**Books are to be returned on or before
the last date below.**

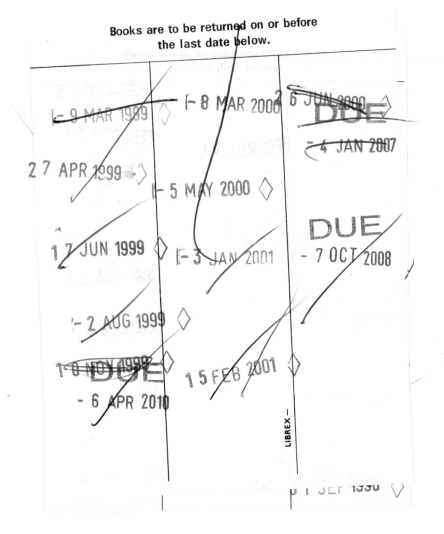

PHOTOVOLTAIC SYSTEM TECHNOLOGY

A European Handbook

Prepared and Edited by

M.S. IMAMURA *and* **P. HELM**
WIP-Munich/Renewable Energies Division
Muenchen, Germany

and

W. PALZ
Commission of the European Communities
Directorate General for Science, Research and Development
Brussels, Belgium

ISBN 0-9510271-9-0

 British Library Cataloguing in Publication Data

Imamura, Matthew
    Photovoltaic System Technology: European
    Handbook
    I. Title
    621.31

    ISBN 0-9510271-9-0

Publication arrangements by
Commission of the European Communities
Directorate-General Telecommunications, Information Industries and Innovation, Scientific and
Technical Communications Service, Luxembourg

EUR 12913 EN
© 1992 ECSC, EEC, EAEC, Brussels and Luxembourg

Published by H S Stephens & Associates on behalf of the
Commission of the European Communities.

H S Stephens & Associates,
Pavenham Road,
Felmersham,
Bedford MK43 7EX, England

Printed in the U.K.

PREFACE

This handbook presents a compilation of photovoltaic system technology in the European Community. It contains comprehensive information and assessments of the present state of knowledge in photovoltaic system design and operation, focusing on stand-alone systems, and provides simple guidelines that should be useful to on-going and future design and operation of photovoltaic systems and plants.

The handbook was prepared within the framework of the Commission of the European Communities' Solar Energy R&D Programme. The cooperation of the owners of currently installed and operating plants and the designers, especially of the original EC "Pilot Plants" and subsequent system applications, was invaluable in the preparation of this manual.

The principal aim of this handbook is to provide a better understanding of how to design, operate, and monitor photovoltaic-powered systems with reliability and cost optimization in mind and to make a more realistic assessment of their capabilities and benefits. Toward this end, a special effort was made in the appendices to describe the design and operating performance of the pilot plants sponsored by CEC/DGXII up to 1990. The original 16 pilot plants installed between 1982 and 1984 were particularly important as a majority of them were stand-alone systems. These plants, ranging up to 300 kW of PV array power with a total capacity of 1.13 MW were built in eight EC-member countries throughout Europe from 1980 to 1989 in the frame of the Commission's (DG XII) second and third Solar Energy R&D Programme. Many of these plants are still operating as of mid-1992 and continuing to provide very useful operational and experimental data.

CONTENTS

Page

PREFACE i

CONTENTS iii

ABBREVIATIONS AND ACRONYMS xi

1. INTRODUCTION 1

   1.1   Status of the Technology: A Brief Overview 1
   1.2   Photovoltaic Power System Types 8
   1.3   Prospects and Strategies for the Future 9
   1.4   The Role of Photovoltaics Among New and Renewable Energy Sources for
        Electricity Production 12
   1.5   PV Activities in Europe 13
         1.5.1   Commission of the European Communities 13
         1.5.2   National Programmes 22
         1.5.3   Industry Programmes 22
   1.6   References 28

2. DESIGN CONSIDERATIONS 31

   2.1   PV Site Architecture and Layout 31
   2.2   System Sizing and Selection 31
         2.2.1   Stand-alone System without an Auxiliary Power Source (Non-hybrid) ... 35
         2.2.2   Stand-alone System with Auxiliary Power Source (Hybrid) 35
         2.2.3   Grid-connected System 35
   2.3   Electrical Load Requirements: Voltage Regulation, Power Quality, Power Profile,
        and Energy 35
         2.3.1   Stand-alone and Hybrid PV Systems 36
         2.3.2   Grid-connected PV Systems 37
   2.4   Plant Use and Operational Modes 37
   2.5   Input Characteristics: The Solar Energy 38

2.6   Characteristics of User Loads 48

2.7   Operating Environment 49

2.8   Safety and Protection 49

2.9   Electromagnetic Interference 50

2.10 Operation and Maintenance 50

2.11 Modularity 50

2.12 Reliability 50

2.13 Lifetime 50

2.14 Expansion Capability 51

2.15 PV System Hardware and Peripheral Equipment 51

      2.15.1   *PV Arrays* 51

      2.15.2   *Energy Storage Devices* 51

      2.15.3   *Power Conditioning and Management* 52

      2.15.4   *DC Bus Voltage and Power Distribution Devices* 54

2.16 Performance Monitoring 55

2.17 References 58

3. PRACTICES AND GUIDELINES 59

   3.1   System Design 60

      3.1.1   *Plant Siting and Landscaping* 60

      3.1.2   *System Configuration and Arrangement* 60

      3.1.3   *Sizing and Selection* 67

      3.1.4   *Lightning and Overvoltage Protection* 75

      3.1.5   *Electrical Grounding* 79

      3.1.6   *Plant Protection and Safety* 80

      3.1.7   *Operation and Maintenance* 81

      3.1.8   *Optimal Utilization of Available Energy* 82

   3.2   Subsystem Design, Performance and Operation 83

      3.2.1   *PV Arrays: PV Modules and Array Configuration* 86

      3.2.2   *PV Arrays: Structures and Foundations* 98

      3.2.3   *Batteries* 108

      3.2.4   *Battery Charge Control/Discharge Protection Methods and Equipment* ...119

      3.2.5   *Battery Operation/Maintenance Methods and Equipment* 124

      3.2.6   *Power Conditioning: DC-DC Voltage Regulation Devices* 129

      3.2.7   *Power Conditioning: Inverters* 133

      3.2.8   *System Control and Power Management* 146

      3.2.9   *Instrumentation and Displays* 148

      3.2.10 *Special Power Supplies: UPS* 149

   3.3   Performance Monitoring 151

      3.3.1   *Monitoring Approaches* 151

      3.3.2   *Selection of Suitable DAS* 153

      3.3.3   *Data Collection Equipment* 153

      3.3.4   *Solar Irradiance Sensors* 165

      3.3.5   *Data Collection Reliability* 170

      3.3.6   *Cost of DAS and Sensor Devices* 172

      3.3.7   *Simplified Method for Determining Available PV Power and Energy* ...172

      3.3.8   *Performance Analysis and Presentation* 174

   3.4   Project Management 174

   3.5   Non-technical Aspects 175

      3.5.1   *Social Benefits Assessment* 175

      3.5.2   *Legal and Institutional Issues* 176

   3.6   Public Information Dissemination 177

   3.7   References 177

4. PHOTOVOLTAIC PILOT PLANTS: OPERATING EXPERIENCES AND LESSONS LEARNED 183

    4.1   PV Plant Characteristics 183

    4.2   Plant Siting and Operational Status 208

    4.3   Photovoltaic Arrays 208

          4.3.1   *Modules and Arrays* 208

          4.3.2   *Array Support Structures and Foundations* 211

          4.3.3   *Cabling, Connectors, and Termination Boxes* 213

    4.4   Batteries 213

          4.4.1   *Problems Experienced* 216

          4.4.2   *Failure Modes and Possible Causes* 216

    4.5   Battery Charge Control and Discharge Protection 218

    4.6   Power Conditioning 220

    4.7   System Control and Power Management 224

    4.8   Lightning and Overvoltage Protection 226

    4.9   Plant Monitoring 226

    4.10 Plant Maintenance and Safety 230

    4.11 References 234

GLOSSARY 235

CONVERSION FACTORS 248

APPENDIX 1   AGHIA ROUMELI PV PILOT PLANT 249

    1   Introduction 249

    2   Plant Designer, Owner and Operator 249

    3   Site Information 250

    4   System Description and Performance 251

    5   Component Description 255

    6   Maintenance 259

    7   Summary of Problems and Solution Approaches 259

    8   Key Lessons Learned 260

    9   Conclusions 260

    10   References 260

    Annex A. Lightning Protection Analysis 261

    Annex B. The New Aghia Roumeli Battery Charge Regulator and Discharge Controller 262

APPENDIX 2   CHEVETOGNE PV PILOT PLANT 264

    1   Introduction 264

    2   Plant Designer, Owner, and Operator 264

    3   Site Information 265

    4   System Description and Performance 266

    5   Component Description 270

    6   Maintenance 273

    7   Summary of Problems and Solution Approaches 273

    8   Key Lessons Learned 274

    9   Conclusions 274

    10   References 274

APPENDIX 3   FOTA PV PILOT PLANT 275

    1   Introduction 275

    2   Plant Designer, Owner and Operator 275

3	Site Information	276
4	System Description and Performance	278
5	Component Description	280
6	Maintenance	285
7	Summary of Problems and Solution Approaches	285
8	Key Lessons Learned	287
9	Conclusions	287
10	References	287

APPENDIX 4   GIGLIO PV PILOT PLANT 288

1	Introduction	288
2	Plant Designer, Owner and Operator	288
3	Site Information	289
4	System Description and Performance	291
5	Component Description	295
6	Maintenance	299
7	Summary of Problems and Solution Approaches	299
8	Key Lessons Learned	300
9	Conclusions	300
10	References	300

Annex A. 1988 Progress Report on Giglio PV Plant 300

APPENDIX 5   HOBOKEN PV PILOT PLANT 305

1	Introduction	305
2	Plant Designer, Owner and Operator	305
3	Site Information	307
4	System Description and Performance	307
5	Component Description	310
6	Maintenance	313
7	Summary of Problems and Solution Approaches	313
8	Key Lessons Learned	315
9	Conclusions	315
10	References	315

APPENDIX 6   KAW PV PILOT PLANT 316

1	Introduction	316
2	Plant Designer, Owner and Operator	316
3	Site Information	317
4	System Description and Performance	318
5	Component Description	322
6	Maintenance	325
7	Summary of Problems and Solution Approaches	325
8	Key Lessons Learned	325
9	Conclusions	325
10	References	325

APPENDIX 7   KYTHNOS PV PILOT PLANT 326

1	Introduction	326
2	Plant Designer, Owner and Operator	326
3	Site Information	326
4	System Description and Performance	328
5	Component Description	334
6	Maintenance	337
7	Summary of Problems and Solution Approaches	337
8	Key Lessons Learned	338
9	Conclusions	338
10	References	338

APPENDIX 8   MARCHWOOD PV PILOT PLANT 339

   1   Introduction 339
   2   Plant Designer, Owner and Operator 339
   3   Site Information 339
   4   System Description and Performance 341
   5   Component Description 345
   6   Maintenance 347
   7   Summary of Problems and Solution Approaches 347
   8   Key Lessons Learned 347
   9   Conclusions 347
  10   References 348
Annex A. Data Acquisition System 348

APPENDIX 9   MONT BOUQUET PV PILOT PLANT 349

   1   Introduction 349
   2   Plant Designer, Owner and Operator 349
   3   Site Information 349
   4   System Description and Performance 351
   5   Component Description 352
   6   Maintenance 356
   7   Summary of Problems and Solution Approaches 356
   8   Key Lessons Learned 356
   9   Conclusions 356
  10   References 356

APPENDIX 10   NICE PV PILOT PLANT 357

   1   Introduction 357
   2   Plant Designer, Owner and Operator 357
   3   Site Information 359
   4   System Description and Performance 359
   5   Component Description 362
   6   Maintenance 365
   7   Summary of Problems and Solution Approaches 365
   8   Key Lessons Learned 365
   9   Conclusions 365
  10   References 366

APPENDIX 11   PELLWORM PV PILOT PLANT 367

   1   Introduction 367
   2   Plant Designer, Owner and Operator 367
   3   Site Information 369
   4   System Description and Performance 369
   5   Component Description 374
   6   Maintenance 380
   7   Summary of Problems and Solution Approaches 380
   8   Key Lessons Learned 381
   9   Conclusions 381
  10   References 381

APPENDIX 12   RONDULINU PV PILOT PLANT 382

   1   Introduction 382
   2   Plant Designer, Owner and Operator 382
   3   Site Information 384
   4   System Description and Performance 385
   5   Component Description 392
   6   Maintenance 394

7	Summary of Problems and Solution Approaches	394
8	Key Lessons Learned	394
9	Conclusions	395
10	References	395

APPENDIX 13   TERSCHELLING PV PILOT PLANT 396

1	Introduction	396
2	Plant Designer, Owner and Operator	396
3	Site Information	398
4	System Description and Performance	399
5	Component Description	402
6	Maintenance	407
7	Summary of Problems and Solution Approaches	407
8	Key Lessons Learned	408
9	Conclusions	408
10	References	408

Annex A. Performance Data from the Terschelling Pilot Plant as of mid-1989 408

APPENDIX 14   TREMITI PV PILOT PLANT 413

1	Introduction	413
2	Plant Designer, Owner and Operator	413
3	Site Information	413
4	System Description and Performance	416
5	Component Description	419
6	Maintenance	422
7	Summary of Problems and Solution Approaches	422
8	Key Lessons Learned	423
9	Conclusions	423
10	References	423

APPENDIX 15   VULCANO PV PILOT PLANT 425

1	Introduction	425
2	Plant Designer, Owner and Operator	425
3	Site Information	425
4	System Description and Performance	428
5	Component Description	432
6	Maintenance	438
7	Summary of Problems and Solution Approaches	438
8	Key Lessons Learned	438
9	Conclusions	439
10	References	439

Annex A. An Application of On-Line Battery Monitoring 440

APPENDIX 16   ZAMBELLI PV PILOT PLANT 445

1	Introduction	445
2	Plant Designer, Owner and Operator	445
3	Site Information	445
4	System Description and Performance	447
5	Component Description	450
6	Maintenance	457
7	Summary of Problems and Solution Approaches	457
8	Key Lessons Learned	458
9	Conclusions	458
10	References	459

APPENDIX 17   ADRANO PROJECT 461

  1   Introduction 461
  2   General Plant Description 462
  3   Technical Plant Data 463
  4   Power Conditioning and Conversion Equipment 466
  5   Experimental Systems 467
  6   Data Acquision System (DAS) 470
  7   First Results and Operating Experience 470
  8   Conclusions 470

APPENDIX 18   FLEXIBLE CABLE-MOUNT ARRAY STRUCTURES 471

APPENDIX 19   PV-POWERED HOUSE IN BRAMMING 474

  1   Introduction 474
  2   Design Considerations 475
  3   Description of the PV house 475
  4   Functional Description 482
  5   Performance Evaluation 482
  6   Conclusions 485

APPENDIX 20   PV-POWERED HOUSE IN MUNICH 486

  1   Introduction 486
  2   PV Array and System Configuration 487
  3   Inverter 488
  4   Results 490
  5   Conclusion 491
  6   References 491

APPENDIX 21   PV-POWERED HOUSE AT VILLA GUIDINI 493

APPENDIX 22   ANCIPA PV PLANT 497

  1   Introduction 497
  2   Water Control and Cleaning System 497
  3   PV Plant Design and Operation 500
  4   Conclusions 502

APPENDIX 23   BERLIN PV PLANT 503

  1   Introduction 503
  2   System Description 504
  3   Components 505
  4   References 508

APPENDIX 24   POZOBLANCO PV PLANT 509

  1   Introduction 509
  2   System Configuration 511
  3   System Sizing 515
  4   Data Monitoring System 515
  5   Plant Performance 516
  6   Conclusions 519

APPENDIX 25   PV APPLICATIONS IN PASSENGER CARS 521

1   General Aspects 521
2   Possible Applications for Solar Cells in Passenger Cars 522
3   Design Considerations 522
4   Test Results 523
5   Conclusions 527

**APPENDIX 26   RECENT ADVANCES IN SOLAR IRRADIANCE MONITORING
DEVICES AND CALIBRATION METHODS** 529

1   Introduction 529
2   Calibration Methods 530
3   Solar Sensors in Use 531
4   Description of the Sensor Calibration Study Task 535
5   Results 537
6   Conclusions 539
7   Acknowledgements 539
8   References 539

**APPENDIX 27   A SIMPLIFIED METHOD FOR DETERMINING THE AVAILABLE
POWER AND ENERGY OF A PHOTOVOLTAIC ARRAY** 541

1   Introduction 541
2   Derivation 542
3   Calculation Example 543
4   Accuracy Improvement Method 545
5   Conclusions 546
6   References 546

**APPENDIX 28   ELECTRICITY PRODUCTION COSTS FROM PHOTOVOLTAIC SYSTEMS
AT SELECTED SITES WITHIN THE EUROPEAN COMMUNITY** 547

1   System Prices per Installed Power Unit 547
2   Calculation of the Annual Energy Production of a PV Generator 548
3   Energy Costs for Selected Examples in Europe 550
4   References 550

APPENDIX 29   RELIABILITY AND AVAILABILITY ASSESSMENT METHODS 551

1   Introduction 551
2   Definition 552
3   Reliability Relationships 553
4   Availability 555
5   Examples of Reliability Calculations 557
6   Conclusions 559
7   References 559

**APPENDIX 30   TRACKING VS. FIXED FLAT-PLATE ARRAYS: EXPERIMENTAL RESULTS
OF ONE YEAR'S OPERATION AT THE ADRANO PILOT PLANT** 561

1   Introduction 561
2   Computer Model Evaluations 562
3   One Year's Operating Results 563
4   Cost Evaluation 564
5   Acknowledgements 566
6   References 566

ABBREVIATIONS AND ACRONYMS

ABB	Asea Brown Boveri AG
A/D	analog to digital conversion
AEG	Allgemeine Elektrizitätsgesellschaft
AFME	Agence Francaise pour la Maitrise de l'Energie *(in Valbonne, FR)*
AGSM	Azienda Generale Servizi Municipalizzati *(in Verona, IT)*
Ah	ampere hour
AM	Air Mass
BE	Belgium
BMFT	Bundesministerium für Forschung und Technologie *(in Bonn, DE)*
BP	British Petroleum
CEC	Commission of the European Communities
CESI	Centro Elettrotecnico Sperimenta Italiano Giancinto Motta, Spa *(in Milan, IT)*
CGC	Comité de Gestion et Coordination
CIEMAT-IER	Centro de Investigaciones Energeticas Medioambientales y Tecnologicas - Instituto Energias Renovables *(in Madrid, ES)*
CPU	Central Processing Unit
CRT	Cathode Ray Tube
DAS	Data Acquisition System
DE, FRG	Deutschland, Federal Republic of Germany or F.R. Germany
DEC	Digital Electronic Corporation
Dfl	Dutch Florin
DG	Directorate General
DIN	Deutsche Industrienorm (standards used in Germany)
DM	Deutsche Mark
EAB	Energie Anlagen Berlin *(in Berlin, DE)*
EC	European Community
ECU	European Currency Unit
emf	electromotive force
EMI	electromagnetic interference
ENEA	Ente Nazionale Energia Alternativa Nucleare *(in Rome, IT)*
ENEL	Ente Nazionale per l'Energia Elettrica *(in Milan, IT)*

ENI	Electriche Nijverheids Installaties
EPROM	Electrically Programmable Memory
EVA	Ethylene Vinyl Alcohol
FD	floppy disk
FI-ISE	Fraunhofer Institut - Institut für Solare Energie *(in Freiburg, DE)*
FL	full load
FP	France - Photon
FR	France
GB	Great Britain
GR	Greece
GTO	Gate Turn Off (a type of thyristor)
HD	hard disk
IEC	International Electrotechnical Commission
IGBT	insulated gate bipolar transistor
I/O	input/output
IRL	Republic of Ireland
IR£	punt (Irish currency unit)
IT	Italy
JRC	Joint Research Centre *(in Ispra, IT)*
K	kelvin
kB	kilobyte
kVA	kilovolt ampere
kVAR	kilovolt ampere reactive
kW	kilowatt
kWh	kilowatt hour
LED	light emitting mode
MB	megabyte
Mftr	manufacturer
MHO	Metallurgie Hoboken Overpelt
MLIT	million Italian Lire
MOSFET	metal oxide semiconductor field-effect transistor
MPPT	maximum power point tracker
MS-DOS	Microsoft Disk Operating System
NF	Norme France (standards used in France)
NL	the Netherlands
Nm	newton meter
NMRC	National Microelectronics Research Centre *(in Fota, EI)*
P&G	Penny & Giles
PC	personal computer; commonly refers to any desktop or laptop computer compatible with the IBM PC's.
PC-AT	PC which uses 80286 16-bit processor that operates at 12 MHz or faster
PX-XT	PC which uses 8086 8-bit processor that operates slower speed than PC-AT
PLC	Programmable Logic Controller
PPC	Public Power Corporation *(in Athens, GR)*
PSP	Precision Spectral Pyranometer (a trade name given by Eppley Laboratories)
PV	photovoltaic
PVB	polyvinyl butane
RAM	random access memory
rms	root mean square
rpm	revolution per minute
SOC	state of charge (battery)
STC	Standard Test Conditions
SW	software
TDF	Télédiffusion de France

THD	total harmonic distortion
TI	Texas Instruments
UPS	uninterruptible power supply
Vac	volt ac
VB	Varta Bloc
Vdc	volt dc
VDE	Verband Deutscher Elektrotechniker. A German organization which tests equipment for public safety and emitted noise.
VOB	Belgian standard for electrical wires and cabling
VVS	Versorgungs- und Verkehrsgesellschaft Saarbrücken mbH *(in Saarbrücken, DE)*
WIP	WIP-Munich/Renewable Energies Division
Wp	watt peak

Chapter 1

INTRODUCTION

1.1   STATUS OF THE TECHNOLOGY: A BRIEF OVERVIEW

Modern Photovoltaics (PV)

The main starting point of modern terrestrial photovoltaics was the scientific congress, "The Sun in the Service of Mankind," held in July 1973 at UNESCO Headquarters in Paris [1]. Since that time, more than three billion dollars have been invested worldwide in the research, development, production, and demonstration of photovoltaic modules and systems. About half the amount came from public funding sources. The largest share of private investment up to 1990 was provided by oil companies but since then semiconductor/electronics companies have taken the lead.

In the late 1980's, the three countries with the largest public support for photovoltaics were the USA, Japan, and F.R. Germany; each of them devoted at least $35 million per year of their national budget to PV research and development. In Germany, the total combined budget from the national (federal) and regional governments steadily increased from about $40 million in 1985 to $100 million annually between 1989 and 1991. The strength of the German effort is relatively recent; it is a direct result of the nuclear accident at Chernobyl, Ukraine in 1986. It is interesting to note that the support for photovoltaics has also been increasing in several other European countries, namely Italy, The Netherlands, Spain, Switzerland, Austria, and Finland.

The Commission of the European Communities (CEC) plays an important role in improving the cooperation on photovoltaic development among its 12 member countries. The Commission provides direct financial support on a cost-sharing basis for R&D and demonstration projects for rural development schemes in Europe's mediterranean areas (Valoren programme). Photovoltaics also play an increasing role in the Commission's development aid agreements, particularly in Africa and the Pacific.

Applications: From Wrist Watches to Central Power Stations

As indicated in Figures 1-1 and 1-2, a wide variety of applications is possible with photovoltaic technology, ranging in power from a few microwatts to tens of megawatts. Photovoltaic power sources have become available in the last few years for many applications and markets around the world. They became popular for consumer products, such as pocket calculators, and many rural applications, notably small residential PV sources and solar water pumps. The total PV capacity produced worldwide as of 1990 amounts to approximately 45 MW.

Consumer products   Water pumping

On Lake Konstanz   Concert at the English Garden in Munich

Near Saarbrücken   At the mountain top

Figure 1-1.   Views of Typical Low-power PV Applications

80-kW VULCANO Plant on the Volcano Island, Sicily

View of the 300-kW PELLWORM Plant on Pellworm Island, Germany. Expansion to 600 kW is scheduled in early 1992.

View of the 300-kW PV Arrays at DELPHOS PV Plant near Foggia, Italy. This plant was expanded to 600 kW in 1991.

Figure 1-2.   Views of Typical High-power PV Applications

General types of applications involving a PV power source have been identified by their end use. For example:

- Agriculture and farming (milk and water pumping)
- Water desalination
- Outdoor lighting
- Recreation
- Warning systems

- Telecommunications
- Rural and remote area electrification (residences and villages)
- Hybrid power plants
- Central power plants

The types of electrical loads which use the power generated by the PV system (i.e., the "consumers"), include dc and ac equipment of varying input voltages. Typical loads are:

- Electric motors (dc and ac)
- Electric lighting
- Refrigerators

- Telecommunications equipment (microwave repeaters, transmitters, etc.)
- Conventional residential and commercial appliances (radios, TV, etc.)

PV Cells

In the solar cells of today, the main types of semiconductor materials used are: single- and poly-crystalline Si, amorphous Si, gallium arsenide (GaAs), copper-indium-diselenide ($CuInSe_2$), and cadmium telluride (CdTe).   Crystalline silicon and amorphous silicon attract most of the development effort. However, with three-fourths of the market supply, crystalline silicon is still the dominant material for commercial PV modules. Newcomers, like $CuInSe_2$ and CdTe along with GaAs, will play only a minor role in the short term.

Silicon Solar Cells: Single Crystal and Polycrystalline

The primary goals of past and future solar cell development are to decrease production costs and to increase the lifetime and efficiency of PV modules. The average sale price for PV modules with silicon solar cells has been reduced by a factor of 8, from about \$40 per Wp in 1973 to \$5 per Wp in 1991.   Correspondingly, many technological changes have been introduced in the manufacturing process. Compared to the early silicon cells in 1973, the thickness has been reduced by more than half, from 500 to 200 μm. The area per individual cell has been increased to 10 x 10 cm and more. Polycrystalline wafers, as compared to monocrystalline, have become increasingly available in the last five years. Complex metal contact structures, such as vapour deposition of TiPdAg, which are employed on solar cells for space applications, have been replaced by simple screen printing techniques.   Integrated production processes in which doping, metallization, and anti-reflection coating are accomplished in only two steps have been developed. Automatic production for wafer processing and parts of the module assembly has been introduced by manufacturers (e.g., BP Solar, TST, Siemens, Helios, HOXAN, and Kyocera).

While degradation and reliability are still problems for amorphous silicon solar cells, crystalline solar cells have clearly demonstrated an extremely high reliability to cperate in all types of outdoor environment with very little degradation. Their life expectancy is well in excess of 30 years. Silicon cell modules employ laminated glass for encapsulation which can be made very thin today (a few mm), and they are sturdy, lightweight, and very easy to transport and install.

In terms of solar cell efficiency, progress has been less visible. A typical commercial silicon cell module of today has a conversion efficiency no higher than a good module of the late 1960's designed for space programmes. Nevertheless, for photovoltaic specialists, the pursuit of higher efficiency will continue to be the most fascinating challenge.

Photovoltaic theory predicts a maximum conversion efficiency of approximately 28%[1] at AM1 intensity (one sun) for a single junction material with a band gap of 1.5 eV. In the laboratory, a maximum efficiency of 22% has been achieved so far, on silicon [2], and InP [3]; very recently, even 23.7% has been announced for GaAs cells [4]. In practice, GaAs cells are only of interest for concentration devices because their cost is excessive for large areas. Under concentrated sunlight, the efficiency of these solar cells is increased; at 100 suns, an efficiency of approximately 28% has been reported with GaAs on silicon solar cells.

Even higher efficiencies can be achieved through the spectral splitting of sunlight. Thus, it becomes possible to optimize solar cells for a specific spectral range and to achieve higher efficiencies in a combined beam splitting system. In practice, two or more junctions are sandwiched together, with each junction optimized for a particular spectral range. These are called the *multi-junction*, *tandem*, or *stacked* cells. For a two-junction cell, the maximum theoretical efficiency is 35%. In theory, for an infinite number of junctions in a series, an efficiency of 54% is possible.

It is expected that the efficiency of commercial modules using Si cells will continue to increase gradually and may reach 17 to 20% within the next several years, as compared with the 10 to 15% level of today.

Thin Film Solar Cells

Table 1-1 gives an overview of some thin-film solar cells which are currently the subject of important development programmes. In the late 1970's, the CdS solar cell was the most popular thin-film cell. Today, development efforts on CdS have been discontinued because it was difficult to increase efficiency beyond 10% and to overcome performance stability problems. The most important thin-film solar cell of today employs amorphous silicon. In mass production for commercial use such as in calculators, the efficiency is not higher than 2 to 4%. In the laboratory, amorphous-silicon solar cells of 6 to 13% efficiency have been produced, but they are plagued with stability problems [5]. In order to overcome the stability problem, the development has shifted from 1989 on to tandem amorphous cells, specifically, amorphous-Si on Si substrates.

Table 1-1.   Achievements in Thin Film Solar Cell Development

Solar Cell Device	Efficiency at STC (%)	References
Poly-silicon on steel substrate, 50 μm thick	12.5	[3]
Amorphous silicon		
- Single junction	11.7*	[5]
- Tandem	12.5	[3,5]
CuInSe$_2$	12.5	[3]
CdS/CdTe	12.8	[3]
GaAs	21	[4]

* unstable

For further cost reduction, thin-film crystalline solar cells are particularly promising. If the thickness of a crystalline silicon solar cell is decreased from 200 to 20 μm, 90% of the material is saved, whereas the efficiency decreases by only 20%. Such a thin-film cell would still have a maximum efficiency of 17% [5]. For many other materials, the optimal thickness of an active layer in the solar cell is in the order of 1 to 2 μm. This is the case for amorphous silicon, GaAs, etc.

---

[1] Unless otherwise noted, all efficiency values are for unconcentrated sunlight intensity of AM1 and 25°C cell temperature.

How to Use This Handbook

Purpose and Scope

The primary purpose of this handbook is to provide useful guidelines to on-going and future PV plant designers and operators. These guidelines are based largely on the design and operating experiences acquired from all PV pilot plants implemented within the framework of the Commission's R&D Programme during the period 1980 to 1991 in the European Community.

Contents and Organization

This handbook presents a compilation of PV system technology in the European Community and overseas. It contains comprehensive information about the present state of knowledge in photovoltaic systems, focusing on currently available hardware and techniques. It is organized and written from system designer and user viewpoints; many of the specific details are left for the reader to pursue in other books and publications.

Chapter 1 provides background information about the state of PV technology. Chapter 2 describes basic factors to consider in the design of stand-alone, hybrid, and grid-connected PV power plants.

Chapter 3 focuses on the basis and rationale for some of the guidelines in terms of the essential topics: systems and subsystems, performance monitoring, etc. Because of the lack of knowledge about batteries operating in the PV environment, this chapter has attempted to cover in depth the topic of battery performance and operation. Also, a special effort was made to provide state-of-the-art information about power conversion devices used in the converters and inverters.

Chapter 4 presents an overview of the original 16 pilot plants and an overall assessment of their operational experiences and lessons learned. The principal topics are organized in the same way as in Chapter 3. Detailed information about these plants is given in Appendices 1-16.

The 30 appendices provide a description of various PV projects installed and technology work performed within the framework of CEC's R&D programmes up to 1991, which are referenced in Chapters 1-4. For detailed operational experiences of individual projects, the reader should refer to the respective appendices.

Terminology Used

To facilitate the reading and understanding of this book, we have defined below a few key terms which are commonly employed herein. For definitions of many other terms, see the GLOSSARY section of this handbook.

System - This is defined herein as two or more subsystems or major components interconnected to form a working unit, such as a power system, a data acquisition system, and application loads such as a desalination system.

PV System (or Plant) - A power system which uses a PV array as power source. For medium to large PV plants, it is convenient to define the plant in terms of the PV power system, auxiliary power source, and facility and enclosures. Furthermore, these plant elements comprise the following:

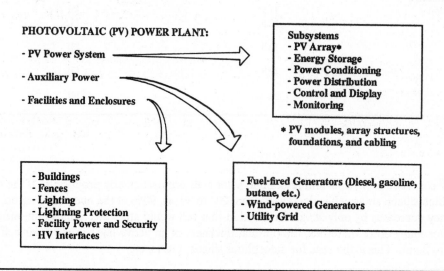

PHOTOVOLTAIC (PV) POWER PLANT:

- PV Power System
- Auxiliary Power
- Facilities and Enclosures

Subsystems
- PV Array*
- Energy Storage
- Power Conditioning
- Power Distribution
- Control and Display
- Monitoring

* PV modules, array structures, foundations, and cabling

- Buildings
- Fences
- Lighting
- Lightning Protection
- Facility Power and Security
- HV Interfaces

- Fuel-fired Generators (Diesel, gasoline, butane, etc.)
- Wind-powered Generators
- Utility Grid

Continued on next page

The three types of PV systems are the stand-alone, the hybrid, and the grid-connected. The basic PV system consists of the following subsystems:

PV array:	The array elements are PV modules, cabling, module mounting structures, and foundations.
Energy storage:	Unless otherwise noted, lead-acid batteries are referred to whenever "energy storage" or "battery" is mentioned, especially in Chapters 1, 2, 4, and in the Appendices.
Power conditioning:	The basic components are the inverter, converter or regulator, and rectifier. The inverter is a dc-to-ac power conversion device. The converter and regulators, which are synonymous, are both dc-to-dc conversion devices; regulator types are partial, full shunt, boost, buck, and boost-buck. The rectifier is an ac-to-dc power conversion device.
Control/display:	The control functions may be implemented with analog or digital circuits. Note that small PV systems may have no manual control provisions or display devices.
Monitoring:	When referring to equipment, this is the hardware for acquisition and storage of measured and processed data; when referring to performance, monitoring means analysis of plant performance data.
Electrical loads:	In the case of a stand-alone system, the consumers are integrated into the plant; otherwise, loads are not considered part of the PV system.

Peak Power - As applied to the PV module, this defines the manufacturer's power output rating in Wp at a solar irradiance of 1,000 W/m^2 and 25°C PV cell temperature (i.e., at Standard Test Conditions). As applied to the PV array panel or array field output, system designers have used the term in two ways:

Installed power rating:	Number of PV modules x the manufacturer's rated module power (disregarding the actual output at the main dc bus)
Array power rating:	Power rating at the array dc bus. The *specification* rating is the value established during the design phase. The *actual* rating is the value determined from actual test of the installed array (on large arrays, I-V data from individual array strings are analyzed to calculate the total array power rating). A common practice among many PV workers is to refer only to the actual rating and erroneously making the design specification and installed rating synonymous.

Commission - An abbreviated reference to the Commission of the European Communities.

Pilot Plant - A PV plant built within the framework of the Commission's R&D Programme. PV plants installed in the Pilot II Programme are also referred to as pilot plants (see next paragraph).

Pilot Programme - All system-level R&D projects are in the "Pilot Programme." The Pilot I Programme included all the original 16 PV pilot plants and the PV-powered houses installed during 1980-1984. Pilot II Programme projects are all subsequent system-level projects implemented afterwards.

JOULE - An acronym for Joint Opportunities for Unconventional or Long-term Energy supply. The JOULE programme is the name given to the Commission's R&D Programme from 1990 to 1994. Directorate General XII administers all research programmes.

THERMIE - An acronym given to the previous programme, *Demonstration*, for the promotion of energy technologies for Europe. The programme is intended to bridge the gap between the R&D stages and the market implementation of technologies.

Concluded

1.2   PHOTOVOLTAIC POWER SYSTEM TYPES

Terrestrial photovoltaic systems can be categorized into three application types: stand-alone, hybrid, and grid-connected. Figure 1-3 shows the block diagrams of typical system types.

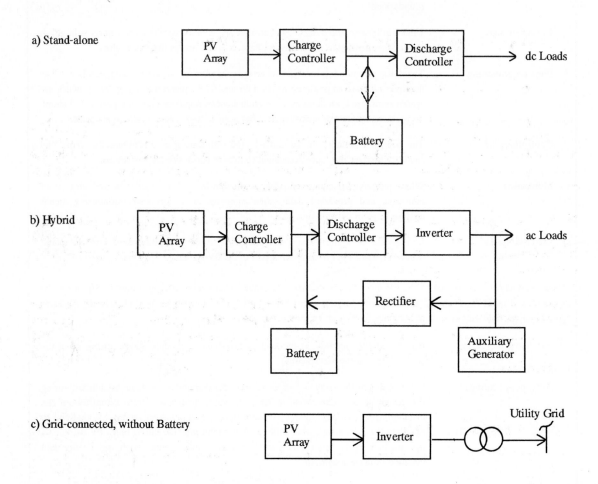

Figure 1-3.   Three Basic Types of PV System Configuration: a) Stand-alone, b) Hybrid, and c) Grid-connected, without Battery

The *stand-alone* systems generally involve batteries and are used in remote locations which have no access to a public utility grid. Most of the water pumping applications above 2-3 kW power consist basically of PV array and inverter(s); note that a PV plant has one or more inverters if it supplies ac power or operates with an auxiliary power source such as utility grid, Diesel, or wind generator.

A *hybrid* system includes a PV array, one or more auxiliary power sources such as wind or Diesel generator, and one or more batteries. Although it requires a more complex controller than the stand-alone or the grid-connected systems, its overall reliability is superior to the other two systems.

The *grid-connected* types are sometimes referred to as cogeneration systems. They normally do not include batteries. Here, the inverter must be capable of accepting the full range of solar array voltage and power excursions, and must be capable of operating at the array peak-power point instantaneously. In this case, the utility network acts as an infinite energy sink and accepts all available power from the PV system. The simplest grid-connected system has a PV array and an inverter as in the case of low-voltage residential grid connection. For high-voltage grid-connected systems (i.e., greater than 220 or 380 Vac), transformers and appropriate power switching and protection devices are included. For any of the grid-connected systems, power factor correction and harmonic filtering devices are essential. However, the grid-interface criteria vary with the utility companies and have yet to be standardized internationally.

Most of the inverters now being used for grid-connected applications incorporate peak-power tracking capability. That is, the inverter controls the PV array output to maintain operation at its maximum power point which changes rapidly with variations in solar intensity and module temperature.

1.3   PROSPECTS AND STRATEGIES FOR THE FUTURE

Accomplishments and PV Market Outlook

Photovoltaic power generation has made tremendous progress in technology and broadened its market over the past few years. There is an immediate commercial market for PV power systems for decentralized power generation. Photovoltaic systems will have an increasing role to play in the coming years by providing lighting and power for telephone stations in remote areas, water supplies, and better health care for dispersed populations around the world.

In the foreseeable future, the trend will be towards both centralized and decentralized (e.g., residential) power production; we anticipate that more and more utility companies in Europe will be involved in installing demonstration/R&D type PV plants with local and national governments actively supporting such projects.

Figure 1-4 illustrates PV market growth over the last 16 years and forecasts a market trend up to the year 2006. In 1974, the market was about 100 kW, mostly for satellite applications. Since that time, the commercial market for terrestrial PV applications has been multiplied by a factor of 250. Figure 1-5 shows the breakdown of major applications in terms of consumers, special projects, and industrial markets.

World-wide Shipments (MWp)

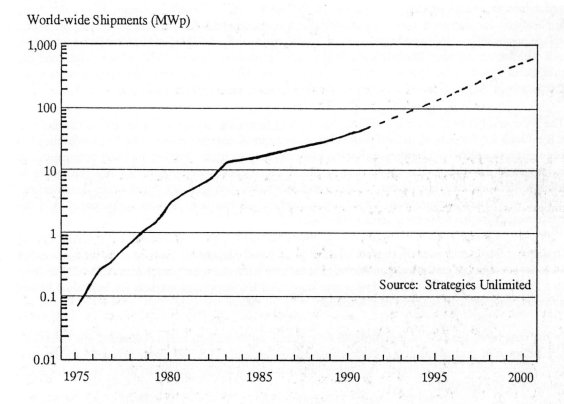

Figure 1-4.   History and Forecast for PV Module Shipments, 1975-2000

Total PV Modules Sold, MWp/Year)

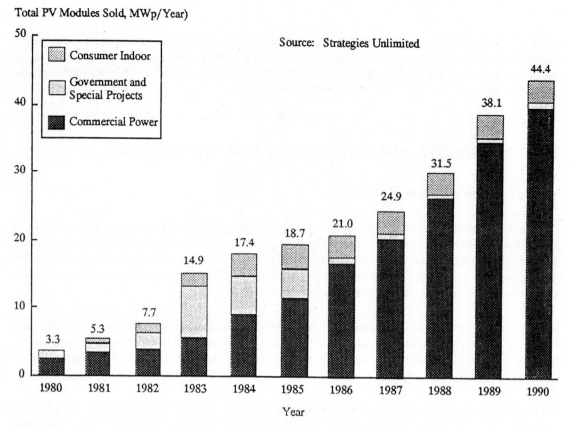

Figure 1-5.   Growth of the PV Market in the Past Decade

As described above, the greater part of the current PV market is in the field of consumer products and telecommunications. A large market of the future will be found, however, in decentralized applications for domestic power in industrialized countries and for rural area development and village power in less-developed countries. It is noteworthy that one-third of the world's population is still without electricity. Photovoltaics is the only technology known today which holds out the promise of a "little power for everyone". It can make available to rural areas the electricity for basic human needs such as lighting, healthcare, water, education, and communications.

The PV market in rural areas is a difficult one, lacking financing. This may be related to the fact that it is difficult for the poor to defend their interests in national energy planning. So far, preference has been given to centralized power production schemes, such as large diesel or hydroelectric plants. It is important that the decision makers of national and regional energy planning become aware of photovoltaics and the possibilities of decentralized power production. Small-scale decentralized electricity is a major impetus for rural development, which can help prevent social unrest and the continuing migration from rural areas to the cities.

In the next five to ten years the market still has to be developed on the basis of a PV module cost of $4 to $6 per Wp. At this price, photovoltaics is definitely the cheapest power source for all needs up to a few hundred Wp. It is in this low-power range that the important markets for individual homes and rural development schemes can be found.

In the long term, the cost of modules will be reduced sufficiently to make *centralized* PV electricity production and grid-connected systems cost-effective as well. This will apply primarily to the sun-belts of the globe, but it is hoped that these opportunities will also spread to areas where the solar resource is less plentiful.

If the current trend of market growth continues, we may expect an overall world market for photovoltaics by the year 2010 of up to 6 GW, mostly for decentralized bulk-power applications.

Cost Goals for PV Module Development

Figure 1-6 gives an overview of PV module prices between 1976 and the year 1991. In 1973 and 1974, the market for modules was extremely small and manufacturers' prices diverged widely. A production cost of approximately $16 per Wp was quoted for instance, by the US company Solarex which was founded in July 1973, and had the capability to simplify the sophisticated technology used for space applications and adapt it to the terrestrial market.

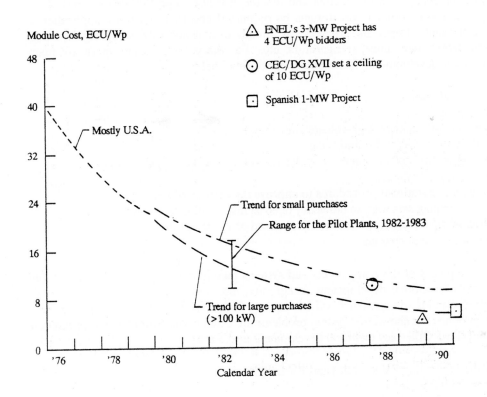

Figure 1-6. Trends in Flat-plate PV Module Cost

In 1987, the cost of producing crystalline silicon modules dropped to approximately $4 per Wp. Note, however, that this is the net production cost to which marketing, distribution, transportation, and profit must be added. Of the $4 per Wp, approximately 40% is for the silicon wafer, another 40% for cell processing and 20% for encapsulation. By simple extrapolation of the classical crystalline silicon technology (i.e., involving wafering from large silicon blocks), a production cost as low as $2 per Wp will be achievable. This is not contested by most PV manufacturers today.

Eventually, it may be possible to achieve even lower production costs for silicon modules. As a director of the German company Wacker Chemical, the world's leading silicon manufacturer [6] recently declared: "The probability to reach the goal set (wafer for less than $0.50/Wp) is increasing..."

There are many thin-film options available as well. We are presently in the middle of an important development process of thin-film solar cells . At the current level of effort, it will take many years to fully explore the various possibilities. A cost goal of $1 per Wp for PV modules by the year 2000 is a realistic scenario today.

Following this strategy, the cost of production will be cut down again by a factor of four in the next 12 years (i.e., the same reduction factor as in the preceding 10 years). To achieve a cost goal of $1 per Wp for PV modules, it will be necessary to maintain or to increase the tremendous development and marketing effort worldwide. As this work has to be related to long-term strategies — which are beyond the immediate interests of industry — the policies and the budgets provided by public sources will be of enormous importance.

Development of System Technologies

For future market penetration, overall PV system optimization will be very important. Not only must photovoltaic modules be cheaper, more efficient and reliable, it is also important that the rest of the system, especially power conditioning and storage, be optimized and that consumer application is particularly energy efficient. Important progress has already been made to develop highly energy efficient fluorescent lamps, televisions, refrigerators, etc. For the years to come, there. are many other possibilities for PV development. Some examples are listed below:

- PV integration into buildings
- High dc voltage systems
- Inverter employing high-frequency, high-voltage switching devices
- Advanced secondary batteries ($Ni-H_2$ and NaS)
- Hydrogen technologies (fuel cell combined with electrolyser and H_2 and O_2 storage)
- Electromagnetic storage (flywheels)
- Battery protection and operational procedures to improve the lifetime of batteries
- Photovoltaic water pumping sets with an efficiency of 80%, instead of 40% today
- Water desalination with 5 kWh per m^3 of seawater, instead of 15 kWh today
- Energy and load management systems

Apart from the development of the components and systems mentioned, it is important to give full attention to the needs of optimal development, integration, test, and operation of all subsystems. Prototype development and field experience will be required for many years to verify long-term field reliability. Past experience has shown that system problems are often underestimated and progress is time-consuming and slow. Other key technical development needs are:

- A systematic method for rating PV modules and PV field
- Sizing of PV array, batteries, and inverters
- Demonstration of long lifetime of PV systems and components
- Energy efficiency improvements of systems (stand-alone, grid-connected, and hybrid)
- Real-time performance analysis and diagnostic methods (for PV array, battery, and inverter)
- Reliable low-cost monitoring systems
- Cost reduction of system and components
- Standardization of hardware, software, and test methods

Non-technical Issues to be Resolved:

- Acceptance of dispersed PV systems in the utility network
- Equitable payment by the utility to the owner of PV systems for PV energy supplied to the utility
- Legal, institutional, and environmental issues

1.4   THE ROLE OF PHOTOVOLTAICS AMONG NEW AND RENEWABLE ENERGY SOURCES FOR ELECTRICITY PRODUCTION

Large-scale production of energy in the form of electricity from new, renewable energy sources is a great challenge. Photovoltaic (PV) and solar thermal technologies have made much progress in recent years as a result of a massive, worldwide development effort which began in the seventies.

In terms of overall installed power capacity, more than 1,500 MW of wind power and over 350 MW of solar thermal plants have been installed and connected to the grid, mainly in California, USA [7,8]. These thermal plants employ parabolic line-focus concentrators and conventional steam turbines to produce electric power which is fed to the local grid.

For grid-connected solar thermal plants in the 30 MW size, the US company, LUZ International Ltd., projects installed costs of $2,200 /kW by the end of this decade. In comparison, wind power today costs approximately $1,500 /kW in units of at least 50 kW and photovoltaics between $6,000 to $8,000 /kW.

The essential differences between photovoltaics and the two other renewables are that both wind turbines and thermal solar power systems come only in relatively large units (i.e., they have a "critical mass") and require certain climatic conditions. Wind power is difficult to exploit economically in regions with wind speeds below 5 m/s yearly average and in turbine sizes below 20 to 50 kW. Solar thermal power plants rely on direct beam radiation which is more abundant in arid regions. They come typically in MW units and require important technological and industrial back-up power sources in the form of conventional fossil-fired turbines to supplement the solar power.

Photovoltaics is the best and least expensive power option today in the small power range up to a few kW. It is suitable for all climates; photovoltaic systems have been installed in all regions around the world, be it at the South Pole or in countries on the equator. Another advantage is photovoltaics' modularity and low need for maintenance. Photovoltaic power has an extra asset which will drive its development in the future. Unlike wind and solar thermal power, it does not necessarily require grid connection: the ground on which PV grows today is the market of decentralized electricity. It is to be expected that the market for all renewable power generation systems will continue to expand in the future as we rely less on oil and nuclear energy, with their respective drawbacks.

1.5 PV ACTIVITIES IN EUROPE

1.5.1 COMMISSION OF THE EUROPEAN COMMUNITIES

In the past decade, the Commission of the European Communities (CEC) has been instrumental in initiating, implementing, and coordinating the photovoltaic (PV) technology activities among its 12 member countries. The Commission has actively pursued PV technology development and utilization since 1975 [9-16]. Their PV activities are now implemented mainly by the R&D, Demonstration (changed to THERMIE in 1990), and International Development Programmes carried out by Directorates-General (DG) XII, XVII, and VIII, respectively. The Commission provides direct financial support on a cost-sharing basis for R&D projects (up to 50%) and Demonstration/ THERMIE projects (up to 40%).

The Joint Research Centre (JRC), a part of DG XII, located in Ispra, Italy, has been actively engaged in monitoring PV plants, qualification testing of modules, and calibration of solar irradiance sensors. JRC has directed the monitoring of all DG XVII PV Demonstration Projects since 1985.

Since 1985, JRC has coordinated the European Working Group on PV plant monitoring. This group consists of about 25 PV technologists who meet at least once a year to discuss and exchange data pertaining to monitoring equipment and plant performance assessment (now mostly on DG XVII programme). JRC has published and disseminated the results of this working group to the PV community in Europe, covering topics such as monitoring criteria and equipment, plant performances and lessons learned, analysis and presentation methods, quality control, etc. [17-25].

R&D Programme (DG XII)

DG XII's PV technology development (R&D) activities fall into two basic categories: 1) solar cell materials and devices, and 2) systems. This handbook deals principally with systems technology.

The system-level R&D projects comprise design and installation of PV plants, subsystem and component development, and special studies. They are grouped into the following time-phased programmes:

Programme	Activities	CEC Contribution (million ECU)
Pilot I, 1980-1984	Original 16 pilot plants and five PV house projects	15.0
Pilot II, 1985-1989	Pilot plant improvement, Concerted Actions, and other technology development projects	3.7
JOULE I, 1990-1992	Follow-on experimentation, Concerted Actions, new applications, and technology development	5.3
JOULE II, 1992-1994	System applications, Concerted Actions, and technology development	Not available

The total contribution by the Commission for the above *system* R&D programmes driving 1980-1992 is about 24 million ECU, which is about one third of the total R&D budget for all PV technology development (the remaining budget is for solar cell materials and devices). Since the Commission provides up to 50% of the cost of these research projects, the total system project cost is at least 48 million ECU. In the case of research institutes, the Commission has normally given 100% financial support for the marginal cost.

Pilot Plants (1980 to 1992)

The most visible activity in the past has been around the 16 "Pilot" PV Plants. All of these plants were installed by the end of 1984 at several sites in Europe (see Fig. 1-7). The key features of these 16 plants are listed in Table 1-2. Chapter 4 presents a summary of the performance from the date of their installation to 1990. Further details of each pilot plant can be found in Appendices 1-16.

Most of the pilot plants were originally intended to operate primarily in the stand-alone mode, but eight of them were modified during the improvement phase (1987-1989) to operate mainly in the grid-connected mode. Collectively, these plants represent the largest group of stand-alone PV plants in the world, and they contain a unique variety of system topology and hardware. With continued improvements, these pilot plants will provide invaluable technology information for a long time into the future.

Ranging from 30 kW to 300 kW and totalling 1.1 MW, the pilot plants were the first major prototypes intended largely for research purposes to encourage the development of new concepts and processes and to stimulate the European industries. To a large extent, these original pilot plants served as the springboard for many national organizations and industries to create markets for photovoltaic applications within the European Community and elsewhere.

Much useful information has already been obtained, but, as expected with research projects, many problems have also become apparent [26]. The pilot plant optimization activities, therefore, were undertaken to improve the technical performance and efficiency of the pilot plants. Many of these plants are actively serving their primary function as a power source. In their secondary role, the plants are excellent for research and technology assessments, experimenting with new components and techniques, and serve as a source of long-term performance reliability data.

PV-powered Houses (1980 to 1984)

As part of the R&D programme, the Commission sponsored five PV-powered residences in Bramming, DK, Munich, FRG, Venice, IT, Sulmona, IT, and Remscheid, FRG. The installation of these PV residences were completed by 1984. Figure 1-8 shows photographs of these installations. (see Appendices 19-21 present brief descriptions of three PV residences.)

50-kW AGHIA ROUMELI Pilot Plant (lower left corner), located on the Southern coast of the Island of Crete in Greece.

PROJECT	PV Power (kWp)	COUNTRY
1. AGHIA ROUMELI	50	Greece
2. CHEVETOGNE	63	Belgium
3. FOTA	50	Ireland
4. GIGLIO	45	Italy
5. HOBOKEN	30	Belgium
6. KAW	35	French Guyana
7. KYTHNOS	100	Greece
8. MARCHWOOD	30	United Kingdom
9. MONT BOUQUET	50	France
10. NICE AIRPORT	50	France
11. PELLWORM	300	Germany
12. RONDULINU	44	France
13. TERSCHELLING	50	Netherlands
14. TREMITI	65	Italy
15. VULCANO	80	Italy
16. ZAMBELLI	70	Italy

Figure 1-7. Locations of Original Pilot Plants and a View of AGHIA ROUMELI Pilot Plant

Table 1-2.   Principal Features of Original PV Pilot (Plants Sponsored by CEC/DG XII) and Several Other Large PV Plants in the European Community

PV Plant (Country)	Array Power Rating (kW)	Primary Application/Loads	System Type*	Array Bus Voltage (Vdc)	Battery Voltage (Vdc)	Battery Capacity (Ah)	Total Inverter Rating (kVA)+	Rectifier Rating (kW)	Converter Rating (kW)	Other Power Sources
AGHIA ROUMELI (GR)	50	Village	S-A	292-353	292-353	1,500	40	---	---	Diesel (40 kVA)
CHEVETOGNE (BE)	40/23	Swimming pool pumps/ lighting	S-A	198-264	198-264	1,500 120	20 x 220	---	---	Grid
FOTA (IRL)	50	Dairy farm pumps	Both	245-315	234-312	300 (x2)	50 (L-C) 10 x 3	---	2	Grid
GIGLIO (IT)	30/15	Refrigeration/water treatment	S-A	90-360 90-350	228-350/ 90-160	2,000 300	7/ 5.7	7	7/ 15	Grid
HOBOKEN (BE)	30	Electrolyzer/pumps	S-A	50-150	22-29	90	---	14	1	Grid
KAW (FR)	35	Village	S-A	292-353	292-353	1,500	40	---	---	Diesel (40 kVA)
KYTHNOS (GR)	100	Village grid	Both	128-192	225-300	2,400	3 x 50	---	25 x 4	Diesel (1 MW) Wind (100 kW)
MARCHWOOD (GB)	30	Grid/DAS	Both	216-295	216-288	400	40 (L-C)	---	---	Grid
MONT BOUQUET (FR)	50	TV/FM transmitter	S-A	191-260	191-254	800	30	---	---	Grid
NICE (FR)	50	Air traffic control equipment	S-A	191-260	191-254	1,500	5	7	---	Grid
PELLWORM (FRG)	300	Recreation centre and grid	Both	230-415	311-415	1,500 (x4)	75 x 2 450 (L-C)	20	---	Wind (3 x 30 kW)
RONDULINU (FR)	65	Village	S-A	80-125	151-202	2,500	50	25	---	Propane-gas (25 kVA)
TERSCHELLING (NL)	50	Merchant marine school	Both	210-273	346-432	250 (x2)	60	20 x 2	2 x 29	Wind (75 kW)

(Page 1 of 2)

* stand-alone (S-A), Grid-Connected (GC).
+ All inverters are self (or force) commutated, except those marked "(L-C)" which are line-commutated types.

Table 1-2. (Concl.)

PV Plant (Country)	Array Power Rating (kW)	Primary Application/Loads	System Type	Array Bus Voltage (Vdc)	Battery Voltage (Vdc)	Battery Capacity (Ah)	Total Inverter Rating (kVA)	Rectifier Rating (kW)	Converter Rating (kW)	Other Power Sources
TREMITI (IT)	65	Desalination/pumps	S-A	128-192	225-300	2,000 250	61**	---	20 x 4	---
VULCANO (IT)	80	Village grid	Both	240-420	240-310	1,500	140 (L-C) 50	10	---	Grid
ZAMBELLI (IT)	70	Water pumping	S-A	200-320	200-270	300 (x2)	40 x 2++	---	15	---
ANCIPA (IT)	1.8	Hydraulic pumps	S-A	24 110	24 110	540 540	--- ---	--- 2	--- ---	DPG
BERLIN (FRG)	9.4	Grid/battery charging	S-A G-C	220	266-350	100	10	---	6 (x2)	Grid
BRAMMING (DK)	5	Residence	S-A	300	194-260	200	10	5	5	Grid
DELPHOS (IT)	300	Grid	S-A G-C	450-600 650-800	450-600 ---	2,400 ---	150 (L-C) 300 (L-C)	--- ---	--- ---	Grid Grid
MADRID (ES)	50 40 10***	Grid	G-C G-C S-A	200-280 280-350 324-450	--- --- 324-408	--- --- 420	50 (L-C) 50 (L-C) 13.5/2	--- --- ---	--- --- 15	Grid
MUNICH (FRG)	2.0	Residence	G-C	11-176+++	---	---	3 (L-C)	---	---	Grid
POZOBLANCO (FRG)	12.6	Milking pumps	S-A S-A	24 106	24 106	1,120 1,600	0.3 3.8 (x2)	--- ---	--- 2	--- Diesel
VILLA GUIDINI (IT)	2.5	Residence	S-A	20-29	26-29	1,600	1	---	---	Grid

Note: Roof-mounted arrays at FOTA, HOBOKEN, MUNICH, and NICE only.

*** 4 - 5.5 kVA, 2 - 3 kVA, 2 - 1.5 kVA, 2 - 15 kVA

++ Variable frequency.

*** 2 kW of A-Si array (Arco G100 modules).

+++ 5 Independent arrays: 11, 22, 44, 88, 176 Vdc nominal.

(Page 2 of 2)

Bramming, DK

Munich, FRG

Remscheid, FRG

Sulmona, IT

Figure 1-8.   PV Houses Co-financed by CEC/DG XII

System Applications and Subsystem Development (1985-1989)

The new system application projects use PV arrays ranging from a few kW to about 50 kW. The largest is a pilot project that started in late 1988 in Saarbrücken, FRG, in which a 50-kW PV array is combined with a 110-kW gas-fired engine. This cogeneration system will provide both electrical and thermal energy to a recently built "Innovation and Technology Centre." (The three installed new plants with at least 2 kWp installed PV power are described in Appendices 17,22,23, and 24.)

The other PV-powered systems are for cable-mounted array structure and PV-assisted car (see Appendices 18 and 25, respectively). Component-level development includes new devices such as the heat pump, 30-kW transformerless inverter, dc pump motors, a monitoring system for crop and forest protection, and domestic appliances. One study was completed on the feasibility of a combined PV/hydroelectric power plant, with the PV array in the MW size. A follow-up project to this study is under way in the JOULE programme to design a 1-MW plant for installation near Toledo, Spain.

Concerted Action Projects (1987-1992)

The Concerted Action Projects comprise seven tasks resulting from common technical problems experienced by the pilot plants. These are identified as: 1) PV array, 2) array structures, 3) battery management, 4) power conditioning, 5) data/plant monitoring, 6) computer simulation and modelling, and 7) social effects study. Appropriate results of these concerted actions are available in CEC reports and publications [27-31].

JOULE I Programme (1990-1992)

The "systems" R&D projects under the new programme, JOULE (Joint Opportunities for Unconventional or Long-term Energy Supply) started in 1989. CEC contributed about 5.5 million ECU for the systems projects which cover the following topics [26]:

System Application

- 1-MW central power station
- Optimized PV/Diesel hybrid system
- Integration of PV into buildings
- Multi-purpose PV plant
- Pilot plant optimization and experimentation

Technology Development and Studies

- Effects of decentralized PV plants on local network and dispatching
- PV/utility interfaces
- Advanced battery charger
- Reliable monitoring system
- External and social costs of energy technologies
- Concerted actions: system development, PV array, battery, power conditioning, and computer modelling

Demonstration/THERMIE Programme (DG XVII)

The Demonstration Programme, initiated in late 1978, is committed to making PV applications a key element in its effort to develop and exploit new energy technologies [32]. It is an integral part of the strategy to achieve the EC's energy objectives, and it aims to promote innovative technologies in the fields of rational energy, renewable energy sources, solid fuels, and carbons. The name "Demonstration" was changed to "THERMIE" in 1990. In its size and range of topics, it is perhaps the most important energy demonstration programme in the world.

The aim of the programme is to provide a means of promoting new technologies in such fields as energy saving, alternative energy sources, the substitution of hydrocarbons and the liquefaction and gasification of solid fuels. Photovoltaics is part of the field of renewable energy sources, along with geothermal, wind, hydroelectric, and biomass.

Projects supported under this programme act as a link between the R&D phase, which sometimes involves a pilot plant, and the later commercial phase. The demonstration phase differs from the R&D and the commercial phases in that the inherent risks are still too high for entrepreneurs to accept without external assistance.

Because of financial risks, there have been instances when technologies or processes have been delayed from entering the market place or even abandoned due to lack of financial backing at this crucial stage.   The Community programme overcomes this problem by providing part of the necessary risk capital, which can then act as a catalyst for involvement by the potential investors on such projects.

For the Community, however, the primary goal is not just to achieve a successful demonstration. Its much broader objective is to encourage the widespread adoption of the demonstrated technology wherever appropriate within the Community. To this end, the Commission uses a wide range of means to disseminate successfully demonstrated technologies, including brochures, press articles, seminars, workshops, conferences and the computer database SESAME.

PV projects are one of the two categories of activities in the energy field. Thermal applications dominated the solar field in the early years of this programme. Since 1983, however, the proportion of proposals submitted for PV has increased to 45% in 1989. About 36% of the solar budget from 1983 to 1990 has been allocated to PV projects.

The number of PV projects authorized as of the end of 1991 is about 106, with an installed PV capacity of about 2,952 kWp at a total CEC contribution of 29.2 million ECU [32,33].   The distribution of the projects according to application type is as follows (also see Fig. 1-9):

	No. Projects To Dec. 1991	Total installed PV Power (kWp)
PV power plants	3	1,360
Houses and villages	44	708
Farms and agricultural	12	185
Pumping and water treatment	10	152
Lighthouses and buoys	6	81
Nature reserves	10	147
Telecommunications	9	137
Warning sign & environmental sensors	12	182
Total:	106	2,952 kWp

International Programmes (DG VIII)

DG VIII implements the development programme for the international community. These activities are oriented towards the developing countries through various cooperative agreements with the countries of the Lomé Convention and several others. The present Lomé III Convention includes most African countries as well as Caribbean and Pacific countries (ACP countries). It was signed between the 66 ACP countries and the 10 EC-member states in 1984/85. Funding for this 5-year programme (1986-1990) was provided by the European Development Fund (EDF), the European Investment Bank (EIB), etc.

Common objectives of the development aid programme of the Commission include self-sufficiency in food, better living conditions for rural populations, the diversification of production and higher productivity. It is in this frame that photovoltaics provides a useful contribution, and an increasing number of projects support self-reliant development activities.

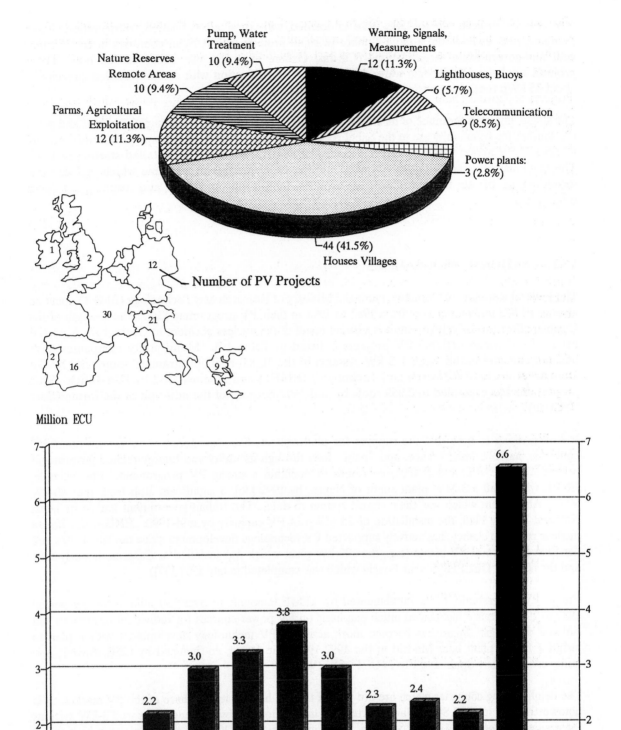

Figure 1-9. 106 PV Demonstration Projects with an Installed Capacity of 2,952 kWp Implemented between 1979 and 1991. Total support from CEC/DG XVII is 29.2 M ECU.

Most of the projects up to 1988 were concerned with rural electrification, refrigeration, water pumping, etc. In addition, a large number of small projects with African countries in combination with non-governmental organizations (Caritas, Oxfam, etc.) are also being carried out. These projects involving photovoltaic water pumping and electrification with a total installed capacity of about 35 kWp receive approximately 1.4 M ECU from the CEC.

The single largest programme of its kind was started in 1989 to install PV pumping systems and small PV sources for community use in the Sahel region of Africa. Over 1,300 pumps powered by 600-W to 3.5-kW PV arrays with a total PV capacity of nearly 2 MW will be installed starting in 1992. The Commission is contributing about 32 million ECU for this programme which includes the development of an infrastructure to support long-term operation and training of local inhabitants.

1.5.2 NATIONAL PROGRAMMES

The Federal Republic of Germany under the Ministry of Research and Technology (BMFT) spent an average of $42 million per year from 1982 to 1991 in their PV programme [34]. The strength of the German effort is relatively recent; it is a direct result of the nuclear accident in Chernobyl. A partial list of the German national PV projects is listed in Table 1-3. Many utility companies have become engaged in the small 1-5 kW systems in the "1,000-roof" programme sponsored by the German Ministry of Research and Technology (BMFT) and administered by KfA-Jülich. This programme has expanded to 2,250 roofs by mid-1992 because of the addition of the former East Germany.

It is interesting to note that the support for photovoltaics has also increased in several other EC countries, namely, Italy, France, and Spain. Italy through its utility and energy-related government organizations, ENEL and ENEA, continues to maintain a strong PV programme. For example, ENEL will install a 3-MW plant south of Naples by 1992 [36], a significant step from the 80-kW VULCANO plant which was their largest system to date. The Italian government had in its latest National Energy Plan, the installation of 25 MWp of PV capacity by mid-1990. ENEA, the Italian nuclear research agency, has actively supported PV technology development since the late 1970's and has financed several PV plants (e.g., the 100-kW plant which was installed at Casaccia in early 1991 and the 600-kW DELPHOS near Foggia which was completed in late 1991 [37]).

France'PV programme, being implemented by AFME, amounts to over $50 million for the period 1982 to 1990. France has placed much emphasis on PV power sources for individual residences and isolated dwellings. Spain has become more active in PV technology development with a plan to install a 1-MW plant near Madrid in the 1992-1993 time frame, co-financed by CEC, BMFT, and utility companies in Spain and FRG (UEF and RWE).

The neighbouring countries which are not part of the EC have a sizable share of the PV market. It is noteworthy that in 1987 Switzerland started the installation of government-sponsored 3-kW systems for residences totalling 1 MW. An innovative grid-connected 100-kW system along a high-speed freeway was installed in 1990 [38]. As of 1990, their total installed PV capacity was about 1 MW (see Table 1-4 [39]). A 500-kW grid-connected plant was erected in early 1992 [40], resulting in about 3 MW total installed capacity in Switzerland as of mid-1992. In Austria, the utility companies with government assistance already have over 27 projects with an installed capacity of over 124 kW (see Table 1-5 [41]).

1.5.3 INDUSTRY PROGRAMMES

Several utility industries in the EC have made a substantial commitment towards the development of PV technology in both residential and central power plant applications. Good examples are RWE and Bayernwerk in Germany.

Table 1-3.   A Partial List of National PV Demonstration Projects in Germany (as of December 1990)

Proj. No.	Plant Owner	Site Location	Application	System Type* S-A	System Type* G-C	Total PV Power (kWp)	Installation Date	Co-financed By CEC
1	Pfalzwerke**	Ludwigshafen	Irrigation	x		18.1	Sept. '88	
2	Pfalzwerke**	Ludwigshafen	Air Ventilation for Pig House	x	x	9.1	Apr. '89	
3	Pfalzwerke**	Kaiserslautern	Cattle Sewage Purification	x		4.7	Sept. '88	
4	Pfalzwerke**	Kaiserslautern	Pig Sewage Purification	x		4.7	Sept. '88	DGXVII
5	VVS, Saarbrücken**	Ensheim	PV-House		x	7.7	May '88	DGXII
6	EAB, Berlin**	Berlin	Heat Pump and Electric Car Charging	x	x	9.4	Dec. '88	
7	City of Ratzeburg	Ratzeburg	Deep Water Aeration		x	22	'91	
8	Stadtwerke Hannover**	Hannover-Messe	Solar Power Station		x	15	July '88	
9	City of Burg	Fehmarn	Hybrid PV-Wind-Thermal Generator		x	140	Mar. '89	
10	Isar Amperwerke**	Brunnenbach	PV-powered Farm House	x		10.4	Apr. '89	
11	Fränkische Überlandwerke AG**	Triesdorf	Agricultural School	x		5	'90	
12	Fränkische Überlandwerke AG**	Triesdorf	Mobile Plant for Feed Production	x		2	'90	
13	Fränkische Überlandwerke AG**	Triesdorf	Cattle Breeding, Automatic Feeder	x		2	'91	
14	Fränkische Überlandwerke AG**	Triesdorf	Pig Breeding, Automatic Feeder		x	6	'90	
15	Fränkische Überlandwerke AG**	Triesdorf	PV for Pond Aeration	x		2	'91	
16	Fränkische Überlandwerke AG**	Triesdorf	PV for Household		x	5	'91	
17	City of Frankfurt	Ffm-Kalbach	Sports Centre		x	5	'91	
18	City of Frankfurt	Ffm-Niederrad	Child Care Centre	x	x	10	Oct. '90	
19	City of Frankfurt	Ffm-Griesheim	Child Care Centre	x	x	10	May '90	
20	Hochschule Bremerhaven (University)	Bremerhaven	PV Pilot Plant for Research		x	1.9	Oct. '89	
21	RWE**	Meckenheim	School and Sports Centre		x	5.1	'91	
22	Solar-Wasserstoff-Bayern GmbH	Neunburg v. Wald	Hydrogen Production		x	482	'90/'92	
23	Jugendwerkstatt Felsberg	Felsberg	24-V-Lighting and Grid Supply	x	x	5	Oct. '90	
24	Solid GmbH	Fürth	Information and Demonstration Centre		x	8.5	Feb. '91	
25	Lechwerke	Wendelstein	Grid Supply		x	30	Oct. '90	
26	HEW	Geesthacht	Pumping	x		65	'91	
27	Stadtwerke Pirmasens	Pirmasens	PV for Lighting and Grid Supply			12	Aug. '90	
28	Stadt Ribnitz-Damgarten	Körkwitz	PV for Sewage Purification Plant		x	250	'91/'92	
29	Schleswag**+	Pellworm	PV-Powered Recreation Centre		x	600+	'91+	DGXII
30	VVS Saarbrücken**	Saarbrücken	Cogeneration, Thermal and PV		x	50	'91	DGXII
31	Weßling	Bochum	Grid-connected PV-Powered Laboratory		x	30	'91	
32	Wasser- und Maschinenamt	Rendsburg	PV for Tunnel Lighting		x	70.5	'91	

Total: 1,898.10

* S-A: Stand-alone; G-C: Grid-connected
** Utility company

+ The original 300-kW system, installed in 1983, was owned by the City of Pellworm. until it was transferred to the local utility company, Schleswag. The expansion of 300 kW, to be installed in late 1991, will bring the total plant capacity to 600 kW.

Table 1-4.   A Partial List of National PV Projects in Switzerland (as of December 1990)

Plant Owner	Site Location	PV Supplier	Total Power (kWp)
EW Kanton Zurich	Aathal	EW des Kantons Zurich	7
B + S Engineering AG	Aeflingen	Bernische Kraftwerke AG	3
Sun-Craft AG	Alpnach	EW Obwalden	2.85
EW des Kantons Thurgau	Amriswil	EW des Kantons Thurgau	2.5 + 2
Hr. Salvisberg	Arch	BKW	2.6
ADEV	Basel	IWB	3
Eisenwaren Brunner	Bassersdorf	EW Bassersdorf	2.5
EW Bern	Bern	EW Bern	2
Ingenieurschule	Burgdorf	Industrielle Betriebe Burgdorf	2.5
CERA	Chêne-Bourg	Industrial Services Geneva	3
TNC	Chur	Industrial Operation Chur	3
EW Davos	Davos	EW Davos	1.0
s'Lotusblüemli	Derendingen	EW Derendingen	7
Fam. Gass	Dittingen, BE	BKW	1.1
Bucher Leichtbau	Fällanden	EW Fällanden	2.5
BEW/TNC Consulting	Felsberg GR	EW Tamins	100
Herr Sprenger	Frauenfeld	EW Frauenfeld	1
W. Schmid AG	Glattbrugg	EW Opfikon/Glattbrugg	5.6
C.U. Brunner	Hinteregg	EW des Kantons ZürichBKW	2.8
Fam. Burn	Interlaken	Industial Operations Interlaken	2.5
W. Haueter	Konolfingen	BKW	3.8
Herr Keller	Kreuzlingen	EW Kreuzlingen	3
Oekozentrum	Langenbruck	Elektra Baselland Liestal	1.2
ADEV	Liestal	Elektra Baselland Liestal	9.3
ADEV Park and Ride	Liestal	Elektra Baselland Liestal	2.4
R. Neukom	Locarno	Società Elettrica Spracenerina	2.4
TISO 15	Lugano	Azienda Industriali Lugano	10
TISO 15	Lugano	Azienda Industriali Lugano	3
Städtische Werke Luzern	Lucern	Städtische Werke Luzern	2.8
Horlacher AG	Möhlin	Aargauisches Elektrizitätswerk	2.5
Solcar	Mönchaltdorf	EW des Kantons Zürich	5.75
Ingenieurschule Biel	Nidau	EW Nidau	1.2
M. Einsenring	Niederuzwil	SAK	3
Städtische Werke Olten	Olten	Städtische Werke Olten	9
ADEV	Outremont	BKW	28
Fa.m. Wenger	Reinach	Elektra Birseck Münchenstein	2.4
ADEV/Solenar	Rheinfelden	Aargauisches Elektrizitätswerk	9
W + S AG	Rohr	Aargauisches Elektrizitätswerk	2.5
Herrn L. Roth	Rüti/Winkel	EW des Kantons Zurich	1.1
Hr. Burkhalter	Rüegsau	BKW	3
SAK, Unterwerk Sargans	Sargans	St. Gallisch-Appenzellische Kraftw. AG	3
Solarmobilgruppe Seedorf	Seedorf	EW Seedorf	2.8
Ch. Gerster	Stetten	EW des Kantons Schaffhausen	3
Fr. Hubacher	St. Gallen	St. Galler Stadtwerke	1
Titlis Bergbahnen	Titlis	EW Nidwalden	2.5
Dr. Albrecht	Uetendorf	Bernische Kraftwerke AG	6
Ch. Leu	Uettligen	Bernische Kraftwerke AG	2.6
Amt für Bundesbauten	Wabern	Bernische Kraftwerke AG	12
Fam. Schulz	Wettswill	EW des Kantons Zurich	2.8
P. Rettenmund	Wynau	EW Wynau	3
Hochbauamt	Zurich	EWZ	20

Total: 323.50 kW

Source: W. Blum [39]

Table 1-5.   A List of PV Projects in Austria (as of December 1990)

Proj. No.	Project	Province	Operational Date	Total Power (Wp)	System Type* S-A	G-C
1	Wetterstation Plattkopf	Tirol	1980	66	x	
2	3 Flutwellenwarnanlagen	Kärnten	1980	66	x	
3	65 Notrufsäulen	Niederösterreich	1982	8	x	
4	Relais-Station	Salzburg	1983	1,200	x	
5	Lanserwiese	Salzburg	1984	1,700	x	x
6	Hochleckenhaus	Oberösterreich	1985	2,000	x	
7	Käserei Baumgartalm	Salzburg	1986	2,400	x	
8	Ybbstalerhütte	Niederösterreich	1987	670	x	
9	Solarzellenteststation	Wien	1987	1,000	x	
10	Solaranlage Gmunden	Oberösterreich	1987	1,300		x
11	Zellerhütte	Oberösterreich	1987	300	x	
12	Hainfelderhütte	Niederösterreich	1988	240	x	
13	Solarkraftwerk am Loser	Steiermark	1989	30,000		x
14	Kesselbachfassung	Tirol	1989	1,500	x	
15	Solarzentrum Kanzelhöhe	Kärnten	1989	1,200	x	
16	Rojacher Hütte	Salzburg	1989	60	x	
17	HTL-St. Pölten	Niederösterreich	1989	20,000	x	x
18	Otto Kandler-Haus	Niederösterreich	1989	240	x	
19	Reichenstein-Hütte	Steiermark	1989	800	x	
20	HTL-Wien X	Wien	1989	10,000		x
21	Hofmannshütte	Kärnten	1989	700	x	
22	HTL-Leonding	Oberösterreich	1990	1,500		x
23	RF-Station Spering	Oberösterreich	1990	3,700	x	
24	Customer operated	Oberösterreich	1990	1,800		x
25	Customer operated	Kärnten	1990	1,000		x
26	Customer operated	Niederösterreich		1,000		
27	Motorway A1 PV-Project (sound barrier)	Oberösterreich	1991	40,000		x
			Total	124,450		

* S-A: Stand-alone; G-C: Grid-connected   Source: Szeless, Verbund-Vienna (1991)

In 1986 RWE Energie AG started a 1-MW PV project which is funded solely by RWE. The first two stages of this project resulted in a 340-kW plant at Kobern-Gondorf in the Moselle Valley near Cologne and a 360-kW plant at Lake Neurath. These plants, pictured in Figure 1-10, started operating in October 1988 and June 1991, respectively. It is rather clear to the visitors to these well-constructed plants that RWE, the largest utility ccompany in Germany, is serious about the potential of PV technology. They also continue to provide valuable service to the PV and renewable energy workers, as well as the local community, by conducting public tours through their PV plants. For instance, in the summer months, it is not unusual for RWE to accommodate over 2,500 visitors per month to the Kobern-Gondorf plant [43].

Bayernwerk, the third largest producer of electricity in Germany, also embarked on a multi-phase project to install and operate PV plants for research and evaluation purposes. In the first phase, they completed a 280-kW plant at Neunburg vorm Wald. This is a unique plant in Europe with hydrogen generation and utilization via operational electrolyzers and fuel cells. Figure 1-11 shows a view of the plant. The sponsors for this plant are of Bayernwerk, BMW, Linde, MBB, and Siemens.

Kobern-Gondorf

Lake Neurath

Figure 1-10.   Views of Kobern-Gondorf and Lake Neurath PV Plants, Owned and Operated by RWE

Water Electrolysis Unit

Pressurized H$_2$ Gas Cylinders

Figure 1-11.   Views of 280-kW PV Plant at Neunburg vorm Wald, Owned and Operated by Solar-Wasserstoff-Bayern GmbH

1.6   REFERENCES

[1] "The Sun in the Service of Mankind," Proceedings of the Photovoltaic Section, published by CNES Paris (1973), W. Palz ed.

[2] M. A. Green, "High-Efficiency Silicon Solar Cells," Proceedings of the 10th European Photovoltaic Solar Energy Conference, Lisbon, Portugal, 6-12 April 1991

[3] W. Freiesleben, "Materials for the PV Market, Present Situation and Future Outlook," Proc. of the 7th European Photovoltaic Solar Energy Conference, Seville, ES, p. 699 (1987), published by D. Reidel, Publishing Company, NL-Dordrecht

[4] S.S Chu et al., "High Efficiency GaAs Solar Cells," 21st IEEE Photovoltaic Specialists Conference, Orlando, FL , USA (1990).

[5] K. Zweibel et al., "Recent Development in Thin-film Solar Cells," 21st IEEE Photovoltaic Specialists Conference, Las Vegas, NE, USA (1985).

[6] Photovoltaic News, Vol. 6, No. 4, (4/1987), P.D. Maycock, Editor.

[7] B.T. Madsen, "Industry Aspects of Wind Turbine Development," International Journal of Solar Energy, Vol. 4, Page 137.

[8] D. Kearney et. al., "Status of the SEGS Plants," Proc. of the 1991 ISES Solar World Congress, Vol. 1, Part 1, Denver, CO, USA, 19-23 August 1991.

[9] W. Palz, "Photovoltaic Power Centre," Series C, Vol. 1, Proc. of the Final Design Review Meeting on EC Photovoltaic Pilot Projects, Brussels, BE, 30 November-2 December 1981.

[10] W. Palz, "Photovoltaic Power Generation," Series C, Vol. 4, Proc. of the EC Contractors' Meeting, Hamburg/Pellworm, FRG, 12-13 July 1983.

[11] M.R. Starr, "An Assessment of the Prospects for Photovoltaics in Europe," Int. J. Solar Energy, 1983, Vol. 2, pp. 35-61.

[12] W. Palz and W. Kaut, "Photovoltaics in the European Community," Proc. of the 6th European PV Solar Energy Conference, London, GB, April 1985.

[13] P. Helm, "Overview on CEC Photovoltaic Pilot Projects," Proc. of the 7th European PV Solar Energy Conference, Seville, ES, October 1986.

[14] W. Palz, W. Kaut, R. van Overstraeten, and G. Willeke, "PV Activities of the European Community," Proc. of the 7th European PV Solar Energy Conference, Seville, ES, October 1986.

[15] W. Palz, "Photovoltaic Power Generation - R&D Programme in Europe," Proc. of the 9th European PV Solar Energy Conference, Freiburg, FRG, September 1989.

[16] W. Palz, "PV Technology Development in EC," Proc. of the 10th European Photovoltaic Solar Energy Conference, Lisbon, Portugal, 6-10 April 1991.

[17] K. Krebs and M. Starr, "First Session of European Working Group on Photovoltaic," Proc. of Meeting in Ispra, IT 14-16 September 1985

[18] K. Krebs and M. Starr, "Second Session of European Working Group on Photovoltaic," Proc. of Meeting at Sophia Antipolis, FR, 10-12 March 1986.

[19] K. Krebs and M. Starr, "Third Session of European Working Group on Photovoltaic," Proc. of Meeting in Madrid, ES, 22-24 October 1986.

[20] K. Krebs and M. Starr, "Fourth Session of European Working Group on Photovoltaic," Proc. of Meeting in Lugano, CH, 1-3 April 1986.

[21] K. Krebs and M. Starr, "Fifth Session of European Working Group on Photovoltaic," Proc. of Meeting in Cork, IRL, 30 September-2 October 1986.

[22] K.H. Krebs, "Quality and Performance Assessment in Photovoltaics," Proc. of the Euroforum - New Energies Congress, Vol. 1, Saarbrücken, FRG, 24-26 October 1988.

[23] G. Blaesser and K. Krebs, "Guidelines for the Assessement of PV Plants, Document A - Photovoltaic System Monitoring," JRC Publications SP-1.87.43, November 1988.

[24] G. Blaesser, K. Krebs, and M.R. Starr, "Guidelines for the Assessment of PV Plants, Document B - Analysis and Presentation of Monitoring Data," JRC Publications SP-1.88.44, November 1988.

[25] G. Blaesser, K. Krebs, and M.R. Starr, "Guidelines for the Assessment of PV Plants, Document C - Initial and Periodic Tests on PV Plants," JRC Publications SP-1.88.45, November 1988.

[26] M.S. Imamura, "PV System Technology Development in the European Community," Proc. of the Euroforum - New Energies Congress, Vol. 1, Saarbrücken, FRG, 24-28 October 1988.

[27] M.S. Imamura and P. Helm, "Concerted Action Project: Monitoring, Array Structures, and Social Effects," WIP-89-9, CEC Contract EN3S-0142-D, November 1989.

[28] G. Beer, G. Chimento, F.C. Treble, and J.A. Roger, "Concerted Actions: PV Modules, Array, and Solar Sensors," CEC Contract EN3S-140-1, Conphoebus Final Report, December 1989.

[29] S. McCarthy, M. Hill, and A. Kovach, "Concerted Actions: Battery Management," CEC Contract EN3S-0137-IRL, NMRC Final Report, December 1989.

[30] J. Schmid and R. von Dincklage, "Concerted Actions: Power Conditioning and Control," CEC Contract EN3S-0141-D, Fraunhofer Institute Final Report, December 1989.

[31] G.T. Wrixon and L. Keating, "Concerted Action Project: Computer Simulation and Modelling," NMRC Final Report, CEC Contract EN3S-0138RL-IRL, December 1989.

[32] W. Kaut, "Status of the Photovoltaic Demonstration Programme of the Commission of the European Communities," Presented at the European PV Monitoring Working Group Meeting, Berlin, FRG, 5-7 June 1991.

[33] Private communication with W. Kaut of CEC/DGXVII, January 1992.

[34] Eisenbeiss, "The German PV Programme," Proc. of the 9th European PV Solar Energy Conference, Freiburg, FRG, November 1989.

[35] K. Wollin, J. Batch, "The German PV Programme," Proc. of the 10th European PV Solar Energy Conference, Lisbon, Portugal, 6-10 April 1991.

[36] C. Corvi, R. Vigotti, A. Iliceto, and A. Previ, "ENEL's 3-MW PV Power Station Preliminary Design," Proc. of the 1991 ISES Solar World Congress, Vol. 1, Part 1, Denver, CO, USA, 19-23 August 1991.

[37] M. Garozzo, C. Messana, A. Previ, and R. Vigotti, "The Italian Programme: Accomplishments and Future Goal," Proc. of the 1991 ISES Solar World Congress, Vol. 1, Part 1, Denver, CO, USA, 19-23 August 1991.

[38] T. Nordmann, L. Clavadetscher, and R. Hächler, "100-kW Grid-connected PV Installation Along Motorway and Railway," Proc. of the 1991 ISES Solar World Congress, Vol.1, Part 1, Denver, CO, USA, 19-23 August 1991.

[39] W. Blum, "An Overview of the PV Grid-connected PV Plants in Switzerland," Swiss Engineer and Architect Society, SiA D049, Photovoltaic Use, Edit. T. Nordmann, 1990.

[40] R. Minder and A. Bertschinger, "The Swiss 500-kW Photovoltaic Power Plant PHALK Mont-Soleil," Proc. of the 1991 ISES Solar World Congress, Vol. 1, Part 1, Denver, CO, USA, 19-23 August 1991.

[41] H. Szeless, "Austrian Utility PV Applications," Presented at the PV/Utility Interface Working Group Meeting, CEC/DG XII, Madrid, ES, 29 January 1991.

[42] U. Beyer, B. Dietrich, R. Hotopp, and R. Pottbrock, "Operating Results of the 340-kWp Plant, Kobern-Gondorf, and Design and Construction of the 330-kWp Plant, Lake Neurath," Proc. of the 9th European PV Solar Energy Conference, Freiburg, FRG, November 1989.

[43] Private communication with U. Beyer of RWE, June 1991.

Chapter 2

DESIGN
CONSIDERATIONS

The principal factors affecting the design and performance of PV systems are solar irradiance, ambient air temperature, electrical load characteristics, system configuration, and characteristics of the three major subsystems, namely, the array, batteries, and power conditioning. The PV system itself can be represented as a transfer function with the solar irradiance as the input and the electrical loads as the output. In addition, there are programmatic considerations such as available budget, schedule, and whether or not the plant is for R&D or commercial use. This last consideration affects the complexity and cost of the monitoring system; for example, a commercial plant does not need expensive data acquisition and storage equipment.

The purpose of this chapter is to delineate the most important factors, given the budget and schedule constraints, which influence the design and selection of the system arrangement and components for the three basic types of PV plants: stand-alone, hybrid, and grid connected. Table 2-1 is a checklist of design considerations for stand-alone and grid-connected plants. Table 2-2 identifies design trade-offs for PV plants in general.

2.1   PV SITE ARCHITECTURE AND LAYOUT

An important requirement for all PV plants co-financed by the Commission is to design and install the array field so that it blends with the surrounding vegetation, ground profile, and landscape. Moreover, many plants situated at remote locations are required to make good use of local materials, rocks, and gravel available on site, mainly to retain or preserve the original appearance of the site and the environment. The key criteria for site selection include solar energy availability, access road, and public visibility.

2.2   SYSTEM SIZING AND SELECTION

Sizing criteria and procedures are quite different for the three types of PV systems defined earlier. In each case, note that optimization criteria are based on cost, battery lifetime, system energy efficiency, and real-time operation.

Table 2-1.   A Checklist of Requirements to Consider for the Design, Performance, Installation, and Operation of Stand-alone and Grid-connected PV Systems

Category/Subsystem	Description
General	- Plant rating - Codes and standards (local and international) - Permits and licences - Reliability and maintainability - Operational and maintenance - Quality assurance provisions - Lightning protection
Site	- Architectural layout/aesthetics and environment - Harmonization with environment and landscape - Soil properties - Solar access and energy availability - Access roads - Site drainage - Public visibility
PV Field	- Array field power rating - Array structures/foundations - Array shadowing - Modularity of array structures, cabling, components, and source circuit - Standardized installation - Expandability (add-on capability) - Circuit isolation & testing - Cabling - Fault detection/isolation - Grounding/field wiring
Inverter	- Inverter rating - Utility power quality (THD, PF, regulation, stability) - Input voltage range - Output voltage and frequency - Maximum array power tracking - Input (dc) ripple filtering - Fault current limiting - Inverter control - Inverter synchronization/control (for multiple inverters) and start-up capability - Over-voltage/over-current protection - Electrical isolation - Data interface (RS232C) - Environmentally protected housing (for outdoor inverters) - Testing
dc Power Distribution	- $I^2 R$ losses - Test and monitoring - Fault detection/isolation/protection - Relays/switchgear
ac Power Distribution	- Fault detection/isolation/protection - Integrated interface to utility grid transmission lines - Minimum $I^2 R$ losses (array field ac) - Reactive power compensation and harmonic filtering - Relays/switchgear - Electrical isolation

Page 1 of 2

Table 2-1.   (Concl.)

Category/Subsystem	Description
Control & Display	- Display panel and meters - Integrated interface to existing control system (e.g., in the inverter) - Automatic control over-ride capability (i.e., manual) - Security and safety display - UPS - Real-time performance display
Data Acquisition (Monitoring System)	- Measurements, special sensors, metering - Solar/meteorological sensors - Portable test equipment - UPS or auxiliary power supply - DAS and PC hardware - Data collection, analysis, and recording software

Facilities	- Control & monitoring room
	- Maintenance room
	- Spare component room
	- Station/utility line switchyards
	- Access roads
	- Water supply/sanitary discharge
	- Switchyard lighting
	- Site drainage
	- Toilets
	- Communication
	- Lighting
	- Lightning protection
	- Security/Safety
	  o Perimeter fencing and lighting
	  o Security alarm system
	  o Limited access areas
	  o Safety signs

Construction, Installation & Acceptance Test	- Standard components & procedures
	- Installation equipment (forklifts, etc.)
	- Field grading/drainage systems
	- Post-installation engineering test
	- Quality control
	  o Receiving inspection/test
	  o Acceptance test
	- Portable equipment for commissioning checkout

Operation & Maintenance	- Solar irradiance sensor cleaning - Array cleaning equipment - Cleaning water/effluent discharge consideration - Component interchangeability - Dispatch of PV power to utility grid - List of spare components - Portable test equipment for sensor calibration, battery cell voltage measurement, etc. - Maintenance/repair manuals

Table 2-2.   A Checklist of Key Design Tradeoffs for PV Power Plants

1. **SYSTEM**

   1.1 **Centralized vs distributed topology**
   - Modular power segments (PV/Inverter/transformer) paralleled at the substation grid
   - Centralized: Array subfields connected to central dc bus; all inverters paralleled (input/output)
   - Size of modular power segment components (PV subfield, inverter, and transformer)
   - Control scheme (on-site and remote)
   - Monitoring and display scheme (real-time and remote data monitoring)

   1.2 **Plant output rating: STC vs NOC and how to validate it on large array field**

2. **PV ARRAY**

   2.1 **PV Array**
   - No. of manufacturers to be involved
   - Power rating: minimum or average
   - Size and no. of modules or standardized panels
   - Interconnection between modules or standardized panels

   2.2 **Array structures**
   - Array orientation/structure
     o Fixed vs moveable (1-axis vs 2-axis)
     o Tilt angle adjustment
   - Foundation: steel vs concrete, 1-pedestal vs 2-pedestal for unit array
   - Module support: standardized unit array structure for one source circuit
   - Panel:
     o Factory vs on-site assembly
     o Panel size (No. modules)

   2.3 **Electrical wiring**
   - dc bus voltage level: between 12 and 1,000 Vdc input to the inverter [+]
   - Power feeders: single- vs bi-polar (plus, minus, and array center tap ground) [+]
   - Array/inverter tie: paralleled or separate input to the inverter
   - Subfield: 1 or mix of PV Mftrs in a subfield relative to inverter connection
   - Module connection: module grouping for a string based on Pm vs Isc
   - Shunt diodes: number and placement within a string vs 1 per PV module
   - Source circuit (i.e., string): standardized or custom-made (i.e., different or same Vm, Pm, Isc, Voc requirements)

3. **BATTERY**
   - Type: Pb-acid vs NiCd
   - Pb-acid
     o Sealed (valve-regulated) vs vented
     o Positive electrode type: tubular or flat pasted
   - NiCd
     o Pocket vs sintered plate
     o Sealed vs vented
   - Allowable depth of discharge
   - Number of batteries in parallel

   - Number of cells in series per battery
   - Rated cell size (Ah)

4. **INVERTER**
   - Power rating
   - Power quality: power factor, harmonics, and voltage and frequency regulation
   - Utility interface: individual or paralleled inverter output
   - Modular size and number of units; and their related control
   - Commutation: self vs line**
   - Control type: PWM or step pulses (6-pulse vs. 12-pulse)
   - Device technology: Transistor, power MOSFETs, thyristor, GTO, or IGBT*
   - Switching frequency: 20 kHz, 50 Hz, or greater
   - Protection and Safety Provisions

5. **ELECTRICAL DESIGN, FAULT PROTECTION, AND GROUNDING**

   5.1 **dc side**
   - Ground-fault detection and correction
   - Shunt diode placement
   - Blocking diode placement
   - Grounding (float or earthground) and ground fault detection

   5.2 **ac side**
   - Inverter protection (dc and ac sides and internal)
   - Utility grid abnormal conditions: allowable time to disconnect
   - Grounding
   - Electrical isolation

   5.3 **Utility-side ac protection**

   5.4 **Grounding (system and equipment)**

   5.5 **Lightning protection**
   - Air terminal and array structure grounding vs just array structure grounding
   - Array structure grounding
   - Active devices (overvoltage/high current units available individually or in modular package)
   - Grounding network

   5.6 **EMI/RFI protection**

6. **QUALITY ASSURANCE AND MAINTENANCE**

   6.1 **Quality control of large number of electrical components**

   6.2 **Test methods:**
   - Manufacturer acceptance
   - Receiving inspection and testing (in the field or staging area)
   - Plant acceptance test
   - Periodic tests

   6.3 **Maintenance**
   - PV array field
   - Power electronics (converters, inverters, etc.)
   - Battery recharging (equalization)

---

*	GTO -	gate turn-off thyristors	
	IGBT -	insulated gate bipolar transistors	
	STC -	Standard Test Condition	
	NOC -	Normal Operating Condition	

** Self-commutation required by stand-alone systems
+ For central power stations and large PV plants, dc bus voltage is usually ≥ 240 Vdc.

2.2.1 STAND-ALONE SYSTEM WITHOUT AN AUXILIARY POWER SOURCE (NON-HYBRID)

The key considerations for the stand-alone system are to satisfy the load demand and to meet the battery operating constraints. The usual procedure is to define a typical or worst-case load power profile at the main load bus and the corresponding solar energy data for one day as the basis for sizing the PV array and the battery. Some designers rely on a weather tape or "test reference year" for the solar radiation. This is a good procedure for engineers with access to a computer program. However, it should be emphasized that a first order sizing can be done quite adequately without a computer program.

Other key criteria that must be specified, all involving the operational constraints to achieve the required battery lifetime, are:

- Allowable battery depth of discharge (DOD)
- Number of sunless days the battery must be capable of supporting
- Length of time (or number of sunny days) to recharge the battery back to the original state (i.e., its initial state of charge before the battery went into deep discharge because of the sunless period).

2.2.2 STAND-ALONE SYSTEM WITH AUXILIARY POWER SOURCE (HYBRID)

The sizing criteria for this hybrid system are essentially the same as for the non-hybrid, except that either the PV array size or the rating of the auxiliary source must first be selected in order to determine the rating of the other power source. Because the PV array is the most expensive item, it is best to start the iterative calculation by assigning an initial power rating to the array.

2.2.3 GRID-CONNECTED SYSTEM

The sizing of a grid-connected PV system, given the peak power rating of the PV array at the main dc bus, involves determining the kVA rating of the inverter, the type and number of PV modules per string, and the total number of array strings in parallel.

The best inverter/PV source configuration for an optimum energy efficiency is multiple inverters in parallel, and operation of only the minimum number of inverters on line such that all inverters operate at, or near their peak efficiency. This simply means that the total kVA rating of the inverters connected on line should be equal to, or slightly less than, the maximum PV power. This sizing criterion implies the implementation of an operational procedure which in turn requires the ability to sense (i.e., calculate) the available PV power and control the on-off switching of the inverters.

However, if a single inverter configuration is desired for non-technical reasons, the inverter power rating should be equal to the actual peak power capability of the array at the inverter input.

2.3 ELECTRICAL LOAD REQUIREMENTS: VOLTAGE REGULATION, POWER QUALITY, POWER PROFILE, AND ENERGY

Electrical load requirements are best considered in terms of two basic types of PV systems, namely the non-grid-connected (stand-alone and hybrid) and the grid-connected. For example, load input voltage, power profile, and energy requirements affect the stand-alone and hybrid PV plants, whereas the ac power quality and ac network safety are the primary requirements of the grid-connected PV plants at the inverter/utility grid interface.

2.3.1   STAND-ALONE AND HYBRID PV SYSTEMS

DC Output Voltage Regulation

The output voltage range is usually dictated by the electrical load equipment. Low-power dc loads tolerate a wide range of input voltage, but the designer must identify the acceptable voltage range for each load so that an appropriate voltage regulation technique or device can be selected. Because most small dc loads are resistive, the steady-state power and voltage values are usually sufficient for selecting the appropriate voltage regulator or regulation approach. For dc motor loads, the transient input current and power characteristics are especially important.

Power Profile and Energy

The total power profile and energy over some increment of time is essential for the sizing of PV array and battery. For system sizing purposes, a worst-case load profile for the worst time period of the year is usually necessary.

In the early design phase, designers very often neglect to evaluate the effects of load management on both PV array and battery size. If discrete night-time and day-time loads can be defined individually, and the system can be designed such that the total battery drain is sufficiently controllable, then the load profile should be determined for these two distinct time periods. The total energy required for this load profile is the sum of the energy consumption during the two periods. In this situation, the cost of a stand-alone system can be reduced dramatically by system sizing and the optimization process (see Section 3.1.3, Chapter 3).

The stand-alone PV system designer should be conservative and aware of the user-load equipment input voltage and power characteristics, and the ratings of power handling components. Based on past experience, it appears that PV applications up to 1 kW continue to rely on 12-, 28-, and 48-Vdc nominal bus voltages because of the need to accommodate commercially available dc loads. Systems in the 1- to 5-kW range use a 120-Vdc bus and above this range, 240 Vdc. At the higher power levels, the consumers are predominantly ac loads, and consequently the key considerations in system design and selection become the availability and cost of self-commutated inverters.

AC Output

Ac load equipment should be defined in terms of the requirements for a particular inverter. The important inverter output parameters are:

- Power demand
- Single- or three-phase (usually the three-phase loads are in the 5 kW range or higher)
- Waveform
- Voltage regulation
- Variable frequency range (such as for pump motors)
- Steady state power
- Transient (start-up) or surge power
- Power factor

Based on the above, other inverter criteria, i.e., its input requirements can be defined with help from the potential inverter suppliers. If wind, Diesel, or other auxiliary power sources are to be operated in parallel, coordination with the generator supplier's application engineers early in the design phase is essential.

2.3.2   GRID-CONNECTED PV SYSTEMS

The load in such plants is the utility network, and the usual assumption here is that the grid is capable of accepting any amount of power from the PV plant. In other words, the utility grid serves as an infinite energy sink. The requirements for the grid-connected system are dictated by the utility company, and each utility may impose a unique set of requirements. The main grid interface criteria which should be checked with the utility are the following:[1]

- Voltage regulation

- Frequency regulation (usually 2% of nominal)

- Harmonic distortion in the operating load range:
  o Total of all current harmonics (usually 5% maximum)
  o Any single current harmonic (usually 3% maximum)
  o Total of all voltage harmonics (usually 5% maximum)

- Power factor and reactive power consumption:
  o Utilities often stipulate a power factor requirement for cogenerators, from 0.9 lagging to 0.9 leading at full load.
  o Reactive power consumption is closely related to the power factor (PF). Typical residential and industrial loads operate with a lagging PF as low as 0.85. Because of this, the utility requires some power factor correction by cogenerating sources to minimize reactive power being supplied by the grid. The inverters used in the PV system consume reactive power and thus, the utility could lose revenue due to real-power line losses.

- Protection and operation criteria such as:
  o Inverter disconnect criteria in the event of a grid failure (loss of voltage), inverter failure, or a ground fault on the dc side
  o Inverter reconnect criteria
  o Adequate safeguard against "islanding" (inability of self-commutated inverters to detect grid shut-down so that they continue to operate and feed power into the grid)

2.4   PLANT USE AND OPERATIONAL MODES

The key considerations in the selection of a PV system configuration, particularly for stand-alone plants, are the intended uses of the plant and its operating modes. Questions and factors which must be assessed are as follows:

1) Will the power output be needed continuously, (i.e., day and night, throughout the day, month, or year,) or only periodically? Are day-time and night-time loads different and predictable?

2) What are the short-term (say, the first 5-6 years) and long-term plans and uses for the plant? Related questions include:
   o Is the plant intended for R&D or commercial/demonstration use only, or R&D initially and commercial use at a later phase?
   o Will the plant hardware be relocated elsewhere at a later date?

3) Is the PV plant to be operated in the stand-alone mode only or will it be later converted to grid-connected operation?

4) Will there be a need to add an auxiliary power source (i.e., wind, Diesel, etc.) at a later date?

5) Possibility of future enlargement of PV arrays and/or batteries.

---

[1] The basic definitions of terms mentioned here and principles of ac theory essential to the understanding of inverter design, selection, and operation are given in Section 3.2.7, Chapter 3.

Of all the preceding questions, the most significant ones are 3) and 4) which affect the choice of inverters — *the ability to work in parallel with another ac power source has become an important factor for many stand-alone systems.* Because parallel operation was not contemplated at the time of project initiation, many projects have incurred significant costs in converting these plants to the grid-connected mode.

Many of the pilot plants were designed only for stand-alone mode because their sites were remote and the possibility of the grid network being extended to that location was not considered. Some of them have been used only at certain times of the year or intermittently during a day (for experimental purposes) so that they have not operated for full energy extraction.

An important lesson learned from almost a decade of plant operation is that PV plants, particularly the high-power ones, should be designed so that they can operate with any ac power source. For some of the stand-alone plants, the reason for converting to the grid-connected mode was simply that "the grid network was finally extended" to the remote site. For others where the utility grid was already available, the reason for conversion was that the battery had degraded significantly, hence the grid-connected mode became a logical alternative.

All future PV plants and likewise those already installed in the range of a few kW should consider the eventual possibility of selling PV electricity to the utility grid.

2.5   INPUT CHARACTERISTICS: THE SOLAR ENERGY

The main purpose of this subsection is to provide a brief review of solar energy terms and radiation characteristics which are important in design, testing, and understanding how the PV systems operate.

Solar Irradiance Terms

Solar irradiance, or solar flux on the earth's surface is best described in terms of its components which are quantifiable and measurable: 1) global normal, 2) direct normal, 3) diffuse (sometimes referred to as scatter or sky radiation), and 4) reflected.   These terms have been generally standardized[2], although some of them are not commonly used (e.g., GNI and DNI).

Global normal irradiance (GNI):	Total radiation on a planar surface perpendicular to the sun. It consists of direct normal, diffuse, and reflected components.
Direct normal (or beam) irradiance (DNI):	Nearly parallel rays that come from the solar disc. This component is GNI minus the diffuse and reflected components.
Diffuse (also scatter or sky) irradiance (DFI):	Global normal radiation on surface (GNI) minus the direct normal radiation.
Reflected irradiance (RFI):	Light reflected from clouds, hillsides, or objects near the ground (e.g., windows or structures).
Global horizontal irradiance (GHI):	Global radiation on a flat horizontal surface.
Tilted-fixed Irradiance (TFI):	Global irradiance on a fixed, tilted surface.

---

[2] See for example ref. [1]

GNI is measured by a "global" sensor mounted on a sun-tracking surface. DNI is usually monitored by a sensor with a light-collimating tube such as the Eppley Laboratories' Normal Incidence Pyrheliometer (NIP) mounted on a sun-tracking surface. This sensor can be constructed very easily, using any light-collimating tube with a 10:1 aspect ratio (length of tube to the maximum width of the sensor) and a global sensor at one end of this tube.

Diffuse and reflected radiation (DFI/RFI) are often defined as the solar flux or energy on a horizontal surface less the DNI component. Note that there is a large difference between the DFI/RFI on a horizontal surface and a tilted surface, depending on the sun's position. The typical instrument for the measurement of the horizontal DFI/RFI is the shadow-ring with a global sensor on a horizontal surface. The diffuse component is the result of radiation reflected or refracted through clouds, aerosols, icy crystals, and dust particles in the atmosphere. (For further discussion of solar irradiance sensors, see Section 3.3.4, Chapter 3.)

Availability of Solar Energy

Radiation from the sun is the dominant source of energy input into the earth's atmosphere. Virtually all the energy consumed by humans has come from the sun — oil, natural gas, and coal are all residues of plants and animals, which are in turn derived from solar radiation. The sun, combined with the rotational inertia of the spinning earth, also causes all weather and climatic changes via the movement of oceans and air. The resulting rain and wind resources power the hydroelectric and wind generators. A belief that solar energy is the most promising of the renewable energy sources is rapidly gaining acceptance.

The sun is about 150 million km from the earth on average, or one Astronomical Unit (1 AU). Sunlight travels at a velocity of 3×10^8 m/s, taking about 8 minutes to reach the earth's surface. The angle subtended by the outer rim of the sun to a point on the earth is 32 min. of arc or 0.5° (see Figure 2-1). Thus, the *direct beam* reaching the earth is almost parallel. The *solar constant* or *flux* received outside the earth's atmosphere at 1 AU has been established and generally accepted as 1,353 W/m^2 since the early 1970's [2,3]. It should be noted, however, that precision measurements of the solar constant have long been desired, but they have become possible only in the past two decades with instruments on satellites (NIMBUS, SMM, ERBS, NOAA, and UARS) in the earth's orbit. Based on actual satellite measurements in the late 1970's, Hickey and Crommelynck have suggested that the solar constant should be in the range of 1,363 to 1,371 W/m^2, depending on the instrument and the time in the solar cycle [4,5]. A key conclusion is that the solar constant varies [6].

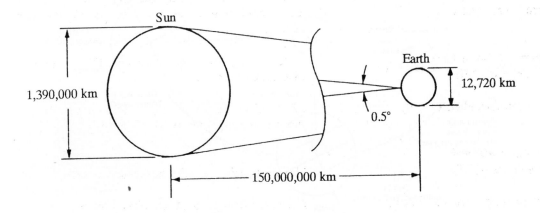

Figure 2-1. Earth and Sun Distance Relationships. Sun's incident rays hitting the earth are nearly parallel (0.5° intercept angle).

As the earth orbits around the sun, other important terms are frequently referred to: e.g., *solar declination* angle, *equinox*, and *solstice* which are also illustrated in Figure 2-2. Since the earth's orbit is elliptical, the sun-earth distance varies with time, and causes the Air Mass Zero (AMO) solar irradiance to vary by ± 3.4% throughout the year. This is illustrated in Figure 2-3. Note that the solar declination varies with time like the variation of the AMO irradiance. Four dates in the year have a particular significance. These are:

- Equinoxes, Spring (21 March) and Autumn (21 Sept); the sun is directly above the equator at noon; day and night duration is exactly 12 hours at any point on the earth's surface.

- Solstices, Summer (21 June) and Winter (21 December); the sun is directly over the Tropic of Cancer at noon on 21 June and directly over the Tropic of Capricorn at noon on 21 December.

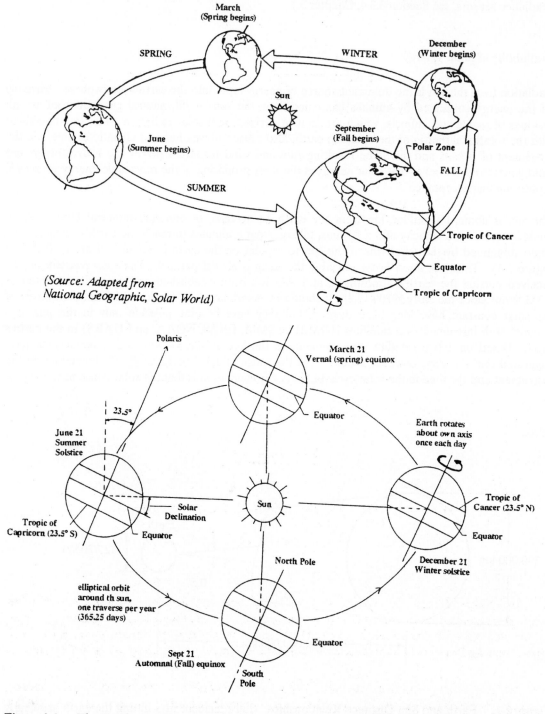

Figure 2-2.   The Earth's Orbit around the Sun with Its N-S Axis Tilted at an Angle of 23.5 Degrees

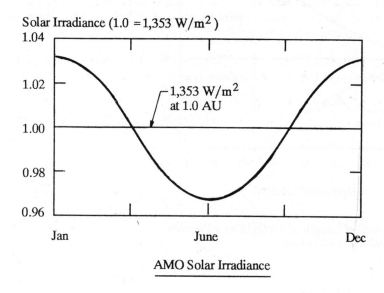

AMO Solar Irradiance

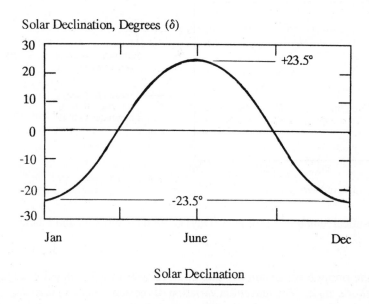

Solar Declination

Figure 2-3.   Annual Variations in AMO Solar Irradiance and
Solar Declination Angle

Also, the daylight duration is a function of the day of year and location. Figure 2-4 shows this relationship for different latitudes in the northern hemisphere. The theoretical sunlight-time per day parameter is accurately predictable on any PC, and it is very useful for performance analysis and operation of the PV plant. This parameter is illustrated in Figure 2-5 for two locations in Europe.

To an observer at a given location, the sun's path across the sky appears different at various times of the year. Figure 2-6 illustrates this profile to an observer standing at the equator and at 40°N latitude. This characteristic is important particularly to the designer of sun-tracking systems.

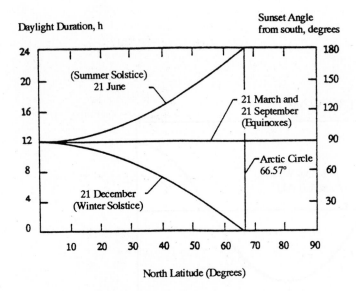

Figure 2-4. Theoretical Length of Daylight as a Function of Latitude and Season

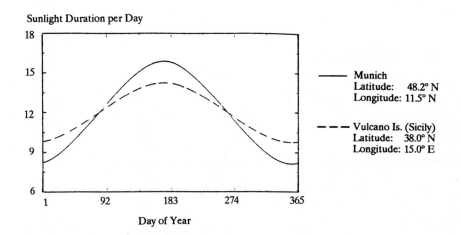

Figure 2-5. Theoretical Sunlight Duration vs. Days at Munich and Vulcano Island

Figure 2-7 shows the position of the sun at solar noon relative to a PV panel oriented to the South and tilted at the latitude angle. The maximum variation of the sun's angle in this case is ± 23.5° (also illustrated in Fig. 2-3). The reason many designers select the latitude angle is to get the most energy for the whole year from the fixed flat-plate arrays. As indicated by Figure 2-7, the average position of the sun angle relative to the plane of the panel occurs at the two equinoxes. The optimum tilt angle is site-dependent, and calculation of this angle requires a solar irradiance prediction computer programme validated with actual site insolation data for the whole year. There is a common agreement that for the higher latitudes the optimum tilt angle is usually 10 to 15° lower than the latitude angle. Hence, a general rule of thumb for the tilt angle for PV panels is to choose an angle which is zero to 15° lower than the site latitude angle.

Solar Irradiance Profile

The solar energy available is highly site-dependent and varies throughout the year. Typical daily profiles of global horizontal irradiance at different times of the year are depicted in Figure 2-8 for a 48°N latitude location (Munich). For the purpose of estimating the output or sizing of a given PV panel, one should make use of the global data on tilted surfaces from the European Atlas of Solar Radiation [1], making appropriate adjustment for the array tilt angle used.

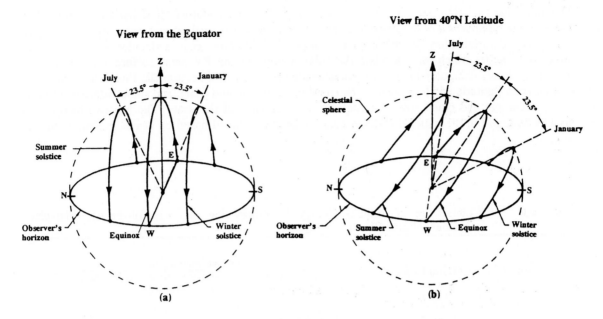

Figure 2-6. Sun's Path Across the Sky to an Observer at a) Equator and b) 40° N Latitude

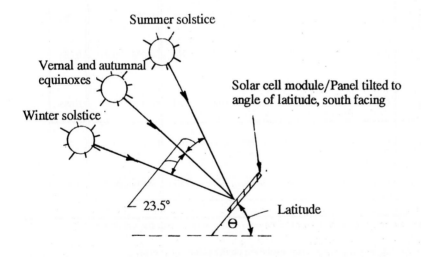

Figure 2-7. Sun's Position at Local Noon on a Fixed South-facing
Surface at Various Times of Year

The important aspects of the tradeoff analysis on different array orientations and tracking configurations are illustrated in Figure 2-9. This figure shows four most commonly used PV array configurations ranging from fixed-tilted to 2-axis tracking types. The solar irradiance profiles applicable to each array type are shown for a given day. For performance comparison and sizing purposes, one must estimate the energy output of each array type for the whole year. One approach indicated in Table 2-3 is to make use of efficiency values for each factor contributing to the array performance. Different array orientations and configurations can then be compared in terms of the overall annual efficiency.

Detailed solar radiation data (hourly or weather tape) is not mandatory for sizing PV systems. But for accurate comparison of different array orientation, sensitivity analysis and optimization of system sizing, and determination of annual energy output and life-cycle cost, a detailed solar irradiance prediction model which can calculate the solar energy on the PV array surface for one year is essential. Many computer models for solar irradiance prediction require the three basic inputs of site location (latitude and longitude), time (time zone, month, and year), and sun conditions in terms of sky clearness and cloud cover for each of the 12 months of the year. The last two factors can usually be found in the national climatic atlas, or simply assumed based on available data at similar sites.

Table 2-3.   Comparison of Annual Energy Efficiency for the State-of-the-Art Flat-plate and Concentrator PV Arrays

Array Energy Conversion Efficiency, $\eta_{pv} = \dfrac{\text{Annual array output}}{\text{Maximum annual solar energy on the sun-tracking surface (i.e., total normal energy)}}$

$$= \prod_{i=1}^{6} \eta_i$$

System Energy Efficiency, $\eta_{sys} = \eta_{pv} \cdot \eta_{iv}$

PV Array Type		Annual Efficiency 1 η_u	2 η_{mod} +	3 η_{mis}	4 η_d	5 η_{sw}	6 η_{fw}	7 η_{pv}	8 η_{iv}	9 η_{sys}
Flat-plate	Fixed-tilted to latitude	0.720	0.107 (0.14)*	0.98	0.997	0.99	0.98	0.073	0.90	0.066
	1-axis tracking, N-S axis horizontal	0.93	0.103 (0.14)*	0.98	0.997	0.99	0.98	0.089	0.92	0.082
	1-axis tracking, N-S axis tilted to declination angle	0.95	0.100 (0.14)*	0.98	0.997	0.99	0.98	0.090	0.93	0.084
	2-axis tracking	0.95	0.100 (0.14)*	0.98	0.997	0.99	0.99	0.090	0.93	0.084
Concentrator	Line focus type	0.85	0.178 (0.22)*	0.97	0.997	0.99	0.99	0.141	0.94	0.133
	Point focus type	0.88	0.204 (0.24)*	0.96	0.997	0.99	0.98	0.167	0.94	0.157

*   Number in parenthesis is rated efficiency (at Standard Test Conditions) of the best state-of-the-art production module available in 1990.

+ $\eta_{mod} = \eta_R [1 - \beta (T_{mod} - 25^\circ)]$ x annual average daily insolation on the plane of the array

Solar Irradiance Spectrum and Sensors

Figure 2-10 illustrates the spectra for three components of solar irradiance, *global normal*, *direct normal*, and *global horizontal* irradiance at AM1 around noon [7]. Note that all radiation components are lower than the AM0 spectrum at all times. This is caused by attenuation of the incoming direct beam by water vapour absorption, aerosols, clouds, etc., in the upper atmosphere. Likewise, global normal irradiance (direct beam plus diffuse) is the highest possible solar intensity at any time during a day as compared to the other radiation components (direct normal or diffuse). The effects of air mass on the solar irradiance spectrum are depicted in Figure 2-11.

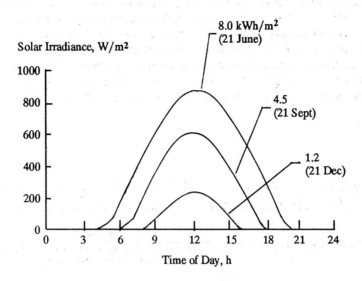

Figure 2-8. Typical Solar Irradiance Profile on a Horizontal Surface at Different Times of the Year at Munich (48° Latitude)

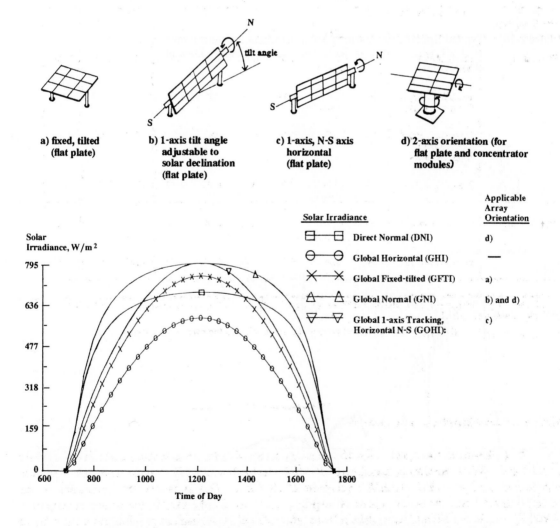

Figure 2-9. A Typical Set of Solar Irradiance Profiles for 15 January at a Low-latitude Site. Note that DNI is the only component applicable to the concentrator PV modules. The rest are for fixed (GFTI) and GHI) and moving (GNI and GOHI) flat-plate module surfaces.

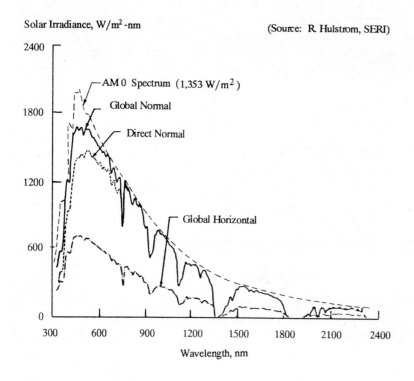

Figure 2-10.   Typical Spectra for Various Solar Irradiance Components at AM1 with AMO as Reference on a "Sunny" Day at Noon Time

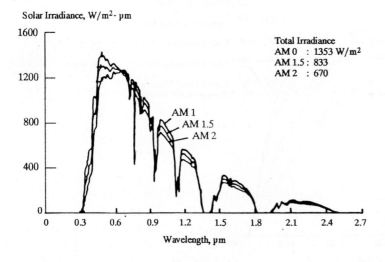

Figure 2-11.   Effects of Air Mass on Solar Irradiance Spectrum. The AMO spectrum is that of extraterrestrial solar irradiance at an average earth-sun distance (one Astronomical Unit)

It is important to understand the two different uses of the term air mass (AM). One is in terms of solar intensity (i.e., AMO intensity is 1,353 W/m^2, AM1.5 is 833 W/m^2, etc.). The other usage is in reference to the spectrum for the calibration of the *reference solar cell* used for solar irradiance measurements to determine the performance of solar cell devices. For this purpose, AM1.5 spectrum is commonly quoted by the PV community when referring to the calibration of reference solar cells or the resulting performance of the PV cells, modules, or arrays.

The AM1.5 spectrum was originally intended to be the reference spectrum for terrestrial PV performance calibration purposes. This spectrum was based on the AMO spectrum given in Table 2-4 [2,3]. The term AM1.5 has been misused or misinterpreted quite often, but its real intent was to standardize the performance rating conditions and to permit detailed spectral mismatch correction on reference solar cells. Thus, testing a PV device to AM1.5 spectrum simply means that the reference solar cell used to measure the solar intensity indoor or outdoor has been calibrated and corrected to the AM1.5 spectrum via computer analysis or transfer of the calibration factor from another reference cell which has been calibrated to the AM1.5 spectrum.

Table 2-4.   Extraterrestrial Solar Spectrum AMO*

λ (μm)	E_λ** (W/m$^2 \cdot \mu m$)	D_λ+ (%)	λ (μm)	E_λ (W/m$^2 \cdot \mu m$)	D_λ (%)	λ (μm)	E_λ (W/m$^2 \cdot \mu m$)	D_λ %
0.115	0.007	1×10^{-4}	0.43	1639	12.47	0.90	891	63.37
0.14	0.03	5×10^{-4}	0.44	1810	13.73	1.00	748	69.49
0.16	0.23	6×10^{-4}	0.45	2006	15.14	1.2	485	78.40
0.18	1.25	1.6×10^{-3}	0.46	2066	16.65	1.4	337	84.33
0.20	10.7	8.1×10^{-3}	0.47	2033	18.17	1.6	245	88.61
0.22	57.5	0.05	0.48	2074	19.68	1.8	159	91.59
0.23	66.7	0.10	0.49	1950	21.15	2.0	103	93.49
0.24	63.0	0.14	0.50	1942	22.60	2.2	79	94.83
0.25	70.9	0.19	0.51	1882	24.01	2.4	62	95.86
0.26	130	0.27	0.52	1833	25.38	2.6	48	96.67
0.27	232	0.41	0.53	1842	26.74	2.8	39	97.31
0.28	222	0.56	0.54	1783	28.08	3.0	31	97.83
0.29	482	0.81	0.55	1725	29.38	3.2	22.6	98.22
0.30	514	1.21	0.56	1695	30.65	3.4	16.6	98.50
0.31	689	1.66	0.57	1712	31.91	3.6	13.5	98.72
0.32	830	2.22	0.58	1715	33.18	3.8	11.1	98.91
0.33	1059	2.93	0.59	1700	34.44	4.0	9.5	99.06
0.34	1074	3.72	0.60	1666	35.68	4.5	5.9	99.34
0.35	1093	4.52	0.62	1602	38.10	5.0	3.8	99.51
0.36	1068	5.32	0.64	1544	40.42	6.0	1.8	99.72
0.37	1181	6.15	0.66	1486	42.66	7.0	1.0	99.82
0.38	1120	7.00	0.68	1427	44.81	8.0	0.59	99.88
0.39	1098	7.82	0.70	1369	46.88	10.0	0.24	99.94
0.40	1429	8.73	0.72	1314	48.86	15.0	0.0048	99.98
0.41	1751	9.92	0.75	1235	51.69	20.0	0.0015	99.99
0.42	1747	11.22	0.80	1109	56.02	50.0	0.0004	100.00

*   This table was adapted from Thekaekara [3]; Solar constant = 1,353 W/m^2
** E_λ is the solar spectral irradiance averaged over a small bandwidth centred at λ.
\+ D_λ is the percentage of the solar constant associated with wavelengths shorter than λ.

Solar energy workers normally do not use these spectral distribution curves in practice; for the PV community, the AM 1.5 condition is generally accepted as a condition for rating the output of PV arrays and modules.[3] Nevertheless, it is still a source of confusion, as related to the calibration of solar intensity sensors and the rating of PV modules and arrays, especially in sunlight testing in which solar intensity can vary widely even under seemingly clear sky conditions.

---

[3] All solar radiation terms discussed previously refer to an integrated value of the solar irradiance spectrum from about 0.3 to 3.0 μm. For theoretical calculations, a terrestrial solar spectrum at air mass 1.5 was first established at the second ERDA/NASA PV Measurements Workshop held in Baton Rouge, Louisiana in 1976.

Another school of thought has led to the most recent attempt to clarify the above discussion for outdoor testing: the procedure is to simply calibrate the reference Si sensors outdoors without relating to the air mass or other atmospheric conditions [8,9]. This procedure has certain benefits because of simplicity, and thus, it is being investigated in DGXII's Concerted Action Project on solar irradiance sensors. This topic is discussed in more detail in Chapter 3, Section 3.3.4.

For outdoor measurements of direct normal and global irradiance, respectively, pyrheliometers and pyranometers are the precision instruments most widely used. They use a thermopile sensor, which, combined with the quartz cover glass, has a relatively flat response between 0.4 and 2.8 μm wavelengths.

All concentrator PV systems have relied on the pyrheliometer (thermopile and Si) reading for direct normal irradiance measurements. For flat-plate PV systems, a mixture of thermopile and Si sensors are being used; some projects (with larger funding) use thermopile types only and others use both types. Silicon-based sensors are commonly used for both indoor and outdoor testing of flat-plate PV modules and arrays, but because of their lower cost, they are becoming more widely accepted by other technologists (solar thermal, meteorology, and environmental). In summary, the silicon sensors, even though they respond only up to 1.1 μm wavelengths, are considered by many experts to be a better sensor for continuous outdoor irradiance measurements because 1) they are much cheaper (3 to 5 times) than thermopile pyranometers, 2) the solar cells used as sensors have little or no degradation with time in contrast to the thermopile device, 3) once calibrated against a thermopile sensor, the Si sensors have demonstrated a 1-year stability better than 2% of the reference thermopile instrument [9] (for further discussion, see Section 3.3.4, Chapter 3).

2.6   CHARACTERISTICS OF USER LOADS

Load Types

Size, cost, and configuration of a stand-alone PV system and its dc bus voltage are strongly dependent on both transient and steady-state characteristics of the loads (i.e., "consumers").

For low-power stand-alone PV applications, only dc loads should be selected because an inverter, and hence its losses, can be eliminated. Individual loads are characterized by their input voltage and power requirements. Dc loads consist of resistive, constant current, constant voltage, or constant power in the applied voltage range. Typical dc loads are lamps, radios, stereo, TV, refrigerators, battery charging, and motors. Most of these loads are designed for use in recreational vehicles (e.g., campers, mobile trailers, and boats). Therefore, the PV system bus and component voltages are typically designed for 12- or 24-Vdc nominal operation.

The magnitude of power consumption and the lifetime of some of the load equipment are sensitive to the input voltage. For example, premature lamp burnout could occur if the lamp is operated at 14 Vdc instead of 12 Vdc.

Fluorescent lamps, power supplies with transformers, and high frequency converters used in microwave oven power supplies comprise a category of ac loads which utilizes induction coupling, resulting in energy transformation.

Transient and Steady-state Loads

The voltage and current inputs to motors vary according to the mechanical torque requirements of the driven loads. The transient in-rush power and current (4 to 6 times the steady-state values) of motors must be considered in system sizing.

2.7   OPERATING ENVIRONMENT

The PV system components installed outdoors must be designed to withstand sand, dust, salt, humidity, rain, snow, hail, wind, and temperature conditions at the site. The other environmental factors of importance are lightning strikes and earthquakes. The protection level required for the affected subsystems is site-dependent. The outdoor temperature variation can be rather wide in some cases (-40 to +50°C) which affects the performance of the solar array and batteries.

The power conditioning devices must provide proper voltage regulation even when solar irradiance fluctuates widely. The system controller must be designed to compensate for temperature variations affecting solar array output and battery life, and must withstand the severe temperature range. Temperature compensation is not commonly used in battery charge controllers, although for lead-acid batteries this is highly advisable to limit excessive battery overcharge, especially if the battery operating temperature varies widely.

For applications especially in coastal regions, special attention should be given to the prevention of corrosion of the electronic parts and terminations via conformal coating or other proven methods. Passive cooling as compared to forced-air cooling of electronic racks is highly desirable for systems in remote locations for two reasons, 1) to minimize entry of salty air which enhances corrosion, and 2) to avoid high temperature problems in case the fan motor fails.

The probability of a direct lightning strike is an important criterion in deciding whether or not lightning protection hardware is warranted. The type of lightning protection for direct or indirect strikes must be evaluated. The costs associated with protection from a direct lightning strike are prohibitively high, so protection only from indirect strikes should be considered in most instances.

2.8   SAFETY AND PROTECTION

The importance of safety and the protection of personnel and equipment cannot be overemphasized. The system must be designed to minimize hazards to operation and maintenance personnel, the public, and equipment. The control subsystem must be equipped with various fuses, built-in fault detection, and protection algorithms to protect the users, the loads, and the PV system equipment. The safety of an operator or technician is of the utmost importance. Personnel must be protected from electric shock by following all available safety practices, such as displaying high voltage warning signs wherever necessary. In general, the system must adhere to the local codes and standards dealing with safety issues.

The battery room should be well-ventilated with forced-air provision. The floor area below the cells should be acid-proof and capable of washing with a water hose; a portable eye-wash station which is self-contained and does not require external plumbing, should be installed in the battery room.

Some of the most important safety criteria are as follows:

1) Electrical components should be insulated and grounded.

2) All high-voltage terminations (>50 Vdc) should be properly covered and insulated.

3) All components with elevated temperatures should be insulated against contact with or exposure to personnel.

4) All moving elements should be shielded to avoid entanglements. Safety override controls and interlocks should be provided for servicing.

5) Structures should be grounded and ground fault relays installed to give warning of ground faults in the arrays or other electrical components.

2.9   ELECTROMAGNETIC INTERFERENCE

All power and control wiring in the PV field should be shielded to prevent any system noise from entering the dc supplied to the inverters. The low level signal circuits should be properly grounded, shielded and separated from the power and control circuits. These signal circuits should be protected from static, magnetic, common mode, and cross-talk noise sources.

2.10   OPERATION AND MAINTENANCE

Operation and maintenance criteria must be thoroughly considered in the design phase of any PV plant. Potential maintenance and test locations should be easily accessible, and components such as PV modules, electronic boards and assemblies, battery cells, etc., readily replaceable. Components subject to wear and damage should be easily serviceable or replaceable. The plant should be capable of being serviced with a minimum of specialized equipment or tools.

It is important to identify and stock critical spare parts, and to ensure that adequate operation and maintenance documents are available from the system designer. These documents should describe basic principles of operational and maintenance procedures, and basic trouble-shooting methods for each major component and field-replaceable part (e.g., printed circuit boards, battery cells, inverter control circuit cards, etc.). They must be written in a manner suitable for training persons unfamiliar with the component. They must be adequate, clear, and of sufficient detail to enable the plant operator's technicians to understand the operation of the system and subsystems, replace failed subsystems or components, and perform post-replacement functional tests.

2.11   MODULARITY

Large PV plants should be configured in a modular form to permit installation and operation of power segments in stages as they are completed. The modular design approach should be considered wherever economically and technically desirable for installation, testing, and maintenance.

2.12   RELIABILITY

Regardless of the size of the PV plant, consideration must be given during both the design and operational phases to achieve high reliability by providing adequate design and operating margins and by utilizing sound engineering practices. Appendix 29 presents the basic reliability principles and procedures. It should be noted that reliability design methods and practices are not standardized in the European Community as they are in the USA. Therefore, one should be careful in the use of reliability estimates specified by the device manufacturers (e.g., MTBFs may be quoted as 500 h by one supplier and 50,000 h by another supplier).

2.13   LIFETIME

The system should be designed to achieve the desired lifetime with nominal maintenance and repair. It has been demonstrated that PV modules can operate in all types of outdoor environment without degradation of their power output (generally less than 5 to 10% over a 10- to 20-year period). Many of the PV module manufacturers have demonstrated the ability of their hardware to meet a 30-year lifetime. The PV power output should not degrade permanently more than 10% from its initial rating over a 30-year life.

2.14   EXPANSION CAPABILITY

Experience has shown that a capability to increase the output power after initial installation and start-up is a key consideration for systems installed in remote, inaccessible sites. In a majority of stand-alone systems, the consumer loads have tended to increase significantly shortly after installation. To allow for system expansion later on, consideration must be given to the following:

- Current-handling capability of the dc power cabling and switchgear
- dc bus voltage level to be consistent with the anticipated addition of array strings or subfields
- Increase in electrical loads
- Availability of land for PV field expansion

2.15   PV SYSTEM HARDWARE AND PERIPHERAL EQUIPMENT

2.15.1   PV ARRAYS

Electrical Power Rating

The PV array converts solar irradiance into dc electrical power at the pre-determined range of voltages whenever sufficient solar radiation is available. The array should be designed to produce the specified rated power at the main dc bus at 1,000 W/m^2 and 25°C cell average temperature. The dc bus is usually the battery bus in stand-alone systems and the inverter input bus in grid-connected systems.

The key considerations in the PV array design are:

- Use of modular building blocks in terms of mechanical arrangement and electrical configuration wherever technically and economically desirable. PV module or panel mounting support structures and wiring should be modularized to permit cost reduction in installation.

- Module electrical isolation capability (should be twice the operating voltage plus 1,000).

- Use of stainless steel nuts and bolts in the array structures to prevent corrosion.

- Environmental requirements at the site, including indirect lightning strikes.

- The array should be designed for a 30-year life with minimum maintenance and repair.

- The number of junction boxes for electrical termination should be minimized.

- Minimization of all power losses, including:
  o Parallel string mismatch (<2%)
  o Series module mismatch (<2%)
  o Cabling (<2%)
  o Series blocking diodes

- Provisions to ensure personnel safety during installation and maintenance of the array.

2.15.2   ENERGY STORAGE DEVICES

Energy storage systems are generally required for stand-alone and hybrid PV plants. However, they increase the cost of the overall system and complicate the power conditioning electronics. Thus, a good understanding of their performance capabilities and operational constraints is essential in minimizing design and operational costs of both battery and power electronics, and in achieving desired battery lifetime.

From the user's viewpoint, the most important factors affecting battery design and operation are depth of discharge (per cycle), temperature, cycle life, number of cells in series, discharge control, charge control, number of sunny days desired to fully recharge the battery, and periodic maintenance (specifically periodic recharging). Figure 2-12 illustrates the effects of depth of discharge (DOD) and temperature on the life of secondary batteries such as Pb-acid (vented type) and Ni-Cd.

Batteries should be designed to meet the anticipated lifetime without any cell replacement. Cycle life is inversely related to DOD and temperature. Thus, a conservative design approach is to allow a low DOD and an ideal temperature range (e.g., 25% DOD and 5 to 20°C operating temperature).

The remaining critical considerations are to 1) minimize inadvertent deep discharges, 2) prevent cell voltage polarity reversals during discharge, and 3) ensure periodic maintenance recharging, especially during continuous low-insolation periods: the greater the DOD per cycle, the shorter the recharging intervals. To do this adequately would require an auxiliary power supply which can provide automatic float charging.

For sizing the lead-acid battery, the following DOD limits[4] for "low" and "high" voltage batteries are suggested to achieve the desired battery life in stand-alone or hybrid PV plants:

Desired Cycle Life (Yrs)	DOD Limit (%) for Sizing Purposes	
	Low-voltage Battery (24-48V)	High-voltage Battery (≥ 120 V)
1	50	40
5	35	25
10	20	15
15	15	10

The above DOD limits are based on conservative estimates for batteries operating in warmer climate regions (say, 40°C maximum ambient air temperature during the summer months). For batteries which will operate in a cooler environment (e.g., 15 to 30°C ambient temperature), the DOD can be increased by five percentage points from the values shown above. An important consideration is to minimize the temperature gradient between cells. This is especially critical for a high-voltage battery which has a large number of cells in series.

In selecting the battery size, the capacity at 10-h discharge rate (C/10) should be used. This provides a design margin because the actual rate in most plants is usually lower.

2.15.3 POWER CONDITIONING AND MANAGEMENT

Power conditioning and management requirements can be grouped into stand-alone and grid-connected categories. The key considerations are efficiency, lifetime, reliability, and degree of autonomy. The principal requirements are:

- Voltage and/or current transformation and regulation
- Control of electrical transients and noise
- Plant protection

- Electrical ground isolation
- Control of internal functions, and in some cases sensing and control of electrical loads
- Battery charge control and discharge protection

---

[4] Assumes C/10 rates in all cases

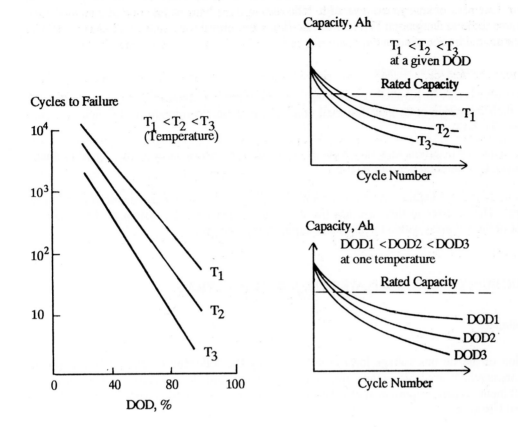

Figure 2-12. Effects of DOD and Temperature on Cycle Life of Batteries. To achieve long life in PV operation, the battery operating temperature should be kept low, and DOD in each cycle should be minimized (ideally 5-20°C and DOD less than 20%).

The power electronics in stand-alone systems may be simple or complex, depending on the power level and the extent of autonomous operation. As the power level goes up, the number of sensing and control functions generally increases, thereby decreasing the system reliability. Power conditioning requirements are more severe for a stand-alone system because of the need for effective battery charge control and discharge protection. Series regulation devices (e.g., dc-dc converters) add to the overall system cost, and should be avoided, especially on high-power plants, if possible.

A critical operational control consideration for stand-alone and hybrid plants is how to prevent the battery or batteries from 1) repeated deep discharges, 2) exceeding the specified DOD limit for normal operation, and 3) preventing individual cell voltage polarity reversal during any discharge period.

Hybrid systems require a large amount of sensing and control because of the need to optimize the overall operation of the system, considering:

- Available power from the PV array
- Battery operating constraints (DOD, low voltage limits, etc.)
- Operating constraints of the auxiliary power source (e.g., Diesel) to optimize its efficiency and life
- Load management criteria

A hybrid system, therefore needs a dedicated computer — typically a process logic controller (PLC) — for real-time monitoring and control functions. A key consideration in the design and operation of inverters is how to achieve high efficiency with varying input and output load, and thereby reduce the heat dissipation in the inverter. There are several ways to do so; a combination of the following should be considered:

- Increase the PV array input voltage

- Use multiple inverters in parallel[5], connecting only the minimum number of inverters on-line to maintain their operation at or near full load condition in order to operate in the high-efficiency region (see Fig. 2-13)

- Use high-frequency inverters with very high efficiency characteristics, such as the digital-synthesis type designed by Fraunhofer Institute [10]

A stand-alone system for higher power (> 1.5 kW) pumping applications may consist of a PV array and inverter. The inverter in this case has the function of operating the system at the maximum power point of the PV array within the allowable input voltage window.

2.15.4   DC BUS VOLTAGE AND POWER DISTRIBUTION DEVICES

DC Bus Voltage

The selection of the dc bus voltage level is influenced by the following considerations: 1) use of standard commercially available equipment, including motors and dc-ac inverters, if ac power is required; 2) input voltage requirements of the user equipment; and 3) minimization of distribution wire size and I^2R losses.

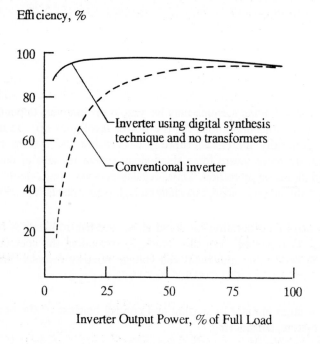

Figure 2-13.   Typical Inverter Efficiency Characteristics

---

[5] KYTHNOS PV plant (see Appendix 7) was designed to operate with one to three inverters on-line. It is one of the very few plants which has demonstrated such parallel-inverter operation.

Small PV systems in the range of 100 W to 1000 W will continue to rely on 12, 28, and 48-Vdc nominal bus voltages as dictated by the characteristics of available loads. Based on past experience, it appears that for PV applications up to 10-kW, the bus voltage should not exceed 120 Vdc nominally. Between 10 and 100 kW, a dc bus voltage of 240 Vdc has proven to be adequate. For higher power levels, 600 Vdc appears to be a practical limit due to switchgear cost and availability.

At the central power station size (several hundred kW and MW), a need to reduce the amount of copper used in power cabling and to increase the power efficiency of the inverter, make it important to apply a high input voltage to the inverter. The high-power semiconductors now available can operate up to several kV's. For this purpose, a bipolar arrangement with centre-tapped array (e.g., ± 400 or ± 500 Vdc) can be considered.

Series-parallel combinations of batteries and solar cell modules can be configured to satisfy almost any voltage and current level ranging from 12 to 1,000 Vdc. The availability and cost of power distribution devices are directly related to the bus voltage. The required output dc voltage level from the standpoint of the switching devices, therefore, should be kept as low as possible.

Switchgear

As noted above, the cost and availability of power distribution devices are directly related to switching voltage. For example, relays and disconnect switches are largely available in the rating of 28 Vdc, 120 Vdc, 250 Vdc, and 600 Vdc. At the same current rating, there is a sharp cost increase for these devices in the last two voltage levels. It is apparent that the dc voltage level for low-power systems (up to a few kW) should be kept in the range up to 120 Vdc to reduce the cost of power switching devices.

Cabling Losses

Large systems in the multi-hundred kW to several MW occupy a large area, hence they are burdened by I^2R losses from dc transmission cables. The multi-hundred kW systems in Europe have mostly stayed within 400-Vdc bus voltage levels, although the DELPHOS PV Plant uses a 800 Vdc bus. The main reasons for preferring the high dc bus voltage are the performance and cost advantages of the inverters and the lower cost of dc cabling. Inverters which use thyristors can operate efficiently at an input voltage over 1,000 Vdc. The drawbacks are the need for special high-voltage insulated cables inside the inverter, the isolation voltage limit of the PV modules, and the availability and cost of switchgear. Therefore, MW-size PV systems tend to use high dc bus voltages in the range of 500 Vdc to 1000 Vdc.

2.16 PERFORMANCE MONITORING

The cost of instrumentation can add significantly to the total cost of the balance of system (BOS). Therefore, to minimize the BOS costs, the data to be collected and displayed must be limited to absolutely essential measurements. Minimum measurements are needed to 1) ensure the basic safety of PV hardware and 2) cover maintenance and operation needs.

Tables 2-5 and 2-6 provide a suggested list of hourly recordings and calculated parameters (daily and monthly), respectively. The designer must review the contractual requirements for plant monitoring, including the allocated budget, when selecting the parameters to be acquired, calculated, and recorded. In the selection of the appropriate data acquisition system (DAS), the designer should compare the monitoring requirements against the capability of the DAS, including its field performance record. The Global and Analytical Monitoring criteria were established by CEC/DGXII for the original pilot plants in 1982 and have been updated periodically afterwards (see ref. [11] for the latest issue).

Table 2-5.   Hourly Recorded Data: A General List for All PV Plants

Parameter Description	Units	CEC Monitoring Guideline		
		Global Monitoring*	Analytical Monitoring*	Advanced Monitoring**
A. Monitoring and System Information				
1. % of data recorded	%			x
2. Total Energy to Grid	kWh		x	x
3. Total Energy from Grid	kWh		x	x
4. System Energy Efficiency	%			x
B. Solar and Meteorological Information				
1. Total POA insolation	kWh/m^2		x	
2. Total HOZ insolation	kWh/m^2	x	x	
3. Avg POA irradiance	kW/m^2		x	
4. Peak POA irradiance	kW/m^2			x
5. Sunlight avg POA irradiance	kW/m^2			x
6. Avg HOZ irradiance	kW/m^2			x
7. Peak HOZ irradiance	kW/m^2			x
8. Avg amb. air temp.	°C			x
9. Lowest amb. air temp.	°C			x
10. Avg wind speed	m/s			x
11. Peak wind speed	m/s			x
C. Plant/System Information				
1. Total energy to grid	kWh	x	x	x
2. Total energy from grid	kWh	x	x	x
3. Energy from auxiliary generator	kWh	x	x	x
4. Energy to dc loads	kWh	x		
5. Energy to ac loads	kWh	x	x	x
6. Hours load connected	h	x		
D. PV Information				
1. Avg array voltage	Vdc		x	
2. Peak array voltage	Vdc			x
3. Avg array current	Adc	x	x	x
4. Avg array power	kW			x
5. Peak array power	kW			x
6. Avg module temp.	°C		x	x
7. Peak module temp.	°C			x
8. Lowest module temp.	°C			x
9. Total array energy	kWh			x
10. Array energy capability	kWh			x
11. Array power capability	kW			x
12. Array energy utilitzation factor	%			x
13. Avg array efficiency	%		x	x
14. Peak array power efficiency	%			x
E. Battery Information				
1. Lowest discharge voltage	Vdc			x
2. Avg discharge voltage	Vdc		x	x
3. Highest voltage	Vdc			x
4. Avg discharge power	W			x
5. Peak discharge power	W			x
6. Avg discharge current (+)	Adc		x	x
7. Peak charge power	W			x
8. Avg charge power	W			x
9. Avg charge current, chg (-)	Adc		x	x
10. Avg cell temp.	°C			x
11. Discharge capacity	Ah			x
12. Charge capacity	Ah			x

*   (see next page)   POA - Plane of array   *(Page 1 of 2)*
** (see next page)   HOZ - Horizontal

Table 2-5.   (Concl.)

Parameter Description	Units	CEC Monitoring Guideline		
		Global Monitoring*	Analytical Monitoring*	Advanced Monitoring**
F. Power Conditioning Information				
1. Avg out voltage, inverter	Vac			x
2. Avg out current, inverter	Aac			
3. Avg out power, inverter	kVA			x
4. Peak out power, inverter	kVA			x
5. Avg out power, converter	kW			x
6. Peak out power, converter	kW			x
7. Total inverter energy	kWh		x	x
8. Total converter energy	kWh		x	x
9. Inverter energy efficiency	%			x
10. Converter energy efficiency	%			x
11. Inverter-on time	h			x
12. Converter-on time	h			x
G. Special Load Information (pumping, etc.)				
1. Avg flow rate	l/s			x
2. Total volume	l			x
3. Total time pump operated	h			
4. Avg current input to pump	Aac			x
5. Avg frequency, pump voltage	Hz			x
6. Avg pressure, pump	Bar			x

* Established originally by DGXII for plants in 1982 and finalized over several years afterwards;   *(Page 2 of 2)*
   now used mostly in DGXVII Demonstration/THERMIE Programme
** Established in 1988 by DGXII for plants which have DAS capable of scanning, processing, and
   recording such parameters

Table 2-6.   Daily and Monthly Performance Data: A General List for All PV Plants, Established by DG XII in 1988

Parameter	Units	Parameter	Units
A. Monitoring Information		**C. Plant/System Information**	
1. Hours of monitored data	h	1. Total load energy/pump energy	kWh
2. Missing data	%	2. Total energy to grid	kWh
		3. Total energy from grid	kWh
B. Solar and Meteorological Information		4. System energy efficiency	%
		5. Plant availability	%
1. Total POA insolation	kWh/m^2	6. Plant capacity	%
2. Daily avg POA insolation	kWh/m^2	7. PV specific yield (kWh/kWp)	h
3. Sunlight avg POA irradiance	kW/m^2		
4. Peak POA irradiance	kW/m^2		
5. Total HOZ insolation	kWh/m^2		
6. Daily avg HOZ insolation	kWh/m^2		
7. Sunlight avg HOZ irradiance	kW/m^2		
8. Peak HOZ irradiance	kW/m^2	**D. PV Information**	
9. Sunlight avg amb. air temp.	°C		
10. Daily avg amb. air temp.	°C	1. Total array energy	kWh
11. Lowest amb. air temp.	°C	2. Daily avg array energy	kWh
12. Peak amb. air temp.	°C	3. Array energy capability	kWh
13. Daily avg wind speed	m/s	4. Array energy utilization factor	%
14. Sunlight avg wind speed	m/s	5. Sunlight avg array power	kW
15. Peak wind speed	m/s	6. Peak array power	kW
16. Theoretical sun hrs.	h	7. Array energy/efficiency	%
17. Actual sun hrs.	h	8. Sunlight avg module temp.	°C
18. Daily avg sun hrs. actual	h	9. Daily avg module temp.	°C
19. Fraction of sun availability	%	10. Peak avg module temp.	°C

Table 2-6.   (Concl.)

Parameter	Units	Parameter	Units
E. Battery Information		3. Daily avg inverter power	kW
1. Daily avg DOD	%	4. Inverter energy efficiency	%
2. Daily avg cell temp.	°C	5. Peak inverter power efficiency	%
3. Recharge fraction, Ah	%	6. Converter efficiency	%
4. Recharge fraction, Wh	%	7. Total converter energy	kWh
5. Number of charge/discharge cycles	--	8. Inverter on-time	h
6. Lowest voltage during discharge	Vdc	9. Converter on-time	h
7. Avg discharge voltage	Vdc	10. MTBF, MTTR*	h
8. Lowest cell voltage	Vdc		
9. Avg charge voltage	Vdc	G. Load Information (pumping, etc.)	
10. Peak voltage	Vdc	1. Flow rate, daily avg	l/s
11. Highest cell voltage	Vdc	2. Total volume, daily avg	l
12. Discharge capacity	Ah	3. Total volume	l
13. Charge capacity	Ah	4. Pump on-time	h
		5. Daily avg current input to pump	Aac
F. Power Conditioning Information		6. Daily avg frequency of pump voltage	Hz
1. Total inverter energy	kWh	7. Daily avg pressure pump	Bar
2. Daily avg inverter energy	kWh;		

*   MTBF - mean time between failure
     MTTR - mean time to repair

2.17   REFERENCES

[1] European Solar Radiation Atlas, "Global Radiation on Tilted Surfaces," Verlag TÜV Rheinland, 1984.

[2] "Solar Electromagnetic Radiation," NASA SP-8005, National Aeronautics and Space Administration, May 1971.

[3] M.P. Thekaekara and A.J. Drummond, "Standard Values for the Solar Constant and Its Spectral Components," Nat. Phys. Sci., 229, 6 (1971).

[4] Private communication with J.R. Hickey of The Eppley Laboratories, Newport, RI, USA, September 1991.

[5] D. Crommelynck, "Solar Irradiance Observations," Science, Vol. 225, pp. 180-181 (1984).

[6] J.A. Eddy, R.L. Gilliland, and D.V. Hoyt, "Changes in the Solar Constant and Climatic Effects," High Altitude Observatory, NCAR, Nature (Vol. 300) 23/30 December 1992.

[7] R.E. Bird, R.L. Hulstrom, A.W. Kliman, and G.G. Eldering, "Solar Spectral Measurements in the Terrestrial Environment," Appl. Opt. 21, 1430 (1982).

[8] G. Beer, S. Guastella, H. Ossenbrink, and M. Imamura, "Concerted Action on PV Modules and Arrays: Investigation of Solar Sensor Calibration Methods," Conphoebus Report Prot. 669, CEC Contract EN3S-0140-1, Nov 1989.

[9] M. Grottke, M.S. Imamura, W. Bechteler, and O. Mayer, "Comparative Assessment of Silicon Cell Pyranometers," Proc. of the 10th European PV Solar Energy Conference, Lisbon, Portugal, April 1991.

[10] A. Goetzberger and J. Schmid, "Recent Progress of PV Systems of Intermediate Power Range," Presented at the 3rd International Photovoltaic Science and Engineering Conference, Tokyo, JAP, 3-6 November 1987.

[11] "Guidelines for the Assessment of Photovoltaic Plants, Document A, Photovoltaic System Monitoring," Issue 4, Commission of the European Communities, JRC-Ispra, June 1991.

Chapter 3

PRACTICES
AND GUIDELINES

This chapter provides a summary discussion of current practices and guidelines in the following areas:

- Design and operation at the system and
  subsystem levels
- Performance monitoring
- Project management

- Documentation
- Non-technical issues
- Public information dissemination

The guidelines presented here are based largely on the lessons learned from the past and on-going CEC programmes. They are intended to highlight only the key areas of interest or concern focusing on large stand-alone systems, so we encourage readers to consult the many excellent publications now available on all types of PV systems.

Of particular importance for sources of new lessons learned and design and operating strategies are the Concerted Action projects sponsored by the Commission in 1987-1989. The first-phase results are available on the following Concerted Action tasks which continued into the JOULE programme (1990-1992):

- PV modules, arrays, and solar sensors [1]
- Battery control/management [2]
- Power conditioning [3]
- Plant data/plant monitoring [4]
- PV array structures and foundations [5]
- Social benefits [6]

We should point out, however, that the PV community in general lacks documentation on actual operating performance and experience. Timely reporting and analyses of failures and mistakes are important steps towards making hardware and software work better at a lower price. But the cooperation of all organization researchers, design engineers, and users is continuously needed to work towards this goal of publicizing their real experiences and feeding the useful performance results to the manufacturers of equipment. Otherwise, the use of the same unreliable and/or costly methodologies and repetition of the same problems will continue in the future.

Early dissemination of information on problems and solutions implies that many of the guidelines and the approaches suggested herein may not yet have been fully validated. We cannot guarantee the correctness of these guidelines in many instances, especially those which involve warranties. The designers or users must also consult other sources of information, and rely on their own experiences to arrive at the best method of solving their problems.

3.1   SYSTEM DESIGN

3.1.1   PLANT SITING AND LANDSCAPING

Plant siting was dominated by environmental and aesthetic concerns. At all plants, much importance was given to the Commission's guidelines to ensure that the PV field blended with the surrounding landscape. As far as possible, local stone and pebbles were utilized for the enclosures and buildings. The resulting array fields did not disturb the skyline. Moreover, the array structures were designed to match the surroundings. The emphasis for the pilot plants has been for stand-alone applications, so many of them were located in rather remote areas in each country. Figure 3-1 shows pictures of the pilot plants, illustrating a few examples of the sites and their landscapes.

3.1.2   SYSTEM CONFIGURATION AND ARRANGEMENT

Types of System Configuration

PV systems as illustrated in Figure 3-2 are often referred to simply as stand-alone, hybrid, and grid-connected systems. At present, the commercial application of grid-connected PV systems is limited by the cost of PV power, which is much higher than that of large conventional power generators. The stand-alone PV systems, although they are more expensive than conventional sources, are penetrating significantly into the world market, replacing more and more small petrol and diesel generators as well as traditional throw-away battery applications.

The configurations for the basic stand-alone systems illustrated in Figures 3-3 to 3-9 are arranged roughly in the order of simple to more complex. An inverter is included in most of them, but it can be deleted if there are no ac power requirements.

The dc systems are generally preferred for small applications up to 2 kWp installed PV peak power. Their main advantages are simplicity and high reliability. Their disadvantage is that it might be difficult to find standard off-the-shelf dc consumer equipment on the market, since common household appliances are designed exclusively for ac grid power supply. However, standard 12/24 V dc systems can feed equipment and components available from the motor vehicle and camping market. A dc system with a battery (Fig. 3-3) represents the standard configuration for most small lighting and electrification applications. During the day, PV energy is used to charge the battery, so that power is available 24 hours a day.

Systems without a battery are extremely simple, with a variable frequency inverter directly connected to the PV array (see Fig. 3-10). They are usually coupled to an external energy storage device (e.g., a water tank in the case of pumping, or eutectic cold storage in the case of refrigerators) since such systems supply power only in daytime. These ac systems without a battery have reached widespread commercial application for PV pumping in a power range from 600 Wp to 3.5 kWp. The variable frequency inverter regulates the pump speed in order to make maximum use of the available PV power. The main advantages of such systems are high efficiency, rugged design, and the possibility to operate many off-the-shelf standard ac pumps.

An optimum arrangement of a stand-alone system must be capable of 1) battery charge control via the PV array or an external back-up power supply, 2) off-line test and maintenance of the battery, and 3) operating the back-up generator in its desirable modes. This is illustrated in Figure 3-11.

VULCANO (Sicily)

LAMOLE (Italy)

RONDULINU (Corsica)

Figure 3-1.   Examples of Pilot Plants Showing Landscapes

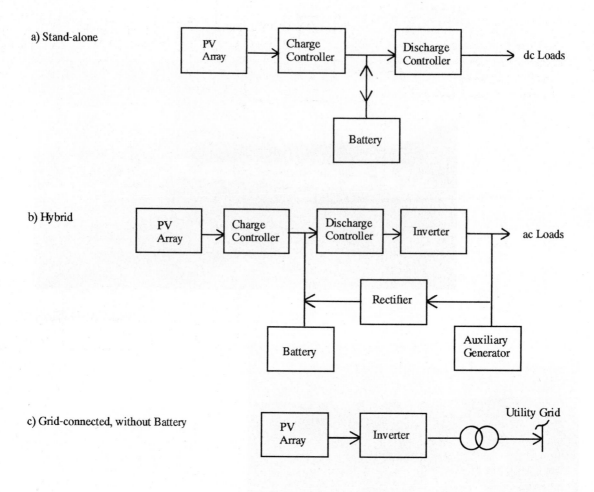

Figure 3-2.   Three Basic Types of PV System Configuration: a) Stand-alone, (PV/ Battery, b) Hybrid, (PV/Battery/Auxiliary Generator), and c) Grid-connected, (PV/Inverter/Transformer).

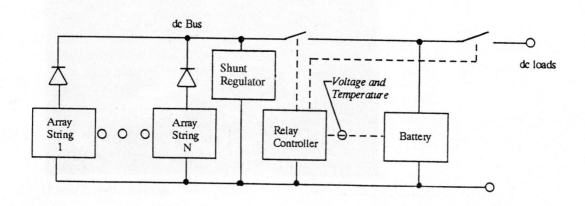

Figure 3-3.   Battery/Discharge Controller via Relay Switching. This arrangement is commonly used in small stand-alone "chargers" which provide both charge and discharge protection functions.

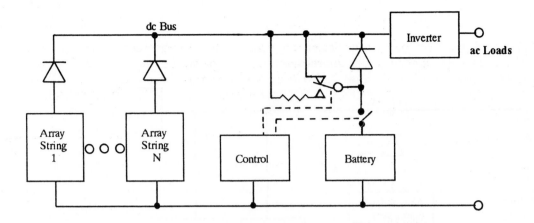

Figure 3-4. Battery Charge Voltage Regulator by On-off Control with Trickle Charging Resistor. This type of battery charge control has been used on large PV plants, but it is not recommended, especially for large batteries.

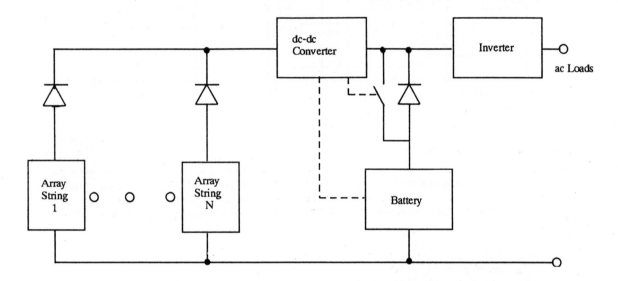

Figure 3-5. Battery Charge Voltage Regulation by Series Converter. The converter is used primarily for battery charge control with or without maximum power tracking; also, the converter may be either the "buck" or "boost-buck" type.

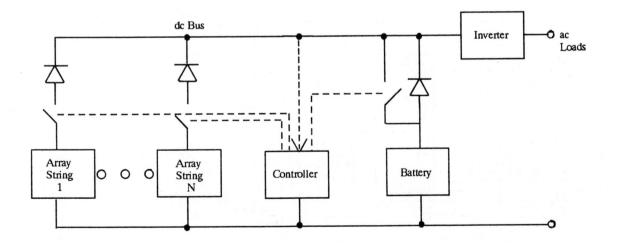

Figure 3-6. Battery Charge Voltage Regulation by Array String Control. This type of voltage regulation has been used very successfully on medium- to high-power plants for dc bus voltage levels up to about 500 Vdc.

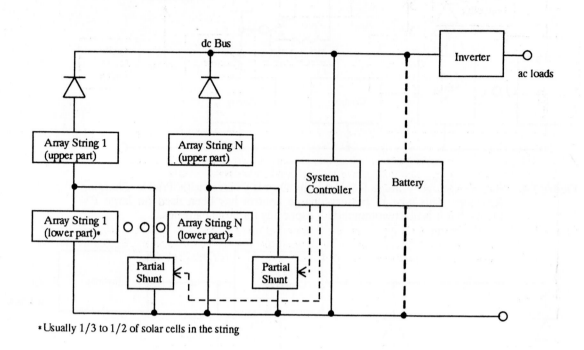

Figure 3-7.   Battery Charge Voltage Control with Partial Shunt (Linear or Digital) Regulation. This configuration is useful for medium- to high-powered PV stand-alone systems. Note that the partial shunt can be a linear analog regulator or an on-off switch (digital), and the system controller can be analog or digital (i.e., μP board or PC).

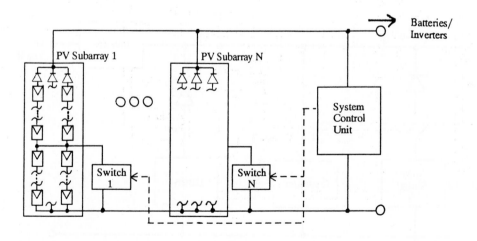

Figure 3-8.   Partial Shunting Approach to dc Bus Voltage Regulation for High Power PV Systems with a Large Number of Array Strings. One subarray shown above may have several strings in parallel, such as in a high power array.

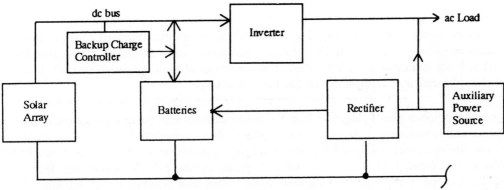

Key Features:
- Permits better battery charge maintenance at low rates which result in higher capacity acceptance
- Allows much better utilization of PV energy as compared to one without the rectifer
- Requires significantly lower PV array size; thus, lower system cost

Figure 3-9. An Optimum Configuration of a Stand-alone PV Hybrid System with an Auxiliary Power Source such as a Small Diesel or Wind Generator. Back-up charger is recommended, but it is optional if the rectifier is reliable.

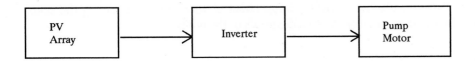

Figure 3-10. PV System without Battery. This type with a variable frequency inverter is commonly used in pumping applications above 2-3 kW of power.

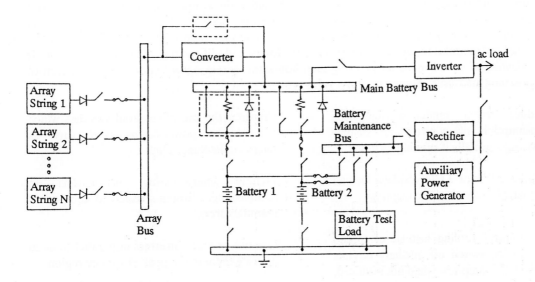

Figure 3-11. A Conceptual Arrangement of a Stand-alone Hybrid PV System for High Reliability Applications. The system illustrated is capable of 1) redundant battery charge control via converter or array string on-off voltage regulation, 2) independent battery connection to the main bus or off-line bus for low-rate or float charging via a rectifier, and 3) off-line battery test for capacity check or battery state of health diagnosis.

Alternating current systems with a battery (Fig. 3-4 to 3-9) represent the standard configuration for larger applications (abvove 1 kWp). In this case, a conventional fixed frequency 50-Hz inverter is used to feed ac loads continuously 24 h per day. The advantage of such systems is that off-the-shelf household ac appliances or equipment can be used as consumers.

Grid-connected PV power plants (Fig. 3-2) feed PV power directly into the grid by means of one or more inverters and transformers. In most larger systems, line-commutated inverters are used for such purposes, equipped with a special PV controller which operates the PV array near its maximum power point. The number of grid-connected PV plants in the 1-5 kW range has increased dramatically in Europe, particularly in Germany, Switzerland, and Austria, within the past three years. This trend is expected to continue in the near future with a growing number of PV-powered residences.

The inclusion of a main battery storage in a grid-connected PV system confers little operational or economic advantage. However, large batteries are used in UPS where a highly reliable power source is required, and/or the utility grid is unreliable. In terms of installed capacity, the UPS market is substantially greater than the PV technology sector. The concept which may gain momentum in the near future is the grid-connected battery system for peak-power and load-levelling uses in the utility network. In such systems, the batteries replace the PV array in Figure 3-2. In one peak-power shaving system in operation in Berlin since 1986, BEWAG has shown that the battery-inverter system is more cost effective than a gas-turbine generator for this application [7]. The system produces 17 MW within a few seconds (after a "brown-out" has occurred). In such cases, grid-connected PV/inverter and battery/inverter plants electrically independent of each other but serving the same grid network can be very attractive in satisfying specific needs of the utility.

General System Design Guidelines

1) Avoid making a PV system more complex than is absolutely necessary. When a battery is added, the system electronics (voltage regulation and control) tend to become complex.

2) Design the total system for the highest power conversion efficiency possible under different sets of operating conditions. To do so, identify the principal operating modes and variables, define individual component efficiencies and loss factors under those operating conditions and then calculate the overall system efficiency using the power flow diagram. For this purpose, a simple sizing equation is very useful (see Subsection 3.1.3). You may want to calculate the cost of the additional PV modules required to compensate for losses and self-consumption in your system. It is worthwhile to spend more engineering man-hours for system optimization and the design of energy-saving measures. Some important rules are:

- Reduce to a minimum the self-consumption of control and display devices, especially if running 24 h/day

- Choose LED or LCD displays instead of control lamps or video displays

- Use natural cooling instead of forced ventilation, switch off auxiliary circuits such as the lamp test when not required.

- Consider the use of solenoid switches which are of the power-saving type, such as magnetic-latching relays

- Include energy-saving instructions in technical specifications issued to component manufacturers

- Use two or more inverters in parallel so as to operate them in the peak efficiency region

3) Avoid depending on ac power for critical control functions. If you do rely on ac power, once the system is "dead" (battery discharged) it cannot be restarted, since the inverter does not have a power source. For this reason, an auxiliary power source such as a separate small PV panel, a separate battery in a UPS, and/or a small, but properly sized gasoline generator may have to be provided.

4) Wherever possible, use dc power for controls, auxiliary devices, solenoids, etc., since battery power (UPS or main battery) is much more reliable than ac power supplied by an inverter, and less power is consumed with only dc loads.

5) Because the battery is the most fragile component in a PV plant, put maximum care into sizing, design, protection, handling, operation, and maintenance of the battery. Carefully consider the need for long-term trickle charging of the battery from a separate power source (wind or generators, utility grid, etc.)

6) Provide short-circuit protection on batteries and incoming utility lines.

7) Do not use an inverter unless it is absolutely essential because of ac loads.

3.1.3   SIZING AND SELECTION

Photovoltaic System Sizing

For satellites and spacecraft operating outside the earth's atmosphere where the solar flux and available energy are highly predictable, it is standard practice to size the PV arrays and batteries on a daily basis in the worst-case eclipse conditions. This is not the case for PV systems in the terrestrial environment. In most parts of the world, full sunlight is not available every day, and often the irradiance is reduced because of inclement weather and constant changes in the sky. For this reason, several approaches to stand-alone PV system sizing have been suggested and used. The sizing models used can be grouped into the following types:

	References
I. Balancing of user and available energy under worst-case conditions in a selected time period or month	Barra, et. al [8] Sharp brochure [9] Saha [10] Imamura, et. al [11,12] Buresch [13] PRC [14]
II. Life-cycle-cost, weather data, annual or monthly basis, and energy balancing	Chapman, et.al [15] Menicucci, et. al [16]
III. Use of "Loss-of-load" probability, weather data, annual or monthly basis, and energy balancing	Evans, et. al [17] Grompous, et. al. [18] Macomber, et. al. [19]
IV. A combination of I and II	Soras, et. al [20]

For each sizing model, two factors affect the level of complexity, namely the input data used and whether or not the mathematical model is based on power flow relationships or on nodal equations (i.e., the calculation of voltage and current at each node). The above types of sizing methods are listed in order of their complexity, Type I being the simplest. We believe that the simplest forms of sizing equations which can be dealt with by non-specialists, without the use of a standard computer are needed by most designers. Therefore, we have attempted to illustrate the basic mathematical equations, based on the simplified sizing considerations and strategies identified herein.

1. Energy Balance on a Daily Basis

The energy balance relationship is the principal basis for the sizing of solar arrays and batteries in stand-alone PV plants. The basic energy balance relationship for any stand-alone PV/battery system for a given day is usually expressed as:

$$E_{SA} = E_N + E_D \qquad\qquad (3\text{-}1)$$

where E_{SA} = Required PV array energy in one sunlight period (i.e., in one day)

             E_N = Energy required to recharge the batteries (to replace the energy supplied by the batteries during the previous night)

             E_D = Energy required by the electrical loads during the daytime

Eq. (3-1) means that the total array energy available in a given day is exactly equal to the energy required during "night-time" and "daytime". In other words, the total load cannot exceed the available array energy if energy balance is to be maintained in a PV/battery system. However, equation (3-1) is *applicable only when successive sunny days are available.* Since this is impossible, we need an energy balance equation over several days which combines sunny and sun-less days. The next subsection describes the sizing approach for this situation which applies to all stand-alone systems. However, in order to derive such multiple-day energy balance relationships, it is easier to start out with the basic energy balance equation on a *daily basis*.

A typical stand-alone PV system comprises a number of array strings and a series converter or a shunt regulator for battery charge control. Also, one or more inverters may be paralleled electrically at the main output bus if ac loads exist. To illustrate the array and battery sizing procedures, we will use the PV system configuration shown in Figure 3-12 to develop the energy balance equation. This system layout is representative of a majority of PV systems. However, the energy balance equation is relatively easy to develop for any other PV system arrangement.

Figure 3-12 shows the power flow diagram for any ac system configuration with no series dc-dc converter, e.g., Figure 3-6 through Figure 3-9. The energy balance equation, eq. (3-1), can be expressed in terms of average quantities of system and component efficiencies and power:

$$P_{SA}\,\tau_D = \frac{P_{NL}\,\tau_N}{K_1 K_2} + \frac{P_{DL}\,\tau_D}{K_1} \qquad\qquad (3\text{-}2)$$

where $K_1 = \eta_{Deg}\,\eta_{W1}\,\eta_R\,\eta_{W2}$

            $K_2 = \eta_D\,\eta_B\,\eta_{W3}$

           P_{SA} = average solar array power required during daylight at the time of installation

           P_{DL} = average electrical load at bus during daytime

           P_{NL} = average electrical load at bus during night-time

            τ_D = daytime duration

            τ_N = night-time duration

         η_{Deg} = total degradation of solar array power output at a given time

          η_{W1} = line loss from solar array source to charger

          η_{W2} = line loss from load regulator to bus

          η_{W3} = line loss from batteries to regulators

           η_B = battery watt-hour efficiency

           η_D = blocking diode efficiency

           η_R = inverter power efficiency

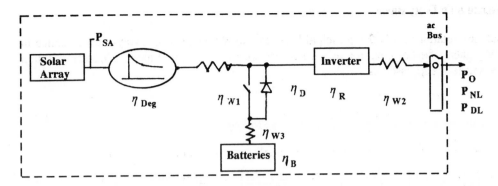

Figure 3-12. Power Flow Diagram for Sizing of PV Stand-alone System. This diagram corresponds to the system configurations given in Figures 3-6 through 3-11 without a series converter.

If we define σ as the ratio of day-time and night-time loads ($\sigma = P_{DL}/P_{NL}$), and substituting this for P_{DL}, in eq. (3-2), the required beginning-of-life array power becomes:

$$P_{SA} = \frac{P_{NL}(\sigma K_2 \tau_D + \tau_N)}{K_1 K_2 \tau_D} \tag{3-3}$$

Note that if P_{DL} and P_{NL} are equal, i.e., $\sigma = 1$, eq. (3-3) simplifies to:

$$P_{SA} = P_O \left[\frac{K_2 \tau_D + \tau_N}{K_1 K_2 \tau_D} \right] \tag{3-4}$$

where $P_O = P_{DL} = P_{NL}$. The term P_O is also the average power output capability of the PV/battery system at the dc bus. This equation for the case of a single average load term is the one most widely used for calculating the size of the solar array. For the case where $P_{DL} \neq P_{NL}$, eq. (3-3) should be used for array sizing purposes.

The required array power in eq. (3-3) can be expressed in terms of the daily average power, P_O, and day/night load ratio, σ, by introducing the following definition of P_O (see Fig. 3-13):

$$P_O = \frac{1}{\tau_O} (P_{NL} \tau_N + P_{DL} \tau_D) \tag{3-5}$$

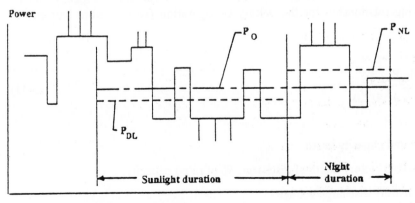

Figure 3-13. Definitions of Average, Day, and Night Power

By letting $P_{DL} = \sigma P_{NL}$, eq. (3-5) becomes:

$$P_O = \frac{P_{NL}}{\tau_O}(\tau_N + \sigma \tau_D) \tag{3-6}$$

Then, eq. (3-2) can be expressed in terms of P_O and σ as follows:

$$P_{SA} = P_O \left[\frac{\tau_O}{\tau_N + \sigma \tau_D} \right] \left[\frac{\sigma K_2 \tau_D + \tau_N}{K_1 K_2 \tau_D} \right] \tag{3-7}$$

Note the differences in arriving at the required array power, P_{SA}. When one assumes a constant load (i.e., $P_O = P_{DL} = P_{NL}$), P_{SA} is determined using eq. (3-4) alone. But, if $P_{DL} \neq P_{NL}$, one must first define the required values of P_O and P_N and then calculate P_{SA} using eq. (3-7). Thus, in this array sizing procedure, eq. (3-6) and (3-7) are involved. The value for P_{DL} is determined if σ and P_{NL} are known.

The total number of modules required in the array field, N_T, is the product of the total number in series per string, N_S, and the total number of strings in parallel, N_P, or

$$N_T = N_S \cdot N_P \tag{3-8}$$

and

$$N_T = \frac{P_{SA} \, \eta_d \, \eta_{fw} \, \eta_{bw} \, \eta_{mm} \, \eta_{bm}}{P_{mo} [1 - \beta (T_1 - 25)] H_1 \cdot 10^{-3}} \tag{3-9}$$

$$N_S = \frac{\text{Maximum dc bus voltage}}{V_m [1 - \beta (T_m - 25)]} \tag{3-10}$$

where all other terms in the two equations above are defined in Appendix 27.

The battery size is mainly dependent upon the night-time (or battery) load, allowable depth of discharge (DOD), and design margin. This battery size in terms of the rated capacity can be calculated from the following relationship for the battery configuration (one or more batteries in parallel):

$$C_R = \frac{N_B P_{NL} \tau_N}{K_2 K_D C_D V_B} \tag{3-11}$$

where: C_R = battery rated capacity in Ah

            C_D = the allowable battery depth of discharge, % of rated capacity

            V_B = average battery discharge voltage

            K_D = design margin

            N_B = number of batteries in parallel

The total battery capacity depends on the number of batteries in parallel (N_B). To achieve the required overall rating, we recommend at least two batteries in any PV plant, even though one battery with larger capacity cells is 10 to 20% less expensive than two batteries containing cells of one half the capacity rating of the single large cells. The technical benefits of implementing two or more batteries are higher reliability through redundancy and the ability to maintain one battery bank while the other battery is working.

2. Energy Balance Over Several Successive Days

When solar energy is not available and the battery is sized to handle the load for a selected number of sun-less days, the daily energy balance equation (eq. 3-1 and eq- 3-7) cannot be used. This number of sun-less days (often referred to as "autonomy days") increases the size of both battery and PV array. Therefore, to reduce the overall system cost, it is best to use an auxiliary source of energy rather than continue to increase the PV array area. (Note that in doing so, the system becomes a "hybrid" plant). This reduces the battery capacity required and consequently the PV array size. For this case, the energy balance equation is:

$$E_{SA} + E_{AUX} = E_N + E_D + E_R \tag{3-12}$$

where E_{AUX} = the additional battery recharge energy required from an additional PV array or an auxiliary power source.

E_R = the total energy supplied by the battery during the sun-less days

and other terms are as defined in equation (3-1).

The additional energy required, E_{AUX}, must equal the demand, E_R.

For a PV-only system (i.e., no auxiliary power source), the power flow relationship for E_{AUX} is:

$$P_{SAR} \tau_{DR} = \frac{P_{NL} \tau_{NR}}{K_1 K_2} \tag{3-13}$$

where P_{SAR} = Average array power during τ_{DR} required to fully recharge the battery represented by E_R

τ_{NR} = Total duration of battery discharge during the sun-less days

τ_{DR} = Sunlight duration allowed to fully recharge the battery after previous sun-less days

The average array power required to maintain the daily energy balance and to recharge the battery after previous sun-less days is the sum of equations (3-2) and (3-10), or

$$P_{SA} \tau_D + P_{SAR} \tau_{DR} = \frac{P_{DL} \tau_D}{K_1} + \frac{P_{NL} \tau_N}{K_1 K_2} + \frac{P_{NR} \tau_{NR}}{K_1 K_2} \tag{3-14}$$

By letting $\tau_{DR} = \gamma \tau_D$, equation (3-14) becomes:

$$P_{SA} + \gamma P_{SAR} = \frac{K_2 P_{DL} \tau_D + P_{NL} \tau_N + P_{NR} \tau_{NR}}{K_1 K_2 \tau_D} \tag{3-15}$$

Since P_{NR} can be assumed to be equal to P_{NL}, and using eq. (3-6) for P_{NL}, eq. (3-15) simplifies to:

$$P_{SA} + \gamma P_{SAR} = P_0 \left[\frac{\tau_0}{\tau_N + \sigma \tau_D} \right] \left[\frac{K_2 \tau_D \sigma + \tau_N + \tau_{NR}}{K_1 K_2 \tau_D} \right] \tag{3-16}$$

and the ratio of total array power required to average load is:

$$\frac{P_{SA} + \gamma P_{SAR}}{P_0} = \left[\frac{\tau_0}{\tau_N + \sigma \tau_D} \right] \left[\frac{\sigma K_2 \tau_D + \tau_N + \tau_{NR}}{K_1 K_2 \tau_D} \right] \tag{3-17}$$

Using the values of the various parameters listed in Table 3-1 in eq. (3-17), the effects of increasing the day/night ratio on the ratio of the required array power to daily average load power with γ and τ_{NR} as parameters are shown in Figure 3-14. Note that the solar array size reduces as the day/night load ratio, σ, is increased. The array size also decreases as the ratio of allowable recharge time to daylight time, γ, increases, and it increases as the number of sun-less days are increased.

Table 3-1.   Recommended Values of Parameters for PV Array Sizing and Design

Parameter**	Value		Parameter	Value
η_{Deg} (10 yrs)	0.95		η_D	0.997 ($\geq$ 220-V bus)
η_{W1}	0.98			0.99 (120-V bus)
η_{W2}	0.99			0.97 (48-V bus)
η_{W3}	0.99		η_R	0.90
η_B	0.80		τ_D	8 h*
			τ_N	16 h*

* These values are for the winter months at a high latitude site (e.g., 40°). The system designer must select the appropriate day and night times of the plant site for sizing purposes.

** see equation 3-2 for definitions.

The equation for determining the required rated battery capacity now becomes:

$$C_R = \frac{P_{NL} T_N + P_{NR} T_{NR}}{K_2 K_D C_D V_B} \tag{3-18}$$

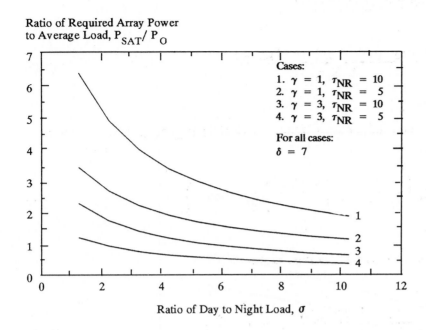

(a) Effects of day/night load ratio

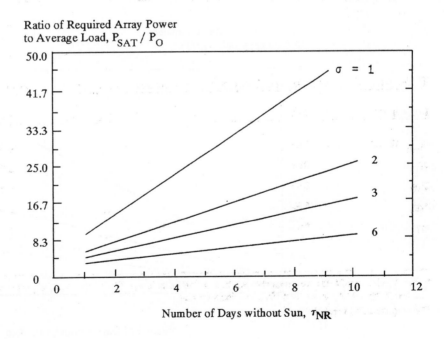

(b) Effects of number of sun-less days

Figure 3-14.   Effects of a) Day/Night Load Ratio and b) Number of Days without Sun on Required Array Power

Figure 3-15 illustrates the effects of battery load and required number of sun-less days ("autonomy" level) on the resulting battery rated capacity, C_R.

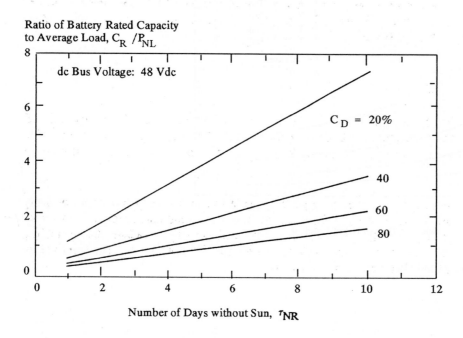

(a) Effects of DOD and sun-less days for 48-V bus

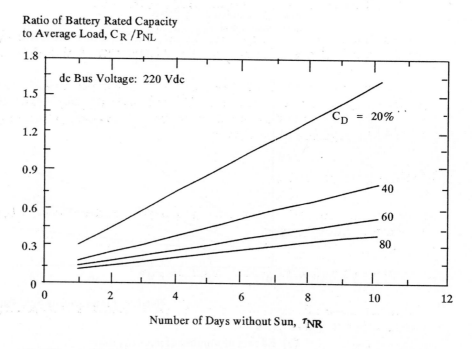

(b) Effects of DOD and sun-less days for 220-V bus

Figure 3-15.   Effects of Battery Depth of Discharge (DOD) and Required Number of Sun-less Days on Battery Rated Capacity for (a) 48-V Bus and (b) 220-V Bus

3.1.4   LIGHTNING AND OVERVOLTAGE PROTECTION

All PV projects in the multi-kW range have provided lightning and overvol'
varying degree. The principal aim in this protection is to reduce the overvoltage ⌐
before it reaches the PV or other subsystem components. The sources of overvoltage
lightning, atmospheric disturbances, switching transients in the power lines, and voltage flick⌐
coming from the grid. However, it is generally agreed that there are no cost-effective protection
methods against direct lightning strikes.

Protection Methods

There are basically three protection techniques in current use: interception of lightning-induced
current surges by means of metallic masts or rods (often called "air terminals"), grounding of PV
array metal structures, and arresting overvoltages by means of semiconductor and other active
devices.

The designers, especially on pilot plants, have generally resorted to a careful grounding of array
structures and connection of surge arrestors between the power lines and earth ground, and across
blocking diodes on each array string.   The local standards usually stipulate the grounding
requirements, including cable sizes for the outdoor components and buildings. The approaches
implemented to protect the PV components are as follows:

1) *Interception via Air Terminals*

- The 9-m rods or "air terminals" located inside the array field are a special lightning protection
  strategy implemented exclusively by Siemens Solar (previously Interatom). The lightning masts
  are connected to the buried ground cables. Siemens Solar has continued this practice at several
  plants in Germany, e.g., at Brunnenbach, FRG (see Fig. 3-16).

- On roof-mounted arrays, the air terminals are usually placed above the highest part of the
  building to attract the electrical discharge (see Appendix 19, BRAMMING PV House).

- Cables strung over the array field (see Fig. 3-17), as well as the array structures, are connected
  to the earth ground via heavy conductors. This is implemented only at GIGLIO PV Plant and is
  not a common practice. However, there has been no reported lightning strike or damage at
  GIGLIO from lightning or overvoltages.

2) *Grounding of Array Structures*

- A general practice with both ground- and roof-mounted arrays is to bond the PV array
  structures to the earth ground.

3) *Suppression via Semiconductors, Fuses, and Reactances*

- The cables from the PV array and sensors are usually protected by connecting them with one or
  a combination of the following devices:

  o Fuses (handle hundreds of kA)

  o Metal-oxide varistors (fast-response but low-current capability)

  o Gas-discharge devices (slow response but very high current capability)

  o Zener diodes (fastest but can handle only low power)

  o Inductive reactance in series

100-kW PV field at Kythnos Island, GR

10-kW PV field at Brunnenbach, Germany

Figure 3-16. Siemens Lightning Interception Rods (Air Terminals) at KYTHNOS Pilot Plant, Installed in 1982 and at BRUNNENBACH PV Plant in 1989. These plants are operated by PPC of Athens and Isar-Amperwerke of Munich, respectively.

76

Three cables suspended above the array field serve as air terminals

View looking north

View looking north

Figure 3-17.   Lightning Interception Cables at GIGLIO PV Plant

Protection Guidelines and Recommendations

The most common practice is grounding of the array structures and the use of metal-oxide varistors as illustrated in Figure 3-18. Some designers advocate the use of varistors across the array blocking diodes. Nevertheless, many PV systems, especially smaller ones, have no overvoltage protection devices, and yet have experienced no apparent damage from lightning strikes. This implies that, in low-risk areas, lightning protection devices could be dispensed with to reduce costs.

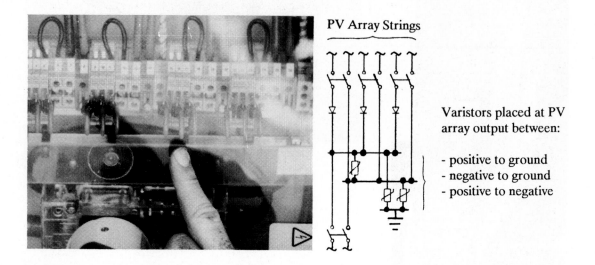

Figure-3-18.   Typical Connection of Varistors for Overvoltage Protection of PV Array Power lines

Figure 3-19 illustrates several methods implemented with the combined protection devices mentioned above which have been introduced into the market recently (1989). This strategy should perhaps be considered only for a highly reliable data acquisition system as the cost of protection devices is very high (about 200 to 400 DM per channel). Because of their proven reliability and low cost, varistors are the protection devices most commonly used.

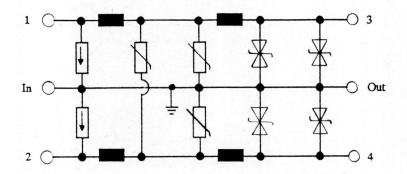

Figure 3-19.   Combined Overvoltage and Surge Protection Device is Now Available (Source: Dehn & Söhne)

On the basis of the experience acquired at the pilot plants and other data, many system experts seriously question the effectiveness of air terminals (metallic rods or overhead cables) for a ground-mounted array field or roof-mounted arrays in relatively open terrain for the following reasons:

- The probability that the air terminals can effectively protect the PV array, inverter, and DAS appears lower than the probability that they may indeed attract lightning to the PV components.

- The cost-effectiveness of air terminals of the type used at the pilot plants is questionable.

- Other protection approaches used have apparently been successful.

The above remarks do not apply to PV plants situated in lightning-prone high alpine regions, as they have experienced frequent damages from lightning. The selection and application of varistors and other high-voltage protection devices should be done in close coordination with the suppliers of these devices. Basic guidelines for lightning protection of a PV array and its power and signal wiring are listed in Table 3-2.

Table 3-2.   Guidelines for Lightning Protection

General:

- Surge arrestors like varistors should have a breakdown voltage greater than the maximum expected voltage in the line they are protecting.
- Consider the use of fuses in addition to surge arrestors.
- Connect the surge arrestors between any two conductors. Make the connection leads as short as possible and avoid unnecessary bends[1].

- For a highly reliable protection against overvoltages, use a combination of the following surge arrestors[2].
  o Varistors
  o Gas-discharge devices
  o Suppressor diodes (such as Spark Gaps and Zener)
  o Inductors
  o Fuses
- Prior to design freeze, always consult the device manufacturer's application engineers.

PV Array Structures:

- Strap all array structures to earth ground with adequate conductors according to local standards.

Power Lines:

- Connect metal-oxide varistors:
  o Between each current conductor and earth ground
  o Across each blocking diode in the array strings

Signal Lines:

- Use only twisted shielded wires.
- Ground the shield on the sensor lines only on one side at the central ground point.
- When grounding the shields, avoid grounding loops.

- Route sensor leads separately from power cables if possible, and in general not close to the high-voltage cables.
- Avoid routing sensor lines in parallel with power lines, and if they have to cross, do it at an angle of 90°.

---

1   Sharp bends result in high mutual coupling inductances which cause very large voltages and current surges.
2   Individual and combined units in a preassembled module are now available.

3.1.5   ELECTRICAL GROUNDING

A key aspect of PV plant electrical circuit design concerns the grounding of PV arrays, electrical equipment, and steel structures. The PV arrays at most of the pilot plants have used floating dc ground, i.e., the PV power circuit and the dc bus are not earth grounded during operation. At shutdown during the night, however, some of the plants (e.g., ZAMBELLI) keep their PV field in the shorted condition, for safety purposes. In spite of the risk of inducing possible hot-spot failures in PV modules, a few plants have adopted such grounding and array shorting approaches for the following reasons:

- Up to six years after plant installation, no specific problems have been reported by the pilot plants mentioned above with a floating ground. Moreover, even in the case of the high voltage PV array at DELPHOS (V_{oc} = 850 Vdc which is the highest in Europe), there has been no report of anyone experiencing an electric shock.

- All PV plants with floating dc ground are monitored continuously for ground faults in order to detect and correct a potentially hazardous condition.

- Grounding of both terminals (+ and - poles) during shut-down of the PV array (as in DELPHOS) reduces the risk of lightning strike damage.

- During operation, accidental contact of a person with a live dc circuit part does not cause an electric shock, since the voltage is nearly zero.

- During shut-down, by grounding and shorting the PV array, the high array voltage is suppressed to nearly the ground level.

- The short-circuit current of the PV array provides a precise morning start-up threshold signal independent of other sensors.

3.1.6   PLANT PROTECTION AND SAFETY

Details on safety practices applicable to electric power plants in general can be found in many commercially available electrical engineering handbooks, as well as standard engineering practice manuals for equipment, cables, and grounding, and their sizing, isolation, and implementation.

It is not commonly known that a current flow of only 30 milliamperes through a human being is considered lethal. However, the human body presents a rather high resistance against electric current flow, hence a high voltage (100-200 volts) is usually required to reach a harmful current level. Nevertheless, in the presence of water (wet skin), the resistance to current flow is reduced substantially, so even a low voltage electric shock may be lethal.

International and most national regulations define 50 Vdc as the maximum limit which the human body can safely withstand. Any voltage above 50 Vdc is considered dangerous, although under certain conditions such as dry air, a person can survive an electric shock of much higher voltage.

The dc power is more dangerous than ac power, because it causes electrolysis of the blood, and the uninterrupted current flow induces clenching of muscles, making it difficult for the individual to free himself once in contact with a live circuit with a sufficiently high voltage.

Safety Guidelines on High Voltage PV Arrays

A PV generator cannot be switched off effectively during daylight. High open circuit voltages are present even under cloudy sky conditions. Theoretically, if all solar cells in the PV array are completely darkened, the array voltage can go to zero. However, it is not possible to fully darken the PV array during daylight time. Accordingly, the only practical way to lower the voltage of a PV string/array is to temporarily ground the array output. All metal casings/shieldings of the plant must also be thoroughly grounded in accordance with local regulations.

If the PV array has been shorted by a switch, the removal of such a short may cause damage to the switch due to arcing between relay contacts upon switch opening. Accordingly, it is advisable to use a temporary shorting device via an appropriately sized circuit breaker that switches off the short before disconnecting it.

Isolated plug-in connections significantly reduce risks during cabling work in PV arrays, as well as the relevant manpower costs. However, reliable weatherproof connections are not readily available.

A few guidelines on the wiring of three or more 36-cell PV modules are as follows:

- Inform personnel of the hazards of making series connections of PV modules; display relevant warning signs and instructions on site.

- If possible, isolate the PV string from the rest of the plant, and mark the relevant isolation switch with a sign warning other personnel not to touch it before the work is finished.

- Ground your working area with a temporary, clamp-on grounding connection.   Check the effectiveness of grounding with a tester. Never work on a circuit part more than a distance of two PV modules away from the next grounding.

- Keep the rest of the string floating (isolated from the rest of the plant and from ground).

- Ensure that no other person is working on the same circuits with which you are engaged. Remember that grounding a given circuit may cause a high voltage build-up at another circuit elsewhere.

- Wear protective gear, such as rubber gloves and rubber shoes, and use specially insulated tools and an isolated footbridge to stand on. During the work, do not hold onto or lean on grounded structures or frame parts.

- When wiring a complete string, first connect a maximum of four PV modules in series with the string isolated from the main dc bus, and then proceed to complete the string cabling.

High Voltage Battery

One of the highest risk areas in a PV plant involves the battery. The general safety precautions for high voltage batteries are listed below:

- The battery room must be properly secured and accessible only to authorized personnel. The risk of death by electric shock must be indicated by large warning signs.

- Due to the danger of an explosive mixture of hydrogen and oxygen gases generated by the cells during operation, safety regulations require flame-proofing of all electrical installations inside the battery room, as well as appropriate ventilation. This includes connections to battery cell terminals, fuses, and switches for cell voltage monitoring.

  CAUTION: If fuses are used in cell voltage sensing lines, place them at least 0.5 m away from the cell terminals (to reduce the possibility of an explosion in the event the fuse burns).

- Safety provisions for personnel protection include the use of insulation and acid-proof foot-mats on the floor below the battery cells and both hand- and eye-wash facilities in case a person comes into contact with the acid.

- Contrary to the PV array, *NEVER SHORT THE ENTIRE BATTERY OR EVEN A SINGLE CELL.* A battery cell is capable of discharging thousands of amperes. The effect of shorting may be disastrous (cell explosion, sparking, and melting of interconnect wiring and termination joints). Use only insulated tools on battery terminals to avoid accidental shorting.

3.1.7   OPERATION AND MAINTENANCE

Discussions of the operation and maintenance of subsystems are given in the relevant subsections of Section 3.2.

The most important requirements for operation and maintenance are adequate documentation and an inventory of spare parts for critical components. In most of the PV plants, the basic operating manuals and instructions were submitted by the designers. These were only for normal plant operation and lacked instructions for trouble-shooting and diagnostic assessment. This has presented serious problems to the owners and operators of several large plants when attempting repair work. When trouble-shooting is needed, a lack of adequate documentation is obviously not only costly, but frustrating to the plant owner and the technicians assigned to look after the plant.

Basic documentation and a spare parts inventory for plant operation and maintenance must be part of any contractual requirement. They must be defined and prepared during the design phase when the designers are still on the project, and not after installation. The reason is simple — the engineers who can best write the required manuals and define spare parts are usually unavailable once the design phase has been completed. Moreover, it is quite difficult for them to return to the project for this purpose.

How to Determine the Needs for Maintenance Manuals and Spare Parts Inventory

The production of maintenance and trouble-shooting manuals presents a very high cost to the project. To minimize this cost, the documentation may be limited to the most critical subsystems, components, and printed circuit cards that are field-replaceable. A systematic approach, in defining these critical parts, is to conduct one or both of the following:

- Failure modes and effects analysis (FMEA). This is best done, first at the system and subsystem levels, then at the major component level (e.g., inverter). For details of FMEA procedures and examples, refer to system reliability text books.

- Mean time between failure (MTBF) assessment of each replaceable part. MTBF is the average of the time periods between failures of the device concerned (e.g., an inverter or a printed circuit board).

The parts or components most prone to fail can be identified by their low MTBF values. The results of FMEA usually indicate the criticality and likelihood of possible failure modes.

Appendix 29 presents a consolidated discussion of reliability assessment methods, including MTBF calculation of electronic equipment or a subsystem. This is written as a guide for PV designers and operators interested in improving the reliability of their system and in developing MTBF data using failure statistics from the plant operation.

Test Equipment

The minimum test equipment needed on site for the maintenance and periodic performance verification of any type of PV plant is as follows:

- For checking the calibration of the solar sensors and monitoring individual battery cell voltages: At least a $4^1/2$-digit multimeter, hand-held type (like Fluke 8062A) or the larger portable 6-digit DVM which is battery operated. This allows 3-decimal place measurements of individual battery cell voltages and solar irradiance sensor outputs (e.g., 2.334, rather than 2.33 on the $3^1/2$-digit multimeter).

- For determining the output and input power quality of inverters: Spectrum (Fourier) analyser and dual memory oscilloscope.

3.1.8   OPTIMAL UTILIZATION OF AVAILABLE ENERGY

Many of the original pilot plants were designed strictly as stand-alone systems. One characteristic of these stand-alone systems is that when the battery is fully charged, the system cannot use all the available array power. To optimize system energy efficiency, the system must be able to consume the maximum energy that is available from the PV array. This requires electrical loads that can be turned on or off, based on available PV power. Since most stand-alone systems does not implement load management for such purposes, they operate at low-power and often produce no power at all.

Long-range planning for post-contract operation was apparently not considered in the early design phase on many stand-alone PV plants. For instance, 6-8 years after the initial startup, many of the original pilot plants designed for stand-alone have converted to grid-connected operation as their main operational mode (e.g., PELLWORM, FOTA, and VULCANO), and others are in the process of doing so (AGHIA ROUMELI and KYTHNOS).

Most self-commutated inverters that have been used in the past for stand-alone systems do not have the capability to switch over to line-commutation mode. The cost of modifying the existing self-commutated unit is generally greater than purchasing a new line-commutated unit, as discovered by several PV projects which converted their systems from stand-alone to grid-connected operation.

To make better use of the available PV energy at existing plants and in future ones, the following strategies could be considered:

- Long-term planning for future operating modes is highly desirable, specifically as related to utility-grid connection.

- For stand-alone systems which are located near a grid network, or where eventual grid connection is planned, an ability to operate in the grid-connected mode is essential. This implies that, if connection to an ac line with other power sources is likely in the future, two options are open:

  o Install an inverter type which has both self- and line-commutation capabilities.
  o Install both types of inverters, self-and line-commutated, one type for stand-alone and the other for parallel operation with the grid.

- In the design of stand-alone PV plants, either with or without batteries, a capability to operate with an external ac power source (wind or Diesel generator, etc. which means a hybrid plant) should be carefully considered. This will allow the stand-alone PV/battery system to be optimized, in terms of both reliability and cost.

- To make full use of PV energy, a capability should be added to calculate in real time the maximum available power and energy of a PV array, especially in PV systems which do not operate at the peak-power point of the array all the time (see Appendix 27).

- Provision of "convenience" loads which can be turned on when ample PV energy is available. This is especially desirable for PV pumping systems without batteries.

Unfortunately, the optimization of available PV energy requires an energy/power management system which increases the complexity of the electronics and reduces the reliability of the plant. Therefore, careful planning and design to implement reliable power management and control strategies are essential.

3.2 SUBSYSTEM DESIGN, PERFORMANCE AND OPERATION

Key lessons learned from the past and on-going programmes are summarized in Table 3-3 for the system and subsystem levels as well as in maintenance and project management categories [21-31]. It should be noted that the discussions pertaining to subsystem hardware in the following paragraphs are limited to commonly available components. That is, the reader should assume that the performance characteristics cited and guidelines suggested are relative to the following types of devices or components:

- PV array: monocrystalline and polycrystalline Si cell modules
- Battery: Pb-acid, vented cells

Table 3-3.   Summary of Key Lessons Learned from the Pilot Plants

PV Array:
- Power degradation on some of first generation modules was attributed to use of PVB material; no apparent degradation of a majority of other module types after seven years
- Careful attention needed on module attachment design and procedures to prevent mechanical damage during and after installation
- On unsealed boxes and metal channels, natural drainage provisions are needed at their lowest gravitational point
- Lighting damage mechanism, protection needs, and methods still not adequately understood; costly protection methods such as lighting rods considered unnecessary
- Cost reduction of array structures possible by reduction of wind load requirements, lowering of tilt angle, and maximization of panel area in a single plane (e.g., large roof-type configuration)

- Module and array power rating method and verification in outdoor testing needs more in-depth assessment and experimental quantification
- Actual module output 10 to 15% lower than specified by manufacturer, thus, for array sizing, designer must allow for the lower module output
- Careful design practices needed in insulation of protection diodes at their mounting interfaces to prevent shorting to structures
- Need further studies on bypass diode placement and number of cells in parallel within a module
- Larger module sizes are very cost-effective, especially for higher power arrays
- Care must be taken in selection and acceptance of varistors (used for lightning protection) as they have caused fires
- Avoid loose cables between modules to prevent wire breakage

Batteries and Battery Protection:
- Immediate attention needed to ensure adequate charge control and discharge protection. Failure to do so can be very hazardous to personnel and equipment, and ultimately could harm stand-alone PV markets
- Charge control: One of the best methods is temperature-compensated constant voltage limit with manually adjustable or selectable discrete voltage limits, with or without SOC parameter for charge temination purposes. If SOC calculator is used, SOC should be used only as a secondary control signal. For high reliability system, a real-time display of average, minimum, and maximum cell voltages in the battery is highly desirable
- Discharge control: Provide manually adjustable or selectable discharge cut-off voltage limits; if at all possible, prevent voltage reversals in the individual cells
- SOC calculator used in eight pilot plants did not work due to a lack of follow-up problem solving; however, SOC math model is unreliable for long-term capacity prediction; even for short-term basis, a reset capability to 100% relative SOC is mandatory
- Cell manufacturers cannot supply battery-level performance data; best sources of these data are from actual PV plants after installation
- Effectiveness and life of H_2 recombination devices (in

vent plugs) are unknown
- Limited knowledge about battery performance in a terrestrial PV operation. Unknowns are:
  o Effects of a large number of cells in series under one control (i.e. battery-level control)
  o Effects of long-term operation in partial state of charge on battery life
  o A reliable model for battery SOC determination implemented in SOC calculation equipment
  o Methods of detecting battery SOC
  o Aging effects on polarization characteristics
  o Cycle life data under different DOD and temperature
  o Proper charge and discharge control as battery ages
  o Effects of air bubbling on electrolyte stratification and battery cell life
  o Effects of using new cells in series with old cells
- Need adequate/timely inspection of battery condition. In addition to electrolyte level, mismatch in cell performance should be assessed by monitoring individual cell voltages towards end of charge and discharge in a given day of operation and comparison with average cells in the same battery string.
- "PV" battery cells are more expensive than other types ($\approx$ 1.0 DM/Ah in FRG)

Power Conditioning:
- Highest inverter efficiencies demonstrated with highest array bus voltages
- To minimize risk, well-proven inverters are better than new designs; avoid custom-designed μP-controlled inverters
- For stand-alone systems with an auxiliary ac power source, inverters with both self- and line-commutation capabilities are often desirable
- Reliable automated operation between PV and auxiliary sources like Diesel or gasoline generators needed to minimize operator intervention

- Design and Tradeoff issues still exist as a function of power level:
  o Buck vs. boost/buck regulators
  o Dissipative converters vs. shunt (full or partial) regulation
  o High-power inverter design approaches for multi-hundred kW applications (e.g., GTO vs. IGTB vs. thyristors; self vs. line commutation)
  o Modular size of inverters
  o Low-cost grid-connected inverters

(Page 1 of 2)

Table 3-3.   (Concl.)

Power Conditioning (Concl.):

- In the event of unreliable utility grid (e.g., "weak" grid or small Diesel power station in the small MW range), a UPS is highly desirable to avoid main inverter shutdowns and also to provide a source of emergency power
- If a transformerless inverter is used in grid-connected operation, pay special attention to safety provisions such as grid-isolation requirements to prevent damage to local consumers (loads) and the inverter
- Forced air (i.e., with fans) vs. passive cooling of power electronics: In coastal regions with salty atmosphere, passive cooling with large heat sinks is more reliable to minimize corrosion in electronic parts and connections. Forcing air flow in the electronic cabinet (especially unfiltered unconditioned air) accelerates corrosion. Also, fan motor failures have occurred frequently, causing heating problems
- Pay special attention to heat dissipation and thermal control in the power electronics cabinet during the design phase, considering low-output power operations in which the converter or inverter is operating at lowest efficiency (highest heat dissipation)

Control Computers and Plant Monitoring Equipment and Devices:

- DAS's cannot be expected to operate autonomously for a long time; the reliability of continuous operation is very low without redundancy
- Reliable meters (amph and Wh) are not available
- Low-cost PC-based systems and reliable data loggers are now available
- Avoid complex costly computers; use commonly available systems (e.g., IBM PCs or compatibles)
- Avoid custom-designed DAS and parts; availability of multiple sources and compatibility with PCs and MS-DOS should be a criterion for selection
- Solar Sensors:
  o Re-calibration should be done on-site;
  o Simple calibration procedures should be used on site by local technicians
- Need basic guidelines on the selection and application of isolation devices, signal grounding, and lightning protection methods
- Do not use one computer for both control and monitoring functions; use separate computers
- Software modification or optimization is expensive; proper documentation of custom-designed software by the original programmer is essential and it should be prepared for possible later modification by another programmer
- Avoid the need for extensive software design; use of commercially available software (like Schlumberger IMPULSE for on-site data acquisition and processing) minimizes time and cost

System Design and Performance Analysis:

- Array bus voltages ranged from about 100 to 800 Vdc; higher voltages lead to following:
  o Higher converter and inverter efficiencies
  o Better Ah capacity from battery due to lower discharge rate
  o Lower battery reliability due to a larger number of cells in series
- A large number of PV cells in series has had no adverse effects on array performance
- Some nomenclatures and methods need to be carefully reassessed, defined, and standardized, e.g.:
  o Array and plant rating definition and method
  o Bus power capability of grid-connected and stand-alone systems (definition does not exist)
  o Method of analysing and predicting PV array capability from operating data
- Development work needed:
  o Integrated sizing of array, battery, and inverter
  o Lightning/overvoltage protection approach
  o System optimization, e.g., pumping applications
  o Lifetime assessment of PV systems
  o Assessment of long-term field performance and reliability .
  o Credible cost analysis methods

Operation and Maintenance/Project Management/Other Issues:

- The project financing organizations (i.e., the customer) should impose documentation, management, plant monitoring and operator training requirements as part of contract requirements. The following should be part of the contract for any PV application:
  o Properly documented hardware and validated software
  o Training of plant users and maintenance personnel by plant designer
  o Spare parts list and inventory during the contract period
  o Operator to maintain and do repairs
  o Operational plan for at least 2 yrs after installation
  o Plant performance analysis for at least two years and documentation of results
- Plant designers should be given the responsibility of user training and maintenance of plant for at least one year after start-up. An alternative is to train the users during the design and installation phases when designers are still on the job
- A concerted effort is needed to promote government utility legislation/policies on the use of dispersed PV plants and payment to the owner for PV energy
- PV utility interface issues need to be addressed, e.g., payment for PV energy at an equitable rate, utility acceptance of grid-connection, etc.

(Page 2 of 2)

3.2.1   PV ARRAYS: PV MODULES AND ARRAY CONFIGURATION

The key factors affecting the electrical performance of the PV array are: 1) solar irradiance; 2) solar cell temperature; 3) solar incidence angle; 4) electrical wiring resistance; 5) voltage drop in the series blocking diodes; 6) accumulation of dirt and other particulates on the module cover glass; 7) electrical degradation of a PV module; and 8) shadowing of individual cells in PV modules.

Basic Output Characteristics of PV Devices

Figures 3-20 and 3-21 summarize the effects of solar irradiance, temperature, and solar incidence angle on the PV device performance in terms of I-V characteristics and key parameters, such as open circuit voltage (V_{oc}), short circuit current (I_{sc}), and maximum power (P_m). The main performance relationships within the normal operating solar irradiance ($< 1,400$ W/m^2 for flat-plate arrays, including reflections) and ambient air temperature range are as follows:

- I_{sc} increases linearly with solar irradiance[1]
- P_m responds similarly to I_{sc}, but for solar cells with a high internal series resistance, P_m will droop slightly with increasing irradiance (above 600-800 W/m^2); the same is true for efficiency
- V_{oc} is a logarithmic function of irradiance
- Both I_{sc} and P_m are cosine functions of the angle of incidence, α

Mathematical Model of a PV Array

An array consists of a number of solar cell module strings or branches connected at the dc bus. The number of modules in series is determined by the desired dc bus voltage level, and the number of strings by the total array power required. Thus, the equivalent circuit of an entire array is a combination of individual solar cell equivalent circuits, interconnecting wire resistances, and the blocking diode.

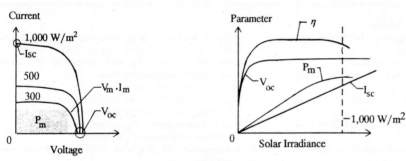

Effects of Solar Irradiance at a Given Temperature

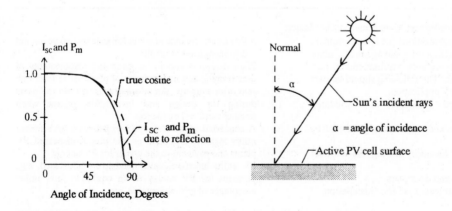

Effects of PV Module Orientation

Figure 3-20.   Effects of Solar Irradiance and Incident Angle on PV Device Characteristics

---

[1] Generally I_{sc} will remain fairly linear with solar irradiance in the low-concentration levels (2-5 suns), even for cells built for 1-sun operation.

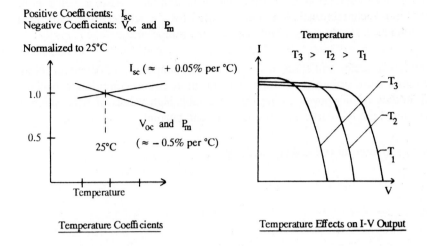

Positive Coefficients: I_{sc}
Negative Coefficients: V_{oc} and P_m

Normalized to 25°C

Temperature Coefficients

Temperature Effects on I-V Output

Figure 3-21.   Effects of Temperature on Si Cells at a Fixed Solar Irradiance. For Si cells I_{sc} increases slightly but linearly with temperature. V_{oc} and P_m decrease with temperature.

A *single exponential lumped-parameter* model of a cell has proven to be adequate for all PV applications involving silicon cells. Some equivalent circuits without the shunt resistance R_{sh} term assume that R_{sh} is negligible, which is valid for most solar cells.

Another widely used mathematical model is *curve shifting* or *translation* from one measured I-V curve to a new one under another set of solar irradiance and solar cell temperature conditions using two algebraic equations. Equations for both the single-exponential and curve translation models are summarized in Table 3-4.

Table 3-4.   Photovoltaic Array Performance Math Models

Curve translation formulae:

$$I_2 = I_1 + \Delta I_{sc}$$

$$\Delta I_{sc} = \left[\frac{L2}{L1} - 1\right] + \alpha(T_2 - T_1)$$

$$V_2 = V_1 - \beta(T_2 - T_1) - \Delta I_{sc}R_s - KI_2(T_2 - T_1)$$

Where:

X_1	=	reference curve (known or measured)
X_2	=	unknowns
I	=	current, A
I_{sc}	=	short circuit current, A
α	=	current temperature coefficient, A/°C
β	=	voltage temperature coefficient, V/°C
K	=	curvature coefficient, Ω/°C
R_s	=	series resistance, Ω
V	=	voltage, V
L	=	solar intensity, W/m²

Single-exponential equation:

$$I = I_L - I_o\left[e^{\frac{K}{T}(V + IR_s)} - 1\right]$$

Where:

I	=	current, A
I_L	=	light generated current, A
I_o	=	reverse saturation current of ideal diode
V	=	voltage, V
K	=	lumped parameters diode constant
R_s	=	series resistance, Ω
T	=	cell junction temperature, absolute

Dark forward characteristics based on single-exponential equation:

$$I = I_o\left[e^{\frac{K}{T}(V + IR_s)} - 1\right]$$

Where the terms are the same as for single-exponential equation

The curve translation method is easier to use, compared to the single exponential model, because the constants α, β, R_s, and K are easier to determine from empirical measurements. Also, it yields more accurate results in the event the I-V characteristic is not exponential (like a "stair-case" curve), or if curve extrapolation is to be done over a wide range of intensity and/or temperature.

Figure 3-22 illustrates the effects of series and shunt resistance on the shape of an I-V curve. Note that the shunt resistance causes a more pronounced slope near the short circuit condition, especially as R_{sh} decreases. A good module of any power rating should exhibit high values of R_{sh}, which is usually indicated by a zero slope of the I-V curve near the short-circuit current.

The shunt resistance, R_{sh} is due to internal cell leakage such as:

- Through p-n junction
- Along the outer cell edges
- n and p metallization path

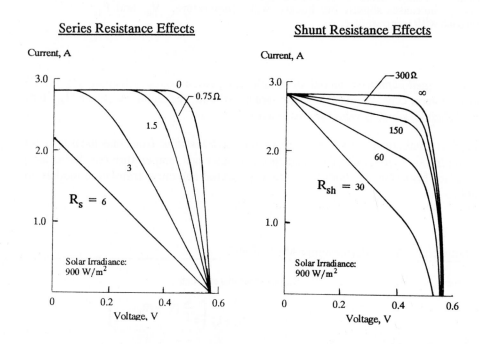

Figure 3-22.   Effects of Internal Series and Shunt Resistances on I-V Curve Shape of a 10 x 10 cm Si Cell. The characteristic curve on the left assumes a cell with very high internal shunt resistance.

Most of the currently available Si cells exhibit high R_{sh} ($> 10^3$ ohms) so that it can usually be neglected in array performance evaluation or computation. Good cells and modules exhibit a flat slope in the I-V curve near the short circuit region. Figure 3-23 illustrates the calculation of R_{sh} for any cell, module, or array.

The effective series resistance, R_s, is in reality distributed in the semiconductor (sheet resistance in the diffused layer and bulk resistance) and its ohmic contacts. The largest contributor is the diffused layer resistance. The effects of R_s are more pronounced in the concentrator cell. When operating at high-illumination intensities, this internal series resistance affects the efficiency of the cell as illustrated in Figure 3-24. Thus, concentrator cells for operation between 50 and 100 suns usually have metallization covering 8 to 10% of the active cell area to reduce its series resistance. Good 100 x 100 mm Si cells have R_s in the range of 0.1 to 0.3 ohm.

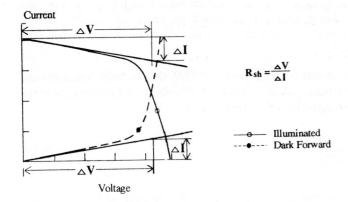

Figure 3-23. Calculation of Shunt Resistance Using Either One Illuminated or One Dark Forward I-V Characteristic of a PV Device.

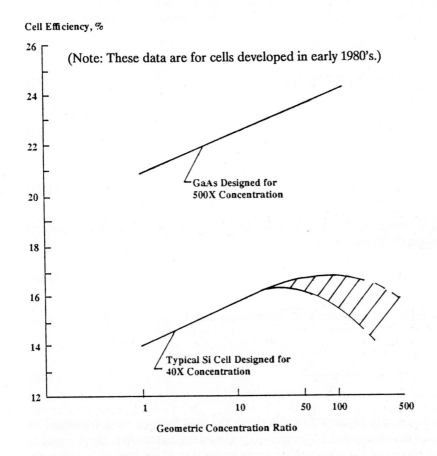

Figure 3-24. Efficiency as a Function of Geometric Concentration at 25°C for Si and GaAs Concentrator Cells

There are two methods of determining the value of R_s in a cell, module, or an array, as illustrated in Figure 3-25. One uses two illuminated curves and the other uses one illuminated and one dark forward curve. Figure 3-26 shows the simple measurement set-up for obtaining the illuminated and dark forward I-V curves CAUTION: These methods are based on single exponential model. Therefore, the R_s determined with these methods apply only to a one-diode equivalent circuit (see Fig. 3-26) only, and not to the 2-diode model.

In the two-illuminated curve method introduced by Rauschenbach [32], the intensity levels do not have to be known. The displacement ΔV resulting from translations of voltage and current axes (see Fig. 3-25) results in the expression for the lumped series resistance:

$$R_S = \frac{\Delta V}{\Delta I_{SC}} \tag{3-19}$$

where ΔI_{SC} is the difference in the short circuit currents obtained at two light levels. Since the maximum power point is the most important operating point on the characteristic curve, the series resistance is usually determined at or near the maximum power region.

In the "dark-curve" method originated by Imamura [33], the equation for the lumped series resistance can be derived as follows.

The loop equations with reference to Figure 3-26 are:

Illuminated: $\qquad\qquad V_L = V_{J1} - I_L RS \tag{3-20}$

Dark: $\qquad\qquad\qquad V_D = V_{J2} + I_D RS \tag{3-21}$

where V_L and I_L are the illuminated voltage and current, V_D and I_D are the dark forward voltage and current, and V_{J1} and V_{J2} are the junction voltages under illuminated and dark conditions, respectively. For a case in which $V_{J1} = V_{J2}$, the difference between the dark and the illuminated terminal voltages is:

$$\Delta V = V_D - V_L$$

$$= R_S (I_D + I_L) \tag{3-22}$$

but,

$$I_D = I_{J2} \tag{3-23}$$

and,

$$I_L = I_{SC} - I_{J1} \tag{3-24}$$

Since the junction voltages are assumed to be equal, I_{J1} is equal to I_{J2}. The substitution of equations (3-23) and (3-24) into equation (3-22) results in:

$$\Delta V = I_{SC} R_S \tag{3-25}$$

Hence, the lumped series resistance term using an illuminated and a dark I-V curve at a given terminal current is:

$$R_S = \frac{V_D - V_L}{I_{SC}} \tag{3-26}$$

This method is graphically depicted in Figure 3-25 which shows the dark forward curve translated by shifting its origin to the I_{SC} of the illuminated I-V curve. At a given terminal current, the separation ΔV between the translated curve and the photovoltaic curve is equal to V_D minus V_L. Using eq. (3-26), the series resistance may, therefore, be determined along the major portion of the characteristic curve for a given illumination level.

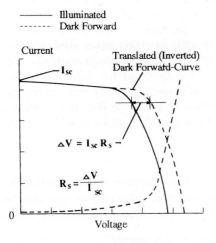

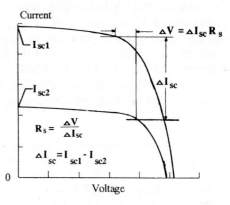

(a) Dark Curve Method Using One Illuminated and One Dark Forward I-V Characteristic.

(b) Illuminated Curve Method Using Two Illuminated I-V Characteristics.

Figure 3-25. Two Methods of Determining Lumped Series Resistance (R_s) of a Solar Cell, Module, or Source Circuit: a) Dark-curve Method and b) Illuminated Curve Method.

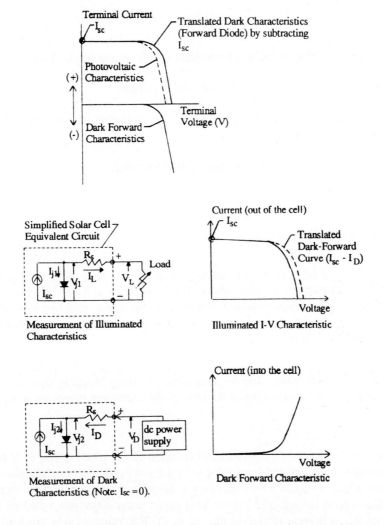

Figure 3-26. Method of Measuring Illuminated and Dark Forward I-V Curves

PV Modules

PV modules available in the European Community are predominantly flat-plate types. They range in power output from about 26 W to 240 W at STC (1,000 W/m^2 and 25°C cell temperature). The nominal operating cell temperature (referred to as the NOCT) is different for different module constructions. A design that yields the lowest cell temperature at a given intensity and ambient air temperature, i.e., the lowest NOCT, usually provides the best specific power and efficiency performance. However, in selecting a module manufacturer, an economic tradeoff should be made in terms of cost per watt at the total array level.

Table 3-5 gives a partial list of higher power PV modules, all flat-plate, currently available in the EC. The latest new entry in the list is BP Solar's "laser-grooved buried-contact" solar cell module which is reported to have a 15% efficiency (17 to 18% cell efficiency). This module is available on a commercial scale as of late 1991.

Figure 3-27 shows the trend in module efficiency. Module efficiency is the product of cell efficiency, cell wiring loss, cell mismatch loss, and cell packing factor, and is generally two to three percentage points lower than the cell efficiency.

It should be noted that improvements are still possible and needed in the PV modules. The key ones are:

- Efficiency

- Single cell vs. multiple cells in parallel, especially for central power station applications

- Placement of bypass diodes, considering the number of cells in parallel

- Standardization of mechanical and electrical requirements, e.g., to permit use of different module suppliers on standard array structures

- Ability to pass "wet" megger testing at isolation voltage up to 3,000 volts.

Module Matching and "Hot-spot" Considerations

A large number of solar cells must be connected in series to provide the desired system bus voltage. For example, a 24-Vdc stand-alone system requires about 80 cells in series and 240-Vdc system about 640 cells. This implies that a performance mismatch can exist in the string, resulting in an effective power "loss" of 1-5%. *Close matching of maximum power-point current* of modules within a *series* string minimizes the string mismatch loss.

Similarly, to minimize the mismatch losses due to *paralleling* of string blocking diodes, the best strategy is to group the strings by *close matching of the sum of the maximum power point voltages* of the modules in series.

Some of the series-connected cells become reverse-biased because of a cracked cell(s) or shadowing of a cell(s). In a single string of cells isolated from the bus by a blocking diode, one fully shaded cell presents a worst-case condition. Figure 3-28 illustrates a simplified analysis of this worst-case condition, showing that the reverse voltage (-15 Vdc or less) could occur even with a 24-Vdc system. Depending on the value of R_{sh}, a reverse voltage in the -10 to -20 Vdc range or less can lead to cell damage from excessive heating.

Table 3-5. List of Several High-Power PV Modules Available in the European Community and Their Features

Manufacturer/ Supplier	Model or Part No.	Cell Type	No. of Cells in Series/Parallel	Performance at STC							Length (mm)	Width (mm)	Weight (kg)
				P_{max} (W)	V_{mp} (V)	I_{mp} (A)	V_{oc} (V)	I_{sc} (A)	Eff. (%)	Fill Factor (%)			
AEG/TST	PQ 40/50	poly	40/1	50	16.8	2.98	22.8	2.98	10	74	1,078	459	7
	MQ 40/52	mono	40/1	52	17.6	2.95	24	2.95	10.4	73	1,078	459	7
	PQ 10/180	poly	15/12	200	7.4	33	9.0	35.3	9.5	76	1,640	1,330	70
BP Solar	BP365	poly	36 (125 x125 mm)	65	16.0	4.06	20.5	4.40	10.3	72	1,188	530	7.5
	BP495	mono*	60	95	30.0	3.16	36.7	3.3	15.0	78	1,188	530	5.9
	BP260	mono*	36	60	18.0	3.33	22.0	3.5	14.6	78	956	431	5.2
	BP2100	mono*	60 (95 x 100 mm)	100	30.0	3.33	36.7	3.5	15.0	78	1,188	530	38.5
Helios	H50	mono	36	50	16.8	2.9	20.5	3.30	11.2	74	1,310	340	5.6
	H100	mono	36/2	90	16.8	5.35	20.5	5.92	11.3	74	---	---	---
Isofoton	M55L	mono	36	55	17.4	3.05	21.8	3.27	12.9	74	1,290	328	5.7
Kyocera	LA 441 J59	poly	44	58.7	20.3	2.89	25.4	3.10	10.2	75	1,216	447	7.3
	LA 441 K63	poly	44	62.7	20.7	3.03	26.0	3.25	11.8	74	1,195	445	7.3
	LA 721 K102	poly	72	102	33.8	3.02	42.5	3.25	12.0	74	1,300	655	14.2
Nukem	Grossmodule	mono/poly	66/3	230	30.5	7.5	37.8	8.2	9.9	73	1,890	1,220	150.0
Photowatt	BP X 47500	poly	36	45	16.5	2.7	20.9	3.0	9.3	72	1,042	462	9.2
	PW X 1000	poly	36/2	91.5	16.5	5.55	21.2	6.08	10.5	71	1,000	873	16.5
R&S	RS M45	poly	36	45	16	2.95	21	3.1	9.4	69	1,006	478	5.9
Solarex	MSX60	poly	40	60	17.8	3.37	21.3	3.65	10.8	77	1,109	502	7.2
Siemens	SM55	mono	36	53	17.4	3.04	21.8	3.27	12.3	74	1,295	332	5.7
	SM144 +	mono	36/4	130	17.0	7.6	21.4	9.00	8.6	68	1,470	1,020	27.0

* Laser-grooved buried-grid cells
\+ Used in several plants (Kythnos, Brunnenbach, etc.) but to be discontinued

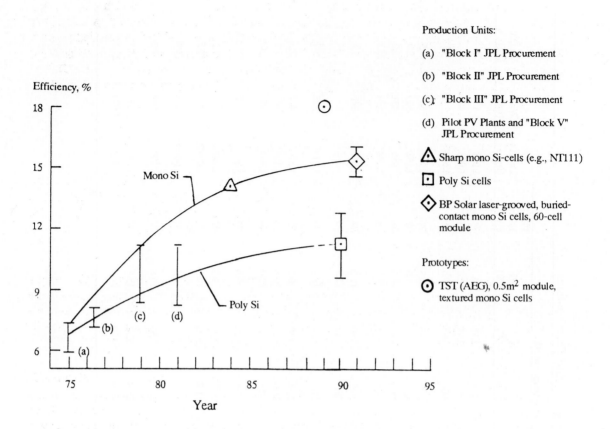

Figure 3-27. Power Conversion Efficiency Trend, Flat-plate Silicon PV Modules. Efficiency at Standard Test Conditions (1,000 W/m^2, 25°C, AM 1.5 Spectrum).

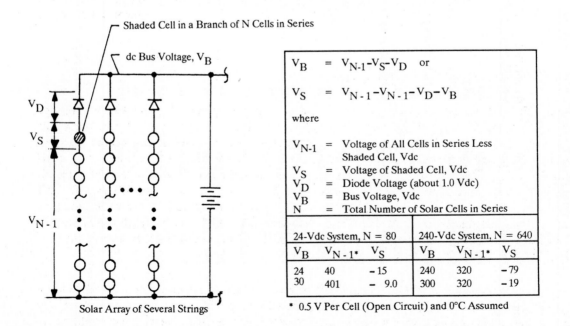

Figure 3-28. Typical Reverse Voltages (Vs) on a Completely Shaded Cell within a String of N Series Cells with No Bypass Diodes

To minimize this "hot-spot" phenomenon a bypass diode is connected in parallel with a number of cells in series. This number varies generally between 18 and 36 cells for flat-plate arrays and as low as four cells in series in a concentrator module [34]. Basic studies on this subject are needed because there has been no systematic method identified to date. Conphoebus and University of Saint Bernard are investigating this problem for the flat-plate arrays under the current JOULE programme (Concerted Action on PV arrays) and the results of this study should be available in mid-1992.

PV Array Configuration: Flat Plate Modules with Stationary and Moving Structures

In large PV systems, especially in central power grid-connected plants, there has been a trend towards one- and two-axis tracking flat-plates, some with light reflectors that result in 1.5X geometric concentration. In sizing or analysing the performance of such systems, it is important to distinguish which of the solar irradiance components are useful to the particular array orientation and the total energy available in various array orientations.

A PV panel which is mounted on a stationary, tilted support structure uses the global irradiance on a tilted fixed surface (see Fig. 2-9, GFTI, in Chapter 2) whereas the two-axis tracking flat-plate uses the global normal component (GNI) which is the maximum possible radiation available to a flat-plate module on a sun-tracking surface. Note that a 1-axis tracker with its N-S axis adjusted periodically to be perpendicular to the solar incidence angle is like a 2-axis sun tracker.

Flat-plate modules, along with their mounting structures for fixed orientation, are commercially available in large production quantities. Flat-plate modules have been used in systems from a few watts to MW size. Practically all PV plants in the European Community (EC) have used fixed stationary array structures, and this trend will continue. What has not been exploited more fully is the pre-assembling of modules at a factory to reduce the in-field assembly and installation cost even for medium-power PV arrays.

In contrast to the practice in Europe and Japan, large central power station PV plants in the USA have resorted to 1-axis and 2-axis sun-oriented structures, but recently the 1-axis structures, both active and passive tracking types, have become more popular (see Fig. 3-29). As the trend towards large grid-connected plants continues in Europe, design options for the EC are expected to include stationary vs. partially and fully tracking flat-plate systems in addition to concentrator vs. non-concentrating flat-plate systems. Further discussion on array structures continues in Subsection 3.2.2.

PV Array Configuration: Concentrating Types

The development of concentrator PV modules is proceeding, mostly in the USA under the direction of Sandia National Laboratories and EPRI. The designs are of two types — line-focus and point-focus Fresnel lenses. The lenses are made of compression-moulded or cast acrylic material. Figure 3-30 shows the first generation concentrator arrays and modules which use these Fresnel lenses. Unlike the flat-plate arrays, the concentrator systems have a much higher potential for greatly increased array efficiency. The advanced point-focus designs of a 200-W size (Sandia Laboratories) achieved 25% module efficiency at 125X using conventional Cz-grown Si cells [35]. The line-focus module built by ENTECH has demonstrated 17% efficiency at 22X, and it is considered to be the only concentrator module design in the commercial production stage as of 1991 [35].

For the concentrators, the smallest building-block element is not the module as in the flat-plate array, but a complete array assembly (e.g., pedestal, PV modules or panels, support structure, module or panel cabling, and a tracking drive mechanism). Moreover, this modular assembly must hold one or more array strings in order for the system to be cost-effective. This is one of the reasons the concentrator systems have a minimum "critical mass" from the standpoint of the smallest building-block assembly.

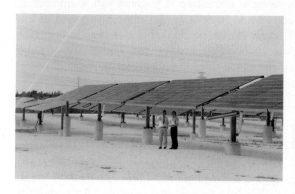

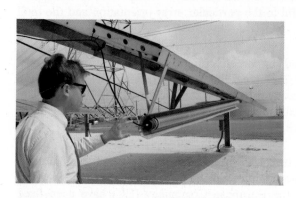

Passive 1-axis trackers (Robbins Engineering) were used on City of Austin Electric Utility
300-kW PV Plant. It was installed in 1986.

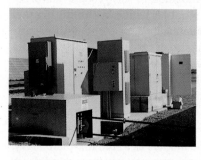

IPC's active 1-axis tracking array started operating in early 1992 at the PVUSA site, Davis, California.
(Photos courtesy of IPC, Rockville, MD.)

Figure 3-29.   One-axis Tracking Array Structures

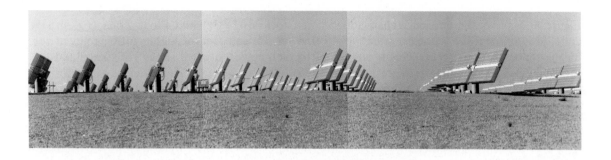

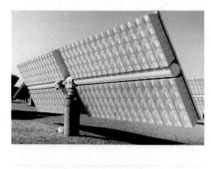

SOLERAS 350-kW Concentrator PV Array (40X Point-focus Fresnel)
at the Solar Village Near Riyadh, Saudi Arabia

300-kW Concentrator PV Array (20X Line-focus Fresnel)
at the 3M Research Center, Austin, Texas

Figure 3-30.   First Generation Concentrator Systems Developed in the USA

Figure 3-31 shows the efficiency characteristics of Czochralski (Cz) and Float Zone (Fz) grown cells designed for 33X concentration. Note that the Fz cells retain higher efficiencies at the higher concentration levels. This is attributed to their lower internal series resistance and the slightly higher open-circuit voltage characteristic of the Fz material. The efficiency data for the Cz cell is based on the large production quantity (> 40,000 of 23-cm^2 cells) delivered for the 350-kW concentrator system installed near Riyadh, Saudi Arabia, by the Martin Marietta Corporation [36].

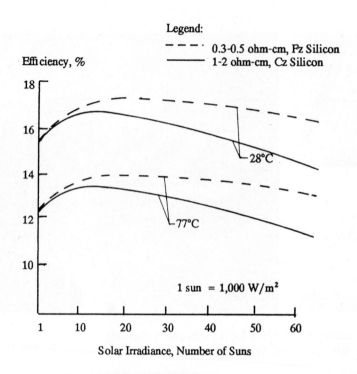

Figure 3-31. Efficiency vs. Solar Irradiance, Illustrating the Efficiency Differences between Fz- and Cz-grown Si. The characteristic shown is for 23.2 cm^2 cells designed for the SOLERAS 33X concentrator modules.

3.2.2   PV ARRAYS: STRUCTURES AND FOUNDATIONS

Structure types

The array structures support PV modules at a given orientation and absorb and transfer the mechanical loads to ground or roof. Practically all array structures utilized in the PV plants in Europe are fixed-tilted and accommodate flat-plate modules.

Figure 3-32 shows examples of roof-integrated or above-roof mounted types at FOTA and HOBOKEN (see Appendices 3 and 5, respectively). Earth-mounted structures are mostly of fixed tilted configuration. Only a small number of the European plants provided a tilt-angle adjustment capability. The structures usually consist of a panel frame support structure, foundations, and attachments. In special PV array structure designs like the cable-mounted configuration (see Appendix 18), a panel frame and/or foundation are not necessary. Unconventional array structures have been implemented, e.g., PV panels for passenger cars (see Appendix 25) where the panel must follow the contour of the roof of the car. The main advantages and disadvantages of roof-mounted versus earth-mounted structures are listed in Table 3-6.

Table 3-6. Comparison Between Roof- and Earth-mounted PV-Array Structures

Array Structure	Advantages	Disadvantages
Roof-mounted	- Low cost - No land consumption	- Bad accessibility - Reduced visual inspection capability - Usually no tilt adjustment capability - Module temperature higher - Not convenient for high power PV plants
Earth-mounted systems	- Good accessibility - Ease of visual inspection - Tilt angle adjustment and tracking capabilities possible - Reduced module heat up due to air venting	- High land consumption - High cost - Problems with module theft

Tracking structures using flat-plate modules have been installed by several test centres and application projects in recent years for experimental purposes (see Fig. 3-33 and 3-34). With the current trend towards large grid-connected PV plants, we can expect some systems to comprise of one- and two-axis trackers with flat-plate modules in the future. It is also entirely possible that concentrator systems which require two-axis sun tracking could also develop rapidly in the EC in the near future for central power plant applications.

Structural Materials

The most common types of structural materials (for panel frames, support members, etc.) are one or a combination of the following:

- Galvanized steel
- Painted steel
- Aluminium

- Wood
- Concrete

The structural material cannot be chosen independently of the foundation type. For example, wooden structures with single pedestal foundations are not usually applicable. Material selection is often dictated by soil characteristics and environmental constraints, including:

- Salty environment
- Humidity
- Temperature

However, the local availability of materials and the capabilities of local manufacturers must also be considered. The main advantages and disadvantages of the different materials are listed in Table 3-7.

Foundation Types

The two principal types of foundations used in medium-power plants are as follows:

Type	Examples

- Strip foundations:
  o north-south oriented
  o east-west oriented
	
	TERSCHELLING (see Appendix 13)
	ZAMBELLI (see Appendix 16)

- Single-pedestal foundations	KYTHNOS (see Appendix 7)

House in Black Forest, Germany

Fota Is., IRL

Bramming, DK

Hamburg, Germany

Ensheim, Germany

Figure 3-32.   Roof-mounted Arrays

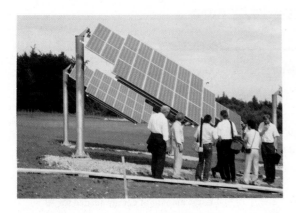

Figure 3-33. Large Passive 1-axis Array Structures at University of Stuttgart's ZSW PV Test Site at Widderstall (near Merklingen, F.R. Germany)

Figure 3-34. Smaller Version of the 1-axis Passive Tracker Carries 8 PV Modules

Table 3-7.   A Comparison of Structural Materials for PV Arrays

Material	Advantages	Disadvantages
Galvanized Steel	- Large variety of profiles available off the shelf - High mechanical strength - Long life expectancy (when zinc coating is not disturbed) - Can be reused (recycled)	- Galvanizing often requires rework (e.g., holes)
Painted Steel	- Large variety of profiles available off the shelf - High mechanical strength - Can be reused (recycled)	- Reduced life expectancy from corrosion - Paint damages during installation
Aluminium	- Corrosion-resistant - Light weight - Can be reused (recycled)	- Large amount of energy needed for Al processing - Cannot be used for all types of structures (e.g., single-pedestal)
Wood	- Inexpensive - Low production energy	- Low life expectancy (depending on type of wood and construction principle) - Movements due to wood expansion and contraction - Loosening of bolts due to wood contraction
Concrete	- Corrosion-proof (if reinforcement is done properly) - Cheaper than galvanized steel	- Aesthetic problems - Cannot be reused - Heavy mass - Applicable only for earth-mounted structures

At some of the plants, structures have been erected without foundations. Wooden or steel poles are rammed into the ground, as in the flexible, cable-mounted array structure at the PV facility at Lamole, Italy (see Appendix 18).

The standard material used for array structure foundations is concrete. Most foundations are cast on site, especially when ground supporting forces are needed, such as single-pedestal foundations, but precast concrete blocks have also been used at several plants (see e.g., NICE, Appendix 10 and VULCANO, Appendix 15). Single-pedestal arrays can also be designed as one continuous array, in which two neighbouring subarrays have one common foundation (see Fig. 3-35).

The continuous-array type was used at KYTHNOS and CHEVETOGNE , and the 2-pedestal type was used at AGHIA ROUMELI. Both types were implemented by RWE in their 340-kW Kobern-Gondorf plant and by the University of Stuttgart at their ZSW PV test facility (see Fig. 3-36 and 3-33, respectively).

Loads on Structures

A wide variety of loads can occur on array structures, and the assessment of these loads has a direct influence on total subsystem costs. The main loadings are:

- Wind load, as a function of:
  o wind velocity
  o tilt angle
  o module spacing
  o ground clearance
  o array aspect ratio
  o location of structure within field
  o natural/man-made windbreakers

- Snow
- Ice
- Dead loads (modules, structure, foundations)
- Live loads (people and machines during installation)
- Earthquakes

102

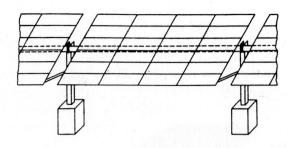

a) Continuous Array

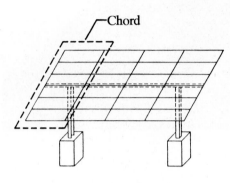

b) Two-pedestal Array

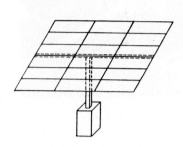

c) Single-pedestal Array

Figure 3-35. Definitions of Array Structure Assemblies Used in This Handbook: a) Continuous Array, b) Two-pedestal Array, and c) Single-pedestal Array.

The optimization of a structure requires a knowledge of array mechanical loading, usually under worst-case conditions. Snow, ice, and earthquake design loads are site-dependent and not considered. The critical structural load turns out to be the result of the expected wind load.

The design reference velocity, which is the basis of calculation, should be taken from on-site conditions if wind velocity recording is performed for a representative time period. If not, data from a nearby meteorological station should be used.

The design reference velocity is input to the wind load equation:

$$F_n = C_n \cdot A \cdot \frac{\rho}{2} \cdot v^2$$

where: F_n = wind load (normal to panel surface)

A = panel area

C_n = normal force coefficient

ρ = air density

v = wind velocity

The coefficient of normal force, C_n, which reflects the geometrical properties of the array, can be derived from either national standards or wind-tunnel tests. In many countries, the national standards do not provide normal force coefficients appropriate to array structure design. The available wind-tunnel test data for flat-plate arrays [37] can close this gap.

Kobern-Gondorf

Lake Neurath

Figure 3-36.   Array Structures at Kobern-Gondorf and Lake Neurath

The influence of tilt angle and aspect ratio is shown in Figure 3-37. A dramatic increase in the wind load can be expected for increasing tilt angles.

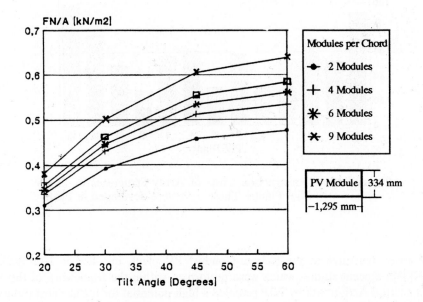

Figure 3-37. Average Net Pressure on Panel Surface as a Function of Tilt Angle and Number of PV Modules as a Parameter. Open flat terrain is assumed.

Array Structure Costs

The main goal in the design and selection of structures is to minimize their cost while ensuring long life. Also, aesthetic measures must be taken to meet the environmental and ecological needs of the site.

The first step in arriving at low-cost structures is to carefully select the PV plant site. Cost-related factors are as follows:

- Cost of land   - Accessibility

- Cost of site preparation   - Distance to consumers

- Soil conditions   - Maximum load to be expected during plant
                                        lifetime (wind , snow, ice, etc.)

The largest cost factors are the materials (structure, panel frame, attachments, and foundations) and labour (workshop activities and on-site installation).

A variety of loads and designs leads to a wide range of structure costs. In a CEC study made in 1982, the cost of structures and foundations for the PV pilot plants ranged from 28 ECU/m² to 100 ECU/m² [38]. The cost of the support structures and foundations on 13 PV Pilot Plants is shown in Figure 3-38. Array structures accounted for 3.5% to 18% of the plant's total budget.

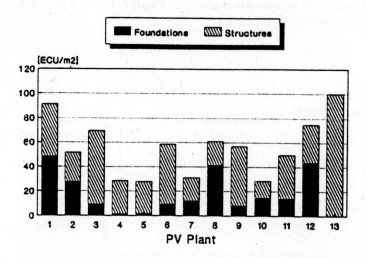

Figure 3-38. Construction Costs of Array Structures in 13 PV
Pilot Plants. These data were developed in 1984.

The cost of array structures at the Kobern-Gondorf PV plant, installed in late 1989, was about
180 ECU/m^2 [39]. Recent studies which were performed within the framework of the CEC's array
structures Concerted Action task by WIP revealed a high potential for further cost reductions on the
KYTHNOS-type array structures and the cable-mounted design [5]. The effects of panel tilt angle on
array cost were also determined (see Fig. 3-39). This study produced the following results for the
conventional and cable-mounted designs,based on 1989 costs:

	Achievable Costs (ECU/m^2)
- Conventional fixed arrays	
o Directly exposed to wind:	54
o Located within the field, and	
protected by the outermost arrays:	36
- Improved cable-mounted design:	25-35

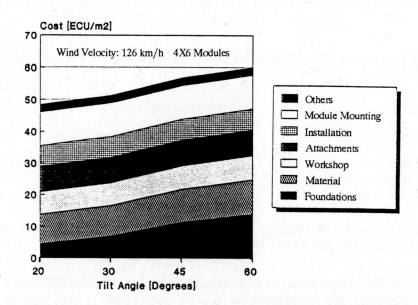

Figure 3-39. Cost Breakdown of a Two-pedestal Array Type Structure
with Fixed Aspect Ratio as a Function of Tilt Angle

Note that no optimum configuration exists for all applications because of non-standardized PV modules and site terrain characteristics.   "Off-the-shelf" structures and foundations are not applicable, in most cases, for large PV arrays. Even if a design exists, it must be adapted to specific site conditions (wind loads, soil conditions, etc.). A further development of state-of-the-art designs or even a brand-new, cost-effective design is to be expected. The benefits of optimization of the existing structure at KYTHNOS are illustrated in Figure 3-40. This optimization led to a cost reduction of more than 37% (1989 cost).

In all cases, the structure selection is determined or restricted by on-site conditions, e.g., single-pedestal foundations should not be used where either the soil is disturbed after ground levelling or the earth's support forces are small, such as in sandy soil. The use of continuous arrays is not advisable whenever the natural slope of terrain must be used.

State-of-the-Art and Future Prospects in PV Array Structures

Currently, the array structures used in Europe are predominantly fixed, tilt-angle designs for flat-plate modules. Experience with tracking systems is limited. A wider dissemination of tracking systems will depend on long-term reliability and energy output advantages demonstrated by existing plants. The trend in the fixed structures is towards single-pedestal structures with galvanized steel because of proven low-cost results.

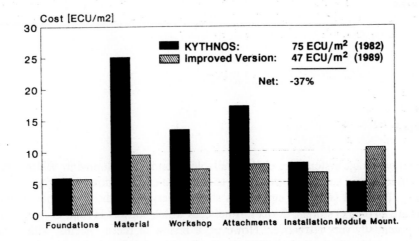

Figure 3-40.   Comparison of Original KYTHNOS Design and Improved Version of Same Configuration (37% Cost Reduction Estimated)

The aesthetics of plant layout will be influenced more and more in the future by social acceptance and environmental considerations. The application of large area modules is limited at the moment because of low demand for central power plants, but a trend towards larger modules can be expected in the 1990's. Optimal preassembling methods for PV panels, either with or without the metal frames on the modules, especially for large plants, will be addressed in the future.

Guidelines for Array Structures

Table 3-8 lists a set of guidelines for PV array structures. They are generally applicable to all sizes of PV array and especially for future low-cost array structures.

Table 3-8.   General Guidelines for Lowering the Cost of Flat-plate PV Array Structures

General

- Minimize the number of foundations	- Use galvanized steel instead of painted steel
- Make full use of PV module frames and minimize the number of module support frames	- Use stainless steel bolts and nuts, especially in regions with a salty environment (to achieve long life, thus reducing cost)
- Use off-the-shelf material (i.e., conventional simple profiles)	- Use preassembled parts to reduce: o In-field assembly and cabling time o Installation time
- Minimize the number of different steel profiles	
- Reduce the number of attachments	

Specific, mostly for large plants

- Make use of natural windbreakers (if any exist)	- Allow for a lower wind load on arrays inside the field
- Consider erecting a fence for both protection and wind breaking	- Use large PV modules to reduce: o Material requirements o Attachments (bolts, nuts, washers) o The number of electrical connections o Field installation costs
- Lower the wind load by: o Minimizing the tilt angle o Minimizing the tilt angle	

3.2.3.   BATTERIES

A wide variety of energy storage systems are either available or under development. Battery systems range from the proven lead-acid and nickel-iron types to the more exotic redox and high-temperature molten salt batteries. Fuel cell (with H_2 and O_2 as stored energy), flywheel, pumped-water, and pumped-air systems are all energy storage devices. Fuel cells which convert H_2 and O_2 gases directly into electricity are potentially very attractive. However, a review of these systems, with regard to availability and cost, quickly eliminates all but the lead-acid, nickel-cadmium, and nickel-iron batteries for PV applications in the immediate future. For these reasons, system designers have selected lead-acid batteries of the vented type for a vast majority of stand-alone systems. The following sections, therefore, cover largely the Pb-acid battery of the flooded type. Furthermore, only the topics of most interest to PV system designers and users are discussed: 1) design and performance characteristics, 2) factors affecting performance, 3) selection of cell type, and 4) maintenance and operation.

Battery Types

The two basic types of batteries are primary and secondary. The secondary types can be recharged, and are thus capable of charge-discharge cycling. For most long-term applications, a rechargeable or secondary battery is required. In Europe, these secondary batteries are commonly referred to as *accumulators*. Non-rechargeable or primary batteries are often used for low-power back-up power supplies, e.g., for clocks, wrist watches, digital memory devices, and calculators.

Tables 3-9 and 3-10 show the status and performance characteristics of the main secondary battery types. The most commonly used secondary battery types are lead (Pb)-acid, and nickel-cadmium (NiCd). The lead-acid battery has been in existence for over 100 years and has been refined to the point that today, it is one of the most inexpensive, most reliable, and most widely used storage batteries. NiCd batteries have a much greater cycle life than the Pb-acid type, but they cost about four to six times more than Pb-acid batteries. Nickel-zinc (NiZn) and nickel-iron (Ni-Fe) are available from only a few manufacturers, and they are not widely used.

The characteristics of two promising batteries are summarized in Figure 3-41 for the sodium-sulphur (Na-S) available from ABB [40] and Figure 3-42 for the nickel-hydrogen (Ni-H$_2$) system. The NaS system operates at a temperature between 300 and 400°C. A limited quantity of this battery which is being developed for car application is available from ABB or Chloride/RWE. The Ni-H$_2$ cell is being developed in Europe for spacecraft (50 to 100 Ah), and in the USA for both space (50 to 100 Ah) and terrestrial applications (100 to 250 Ah). This system is attractive especially for the space application because it has about 50% of the mass of the sealed NiCd battery cell. In NaS and sealed NiCd batteries, safety and proper charge control are critical design and operational issues.

The Lead-acid Battery

Lead-acid batteries are classified roughly in terms of the following applications:

- Starting, lighting, and ignition (SLI) (automobiles and aircraft)
- Emergency power supplies (UPS, telephone substations, etc.)
- Stationary or industrial (utility load levelling, submarines, and railways)
- Electric car (battery-powered passenger cars and automated remote control movers)
- Solar or PV energy storage (stand-alone and hybrid PV systems)

Emergency power and automobile batteries are used only for a short duration at a fairly high discharge current and are placed in float charge condition immediately after discharging. They can also be classified as shallow discharge (low DOD) or shallow-cycle batteries and consequently are not normally recommended for PV applications.

The stationary and electric car battery types are designed for deep-discharge, high cycle life operations, such as utility load leveling, submarines, and railways. Because of their low cost, proven reliability, availability, and well-established technology, together with high life-cycle and deep-DOD capabilities, these battery types are often used in PV applications.

Construction Features of Vented Pb-acid Cell

The lead-acid cell contains N positive and N + 1 negative plates which are made of a porous form of lead, and lead peroxide (PbO$_2$). These plates are immersed in a diluted solution of sulphuric acid (H$_2$SO$_4$), which acts as the electrolyte. When the battery cell discharges, the sulphuric acid reacts with the lead materials to form lead sulphate, and water is produced which dilutes the electrolyte. When the battery is fully discharged, the specific gravity of the electrolyte approaches 1.12.

Table 3-9.   Status and Availability of Major Secondary Battery Systems

Type	Availability/Application	Cost, 1990 (US$/kWh)	Individual Cell or Battery Sizes Availlable/Developed	
			Ah	kWh
Conventional Battery Systems:				
- Lead-acid, vented*	- Widely available and used in telecommunication and transportation fields	80-120	50-10,000	0.2-10
	- Main type used in PV application			
	- Available throughout the world			
- Lead-acid, sealed*	- Widely used in automotive vehicles and telecommunication applications	100-160	60-6,000	
- Nickel-cadmium				
o vented	- Widely available and used for stationary, car, telecommunication, and UPS	700	50-600	
o sealed	- Mainly used in small units (AA and D size)	100-500	10	
	- Spacecraft	> 5,000	50	
- Nickel-iron	- Sources limited	500		
Advanced Battery Systems:				
- Fuel cell o H_2-O_2 o Natural Gas				
- Lithium-metal sulphide	- Advanced state of development	20-30		
- Nickel-hydrogen				
o Terrestrial application	- Advanced state of development	2,000	100	15
	- Sources limited			
o Space application	- Used in Comsats since 1982	> 25,000	50	
	- Sources limited			
- Sodium-sulphur	- Near full-scale production by ABB (BBC) for cars	400-1,200	---	32 (50 kW)
	- Sources limited		---	
- Zinc-bromine	- Advanced state of development	500-1,500	---	20

*   The "vented" type has free liquid electrolyte and a vent hole (unsealed). The "sealed" type has its electrolyte suspended in the separators or in gel form such that there is no apparent free electrolyte. The latter is called "maintenance-free" because no water needs to be added. They are partly sealed with a one-way valve, so they are often referred to as "valve-regulated batteries" by telecommunication equipment users.

Table 3-10. Technical Characteristics of Secondary Battery Systems

| Type | Nominal Voltage (Vdc) | | Operating Temperature (°C) | Energy Density (Wh/kg) | Cycle Life* |
	No-Load**	Discharge			
Conventional Battery Systems:					
- Lead-acid					
o vented	2.12	1.6-2.0	-10 to 50	15-30	800-2,000
o sealed	2.12	1.6-2.0	0 to 50	10-30	500
- Nickel-cadmium					
o pocket-plate, vented	1.3	1.0-1.2	0 to 50	18-45	> 5,000
o sintered-plate, sealed	1.3	1.0-1.2	-5 to 40	10-20	1,000
- Nickel-iron	1.4	1.2	-10 to 50	22-45	> 2,000
- Nickel-zinc	1.85	1.5-1.65	-10 to 60	60-90	250-350
Advanced Battery Systems:					
- Fuel cell, H_2-O_2	1.3	1.0-1.2	-20 to 50	NA	> 2,000
- Lithium-metal sulphide	3.7	1.5-3.0	400-500	100-225	1,000
- Nickel-hydrogen	1.4	1.2-1.3	-5 to 50	44-60	> 3,000
- Sodium-sulphur	2.5	2.2	300-400	120-250	900-2,000
- Zinc-bromine	2.2	1.4-1.7	-10 to 50	65 to 75	600-1,800
- Zinc-chlorine	2.2	1.6-1.9	-10 to 50	60-90	500-800

* Usually at 60-80% DOD; significantly higher at lower DOD
** Open-circuit voltage at 25°C

While charging, the process is reversed; the lead sulphate at both plates is changed back into sponge lead and lead peroxide, and H_2SO_4 is produced, hence the specific gravity rises. When fully charged, the specific gravity is 1.28 at 25°C. "Gassing" occurs as the charge progresses due to the evolution of hydrogen and oxygen which bubble up through the electrolyte. Theoretically, this gassing starts when all active materials in the positive plates have been fully utilized or charged, so that they are no longer capable of converting the charge current into electrochemical energy.

The reversible chemical reaction in a lead-acid battery is:

$$PbO_2 + Pb + 2H_2SO_4 \rightleftharpoons 2PbSO_4 + 2H_2O$$

(charge)   (dicharge)

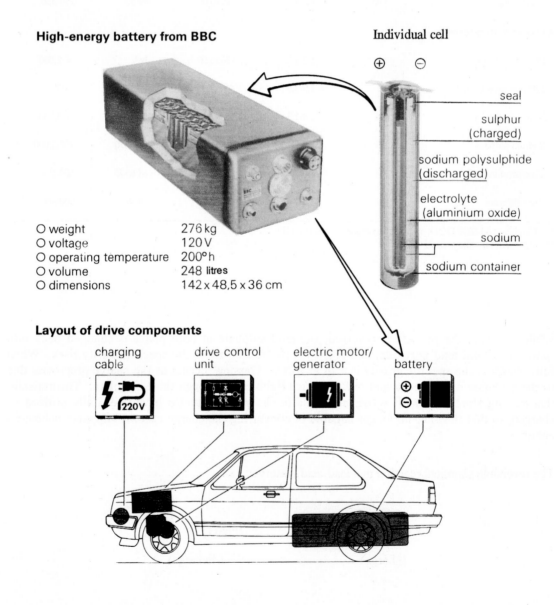

High-energy battery from BBC

Individual cell

$\oplus$ $\ominus$

seal

sulphur
(charged)

sodium polysulphide
(discharged)

electrolyte
(aluminium oxide)

sodium

sodium container

O weight	276 kg
O voltage	120 V
O operating temperature	200°h
O volume	248 litres
O dimensions	142 x 48,5 x 36 cm

Layout of drive components

charging   drive control   electric motor/
cable   unit   generator   battery

220V

Figure 3-41. Sodium-sulphur Battery Being Used in Experimental Cars (Courtesy of ABB/BBC).
The above battery contains 60 cylindrical cells connected in series.

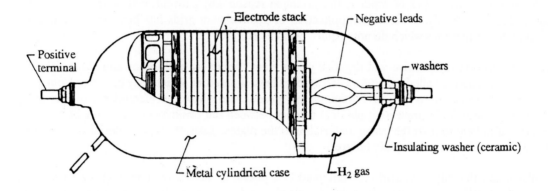

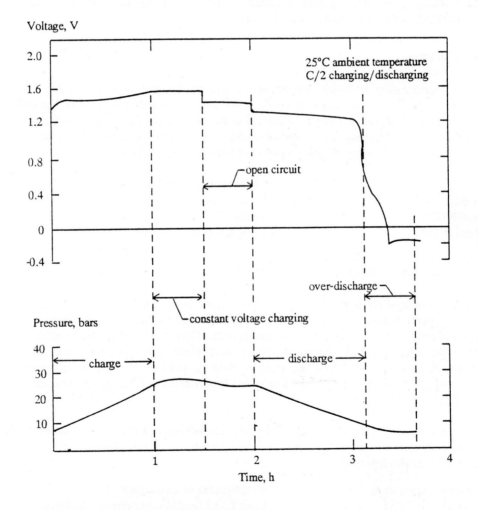

Figure 3-42.   A Typical Nickel-hydrogen Cell (Cylindrical Vessel Configuration Used in Space Programme) and Its Operation Characteristics. A 28-Vdc battery requires about 24 of these cells.

The construction of a typical vented cell is illustrated in Figure 3-43. The pastes used in the cell are structurally weak, and are poor conductors of electricity. In order to attain an even distribution of current in the paste and to support the paste structurally, a grid made from metallic lead is used. The grid consists of a lattice structure, cast from the lead, which holds the paste in the spaces between the lattice plates. Antimonial lead has been the traditional material for lead grids, although recently other alloys, notably calcium-lead alloys, have been used for grid material. These alloys have been incorporated primarily to reduce the decomposition of the water in the battery while charging via electrolysis. Electrolysis of water is the principal reason why conventional batteries must have water added periodically. The major application of lead-calcium grids has been in "maintenance-free" automotive batteries which do not require periodic addition of water .

In order to prevent the positive and negative plates from contacting each other and short-circuiting the cell, thin insulators called separators are used. These separators, which may be made from glass, plastic, wood, asbestos, rubber, or fibreglass, are porous and allow the electrolyte to pass freely between the plates while preventing physical contact between the positive and negative plates. The separators also help to hold the active material onto the plates. Loss of active material from positive plates is a common cause of battery failure.

The plates and the sulphuric acid are contained in a plastic or rubber container. Polypropylene is often used for cell cases because it is light, chemically inert, and strong, but other materials that are nonporous and resistant to sulphuric acid may be used.

Transparent Cell Cases

An important consideration in monitoring the operation of the battery is the ability to visually examine the amount of plate materials which flake off and settle at the bottom of the cell. The amount of these plate materials or residues collected at the bottom gives an indirect indication of the condition of the cell. For this purpose, the use of transparent cell cases is extremely useful, and it is a standard practice of many European manufacturers of battery cells, such as:

- FIAMM (IT)   - Oldham-France (FR)
- Hagen (FRG)   - Sonnenschein (FRG)
- Hoppecke (FRG)   - Tudor (IT)
- LYAC Power (DK)   - Tungsten (UK)

Failure Modes and Factors Affecting Battery Cycle Life

The basic failure modes of the battery are *open-circuit*, *short-circuit*, and *low capacity* of any one cell, and intra-cell connection open or exhibiting high resistance due to corroded terminals. The most common types of failure are shorted and low-capacity cells. Corrosion of the positive grid, shedding of positive active materials, and growth of the positive plate (see Fig. 3-44) are all possible contributors to the failure mode. However, available data are not too clear for us to determine the precise cause and effect relationship. Some controlled experiments and testing are thus needed and useful in every installed PV plants.

The low-capacity cells are usually on the verge of going into voltage reversal during battery discharge or turning into a shorted condition. Factors which enhance or accelerate this type of failure in PV plant batteries include the following:

- High operating temperature   - Stratification of electrolyte
- Many cycles at high DOD's   - Long stand time in open-circuit condition
- Periodic deep discharges   (i.e., without "exercising" or float charging)
- Long stand time while discharged   - Failure to apply low-rate recharging to fully
- Large number of cells in series   charge the battery periodically.

The above factors all affect the performance and life of any rechargeable battery. These factors are briefly discussed below. Additional discussions of specific items relating to charge/discharge control and battery operation are given in sections 3.2.4, 3.2.5, and 3.2.6, respectively.

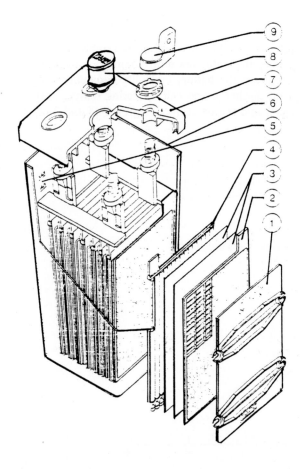

1) Polystyrene buffer plates

2) Negative plates, pasted

3) Separators - microporous material between each positive and negative plate

4) Positive plates - pressure cast from low antimony alloy

5) Electrolyte - Diluted sulphuric acid having a specific gravity of 1.25 at 25°C

6) Cell containers - Injection moulded transparent styrene acryloni-trile (SAN)

7) Cover - moulded plastic material

8) Vent plugs - Allow venting of internal gas during charging

9) Terminal, lead alloy (+ and -)

Figure 3-43.   Construction of a Typical Vented Pb-acid Battery Cell (shown for FIAMM cell)

a) View of the electrode stack from the bottom showing the elongation of the positive plate material (Courtesy of NMRC).

b) A sketch illustrating the positive plate growth phenomenon

Figure 3-44.   Growth of the Positive Plate Not Only Reduces Electrical Contact between the Metal Grid and Active Material, but also Accelerates Loss of Active Material. It is known to result in sufficiently high mechanical pressure against the container walls to crack the case.

1) *Depth of discharge, number of cycles, and temperature* — Major attention is usually focused on DOD and temperature because they are the principal parameters most commonly used by cell manufacturers and power system engineers to estimate the battery cycle life capability. Figure 2-12 (Chapter 2) illustrates the effects of depth of discharge and cell operating temperature on cycle life. Cycle life is an inverse function of DOD and temperature. Figures 3-45 and 3-46 are typical cycle life data available from Varta.

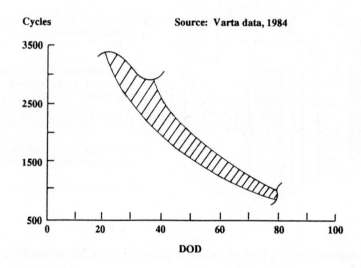

Figure 3-45.   Reported Cycle Life as a Function of DOD for Varta Pb-acid Battery Cell

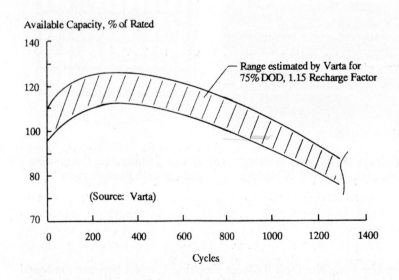

Figure 3-46.   Variation in Available Capacity with Cycling, as Estimated by Varta

The capacity of any secondary battery is generally known to degrade faster as battery operating temperature and/or depth of discharge increases. Therefore, in order to minimize cell replacement costs, a general approach to handling and operating the batteries is to minimize the overcharge and the battery temperature, and limit the number of charge/discharge cycles. The basic strategy in recharging the battery is to minimize the overcharge level. The principal difficulty is in detecting the "full-charge" level.

A word of caution: Field experience to date indicates that the available life data from cell manufacturers are too optimistic. High-voltage batteries have been failing 1-3 years after they were installed. There is insufficient actual cycle life test data at the cell and battery level, so be careful about the reliability and absoluteness of any cycle life data you collect; inquire about the test samples, the number of cells in series per test sample, test regimes, temperature and durations of charge and discharge in a given cycle. Cell-level or 3- to 4-cell group cycle test data cannot be extrapolated to the full (e.g., 108-cell or 250-cell) battery performance.

2) *Charge/discharge control technique and operating mode* — Charging is a critical operation which directly affects the useful life of the battery. The primary objective of a charge control system is to charge the battery efficiently while avoiding the detrimental effects of excessive overcharging. Figure 3-47 shows the typical charging profile of vented Pb-acid cells.

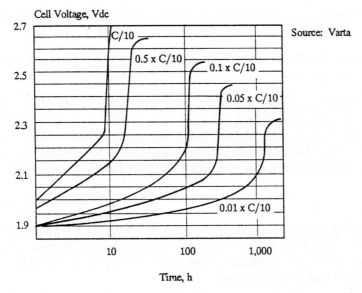

Figure 3-47. Charging Profile of New Varta Battery Cell at Different Constant Rates

The evolution of O_2 and H_2 gases in a cell mainly occurs during charging, but may also occur during a normal battery discharge. Vented Pb-acid cells can withstand a moderate amount of overcharging since the gases evolved, O_2 and H_2, can escape through the vent plug. However, both the oxygen evolution and recombination reactions are exothermic, and the resulting increase in battery temperature complicates the charge control problem. In addition, the increased cell temperature increases the rate of degradation of both electrodes and separators, and reduces the battery life. Moreover, from a system standpoint, *excessive* overcharging is generally undesirable, since it increases the cell temperature and represents an efficiency loss which must be compensated for by the solar array. This is not to say that overcharging is undesirable; some overcharging is mandatory to fully charge the cell plates.

117

3) *Periodic maintenance recharging* — Most stand-alone PV plants tend to operate for several days or weeks without adequately recharging the batteries because of lack of solar energy. We suspect that this lack of timely maintenance recharging during bad weather is a key contributor to the curtailment of cell life. A rule of thumb criterion is to provide maintenance-type recharging every 3 to 7 days, especially during cloudy days. To do so, an auxiliary power source (e.g., a small Diesel generator and rectifier with a constant-current output) is necessary. More costly alternatives are to substantially oversize the PV array and implement a load management system that limits the amount of battery energy used between recharges (i.e., minimize the allowable DOD).

Others have suggested use of several PV modules (properly connected to provide the required voltage) strapped across the battery to provide an independent source of low-current charging.

4) *Number of cells in series and effects of deep discharges* — The number of cells in the battery string is a very important consideration in designing the charge control and discharge protection schemes. For a 120-Vdc bus system, normally 60 battery cells are connected in series. The 340-V system requires at least 170 cells, and so on.

The performance mismatch in cells within a battery is attributed to three sources:

- The inherent difference in cell capacity at the time of cell manufacture (due to plate capacity differences)

- Different rates of degradation among the cells

- Temperature gradient among the cells (e.g., in double- and triple-deck cell mounting racks, the highest cell stacks closest to the ceiling usually operate at a slightly higher temperature).

In high voltage batteries, cell mismatch increases with cycling and ageing, and this can severely affect the life of the battery string.

Cell voltage reversals in low-capacity cells can easily occur during any discharge period, and are worse during deep discharges and as the cell ages. This failure mode is caused by the higher-capacity cells in the battery string driving lower-capacity cells into reverse-polarity as the discharge continues. This reverse-polarity condition is known to be more detrimental to cell life than excessive overcharging of cells, so it is essential that such conditions be prevented.

When one or more cells are in the shorted condition, the remaining cells in the battery will be subjected to a more severe condition while the battery is charging or discharging. During charging, the remaining normal cells will receive a greater overcharge because the effective cell voltage limit will be higher than that of the battery with no shorted or weak cells.[2] Likewise, during the discharge period, the remaining normal cells can easily be subjected to a higher DOD because of the resulting higher discharge rate.

5) *Discharge control criteria* — A lack of adequate provision to terminate battery discharging can cause battery failure or curtail its life. Some of the common strategies used to prevent deep battery discharge and to extend battery life are to:

- Cut-off battery discharge when a predetermined battery low-voltage limit is reached, e.g., 1.9 or 1.8 V per cell average.

- Minimize deep discharges by limiting the ampere hours delivered in a given day or discharge period (e.g., 20% DOD per cycle or per day)

- Use the calculated SOC to terminate the battery discharge at a certain preset SOC value, e.g., 70%.

---

[2] The severity of the cell mismatch condition is a function of the battery voltage (V_b), the number of cells in series (N_t), and the number of shorted cells (N_s). The average cell voltage (V_c) is: $V_c = V_b / (N_t - N_s)$. The problem becomes worse for a smaller battery, or lower N_t. Consider, for example, a 48-V battery with 24 individual cells and two dead or shorted cells that are not detected. The constant voltage charger set at 56.4 Vdc limit (2.35 V per cell) will cause the normal cells to be charged at 56.4 / (24 - 2), or an average 2.56 V per cell instead of 2.35 V.

The last two methods are not commonly used because of the added cost of electronics and the lack of an accurate, reliable Ah integrator or SOC calculator.

Typical discharge profiles at several rates and the effects of discharge rate on the available capacity of a cell (given an ideal charge condition) are illustrated in Figures 3-48 and 3-49, respectively. Operating temperature is also a critical factor. Capacity drops rapidly below 0°C and above 35°C, and its effect on battery cycle life is depicted in Figure 3-50.

Actual field operating experience and performance data on charge/discharge control methods in use by many operating PV plants would be valuable, especially with regard to the DOD and SOC used as control parameters.

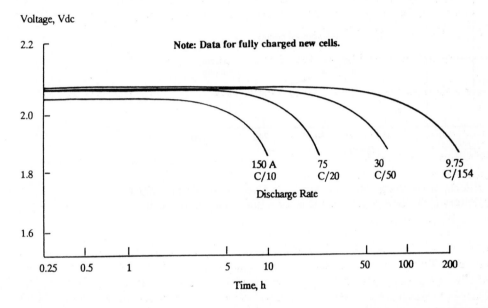

Figure 3-48.   Voltage Discharge Profile at Several Discharge Rates for Varta 1,500-Ah Vented Pb-acid Battery Cell. These characteristics are for a fully charged new cell.

3.2.4   BATTERY CHARGE CONTROL/DISCHARGE PROTECTION METHODS AND EQUIPMENT

Charge control and discharge protection criteria implemented in the PV system play a key role in extending battery life. However, they have received little attention in the past. Our knowledge of their effects on battery life is still largely empirical and they are not very well understood. Unfortunately, this situation will probably persist for some time. One reason for this is that it is very costly to run a cycle life test programme in a solar/PV operating environment, and therefore adequate actual cycle life data as a function of key operating variables (e.g., DOD, temperature, recharge time, and available energy, all of which fluctuate with time) are not available even at the cell level.

Cell-level testing is simpler than battery testing, but the resulting cycle life data from individual cells cannot be extrapolated to battery-level performance. The cell manufacturers usually do not conduct system-level tests for different users. Generally speaking, therefore, system designers should rely on the manufacturers for cell-level data and not for battery-level data. Battery-level data should be gathered in the installed PV plants, especially on high voltage batteries (200 to 500 Vdc).

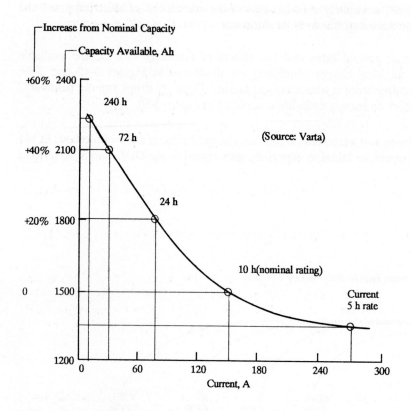

Figure 3-49.   Effects of Discharge Rate on Available Capacity of Varta 1,500-Ah
Battery Cell. This characteristic is for a fully charged new cell.

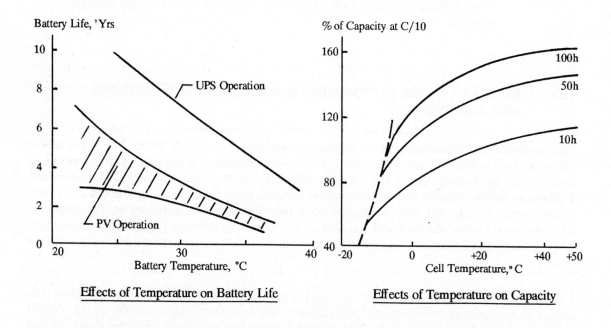

Effects of Temperature on Battery Life   Effects of Temperature on Capacity

Figure 3-50.   Effects of Temperature on Pb-acid Cell Cycle Life and Capacity (Data Source: Left
chart is author's estimate based on available data; right chart is from Varta)

120

Many battery failures have occurred in the past because the charge control is at the battery level and is usually not designed to respond to individual cell performance mismatches and/or degradation.

Charge Control Methods: Battery Level

Most charge control systems in use today can be classified as "battery-level control based on battery level parameters". In the battery-level control method, the cells of a battery are charged as a group with no provision for the use of individual cell parameters, and charging is based on battery parameters such as terminal voltage, temperature, and/or state of charge. Practically all PV plants use battery-level charge control with some capability to adjust the charge voltage limit manually.

The most common techniques for charging large batteries used in PV systems can be categorized as follows:

1) Constant charge voltage limit with temperature compensation, and charge current limited by the PV array capability (see Fig. 3-51).

   In this approach, the battery charges at a current determined by the array capability and load demand. When the battery reaches the preset voltage limit, the charge current tapers off. The battery continues to charge until the PV power is lower than the minimum necessary to sustain charging. The upper voltage limit is usually limited to the "gassing" level between 2.3 and 2.4 Vdc cell voltage.

2) Two-step temperature-compensated voltage limit

   This method is intended to recharge the battery quickly when solar energy is available while reducing overcharge to a minimum. During low DOD operation, the battery is charged using the lower charge voltage limit to minimize overcharging. In this mode, as illustrated in Figure 3-52, the charging voltage is initially 2.45 Vdc/cell. It then switches to a float voltage of 2.35 Vdc/cell to maintain the battery in a "trickle" or float-charge condition until cut-off by low solar array output.

3) Preset charge voltage limit cut-off followed by trickle-charge or zero-current mode

   This is a variation of method 1) above. In this mode, the battery is allowed to charge up to the preset voltage (e.g., 2.35 or 2.40 V per cell). After this point, the charge current is either completely terminated or dropped to a very low level, or "trickle charge" condition.

In combination with the above three methods, some systems employ battery state of charge (SOC) or Ah limit signals to terminate charging. The SOC calculator basically consists of input and output ampere-hour integrators and a charge return factor to allow for battery inefficiency. In practice, most SOC equipment has not been at all effective. At the few projects where these instruments work adequately, the reason for their success appears to be that they have ample provisions to periodically reset the instrument to the "100%" SOC.

The most thorough monitoring and control of battery operation, which we characterize as battery-level control with partial cell-level monitoring, has been done by Jydsk Telephone [40,41]. Detailed battery information such as 3-cell group voltages (108-cell battery), specific gravity of several cells, electrolyte levels, battery voltage and current; and status of the electrolyte bubbling system (air pumping) is routinely monitored (printed automatically) at their home office about 200 km away (see Appendix 19). In their case, the battery is an important part of the reliability of the remote telephone switching substations in Denmark and in Iceland.

For moderate size stand-alone or cogeneration systems in which batteries are absolutely essential, the optimum system configuration and operational strategies are as follows: 1) Configure the system as shown in Figure 3-9 and recharge batteries from the PV array or other available power sources; 2) Minimize battery discharge to absolutely essential loads which would also minimize the size of the auxiliary charger; 3) Mechanize a load management scheme to shift night-time (i.e., battery loads) to day-time loads as much as possible. This configuration achieves the desired objectives of reducing PV array and battery sizes, simplifies the power conditioning and control requirements, and significantly lowers the overall system cost.

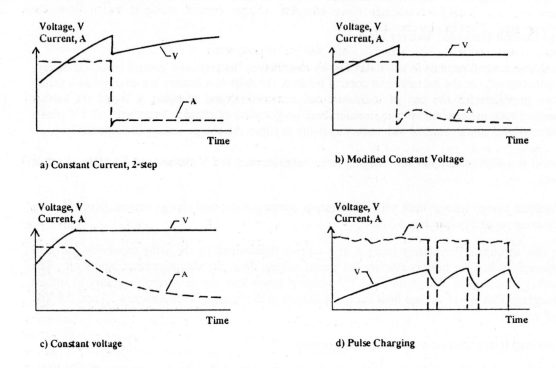

Figure 3-51. Battery Charging Methods Illustrated by the Resulting Battery Current Profiles: a) Constant current after the battery reaches its upper voltage limit, b) Modified constant voltage where the upper voltage limit is re-set to a slightly lower value after the battery reaches the first upper voltage limit, c) Constant voltage after the battery reaches the pre-set upper voltage limit, and d) Pulse charging.

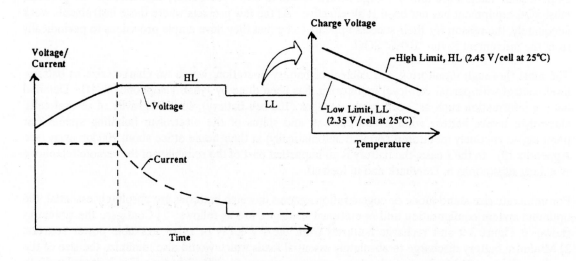

Figure 3-52. Voltage/Current Profile for Constant Voltage Charging with Temperature-compensated Voltage Limits

Charge Control Method: Cell-Level

There are two other, increasingly more complex, charge control methods which have been demonstrated. They can be classified as:

- *Battery level control based on individual cell and battery parameters*
- *Cell level control based on individual cell level monitoring*

Cell-level parameters can be one or a combination of cell voltages, specific gravities, and electrolyte level in addition to common parameters like battery voltage, current, and temperature.

The use of individual cell voltages to terminate battery charge is an example of battery-level control based on cell-level monitoring. There is no large-scale demonstration of this cell-level control technique reported in open literature.

Cell-level control refers to the ability to switch an individual cell into, or out of, a battery pack. This type of cell-level controller is illustrated in Figure 3-53 [42,43]. Cell-level monitoring and control may appear to be an optimum form of charge control because it theoretically eliminates uncertainties due to cell divergences.   An additional potential advantage of cell level control in the battery configuration is the ability to carry spare cells in the battery pack and switch them into the series string as other cells fail.   However, the cell-level controller has been rarely used because of its complexity, low reliability, and cost. If the application has a very high reliability requirement and/or the cost of an individual battery cell is substantially greater than the cell-level control electronics, then the cell-level protection system may be justifiable. The best compromise appears to be *cell-level voltage monitoring combined with a battery-level control*, which provides greater reliability of battery protection at an affordable cost. Other measurements such as specific gravity and electrolyte level can be considered for research purposes, but they are not reliable for control purposes on a routine basis for commercial application.

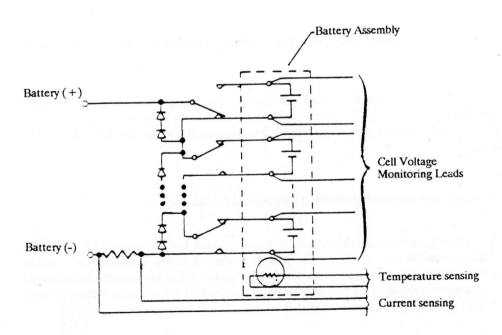

Figure 3-53. Individual Cell-level Controller. This configuration allows individual cells to be connected into and out of a battery string for charge/discharge control purposes.

3.2.5   BATTERY OPERATION/MAINTENANCE METHODS AND EQUIPMENT

Operational Problems Common to All Batteries

The most important hardware problem facing stand-alone PV plant operators is how to prevent catastrophic and premature failures by operational measures. Cell failures experienced fall into two basic types, catastrophic and non-catastrophic. Catastrophic failure is one in which the cell exhibits an electrically open circuit or a short-circuit condition very shortly after a sudden event like cell explosion and/or cracking of the cell case, accompanied by loss of electrolyte. In a non-catastrophic failure, the cell gradually fails by shorting, or there is a progressive or "graceful" degradation of its available capacity after many cycles.

The operators of all stand-alone PV plants will be confronted with questions such as:

- How can we best operate and maintain the battery?

- What data are essential to future operation and maintenance of the battery? How and where can we get them?

- How can we extend the life of the battery cells? Can we operate without the battery?

- What is the best way to determine the available capacity of a battery while in normal cycling mode?

- How can we detect the "graceful" degradation in capacity among different cells, and when do we replace them?

- When new cells are put among old cells, what is the effect on the life of both new and old cells in the string?

- Should we install new cells in the entire battery if the number of bad cells exceeds a certain percentage within that string? What percentage?

- What is the total cost of cell replacement (including hardware, shipping, installation, equalization, and checkout)?

Some of these questions must be considered during the plant design phase. Past projects co-financed by the CEC have already provided good answers and very useful lessons have been learned [44]. A number of CEC's JOULE projects under way will be investigating the above problems during the 1990-1992 time period.

Battery State of Health Assessment via Individual Cell Voltage Monitoring

One concept for battery state-of-health determination was developed by WIP under CEC's JOULE programme. Their approach is to determine the relative mismatch in the individual cells and to detect and identify, in real time, weak cells which need recharging or other preventative maintenance actions. The measurements required for this analysis are the individual cell voltages, battery voltage, and battery current (specific gravities are not used).

Figure 3-54 shows the basic hardware configuration for data acquisition and processing. PC1 performs data acquisition and processing, and PC2 does the calculation and evaluation functions. Remote monitoring from off-site offices is possible via modem/telephone network.

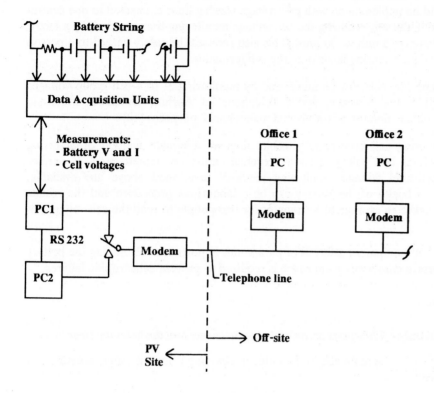

Figure 3-54. Block Diagram of Real-time Battery Monitoring System at the ZAMBELLI PV Pumping Station. It measures battery voltage and current, individual cell voltages, and cell temperature. Remote monitoring capability via modem is included to permit manual intervention of battery operation in case critical situations arise.

In the above system, the cell voltages are scanned during charge and discharge and performance mismatch is evaluated at a given time towards the end of discharge, end of subsequent charge, and during the open-circuit condition. The following parameters are calculated and sorted by a dedicated PC ("PC2" in Fig. 3-54):

- Average cell voltage at a given time at the end of discharge (EOD)
- Cells with lowest voltages at EOD
- Cells with highest and lowest cell voltages at a given time during the charging period
- Cells with lowest open-circuit voltages at a given time
- Standard deviations of cell voltages for selected key events
- State of charge of the battery

Based on the above data, the operator makes a "quick-look" analysis to arrive at the overall state of health of the battery, along with the identity and number of the weakest cells in the battery string. The most important improvements over the existing battery systems are 1) the ability of the operator to prevent voltage polarity reversal of any cell, especially during deep discharges, and 2) the capability to detect severe cell mismatch and initiate some maintenance recharging (equalization), both in real time.

The first prototype of this real-time battery monitoring system was installed at ZAMBELLI PV pumping station near Verona, Italy in September 1991. A similar battery monitoring system was installed at BRUNNENBACH PV/Diesel hybrid plant in early 1992 (as part of a collaborative effort between CEC and BMFT projects).

The cost of such a system for automatically monitoring individual cell voltages in a battery string goes up with the number of cells in this string. As an alternative low-cost approach, manual scanning of individual cell voltages could be implemented with cell voltage sensing lines connected to one central area outside the battery room, thereby reducing the cell voltage monitoring time required by a factor of about six for a 108-cell battery (5 min vs. 30 min). This also reduces the safety hazard to a person attempting to attach a pair of bare sensing leads onto the cell terminals.

The number of measurement channels can be minimized by monitoring 3- to 4-cell group voltages rather than individual cells. For instance, Jydsk Telephone in Aarhus, Denmark effectively implemented 3-cell group voltage monitoring for normal maintenance purposes [40].

When only battery-level sensing and control is possible, then an adequate method of detecting undesirable conditions such as cell voltage polarity reversals must be implemented. If neither automated nor manual cell-level voltage monitoring methods mentioned above are available, determination of the state of health of the battery can be a labourious procedure and the results imprecise. For example, it takes an average of 15 minutes for two people to read the cell voltages of all 108 cells in a battery.

Remember that personnel and equipment safety is of paramount importance. Extending the battery life is secondary, nevertheless, a conservative approach is required to prevent catastrophic failures.

How to Avoid Cell Explosions

1) *Periodically check the electrolyte-filling ports or the vent plugs to ensure that the holes are clear.*

   If the vent plug is not blocked, the main effect of continued charging is loss of water; pressure will not build up inside the cell..

2) *Terminate charging whenever there is a severe imbalance in the cell voltages or if any cell exhibits an abnormally large deviation from the mean cell voltage.*

   It is essential to check the condition of individual cells by analysing the cell voltages during normal charging and discharging. Low-capacity cells are easy to detect during deep discharge operations as they will normally exhibit lower-than-average cell voltages, especially towards the end of the discharge period and higher voltages during charge. The cell voltages will clearly identify any shorted or very low-capacity cells. Any "dead" cells should be removed from the string, and "weak" cells recharged separately from the pack. Sometimes the "dead" cells can be rejuvenated by replacing the electrolyte and subjecting the cells to formation type cycling. AGSM has attempted to do so on the ZAMBELLI battery, but they concluded that internally shorted cells cannot be reconditioned [45].

How to Extend the Battery Cycle Life

In order to get the longest life from a Pb-acid battery working in the PV system/solar environment, it is not enough to "size it properly," "apply proper charge control," and "do maintenance checks." Both automatic and manual procedures at the system level are essential to achieve long battery life. Listed below are a set of guidelines not only to reduce safety hazards arising from catastrophic failures but also to reduce the cost of battery cell replacements.

1) *As far as possible, keep the battery in the fully charged state, avoid operating it in a partially-charged state, and minimize long periods of open-circuited condition.*

   Note that this applies only to the Pb-acid system. It does not apply to Ni-Cd or Ni-Fe batteries, both of which require complete discharge of cells for full reconditioning.

   The best way to satisfy this guideline is to periodically apply low-rate charging at the battery level or individually on low-capacity cells every three to seven days. This is often referred to as "battery cell equalization." This maintenance recharging can be done by constant voltage "float" or applying a low constant charge current. For high voltage batteries, a low-rate constant-current charging rather than float charging at the battery level is better to minimize cell voltage divergence while charging.

The idea behind the battery equalization charge is to allow the low-SOC cells in the string to come up to the same SOC as the high-SOC cells in view of the fact that the latter will not accept any more charge. For vented Pb-acid cells, the excessive overcharging and higher rate of water loss in the high-SOC cells during this equalization charge is not considered detrimental if done only occasionally. Battery-level equalization is permissible mainly for the vented cell types, but with some caution — do this at a low charge rate which allows higher charge acceptance and avoids excessive bubbling in the electrolyte, resulting in loss of water and higher cell temperature.

However, trickle charging cannot be applied properly and systematically every few days with only the PV array which produces no power whenever the sun is obscured. The most practical way is to provide a generator (Diesel or propane gas) with a small rectifier on site for the purpose of battery maintenance recharging. The same system, illustrated in Figure 3-55, can be used for cell-level equalization charging. It is cheaper to implement this approach during the design phase, and not after plant start-up.

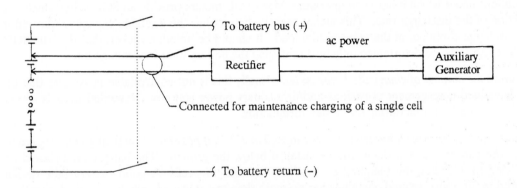

Figure 3-55.   Portable Small Rectifier for Trickle-charging of Small Number of Cells in Series (1 to 6). The rectifier should be capable of supplying low-current (C40 to C/100) at low voltages (3 Vdc for a single cell). This method of maintenance charging of a single cell in a series string of cells must be done with the battery open-circuited as shown. An acceptable alternative to the auxiliary generator and rectifier is to uise a number of PV modules in parallel with a series blocking diode.

In some cases, a separate PV panel or modules connected directly to the battery terminal via a series blocking diode may be just as effective for float charging purposes (upper battery voltage limit is uncontrolled; battery input current is limited by the short-circuit current of the PV panel). However, this alternative method of battery recharging is not as reliable as other methods.

2) *As dictated by the cycle life desired, limit the battery discharge per cycle to a pre-defined maximum allowable DOD; avoid frequent deep discharges (or high DOD's) other than controlled deep discharges for capacity test.*

This guideline must be implemented or provided for by appropriate battery and PV array sizing, and system load management during the design phase. The load management procedures should be verified during the post-installation checkout.

Battery life is an inverse function of DOD and temperature. By limiting the DOD to about a 20% average per cycle, a 5-year battery life without cell replacement may be achievable. As the number of cells in series increases, e.g., in high voltage batteries, these two considerations become more critical.

Many designers have sized the battery for 50 to 80% permissible DOD per cycle. At 80% DOD/cycle, it would be highly unlikely for the battery to last more than one year without losing any cells.

Comments on Discharge Protection Approaches: Knowledge of the available capacity is a prerequisite for proper battery maintenance and control of system operation. However, the battery discharge protection must be designed to use a simple positive cut-off signal such as low-voltage threshold for primary control. SOC and DOD signals, if available, should be used only as secondary control parameters because they are not completely reliable.

Caution on Use of State of Charge (SOC) Meters: If the battery has discharged to 2.0 V/cell at a very low rate (say, C/50 or lower), this could mean the battery has actually discharged 100% of its actual capacity at C/10 rate, even though the battery did not reach its pre-set low cut-off voltage (e.g., 1.8 Vdc/cell). If an SOC meter is available, and it displays "60%," the meter simply tells you that on the last discharge pass, the depth of discharge was 40% of the rated capacity, and it does not mean that the battery still has 60 % of rated capacity remaining.

If further discharge results in the battery reaching its low voltage limit, and the meter still indicates a large SOC (say, 35%), it means that the SOC meter calculation is in error. Moreover, it does not necessarily mean that the battery has been undercharged in the last few cycles — it may have naturally reached a steady-state *operating* capacity under the cycling duration and conditions under which it has been operating. Many SOC meters provide an SOC value which is a function of the discharge rate. This means that the battery has the available capacity indicated if the discharge continues at the same system load. It should be noted, however, that the true SOC of the battery can be determined only by discharging the battery to its low voltage limit.

3) *Minimize the number of charge-discharge cycles.* Usually this is not controllable in systems that do not have load management provisions and/or auxiliary power sources. Nevertheless, it is a very important consideration during the design formulation.

4) *Operate the battery in the lower temperature range (0 to 25°C), if possible.* Practical guidelines are to place the battery under a shade and/or install it below the ground. The battery rack should be designed to minimize the temperature gradient between the cells. A good battery rack design located in the above-ground container has no more than two stacks of cells in a vertical direction. Attempts to keep the battery cool (less than 35°C) will generally improve the reliability of the battery, and consequently the whole PV plant.

5) *Periodically check for severe imbalances in relative SOC of all cells in the battery.* This should be done by monitoring individual cell voltages during charge and discharge and comparing against average cell performance. The specific gravity measurements should be used only as back-up data. It is suggested that the monitoring period should be every 3 to 6 months, depending on the system reliability, or at least at the start of summer and winter months. Towards the end of a charging period, cells which exhibit low voltages (e.g., 2.250 V vs. 2.350 V average) may be undercharged, have low capacity, or have developed high resistance shorts.

The best indication of performance mismatch in terms of high- and low-capacity cells can be seen towards the end of a given discharge period, especially during a deep discharge. The low-SOC cells will generally exhibit lower voltages than the average of all cells towards the end of discharge, and they will often be driven into reversed polarity as battery discharge continues.

SOC Calculator

The SOC calculator is intended to determine the battery capacity available at a given time during regular battery operation. It can be used either for monitoring only, or for both monitoring and control. The device is often described as SOC "measurement" equipment, however, the SOC parameter is calculated by a pre-defined logic and a math model stored usually in an EPROM, and is not a directly measured quantity. Typical input measurements are:

- Battery voltage and current
- Battery cell temperature

and typical calculated output parameters are:

- SOC in % of battery rated capacity
- Depth of discharge (DOD) in % of battery rated capacity

The equipment often displays the above input and output parameters (see Fig. 4-25, Chap. 4). If used for control purposes, the equipment provides the signal to 1) terminate the battery discharge whenever the battery SOC reaches a preset minimum limit, e.g., 20%, and/or 2) terminate or reduce battery charging whenever the battery SOC reaches 100% or nearly 100%.

3.2.6   POWER CONDITIONING: DC-DC VOLTAGE REGULATION DEVICES

Power conditioning devices used in the PV power systems are of the following types:

- dc-dc voltage regulation circuits
- dc-dc voltage and power converters
- dc-ac inverters
- ac-dc rectifiers

This subsection covers the first two types, namely, the voltage regulation circuits such as the series switch, shunt regulators, and converters. Subsection 3.2.7 covers the inverters and subsection 3.2.10, the rectifiers as part of the UPS. The voltage conversion and regulation circuits may be simple or quite complex, depending on the extent to which battery and load management strategies are implemented. Transferring the load bus from the solar array to the battery and vice versa is fully automatic, based on the pre-set voltage levels and thresholds for the PV array and the battery.

Voltage Regulation via Series Switch or Shunt Regulators

Voltage regulation can be accomplished in several ways. In the low-power PV systems, a widely used approach is a simple on-off switch in series with the array and/or a full shunt regulator (see Fig. 3-3). The system with only a series switch is normally designed to tolerate a wide bus voltage range. The switch in this case simply prevents both overvoltage and undervoltage condition of the battery, and the system uses the current-limiting characteristics of the PV array.

The full shunt regulator controls the upper voltage limit by shunting part of the current from the PV array, causing the array voltage to vary but remain within the design range. This scheme is relatively easy to implement, but it does require some method of dissipating the heat generated by the shunted current.

Voltage Regulation via Array String Switching

Another voltage regulation scheme used quite often in medium-power PV plants is a series string on-off control (illustrated in Fig. 3-6). The control electronics, usually a computer, adjusts the number of array strings connected to the bus such that the array bus voltage remains within a pre-set range. In order to reduce the large step-voltage changes during the switching operation, the number of array strings needs to be in the order of at least six to ten.

Partial Shunt Regulators

For higher-power PV systems, the partial shunt techniques illustrated in Figures 3-7 and 3-8 are well known, but they have not been applied very often. In this method, the power segments are switched in and out sequentially as the difference between the required load and array power varies within the pre-determined range of voltage regulation. Note that a digital switch (e.g., mechanical or solid state) can substitute for the linear partial shunt shown in Figures 3-7 and 3-8. The advantage of the digital switch is that it has no dissipative losses.

Power Converters

The power converters are dc to dc regulation devices placed in series with the PV array. Figure 3-56 shows simplified diagrams of the three basic circuits used in power conversion. The "buck" converter regulates the higher input voltage to a lower value. The "boost" converter provides a higher output voltage than the input. The "buck-boost" converter can provide both lower and higher output voltages.

129

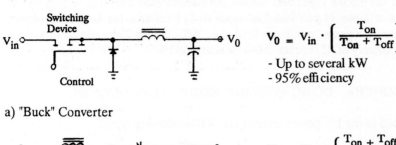

$$V_0 = V_{in} \cdot \left(\frac{T_{on}}{T_{on} + T_{off}} \right)$$

- Up to several kW
- 95% efficiency

a) "Buck" Converter

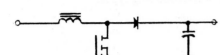

$$V_0 = V_{in} \cdot \left(\frac{T_{on} + T_{off}}{T_{off}} \right)$$

- Lower efficiency than buck regulator
- 90-92% efficiency

b) "Boost" Converter

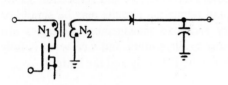

$$V_0 = V_{in} \cdot \left(\frac{T_{on} N_2}{T_{off} N_1} \right)$$

- Lowest efficiency and lowest reliability because of higher switching current and higher voltage stress
- 90% efficiency

c) "Buck-boost" Converter

Figure 3-56.   Basic Circuits for dc-dc Power Conversion: a) Buck converter regulates the input voltage to a lower value, b) Boost converter converts the input voltage to a higher value, and c) Buck-boost converter regulates the input voltage such that its output voltage can be higher or lower than the input voltage.

All of these circuits make use of a high-frequency switching regulator which usually operates at 20 kHz or higher to minimize the size and weight of the magnetic (i.e., transformer) components and capacitors.   The switching regulator converts the dc input voltage to ac output voltage as an intermediate step, and with the use of a transformer, steps the voltage up or down, and then reconverts the ac voltage to dc output.

The basis of all switching regulators is the *pulse-width-modulation* (PWM) technique.   The principle of PWM has been well known for decades.   A patent was issued in 1944, but it was not until the early 1960's in the space programme that it began to be widely applied to power electronics.   Today, the PWM technique is used extensively in both dc-dc converters and dc-ac inverters.

The PWM is a linear modulation process where the duration of each pulse in a train of pulses is made proportional to the instantaneous value of the modulation signal.   The pulse width is determined by the ratio of power switch on-to-off times.   This process effectively maintains a relatively constant output voltage independent of variations in the input voltage.

The power switches in the low-to-medium power converters can be bipolar transistor, power MOSFET, or thyristors.   Transistors and MOSFETS rather than thyristors are normally used with PWM because of their faster switching time.

Table 3-11 lists a family of efficient converters manufactured by Siemens Solar, ranging from 6 to 28 kW.   They are designed to operate with 12 to 140 Pb-acid battery cells in series.   A typical efficiency curve for the smallest 6-kW unit is shown in Figure 3-57.   If you choose to use an "off-the-shelf" converter such as one of the sizes listed in Table 3-11, you must consult with the converter manufacturer to check its compatibility with the rest of the system.

Table 3-11. A Family of Converters from Siemens Solar Designed for Use with Vented Pb-acid Batteries

Parameter	Unit	Solarconverter Model			
		SC24	SC48	SC220	SC280
Electrical:					
- Rated Power	kW	6.0	8.8	24.0	28.0
- Efficiency, Full Load (FL):	%	92	94	97	97
20% FL:	%	94	96	98	98
- Array input, nominal voltage	Vdc	51	34	160	220
- Open circuit voltage	Vdc	25.4	51	223	297
- Battery output, nominal voltage	Vdc	24	48	220	280
- End-of-charge voltage	Vdc	28.2	56.4	253.8	323.4
- Maximum current	A	249	144	116	106
Mechanical:					
- Dimension:	mm	758 x 595 x 200 (all models)			
- Weight:	kg	50	50	60	60
Number of 2-V battery cells:		12	24	110	140

Source: Siemens Solar Commercial Brochure

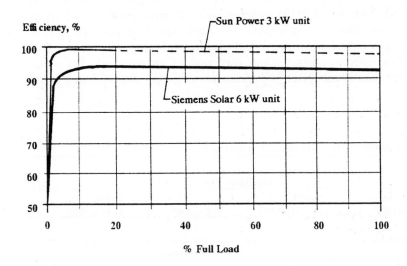

Figure 3-57. Efficiency Characteristics of Typical dc-dc Converters

Comparison of Voltage Regulation Approaches

Briefly summarized above are the basic features of several voltage regulation and control techniques used in the stand-alone PV systems. We must keep in mind that their most important tasks are to minimize excessive overcharging and to prevent batteries from discharging below a pre-defined level. The latter can be implemented by the battery state-of-charge calculator, depth-of-discharge indicator, or simply by a selected battery low-voltage limit. In any of these approaches, a conservative limit should be established to allow for cell mismatch and number of cells in series.

Table 3-12 summarizes a comparison of several methods currently used for charge control and load voltage regulation. A 300-W PV system is used to illustrate the power dissipation differences; however, this chart is applicable to very high-power PV plants. The principal differences between the regulation methods can be summarized as follows:

- *Series-relay* control method simply disconnects the PV array after the battery has reached its prescribed upper voltage limit. It is the most economical method and applicable for low-power systems; however, it is deemed too coarse for efficient charge control as well as adequate utilization of array capability.

- *Full shunt* handles excess current from the PV array, which can generate a large amount of heat. It is effective only for low-power systems but economical.

- *Series* regulator or converter accepts a wide range of input voltage and provides a good output voltage regulation. A drawback to all series converters is it presents a power dissipation in the series elements and can be costly. However, it is probably the best approach for a reliable battery charge control.

- *Array string on-off* control sequentially disconnects or adds the array strings from the dc bus. This method is usually implemented by the computer which may be a drawback for smaller systems. It is applicable for any size of PV array, but limited to systems with dc bus voltages up to about 500 Vdc because of the availability and cost of series switches.

- The *partial shunt* regulators have been used widely from 50 W to 2-kW systems in communication satellites [46], where minimum power dissipation and high-system efficiency are critical drivers. The approach can be implemented with analog (linear shunt) or digital circuit (switch) as illustrated in Figures 3-7 and 3-8, respectively. This type has been applied, but infrequently, in terrestrial PV systems, ranging in power from 200 W to 350 kW [47,48]. The main drawback to the partial shunt systems is the need to install the shunt switches near the array panels or subarrays, along with the controller (error amplifier or a computer). An alternative is to place the control electronics in the central building or a shed and install signal wires from the controller to the shunt switches out in the field. This approach has worked very well in an extremely hot desert environment in Saudi Arabia for the 350-kW plant [48] which was originally designed as a stand-alone system.

Table 3-12.   Comparison of Various Array Regulation Approaches

Parameter	Regulation Type			
	Series Switching[1]	Linear Partial Shunt[2]	Full Shunt[3]	Partial Digital Shunt[4]
Adaptability	Special design for each system	Wide power range	Special design for each system	Wide power range
Efficiency at Full Load	75-85% (48-V system) 90-95% (> 300-V system)	100%	100%	100%
Redundancy	Yes	Yes	No	Yes
Power Dissipation	Low at full load, 15 W*	Medium at no load, 55 W*	Very high at no load, 300 W*	None
Cost	High	Low	Medium to very high	Medium

* Wattages are examples for a 300-Wp PV system; they are listed for comparison purposes.
1 See Fig. 3-5 for series converter and Fig. 3-6 for array string control.
2 See Fig. 3-7.
3 See Fig. 3-3.
4 See Fig. 3-7 or 3-8; voltage regulation may be coarser than analog linear shunt regulator

3.2.7   POWER CONDITIONING: INVERTERS

Inverter Functions and Operating Principles

The inverter converts dc to ac power. There are basically two types of inverters — rotary machine and static solid state. Because of the rapid progress in power semiconductor devices, the rotary type is being replaced by the solid-state inverter. For this reason, only solid-state inverters are considered here. The inverter's main functions are:

1) *Inversion* of dc voltage into ac (thus, the word *inverter)*

2) *Waveshaping* of the output ac voltage. The ideal ac waveform is sinusoidal. For resistive loads which can use both ac and dc power supplies the waveform is not critical. For inductive loads (e.g., motors) the ac waveform supplying the motor must be close to sinusoidal because the maximum efficiency of rotation (i.e., mechanical energy) is produced by the fundamental 50 Hz sinewave. Any higher harmonics included in the ac waveform supplying the motor do not produce rotation, but they cause I^2R heating in the copper coils and the body of the motor, and very often, they are responsible for degradation and destruction of the coil insulation.

Static inverters use power semiconductor switches which operate at the cut-off and saturation mode (on-off), so the output of the inverter is square pulses.[3] A square waveform can be converted to a sine waveform using a power filter. The filtering of harmonics closest to the fundamental requires expensive, bulky, and energy-consuming capacitors and inductors, as well as costly switchgear, if used. A key objective is to design inverters whose output waveform has as few higher harmonics as possible and further away from the fundamental, by controlling the power semiconductor switches in a special way, rather than producing a square waveform at the output of the inverter and filtering it with power filters.

3) *Regulation* of the effective value of the output voltage. This is accomplished through pulse-width modulation (PWM).

Since in a PV system the input dc voltage to the inverter varies, the output load current, and consequently, the forward voltage drop of the power semiconductor switches also vary. This means that the amplitude of the produced pulses at the output of the inverter is not constant. So, in order to maintain a constant effective value of the output voltage, the width of the output pulses must be modulated (narrower or wider). Pure square-wave inverters cannot have output voltage regulation as they do not have the ability to modulate the width of the square wave, so some designs accomplish the output voltage regulation via switching to the appropriate power transfer taps through power transformer tap changing (see Fig. 3-58).

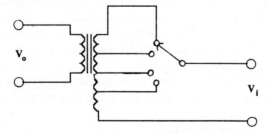

Figure 3-58.   Output Voltage Regulation by
Tapping the Secondary of the
Output Transformer

---

[3]   Power semiconductor switches operate at cutoff or saturation region (off or on). These two modes of operation are the most efficient ones, because at cut-off, no current flows through the switch (i.e., Power $= V \cdot I = 0$ as $I = 0$) whereas at saturation there is no voltage across the switch (i.e., Power $= V \cdot I = 0$ as $V = 0$). In reality, at saturation (on) mode of operation, there is always some voltage drop across the switch, which is called conduction voltage drop. The higher it is, the higher are the conduction losses. This is the main reason why IGBTs are used. Power MOSFETs are very reliable and very fast semiconductor switches. Their main disadvantage is that the high voltage devices present high voltage drop when they are conducting (7 V for a 1,000-V/10-A device). Bipolar power transistors present lower voltage drop (3 V for a 1,000-V/200-A device), but they are not as fast as the power MOSFETs. New generation IGBTs are 1,000-V/300-A devices with only 3 V conduction voltage drop. So IGBTs combine the advantages of the power MOSFETs with those of the bipolar transistors.

A transformer is not necessary for any of the above-mentioned tasks (inversion, waveshaping, and regulation), and it can be eliminated from an inverter's design. Transformers along with inductors are the most bulky, noisy, and expensive components of any power conversion devices. The price of transformers, capacitors, and inductors essential for power factor correction and harmonic filtering will continue to rise because of increasing labour and material costs, whereas the price of the electronics, in particular, semiconductor switches, will continue to fall. Transformers offer the following advantages:

- They provide *galvanic isolation* between the PV system and the grid or ac loads to provide more safety.

- They permit *flexibility in* selecting the PV array and battery voltages, or the battery voltage, irrespective of the output voltage. This is accomplished through the transformation ratio of the transformer. Without a transformer, a special PV array wiring is required (for example, the Fraunhofer Institute's inverter, described in Appendix 20).

- They permit *grounding* of the PV array for safety purposes.

4) *Operation at or near the array peak power point,* with a capability to accept a wide range of array output voltage. The efficiency of a typical inverter varies with its output power as shown in Figure 2-13 (Chapter 2). The significance of this characteristic is that the total PV system energy efficiency in a given day can be very low if the inverter operates at a fraction of its rated load during that day. Two ways of operating the system at high efficiency are to: 1) Design and operate a 1-inverter system such that the inverter operates only when the load is high enough to attain an inverter efficiency of, say, 90% or more, or 2) Use multiple inverters of the same rating to cover the full range of load power requirements, and connect one or more inverters on-line so as to operate the inverter(s) in the high efficiency region.

There are other operational issues such as automation of the basic control actions (e.g., sun-up and sun-down, start-up and shut down) and protection in the event of utility power shut-down or grid failure. The control issues are easily solved electronically. It is very simple to implement automatic start-up and shut-down procedures. Some inverters use the short-circuit current signal from the array for turning on and the array open-circuit voltage for shutting down. In the event of an interruption of the utility power, the inverter is arranged to turn off within the next half cycle of the operating frequency.

The simplest and the least expensive static inverter has a square-wave output (see Fig. 3-59). It consists of a square-wave generator, voltage regulation circuit, and a tuned filter. A square wave output can be obtained from a dc source using four solid-state switches illustrated as S_1 through S_4 in Figure 3-60. This figure represents a single-phase bridge circuit using power semiconductor switches. Inversion is accomplished by alternately closing switches S_1 and S_4 and then S_3 and S_2 every 10 ms. This is repeated 50 times every second which produces an ac current flow through the transformer. Basically, the same principle is used in generating the other step-pulse sine waves shown in Figure 3-61.

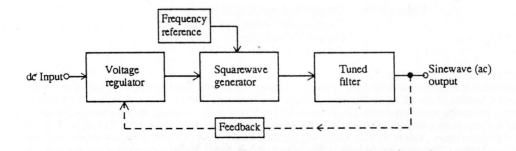

Figure 3-59.   Square-wave Generator Used as an Inverter

134

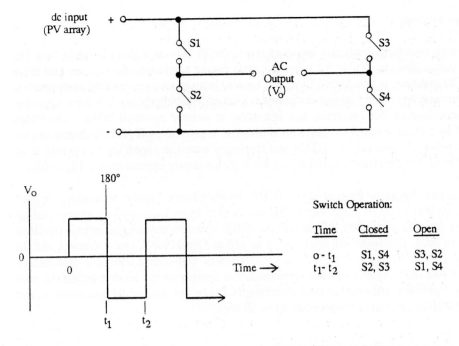

Switch Operation:

Time	Closed	Open
$0 - t_1$	S1, S4	S3, S2
$t_1 - t_2$	S2, S3	S1, S4

Figure 3-60.   Generation of a Square-wave ac Output from a dc Source

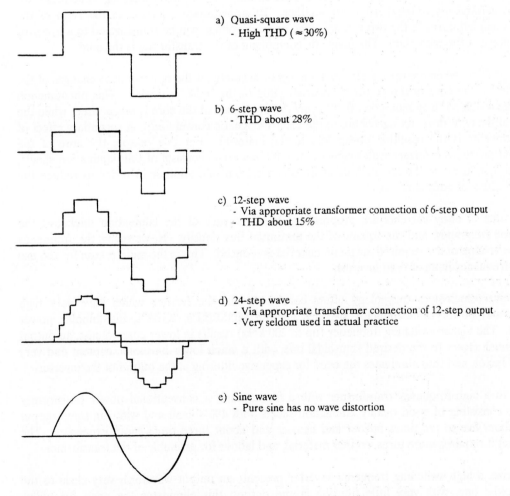

a) Quasi-square wave
   - High THD ($\approx 30\%$)

b) 6-step wave
   - THD about 28%

c) 12-step wave
   - Via appropriate transformer connection of 6-step output
   - THD about 15%

d) 24-step wave
   - Via appropriate transformer connection of 12-step output
   - Very seldom used in actual practice

e) Sine wave
   - Pure sine has no wave distortion

Figure 3-61.   Types of Inverters According to Their Output Waveforms

Power Semiconductor Devices

The technology for high switching-frequency inverters (typically 20 kHz or higher) is made possible by the switch-mode power devices. Table 3-13 lists the main power semiconductors in use and their key characteristics. Transistors, power MOSFETs, and bipolar transistors predominate as the power switches in low-power inverters. The high-power inverters generally use thyristors. Multiple thyristor bridges in HVDC transmission line terminals are operating at several hundred MWs. Thyristors cannot be turned off by a control signal. A thyristor derivative, a gate turn-off (GTO) device can be turned off by a gate pulse. In general, the GTOs and thyristors have the capability to operate near the 5 kV, multi-MW level, but they have limited ability for high-frequency operation (see Fig. 3-62).

The device which recently began to emerge is the IGBT, insulated-gate bipolar transistor. In late 1990, the IGBT was capable of up to 1,000 Vdc at 300 A. A year later, the suppliers announced an increase in the device capability to 1,200 Vdc at 600 A. IGBT inverters capable of several hundred kW will be available in 1992. Thyristors will continue to be utilized for HVDC, motor control, etc., in the utility and transportation (train) sectors, but the application of high-speed IGBTs is expected to grow at a much faster rate than any other power semiconductor devices on the market within the past decade. GTOs can be used for self-commutated inverters, but they are limited to operation at less than 1 kHz while the IGBTs can run at frequencies up to 50 kHz.

Conventional and High-frequency Power Transformers

Power transformers are electromagnetic devices which transfer electrical energy from one electric circuit, called the primary winding, to another electric circuit, the secondary winding, while these two circuits are galvanically isolated from one another. By appropriate selection of the ratio of the number of complete turns in the two windings, the primary voltage can be transformed to a lower or a higher voltage at the secondary. The magnetic component of the transformer is the core.

The major problem when converting electrical energy to magnetic, is the opposition to changes of the magnetic flux at the core of the power transformer, due to the "eddy" currents. This phenomenon results in the so-called "magnetic skin effect" (only the outer part of the core is magnetized, while the inner part of the core does not contribute to the electromagnetic conversion). A secondary effect of the eddy currents is the resulting power loss in the material. For this reason, the core of the conventional power transformer with magnetic metallic materials consists of thin lamination sheets, rather than a solid mass, the laminations being isolated from one another in order to reduce the undesirable effect of eddy currents.

The losses due to eddy currents are proportional to the square of the lamination thickness, the square of the frequency, and the square of the maximum flux density. However, as the frequency goes up, the volume of the required magnetic material is reduced. This is the main reason for the use of 400-Hz frequency in an aircraft inverter.

Modern power electronics technology offers core materials like ferrites which have very high resistivities and power semiconductor switches (power MOSFET's, IGBT's, fast bipolar power transistors). The higher switching frequency (up to 200 kHz) results in lower core volume, an output waveform much closer to the desired sinusoidal one, with a much lower harmonic content and very high power factor, and thus eliminates the need for expensive filtering at the output of the inverter.

Compared to a high-frequency transformer with a ferrite core, a conventional power transformer with a core consisting of good quality lamination sheets with 4% silicon and with the same output power, is about five to ten times bulkier and heavier and about three times more expensive. The reason is that it requires more turns, copper material, and labour for assembly of the transformer.

To summarize, a high switching frequency inverter presents an output waveform very close to the pure sinusoidal one, with very little filtering at the output; this eliminates the need for bulky, expensive, and energy-consuming power filters, along with switchgear. Finally, the noise emitted by a high frequency inverter is inaudible, because it is beyond the human ear's frequency range.

Table 3-13. Comparative Data on Various Power Semiconductor Devices

Parameter	Device				
	Thyristor or silicon controlled rectifier (SCR)	Gate Turn-off Thyristor (GTO)	Bipolar Power Transistor	Power MOSFET	Insulated Gate Bipolar Transistor (IGBT)
Symbol					
Current control	When a short pulse is applied at gate G, current flows from anode to cathode (see Note 1)	Same as thyristor but it can be turned off by a negative voltage gate signal.	When base signal at B is continuously applied, current flows from C to E (no current if base signal is removed)	Same as transistor (note 3)	Similar to Power MOSFET
Working Voltage	up to 3,000 V	up to 3,000 V	1,000 V	up to 1,000 V (note 4)	up to 1,000 V
Working Current	10,000 A	1,000 A	200 A	100 A (note 5)	300 A
Forward Voltage Drop	1-2 V	4 V	3 V	2 V for 50-V/100-A device, 7 V for 1,000-V/10-A device	3 V
Switching time	20 μs (for fast inverter type)	20 μs	1 μs	100 ns	200-600 ns
Cost and power handling capability	Low cost; few MW	Low, but higher cost than thyristor (see Note 2); 1 MW	High cost; 200 kW	Higher cost than transistors; up to 20 kW	Same cost as transistors; up to 300 kW

Notes:
1) Current flow stops below a minimum level "holding" current. If the thyristor controls ac current, current flow stops every half period and a new gate signal is required every half period to turn the thyristor "on". If the thyristor controls dc current, any gate signal cannot stop the current flow, and an external commutation circuit is required to switch the thyristor off.
2) Comparable to thyristor if external commutation circuit, auxiliary thyristors, commutating capacitor, inductance, etc., are considered.
3) No base (gate) current is required (i.e., voltage-controlled devices).
4) Efficient up to 200 V (high conduction losses over 200 V).
5) Up to 10 A for 1,000-V devices, 100 A for 200-V devices.

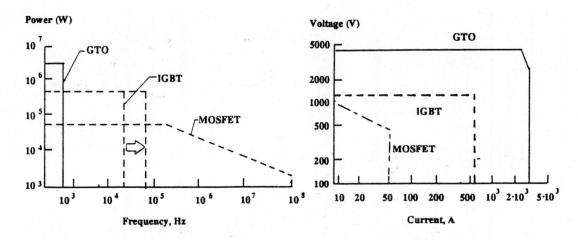

Figure 3-62.   Power Semiconductor Device Characteristics as of 1991.

Inverter Types

The inverters now used in PV applications fall into two main categories: *self-commutated and line-commutated*. They are also referred to as "*voltage-fed*" and "*current-fed*" types, respectively, by the inverter designers. The first can operate independently, being activated solely by the input power source (PV, wind, etc.). The line-commutated type can function only when the output ac line voltage is present. Note that the source of this voltage could be the utility grid or any other power source which is tied to the inverter output. This external power source could therefore be a UPS inverter, Diesel generator, wind generator, etc.

Figure 3-63 shows two basic types of low-power single-phase inverters, classified as *bridge* and *push-pull* inverters. The power switches can be transistor types (MOSFET, Bipolar, IGBT) or thyristors. For higher power (>5 kVA), a 3-phase bridge configuration is usually used.

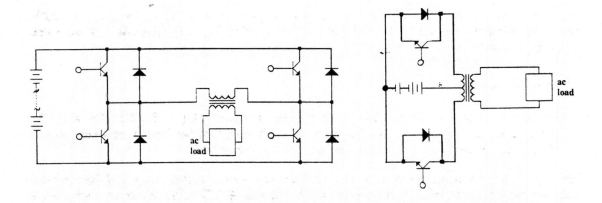

Figure 3-63.   The Two Most Common Single-phase Low-phase Inverter Configurations. The battery can be replaced by PV array as the dc power source. The 3-phase configuration of bridge inverter with IGBTs can go up to 300 kVA using switching devices available in 1991.

Two types of low-power inverters are available in Germany with and without transformers. A transformerless inverter invented by the Fraunhofer Institute [49] uses digital synthesis to generate a near-sinusoidal waveform (see Fig. 3-64). The Sun Power inverter operates at 20 kHz to achieve a

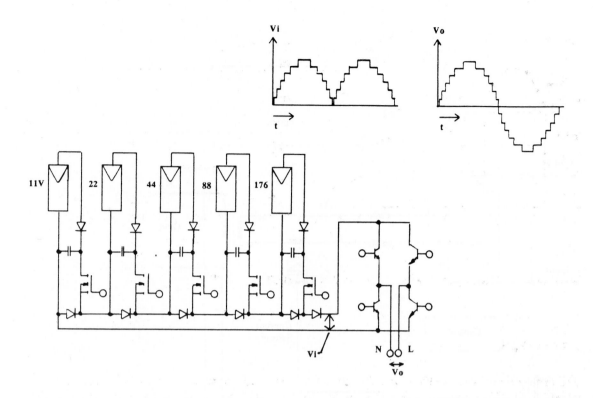

Figure 3-64. Transformerless Using Digital Synthesis, Designed and Marketed by Fraunhofer Institute, Freiburg, FRG.

good sinusoidal output (see Fig. 3-65). Their 3-phase version is illustrated in Figure 3-66.

The current-fed inverter requires line voltage for commutation (thus, it is referred to as a line-commutated type). It operates at a generally poor lagging *power factor*, has a high harmonic distortion at its ac output, and is susceptible to commutation faults during ac line failures. It is, however, cheaper than the high-frequency types. Two of its main drawbacks, power factor and harmonic injection, are usually reduced by appropriate correction with capacitors and L-C filtering. Also, its *ripple current* injection to its input dc terminals can be controlled.

Thyristor-based inverters, in particular, 6-, 12-, and 24-pulse systems are available in the multi-hundred kW to MW range. The IGBT inverters in the 40 kVA range are available in Europe, but systems up to 200 kVA can be built.

Inverter Comparison

Single-phase inverters are usually made for low-power applications (up to 5 kW). Above 5 kW, 3-phase inverters are common. Generally, the inverter efficiency is higher for inverters with low no-load losses, and the efficiency goes up as the input dc voltage increases.

For single-phase inverters, with square-wave or quasi-square-wave output, the cost of the required output power filter is almost the same as the inverter (about 1,000 DM for a 3-kW inverter). For a 3-phase inverter, with square wave or quasi-square wave output, the cost of the output filter is considerably lower than that of the inverter (about one-fifth). For example, a 50-kW 3-phase inverter with quasi-square wave output could cost about 100,000 DM, and its output filter about 20,000 DM.

The voltage-fed (self-commutated) inverter has an inherent stand-alone operating capability and, with proper design and control, operates at a high-power factor. Most self-commutated inverters exhibit low harmonic distortion. They have been manufactured in versions ranging from a few hundred watts to several hundreds of kW, at dc voltages ranging from a few tens of volts to about 500 Vdc. The active switching devices used, in contrast to the line-commutated inverter which uses

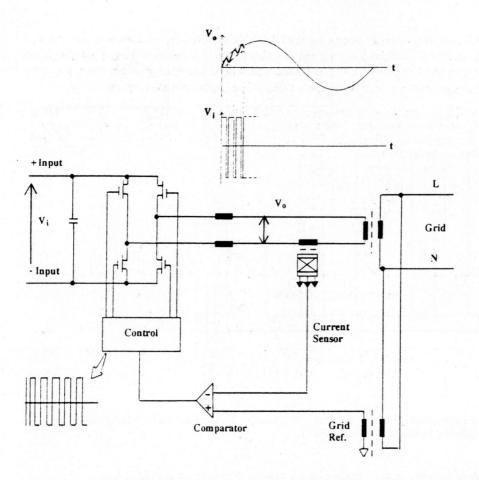

Figure 3-65. Self-commutated High Switching Frequency Full-bridge Single-phase Inverter Using Power MOSFETs for Grid-connected Operation. Sun Power 3-kVA unit is an example of this configuration.

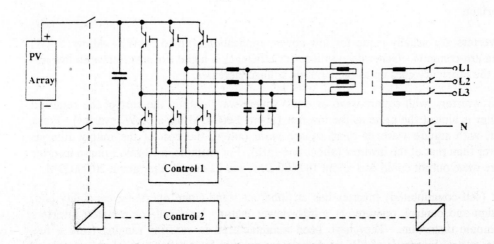

Figure 3-66. Self-commutated 3-phase High-frequency (20 kHz) Inverter, PWM using Power MOSFETs for Grid-connected PV Systems (Source: Sun Power, Frankfurt, FRG)

predominantly thyristors, may be transistors, power MOSFETs, or thyristor relatives such as the gate-assisted turn-off device (GTO). Transistor versions have been limited to no more than a few kW, at dc voltages no higher than 400 Vdc. Now, IGBT types can be used for inverters up to 200 kVA, but their capability to handle larger currents is growing more rapidly than any other semiconductor device.

Table 3-14 is a partial list of small power inverters available in the EC for use in PV plants. Note that a majority of low-power inverters uses PWM with high switching frequency, and therefore good sinusoidal waveforms are possible (i.e., high PF and low THD). Table 3-15 is a set of specifications for a typical high-frequency inverter.

Table 3-14. A Partial List of Small Power Inverters Available in the EC in 1991

Manufacturer (Country)	Rating (kVA)	Commu-tation Type	Waveform	Power Switching Devices	Pulse Control Method	Nominal Voltage Input (Vdc)	Output (Vac)*	1990 Unit Price (DM)	DM/VA
AEG/TST (FRG)	1.4	line	sinusoidal	MOSFETs	PWM (20 kHz)	100	230	4,591	3.06
Fabrimex (FRG)	3.0	line	sinusoidal	MOSFETs	PWM	48	220	9,386	3.12
FI-ISE (FRG)	3.0	line	sinusoidal	MOSFETs	PWM	+	220	9,000	3.00
Grosmann (DK)	1.0	line	sinusoidal	MOSFETs	PWM (20 kHz)		220	8,800	8.80
	0.5	line	sinusoidal	MOSFETs	PWM (20 kHz)		220	7,600	15.20
Hardmaier (FRG)	3.0	line	sinusoidal	MOSFETs	PWM (50 kHz)	96	230	NA	
IBC (FRG)	1.8	line	quasi-square	Thyristors	2-pulse	200	230		
	1.5	self	square wave	Thyristors	2-pulse	100	220	3,285	2.19
Laschek (FRG)	1.0	self	quasi-square	Thyristors	2-pulse	28	220	4,000	4.00
MSE (FRG)	1.0	line	sinusoidal	MOSFETs	PWM (50 kHz)	24	220	NA	
Respect (GR)	2.0	self	sinusoidal	MOSFETs	PWM (50 kHz)	48	220	5,000	2.50
Siemens (FRG)	1.5	line	sinusoidal	Thyristors	6-pulse	250	230	4,355	2.89
	2.5	line	sinusoidal	Thyristors	6-pulse	250	400	5,125	2.05
	5.0	line	sinusoidal	Thyristors	6-pulse	250	400	NA	
SMA (FRG)	1.5	line	sinusoidal	MOSFETs	PWM (50 kHz)	100	220	3,985	2.64
Solar Diamant (FRG)	1.7	line	sinusoidal	Thyristors	50 Hz	175	230	4,066	2.39
Sun Power (FRG)	1.2	self	sinusoidal	MOSFETs	PWM (20 kHz)	165	230	7,900	6.58
	3.0	self	sinusoidal	MOSFETs	PWM (20 kHz)	240	230	9,900	3.30
Trace (FRG)	2.0	self	square wave	MOSFETs	2-pulse	24	230	2,380	1.19
	0.6	self	square wave	MOSFETs	2-pulse	12	230	948	1.58
UfE (FRG)	1.6	line	sinusoidal	MOSFETs	PWM (100 Hz)	68	230	4,591	2.88
	0.4	line	square	Transistors		32	230	1,150	2.88
Victron (FRG)	1.0	line	sinusoidal	MOSFETs	PWM (50 kHz)	80	220	3,500	3.50

* All 1-phase except those above 230 Vac
+ 5 separate voltage sources
NA Not available

141

Table 3-15.   Specifications of Typical Low-power High Switching Frequency Inverters (Sun Power)

Parameter	Capability	
	1 kVA unit	3 kVA unit
Input voltage, nominal (Vdc)	165	240
PV Array	10 modules in series	15 modules in series
Input release voltage (Vdc)	> 175	> 262
Input cut-off voltage (Vdc)		
- Minimum	145	215
- Maximum	250	375
MPPT range (Vdc)	150-175	225-265
Operation mode		
- Mode 1	Stand-alone	Stand-alone
- Mode 2	Grid-connected	Grid-connected
	Maximum power point tracking	Maximum power point tracking
Output voltage (Vac)	220	220
Wave form	Sine	Sine
Frequency (Hz)	50	50
Frequency stability	± 0.5%	± 0.5%
Harmonic distortion (%)		
- Stand-alone	1-2	1-2
- Grid-connected	3-4	3-4
cos φ	0.95 minimum	0.95 minimum
Automatic grid-disconnect	Grid voltage < 190 Vac	Grid voltage < 190 Vac
	Grid voltage > 260 Vac	Grid voltage > 260 Vac
- Minimum voltage (Vac)	< 190	< 190
- Maximum voltage (Vac)	> 260	> 260
Inverter power efficiency (%), at fraction of rated power:		
- 0.1	80	90
- 0.5	87	94
- 0.7	87	93
- 1.0	85	92
Power consumption at idle operation mode	1.5% Pnom	1% Pnom
Status indication	Input OK	Input OK
	Output OK	Output OK
	Inverter ON	Inverter ON
	Fault	Fault
Protection devices		
- Input	Undervoltage lockout	Undervoltage lockout
- Output	Overvoltage lockout	Overvoltage lockout
Ambient air temperature (°C)	0-40	0-40
Noise (dB)	30 maximum	30 maximum
Weight (kg)	40	60

Tables 3-16 and 3-17 list medium- to high-power inverters available in the EC. Recently, several multi-hundred kW plants have installed or planned the use of thyristor-based 6-pulse inverters [50]. By connecting the outputs of these 6-pulse inverters into Star and Delta configurations at the primary

Table 3-16.   A List of Medium-power Inverters Used in the PV Plants in the European Community (Installed as of mid-1991)

PV Plant (Country)	Inverter Manufacturer	Switching Device	Total Inverter Rating (kVA) [+]	Array Power Rating (kW)	Input Voltage (Vdc)	Efficiency at Full Load (%)
AGHIA ROUMELI (GR)	Jeumont Schneider	Transistor	40	50	292-353	93
CHEVETOGNE (BE)	ETCA	Thyristor	20 x 220	40/23	198-264	95
FOTA (IRL)	AEG	Thyristor	50 (L-C) 10 x 3	50	245-315	88
GIGLIO (IT)	Silectron	Thyristor	7/ 5.7	30/15	90-360 90-350	95 95
KAW (FR)	Jeumont Schneider	Transistor	40	35	292-353	94
KYTHNOS (GR)	Siemens Solar	Thyristor	50 x 3	100	128-192	94
MARCHWOOD (GB)	AEG	Thyristor	40 (L-C)	30	216-295	92
MONT BOUQUET (FR)	Aérospatiale	Thyristor	30	50	191-260	90
NICE (FR)	Aérospatiale	Thyristor	5	50	191-260	90
PELLWORM (FRG)	AEG	Thyristor	75 x 2 450 (L-C)	300	230-415	91.5
RONDULINU (FR)	Aérospatiale	Thyristor	50	65	80-125	93
TERSCHELLING (NL)	Holec	Thyristor	60	50	210-273	87
TREMITI (IT)	CO.EL.	Thyristor	61 [*]	65	128-192	94
VULCANO (IT)	Marelli (140) Borri (50)	Thyristor	140 (L-C) 50	80	240-420	90 90
ZAMBELLI (IT)	Silectron	Thyristor	40 x 2 [++]	70	200-320	97
BERLIN (FRG)	Siemens Solar	Thyristor	10	9.4	220	92
BRAMMING (DK)	Grossman	MOSFET	10	5	300	92
DELPHOS (IT)	CO.EL. (300)	Thyristor Thyristor	150 300 (L-C)	300 300	450-600 650-800	95 92
MADRID (ES)	JEMA JEMA Victron	Thyristor Bipolar MOSFET	50 (L-C) 10 (L-C) 3	50 40	140-400 140-400 83-110	88 88 90
MUNICH (FRG)	FI/ISE	MOSFET	3 (L-C)	2.0	**	95
POZOBLANCO (ES)	Ambar (0.3) Jema (3.8) Grundfos (3.8)	Bipolar Bipolar	0.3 3.8 (x2)	12.6	24 106	
RAPPENECK (FRG)	FI/ISE	MOSFET	3	3	**	95

[*]   4 - 5.5 kVA, 2 - 3 kVA, 2 - 1.5 kVA, 2 - 15 kVA
[**] 5 independent array circuits: 11, 22, 44, 88, and 176 Vdc nominal.
[+]   All inverters are self (or force) commutated, except those marked "(L-C)" which are line-commutated types.
[++] Variable frequency.

Table 3-17. List of Other High-power Inverters Available or Being Developed in the EC for Grid-connected Application

Manufacturer (Country)	Rating (kVA)	Commu- tation Type*	Power Switching Devices	Control Method	Efficiency (%) 10% FL	Efficiency (%) 100% FL	Voltage Input (Vdc)	Voltage Output (Vac)[++]
CO.EL (IT)	100	line	Thyristor	6-pulse	91.5	96.5	500	440
JEMA (ES)	100	self	Thyristor	12-pulse	83	94	550	380
Siemens	330	self	SIPMOS	PWM	92	94.5	440	380
(FRG)	500+	self	IGBT	PWM	92.5	95	440	380
	100	line	Thyristor	6-pulse	91	93	440	380
	500	line	Thyristor	12-pulse	92	94	550	380
SMA (FRG)	350	line	Thyristor	6-pulse	95+	98+	800	830
Sun Power (FRG)	40	self	IGBT	PWM (20 kHz)	90	95	420	400
	100	self	IGBT	PWM (20 kHz)	92	96	500	380
TST/AEG (FRG)	1,000	line	Thyristor	12-pulse	92	94	500	380
	300	line	Thyristor	12-pulse	92	94	900	380

*   self-commutated units are grid-synchronized to be connected to the utility grid
+   without filtering, transformer, and power factor correction capacitors
++   3-phase

winding of the transformers as illustrated in Figure 3-67, a 12-pulse sine wave can be obtained. The 12-pulse inverter results in a lower total harmonic distortion as compared to a 6-pulse inverter (29.7 vs. 15.1%). Similarly, a 24-pulse inverter with even less harmonic distortion can be achieved by using a combination of 12-pulse inverters in conjunction with an appropriate transformer configuration.

Thyristor versions have dominated the higher power implementations at low frequencies. Transistorized types are used in high frequency (>20 kHz) applications. Generally, because of device and/or commutation circuit costs, they are more expensive than the current-fed type. It is likely that the voltage-fed self-commutated inverters will dominate the market for residential-size applications, depending on the particular codes and standards imposed by the host utility with regard to the inverter power quality.

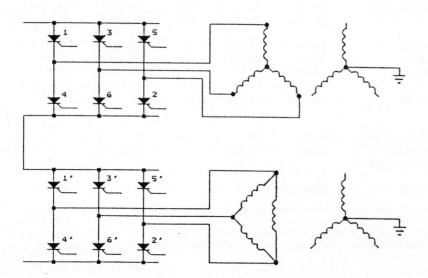

Figure 3-67. 12-pulse Inverter Consisting of Two 6-pulse Bridges in Series. The 12-pulse "sine" wave results from Star/Delta connection of the transformer primary.

Power Conversion Efficiency

The efficiency of all inverters is affected not only by the losses produced by the active switching converter but also by the losses introduced by the passive components such as transformers, reactors, (i.e., inductors), and capacitors. Thus, the overall system conversion efficiency of the inverter chain (input filters, power switches, output filters, and transformer) is more appropriate for inverter comparison purposes than the individual inverter efficiency. If you are comparing the efficiency of inverters such as thyristor- and IGBT-based units, you must inquire about the power quality (THD and PF) at the stated efficiency.

Current-fed inverters generally have full-load efficiencies ranging from 90 to 94% for low dc input voltages (< 220 V) to 90-95% for the high dc input voltage ($\approx$ 400 V), systems. Voltage-fed inverters have full-load efficiencies in the 90-95% range, the higher range for higher power units, especially as input voltage increases. Since their fixed losses are usually greater than their resistive losses, they exhibit steadily deteriorating efficiency as the load is reduced.

Power Factor and Harmonics

The two key technical issues at the ac interface are power factor and harmonic distortion. In general, equipment operating at significantly less than unity power factor is technically not completely acceptable for grid-connected applications. It is not suitable because inverter VAR demand usually affects system voltage regulation which can seriously degrade service to other consumers and communication loads; moreover, it is undesirable to the utility because it cannot reasonably be expected to bear the costs of supplying reactive power while it is supplying no real power, or perhaps buying real power from the cogenerator.

The VAR demand of both loads and inverters may become acceptable if the utility charges money to the owner of the cogenerator for reactive power consumption. Current regulations do not permit this for residential customers. Thus, unity power factor both at the utility connection as well as at the converter output is ultimately desired by the utility.

The harmonic content of the PV inverter output is difficult to assess and specify because 1) not much information is available, 2) the effects of harmonic distortion on utility equipment and consumer loads, including harmonic propagation into the grid network [51], are not well understood, and 3) it is not easy to measure. The European regulations do not yet impose undue cost and efficiency penalties on the inverter because of their belief that the current harmonic specifications have a high probability of not creating problems with existing utility equipment or other consumer loads. A current specification, rather than voltage, is important because it is the harmonic current that causes problems, and because it is in some respects easier for an equipment manufacturer to demonstrate compliance with a current-injection specification. This specification is usually given in terms of the Total Harmonic Distortion (THD). The magnitude of any of the harmonics, called the Single Harmonic Distortion (SHD) is also an important parameter for utilities. In Europe, these are usually specified as 5 and 3 % maximum, respectively.

A few organizations have instrumented their PV plants to investigate the harmonics on the inverters [52, 53]. In this regard, low-cost reliable equipment to make such measurements are highly desirable.

DC Interface

The key technical issues at the dc interface are fault detection and clearing, operating voltage ranges (absolute and relative), and ripple current injection from the inverter output to the array input. into the PV array. Moderate ripple current levels, up to 10% peak-to-peak, have an insignificant effect on array performance and levels up to 20% peak-to-peak may well be permissible. Consideration of array terminal voltage variations with changes in PV module temperature, solar irradiance. and loading indicates that a dc operating voltage range of 1.7 to 1 times the nominal value should certainly be adequate to ensure negligible loss of energy production. It may be possible to specify a narrower range, but there are cost risks other than energy loss associated with doing so. The cost and loss penalties for this range are not too severe in the PV application.

AC Interface

At the ac interface of a line-commutated inverter, no direct control of the inverter is required (or possible). Control of VAR demand and, perhaps, harmonic injection, is accomplished by means essentially independent of the inverter and its control. For a self-commutated inverter, this is not the case. The inverter's ac voltage magnitude must be controlled independently of dc side conditions if VAR demand (or supply) is to be controlled, and control of real power flow is accomplished by adjustment of this voltage's phase with respect to the grid. Thus, the most important dc side parameter, PV array/power, is controlled at the ac interface of a self-commutated converter.

Radio Frequency Suppression

Due to the fact that more and more inverters operate at higher switching frequencies of 20 kHz or higher using PWM, lower harmonics and better power factors are achieved; however, they cause some interferences in RF region. This trend is noticeable on residential inverters operating at lower ac voltages.

The harmonics of the inverter switching frequency can interfere with frequencies used by telecommunication equipment (radio, TV, telephone). To avoid such interferences the inverters usually suppress those harmonics by appropriate filtering and shielding

Electrical Isolation

Another ac interface issue of considerable importance is the need or lack of an isolation transformer between the inverter and the utility. Isolation may be mandatory if safety needs are considered. First, all provisions of the various codes in the EC countries addressing the safety of residential installations may not be met unless isolation is provided. Second, consumer and utility fault clearing devices are incapable of safely interrupting dc fault currents, therefore, such a possibility must be considered. Third, a converter fault with dc content is likely to saturate the distribution transformer and cause disruption of service to other consumers, both those on the same transformer and those on the same feeder since some primary fault clearing devices will then operate. All of these are still areas of concern to the utilities.

Fault detection and clearing at the array-inverter interface is difficult because of the low short circuit current capability of the source and the high cost and poor service life of dc fault interrupting devices like switchgears. However, since the array does not harm the inverter, the inverter is often used to clear most dc-side faults. For those which cannot do so, or for catastrophic inverter failures, a deliberate short across the array terminals is placed via a short-circuiting switch. One pilot plant has done so as part of normal shut-down operation [45].

3.2.8   SYSTEM CONTROL AND POWER MANAGEMENT

Real Time Monitoring and Control Functions

In the case of an industrial stand-alone PV system, the basic monitoring and control functions required are usually the following:

- Night/day operation of PV system
- Battery charge regulation
- Battery protection against excessive overcharging and deep discharging
- Specific controls related to electrical loads (e.g., switches for a pumping motor and load on-off)
- Monitoring and control of critical performance parameters (checkout and trouble-shooting)

In the case of a grid-connected PV power plant without a battery, the following functions are usually implemented:

- Night/day operation of PV system
- Inverter output and input regulation and protection
- Grid black-out or brown-out sensing and automatic reset/restart following grid power recovery
- Monitoring of entire plant performance and meteorological data (for on-site use)

For each of these functions, specific devices or circuits which can perform the function independently from other devices, are usually available in the market. A central process controller performs the automatic control functions, but an ability to override any critical function manually should be provided.

Obviously, for an unattended plant, the system operation must be automated. This automation can be implemented by analog or digital approach. The latter is advisable if 1) flexibility is required in changing plant configurations and power management strategies and 2) its reliability can be assured.

In the control system of any industrial plant involving a central computer, the three hierarchies of control are:

- Process instrumentation or sensors measure the value of process parameters and transmit this information to a signal conditioning/conversion device, subsystem component, process controller or data acquisitioning system, or else an indicator for the human operator.

- Individual components perform specific control functions not requiring high level coordination with other plant components (e.g., battery charge regulator, inverter regulator, etc.) based on the sensor outputs.

- The central processor implements the automatic operation of the entire plant (i.e., provides total system control), and provides the interface to the human operator. A data acquisition system or a data logger, which collects and records the meteorological data and plant performance is also classified under this level.

These functions are implemented by a central processor residing in the process logic controller (PLC). The industrialized PLCs are now widely available and used, like the Siemens SPS. They are known to be very rugged and reliable. Because of the large number being marketed, the cost of the PLCs has been decreasing.

Centralized vs. Distributed Control

In the early days, when the first expensive automatic process controllers became available on the market, designers tended to concentrate all data acquisitioning and control functions of an installation into one central computer. It was believed to be cheaper and also simpler, since powerful software was expected to do all the work.

Frequently, however, the result was that the entire installation became dependent on the correct operation of a very sensitive and unreliable plant component, the process controller, and no one was able to operate the plant manually without the computer. And even worse, if the computer failed, there was no way to find out what had happened because all data were lost. Even a fault or alarm from a single measurement required by this controller could cause a shutdown of the entire plant.

Nowadays, designers tend to adopt a distributed control architecture, i.e., all control functions are split-up into single and independent functional blocks, which are implemented by front-end control devices installed inside or near the equipment in question. Coordination between different units is assured via shared process parameters (e.g., battery voltage, dc bus voltage, etc.). The functions of the central automatic process controller, if required, should be only to supervise and implement the operation of the overall plant.

Furthermore, plant monitoring equipment for collecting and recording data, except for certain sensors, is now usually separated from the system control and display subsystem for reliability reasons.

Guidelines for Control and Display Subsystems

When designing a control system for a PV system, the following guidelines should be considered:

1) Provide a reliable power supply, such as a UPS, to crucial instrumentation and controls. The dc power from the main battery or a separate battery with its own charger is one of the best solutions; use ac power only if the battery solution is not feasible.

2) Reduce the power consumption of instruments and controls. Their impact on the overall energy balance of a PV plant is significant, since usually they are kept operating 24 hours per day.

3) Provide for appropriate protection of electronics.

4) Overvoltage protection should be provided on power supply and signal lines. All signal wires should be isolated from power circuits. All equipment shielding and negative polarity cables should be properly grounded.

5) Provide for redundancies to assure plant operation in case of failure of single instruments or signals (e.g., solar sensors).

6) Design for worst-case temperatures and foreseeable tolerances in order to compensate for bad calibration of instruments and signals.

7) Implement system diagnostic features by comparing the controller output and the corresponding feedback from plant.

8) Render the system "foolproof" and "fail-safe" by adopting safety interlocks wherever possible.

9) Provide a manual override capability for each critical control function implemented automatically.

3.2.9   INSTRUMENTATION AND DISPLAYS

In stand-alone PV systems, internal energy consumption is an important design factor. Thus, controls, instruments, etc. should be designed in such a way that their consumption is minimized by the use of low-power devices (e.g., LCD's for displays, laptop computer, etc.), or switching them off when they are not needed.

Such energy saving may be effected, for example, by designing the power circuits to indicators and display devices (e.g., LED's) in such a way that they give an indication (and consume power) only when the human operator activates the push-button switch.

The minimum essential instrumentation and controls required for a high-power PV system with high reliability should be aimed at monitoring caution, warning, and alarm conditions and displaying key parameters such as:

- Solar irradiance (in W/m^2)
- PV array output power (in kW)
- Battery voltage and current
- Operating state of switches, especially load power and critical controls
- Battery state of charge (SOC)

- Indicator allowing the user or operator to distinguish the following conditions:
  o Normal battery charging and discharging
  o Dangerous battery overvoltage and under-voltage conditions

The battery SOC calculator was discussed in Section 3.2.4. Suffice it to re-state that a reliable SOC meter with a resetting capability (to 100% relative SOC) can play a key role in some plants.

For the battery indicator, either four LED's may be used to indicate a discrete condition, or an analog voltmeter with coloured voltage ranges on a scale that shows danger zones. PV systems without a battery may dispense with instrumentation altogether. For example, a directly-connected PV pumping system requires only load-specific controls, such as a manual on/off switch and a level switch to avoid running the pump when the well is dry.

3.2.10 SPECIAL POWER SUPPLIES: UPS

An uninterruptible power system (UPS) provides suitable ac power to critical loads without interruption or discontinuity in voltage during temporary loss of power from the primary source. The primary power source can be the utility grid or independent power generators (PV, wind, Diesel, etc.). Problems which occur in the prime power source can be transient or steady-state conditions such as:

- Voltage surges and spikes

- Instability in voltage or frequency

- Sustained low voltage

- Power blackouts and brownouts

Equipment items which are susceptible to malfunction and are often damaged by the above conditions are computers and voice, video, and data communication lines. Circuits and devices performing critical control functions in power plants, in general, must have a back-up power source. There are several ways to handle the above problems, e.g., the use of surge arrestors, isolation transformers, capacitor banks, voltage regulators, and UPS's. However, only a UPS is designed to provide a continuous, or nearly continuous, power of an acceptable quality. Depending on the required availability of the power plant, most PV power systems should provide a back-up power source and/or a UPS for control and protection equipment, and often monitoring sytems.

Today's UPS designs are classified as *off-line*, *hybrid*, and *on-line*. They may differ somewhat in their functional features, but the basic operation is nearly the same. Figure 3-68 shows the three basic topologies.

The *off-line* UPS, also known as a standby power source, is the simplest and cheapest. During normal operation, the UPS routes ac power from the grid directly to the load while trickle-charging the battery. Whenever the grid power drops below a designated value, a detection circuit switches the battery power to the inverter. This is possible because the inverter is normally off. The only significant advantage of the off-line UPS is low cost.

The hybrid UPS operates primarily by passing utility power through to the load, and is designed to eliminate the voltage glitch or inverter output voltage drop by using the energy stored in the capacitor during the grid-to-inverter switching operation. However, this automatic switchover is not trouble-free and often exceeds the hold-up capability of the electronics or the load.

The *on-line* UPS is designed to operate with the rectifiers and inverters on, so no power switching is done. It generates ac power via the inverter while charging the battery. A drawback has been the higher price.

Many information and communication system loads are powered by switched-mode converters which draw power in the form of non-linear repetitive, high-peak current. This current is typically more than twice the linear rms current. The ratio between the non-linear repetitive peak current and the linear rms value is called the *crest factor ratio*. High crest factor and low THD (less than 5%) are the key technical features to look for in selecting the UPS. Inverters operating with PWM or high frequency usually generate a sine wave output and usually the lowest THD. Thus, if the manufacturer is not sure of the THD level, simply ask for photographs of the output waveform at full load when the inverter is drawing power from the grid and also from the battery (waveforms can be quite different when operating from these power sources).

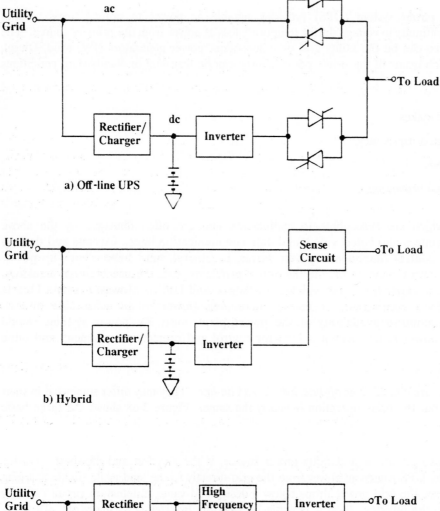

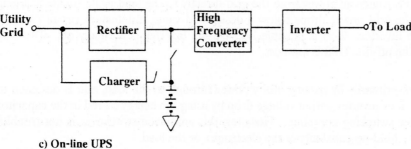

Figure 3-68. Block Diagrams of Basic UPS Configurations: a) Off-line, b) Hybrid, and c) On-line. The only significant advantage of the off-line unit is low cost.

3.3   PERFORMANCE MONITORING

All PV projects co-sponsored by the CEC have some monitoring requirements. For the CEC Demonstration/THERMIE projects at least two years of plant monitoring are required. This section discusses monitoring requirements, methods, and the common format for reporting information from the PV plants. In light of all the lessons learned, there is a continuing need for the development of effective plant monitoring and for the dissemination of plant performance data.

3.3.1   MONITORING APPROACHES

There is a certain minimum level of monitoring required by the on-site technician to assure proper start-up, normal operation, and shut-down sequences, and to assure personnel safety. This type of monitoring must be implemented by the instrumentation, display, and warning devices integrated in the control/display subsystem in the PV plant. Monitoring as described herein deals more with the collection, recording, and transmission of data for archival storage and off-site analyses of plant performance. However, if properly planned and designed, the latest DAS hardware available combined with a PC-AT can serve as a back-up to the on-site display system for use on site. Many of these systems have been installed within the last three years and are operating very successfully in Europe [54].

Monitoring for subsequent analysis and data archiving requires special data acquisition equipment which, for reliability purposes, must be separated from the plant control and display system. The exception is perhaps the sensors. Sensors such as those for the current shunts, voltage, and solar irradiance, can be shared by the plant control and separate DAS subsystems. For this purpose, different types of data collection or monitoring were established as listed in Table 3-18.

In defining the different types of measurements to be recorded, it is important to consider different sizes of PV plants, subsequent analysis required to satisfy the aims listed in Tables 2-5 and 2-6 (in Chapter 2), and the equipment and devices currently available on the commercial market. The last item is the most critical consideration. The relative applicability and characteristics of the different types of monitoring can be identified as follows:

Type I: **Monitoring of Demonstration or Commercial Plants with Few or No Scientific Goals and Minimum Evaluation**

- Measurement and/or recording of energy parameters only to give a rough indication of the PV system performance

- Use of simple data loggers or energy meters to record the energy data on an hourly basis, or only one set of data per day to save storage memory. (Note: Manual data taking on a permanent basis is not recommended, except for very small low-cost PV systems, even for user-motivational purposes.) Reliable energy counters are necessary for this type of monitoring to be effective.

Type II: **Monitoring of Plants without In-depth Performance Evaluation**

- Only voltages, currents, and inverter output/utility energies (energy computation done with an off-line PC using V and I to avoid the cost of energy counters)

- Minimum number of kWh counters to reduce costs

- Use of a simple data logger like the Campbell Scientific, Delta-T, or Lambrecht with or without modem data transfer

Table 3-18.   Classification of Data Measurements and Recording on Typical Stand-alone and Grid-connected PV Plants

Measurements	Type I	Type II	Type III	Type IV
		Monitoring Classification		
Solar and Meteorological				
- Solar Irradiance (power)				
o Plane of array		x	x (p)	x (p)
o Horizontal		x	x (p)	x (p)
- Insolation (energy)				
o Plane of array	x			x
o Horizontal				x
- Ambient Temperature		x	x	x
- Wind Speed		x	x (p)	x (p)
PV Array				
- Voltage and Current		x	x	x
- Power			x	x (p)
- Energy	x			x
- Temperature, PV module			x	x
Battery				
- Voltage and Current		x	x	x (p&m)
- Power			x	x
- Energy	x			x
- Cell Voltages				x
- Cell Temperature		x	x	x
Inverter/Converter				
- Output Voltage and Current		x	x	x (p)
- Output Power			x	x
- Output Energy	x			x
Auxiliary Power Sources				
- Output Energy	x	x	x	x
Rectifiers (Battery Charging)				
- Output Voltage and Current		x	x	x
- Output Power			x	x
- Output Energy	x			x
Utility Grid				
- Energy to grid	x		x	x
- Energy from grid	x		x	x

Note: All recorded values with check marks (x) are averages for each duration of scanning and recording time, except for discrete (digital) data, e.g., energy, peaks, and minima.
(m) Minimum value during the averaging interval recorded.
(p) Peak value during the averaging interval recorded.

Type III: Monitoring of Plants with an In-depth Performance Evaluation

- All of Type I and II measurements plus:
    o Ambient air temperature
    o PV module temperature
    o Wind speed (including peaks)

- Use of either a sophisticated data logger like the RDE, or Australian design, or a PC-based system, with or without modem data transfer

Type IV: **Monitoring of Plants with an In-depth Analysis of Plant Performance and of Recorded Data**

- All of Type III measurements plus:
   o Energy counter data for redundancy and accuracy
   o Peak values of solar irradiance, wind speed, PV power output, and minimum and peak battery voltages
   o Individual battery cell voltages (Note: Other battery cell parameters such as specific gravity or electrolyte level may be included)

- Display of on-line real-time data on PC/video monitor

- Storage of real-time calculated parameters (power, energy, averages)

- Scanning of peaks and minima

- Use of PC-based system with or without modem data transfer

3.3.2 SELECTION OF SUITABLE DAS

To select which type of monitoring is the most suitable for a particular PV plant, it is important to consider the following:

- Technical requirements for monitoring (see Tables 2-5 and 2-6) and budget allocated for the monitoring equipment hardware and installation

- Basic goals of the PV plant or the project itself; both short-term and long-term operational plans

- Availability of DAS equipment and devices and related computer software on the commercial market and their cost

- Off-site analysis requirements. (Note: If both measured and calculated parameters are not properly designed and programmed for recording, you could end up spending much more time and money for this off-site analysis later on).

3.3.3 DATA COLLECTION EQUIPMENT

Monitoring equipment generally consists of a serial chain of sensors, signal conditioning devices, the data acquisition system (DAS), and data transmission/receiving devices as illustrated in Figure 3-69. We have defined this whole chain as the data collection system (DCS). The heart of this chain, the DAS, is the most critical part. The DAS itself has several elements (see Fig. 3-70), each of which can cause a failure in the data collection process.

For the purpose of this handbook, we have classified DAS's into "PC-based" systems and "data loggers". The PC-based systems are those which use an IBM PC, or its compatibles. A data logger is thus any DAS which does not use the PC for on-line data collection and storage.

We have limited the PCs specifically to the IBM/compatibles simply because in the foreseeable future it is very likely that they will dominate the market for data acquisition applications. In fact, all data loggers described herein make full use of the PC not only for the initial set-up and programming of the input data, but also for subsequent data analysis and presentation of results after retrieving the stored data.

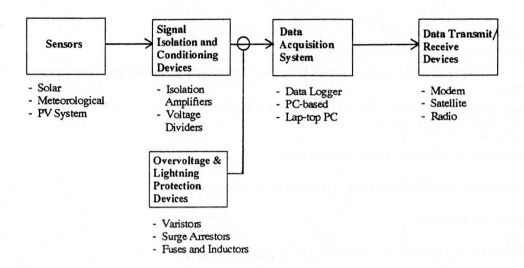

Figure 3-69.   Elements of a Generic Data Collection System

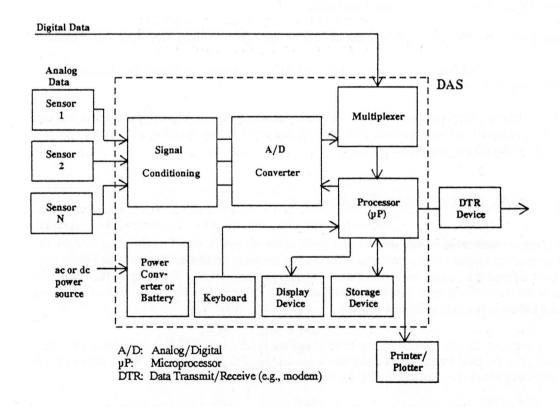

Figure 3-70.   Typical Data Acquisition System (DAS) and Its Interface. Printer and plotter are usually not integrated into the DAS. Also, signal conditioning circuits are often provided separately (e.g., isolation amplifiers and voltage dividers).

154

Figure 3-71 is a block diagram showing a more advanced version using two PCs which is intended for real-time analysis.

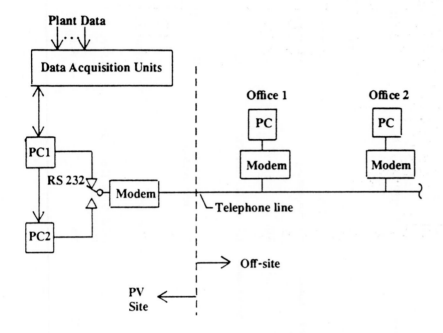

Figure 3-71. A More Advanced Monitoring Configuration for Real-time Analysis. PC1 is used for routine data acquisition, processing, and storing while PC2 does engineering and diagnostic calculations.

The state-of-the-art hardware and software for the principal elements of the DCS, namely the DAS's, PCs, signal conditioning devices, power and energy meters, some sensors, and modems are described in the following paragraphs. These are the main components which usually present the greatest problems to designers.

Table 3-19 lists the principal characteristics of most of the recently available DAS on the market and installed in the PV plants. The Concerted Action group sponsored by the Commission has evaluated the following monitoring equipment:

DAS/Datalogger	Evaluated by
AEG DAM 800	AEG (TST), WIP, JRC
Schlumberger IMP and PC-AT	WIP, JRC
RDE	WIP, JRC
HP	PPC, Fraunhofer Institute
Lambrecht	WIP
Delta-T Logger	WIP
Datataker	CIEMAT-IER
Campbell Scientific, 21X or CR10	WIP, Hyperion, Swedish Institute of Microwave Technology

155

Table 3-19. State-of-the-art Data Acquisition Systems

Parameter	Manufacturer and Model*			
	AEG DAM 800 (Wedel, FRG)**	Schlumberger IMPs with PC (Farnborough, GB)	rd Electronic (RDE) with PC (Eching, FRG)	HP 3421 A (all EC countries)
No. Channels				
- Analog, single-ended	- 16 or 32	- 10 to 600	- 8 to 206	- 30
- Analog Differential	- 8 or 16	- 10 to 600	- none	- 16
- Digital	- 2	- 10 to 600	- 8 to 200	- 2
Input Levels				
- Analog	- 0 to 10 Vdc - 0 to 20 mA	- 0 to 12 Vdc (IMP A) - 0 to 2 Vdc (IMP B) - 0 to 500 Vdc (IMP C) - 0 to 20 mA	- 0 to 10 Vdc - 0 to 20 mA	- 0 to 10 Vdc
- Digital	- 0 to 5 V/TIL	- 0 to 5 V, 0 to 15 V (selectable) 49 kHz	- 0 to 5 V/TIL	- 0 to 5 V/TIL
Scan Rate				
- Channel	- 20 ms	- 20 ms	- 2 ms	- 50 ms
- System	- 5 to 120 s	- 1 s to 24 h	- 1 s	- 1 s
Averaging/Storage Frequency	- 1 h	- 1 s to 24 h	- 1 s to 24 h	- 10 s to 1 h
Calculated or Scanned Values	- 5 s and avg	- 1 s, min, max, avg, and integrals, plus Intel Math	- min, max, and avg	- min, max, and avg
Real Time Calculation Between Measurements +	- 2 ch	- 64 ch	- 2 ch	- 2 ch
Programming Set-up	- Very easy with video terminal & keyboard, SW written in CP/M (not MS-DOS)	- Very easy with IMPULSE SW and PC	- Easy with PC (RDE SW)	- Easy
Memory Storage	- 720 kB (3.5 in FD), - 360 kB (5.25 in FD), - 256 kB (RAM), or - 512 kB (IC-Card)	- 40 MB or more with PC	- 32 to 128 kB (RAM) - No automatic readout	- HP 9114B FD
Operating Temperature	- With RAM Floppy, -20 to 70°C; with FD, 5 to 45°C	- -20 to 60°C	- -20 to 70°C	- -20 to 60°C
Power Source Voltage	- 24 Vdc or 220 Vac	- 12 Vdc or 220 Vac	- 220 & 30 Vac, 12 & 7 Vdc, - Internal battery	- 12 Vdc
Power Required	- 25 W	- 10 W with Grid Plus PC - 30 W with other PC	- 4 W	- 6 W avg/day

* All listed have PC interface via RS 232 C, except HP 71 B

+ Between two channels, e.g., $V \times I = P$

** Renamed TST in 1989

(Page 1 of 2)

Table 3-19. (Concl.)

Parameter	Manufacturer and Model*			
	Lambrecht Datenlogger ADLAS (Göttingen, FRG)	Campbell Scientific, 21 X (Loughborough, GB)	Delta Logger with PC (Cambridge, GB)	Datataker of Data Electronics, Australia with PC (Madrid, E)
No. Channels				
- Analog, single-ended	- none	- 16	- 60	- 46
- Analog Differential	- 7	- 8 (up to 192**)	- 30	- 23
- Digital	- 7	- 2	- 2	- 8
Input Levels				
- Analog	- 0 to 1.5 Vdc	- +5 Vdc	- 0 to 2 Vdc	- ±25 mV, ±250 mV, ±25 Vdc, or greater by internal attenuator circuit
- Digital	- 0 to 5 V/TTL	- +20 Vdc, 250 kHz	- 0 to 5 V/TTL	- low V and high > 4 V
Scan Rate				
- Channel	- 1 s	- 37 ms	- 50 ms	- 25 ms or 5 µs
- System		- 5 s to 24 h	- 1 s to 24 h	- 1 scan/sec to 1 scan/year
Averaging/Storage Frequency	- 1 s to 1 h	- 5 s to 24 h	- 1 s to 24 h	- 10 min to 24 h
Calculated or Scanned Values	- min, max and avg	- 1 s, min, max, avg, and integrals	- 1 s, min, max, and avg	- min, max, and avg
Real Time Calculation Between Measurements +	- none	- 2 ch	- 2 ch	- 2 ch
Programming Set-Up	- Easy with two-point keyboard	- Easy with PC (Campbell Scientific SW)	- Easy with PC (Delta SW)	- Easy
Memory Storage	- 1 MB IC-Card	- 30 kB (RAM)	- 16 to 128 kB (RAM)	- 24 to 32 kB (ROM), 16 to 24 kB (RAM)
Operating Temperature	- -20 to 70°C, With FD, 5 to 45°C	- 0 to 60°C	- -20 to 70°C	- -20 to 50°C
Power Source Voltage	- 12 Vdc, internal battery	- 9.6 to 15 Vdc, internal battery	- 7.5 to 15 Vdc, 220 Vac, internal battery	- 110 to 240 Vac, 11 to 18 Vdc
Power Required	- 0.5 W	- 0.5 W	- 1 W	- 1.4 W

* All listed have PC interface via RS 232 C
** With 32-channel Mux Card
+ Between two channels, e.g., $V \times I = P$

(Page 2 of 2)

The most important operating features of the most widely used equipment are briefly summarized below.

PC-based DAS

One of the best arrangements for data acquisition and processing is a PC-based system. It basically consists of a "front-end" electronics PC and simple software to process, calculate, and store specified engineering parameters. The front-end provides the interface between the sensors and the PC. Any of the following commercially available data loggers and A/D conversion equipment can serve as the front-end electronics:

- AEG DAM 800   - Lambrecht

- Schlumberger IMP and PC-AT   - Delta-T Logger

- RDE   - Datataker

- HP   - Campbell Scientific, 21X or CR10

Figure 3-72 shows the basic arrangement of a PC-based design, illustrating the modem connection for data transfer, which is an option. The PC can be a desktop or a laptop; the laptop is better for low-power remote applications. The other main parts consist of Schlumberger IMP/S-net card and Analog Devices isolation amplifiers. Key features of the design are summarized as follows:

Positive Features ──────────────────────────────────

- Laptop vs. desktop: Laptop is more rugged than PCs and consumes much less power (e.g., 5-7 W with Toshiba 1600); desktop is cheaper and preferable whenever temperature-controlled shelter is available.

- Different software languages are available to run the S-net card (Fortran, Pascal, Basic).

- The S-net card contains its own microprocessor. This enables background processing to do mathematical calculations or logical operations on one or more channels.

- Modem control is possible (e.g., Carbon Copy or Procomm software).

- It has excellent on-line display function on measured and calculated parameters.

- One IMP has 20 channels which can interface with a PC over a 2-wire digital line. Up to 30 IMP's can be hooked on this digital line (600 channels maximum); easy to transport and install.

- All channels can be manipulated with the complete Intel-mathematics contained in the S-net card.

- Up to 64 channels can be combined to get one computed parameter which can be stored.

- With a hard disk there is virtually no limitation on memory storage capacity.

- 16 different timers are available to allow simultaneous operation of display, printer, modem, different storage media like HD, FD, and Directories.

- User-friendly prefabricated software, RTM or Impulse, is very reliable and easy to program.

- Not easily affected by electromagnetic interference because of the use of a digital bus. (This bus over a 2-wire twisted shielded cable can be as long as 1 km, which is a big plus for PV plant application where an electrically noisy inverter is used.)

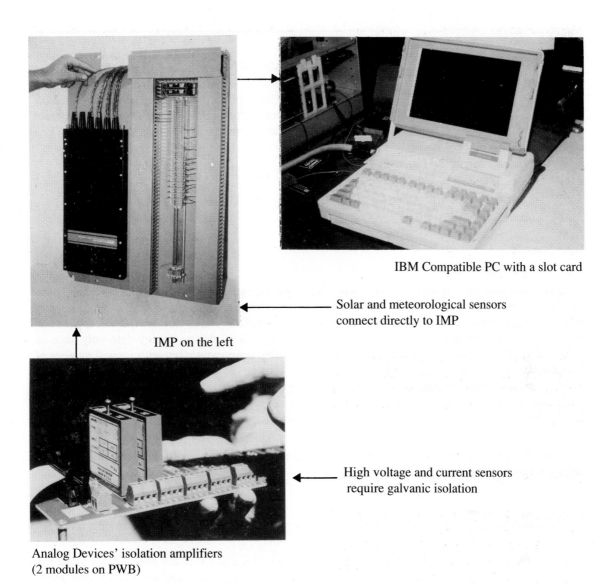

IBM Compatible PC with a slot card

Solar and meteorological sensors connect directly to IMP

IMP on the left

High voltage and current sensors require galvanic isolation

Analog Devices' isolation amplifiers
(2 modules on PWB)

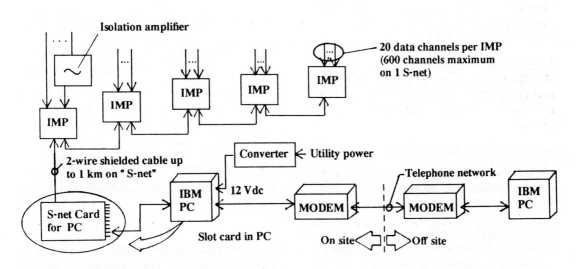

Figure 3-72. PC-based System Designed and Installed by WIP is Operating in Over Fourteen PV Plants. It uses PC-AT and Schlumberger's IMP and RTM software which have demonstrated a very high reliability over two years in a variety of environments. It represents the best commercial state-of-the-art data acquisition system for on-site and remote monitoring.

Negative Features ———

- Temperatures under minus 5°C cause problems with the Laptop PC.

- Relative humidities below 70% are not advisable.

- XT-Laptops are not useful (no standard real-time clock card available and different incompatible address on the bus).

- Most AT-Laptops need an external interface box to be able to install the S-net Card. These boxes are only available with ac-connectors; for dc they need an external dc/dc converter with +12 V and +5 V.

Data Loggers

Smaller lower-cost data acquisition units have their own microprocessors and limited storage space for data. Most of the commercially available and custom-designed units have been found to be very versatile and rugged. Simple data loggers are probably the best choice for low-power application with a very small amount of on-site storage requirement, such as storing energy data only once or twice a day. For projects that require hourly data recording continuously but with a small number of measurements, data loggers combined with periodic transfer of data via modem provide a very reliable and cost-effective approach. There are several data loggers which have proven to be very adequate. It must be noted, however, that these data loggers do not have the capability of the PC-based system in terms of on-line (i.e., real time) and off-line processing.

To illustrate the characteristics of a typical data logger, the key features of two Campbell Scientific units are summarized below:

Positive Features (Campbell Scientific 21X and CR10) —————————————————————

- Easy to hook up, set up, and operate

- Easy to program with PC and CS software (PC208)

- Stores engineering units

- On 21X only: individual channel can be scanned on a small LCD display supplied with the unit

- Both 21X and CR10 are excellent for small PV systems; however, multiple 21X's can serve large PV plants.

- 16 single-ended analog channels on 21X

- One of the lowest-cost data loggers (retail price: £ 1520 for 21X and £ 870 for CR10 from GB).

Negative Features (Campbell Scientific 21X and CR10) —————————————————————

- On-line display of all parameters not possible without PC

- Needs PC to program it
  o Battery consumption is high, if many channels are scanned at a fast rate
  o Very sensitive to high common-mode voltage range (cannot accept higher than 5 Vdc input signal on 21X and 2.5 Vdc on CR10)

- If battery energy is depleted, all stored data are lost.

- Low internal memory (19,000 data words); to store or collect a large amount of data, a modem or a PC with HD must be used.

Personal Computers (PCs)

A computer is used in two ways in plant monitoring, 1) as part of the DAS to receive, record, and/or transmit data over a modem, and 2) to retrieve data from the data logger or another on-line computer. The key considerations and tradeoffs in the selection of the computer are as follows:

1) It must have the capability to interface to an external device such as:

    - Modem
    - Printer
    - Plotter

2) At least one RS 232C interface connector is essential; two are necessary for both modem and connection of another PC.

3) IBM PC vs. Apple/Macintosh PC: Generally, the Apple PCs are preferred for desk-top publishing and report preparation. A distinct disadvantage with the Apple PC is that they do not have an A/D slot card, so they cannot be readily used for data acquisition purposes like the IBM PCs. For this reason, the IBM PCs or their compatibles should dominate the data acquisition application for a long time to come.

4) AT vs. XT: The AT has a 16-bit processor and the XT an 8-bit system. The XT is not capable of multi-tasking for background processing (modem transfer for instance) and has serious problems with commercial software required for data collection and analysis purposes. Moreover, it has become almost obsolete. Thus, in general, AT should be selected for most on-line DAS applications. Its cost is not much higher when desired options are added to the XT.

    On the other hand, the XT is cheaper, so there will be a continual market demand for other uses. For data acquisition purposes, the XT machine (e.g., Toshiba 1000 or 1200) is useful as a portable unit which can be carried to the field for data retrieval purposes.

5) Memory storage: Hard disks (HDs) are now very common with typical sizes in 20, 40, 80 MB, and so on. Either 20 or 40 MB are sufficient to take care of most of the data storage needs of the PV plants. The other most common memory devices are the 3.5- and 5.25-in floppy disks (FDs). Both types should be requested when purchasing a PC as commercial software is available in both sizes. Table 3-20 lists the main characteristics of other devices, including bubble, CMOS RAM, and RAM (or IC) card. The FDs and IC cards are the best storage devices.

6) Laptop vs. desk-top: The main reasons for the laptop for on-site DAS use are its low power consumption and its ability to operate in a harsh environment. The laptops available for remote application must be able to operate with a 12-Vdc power source and have a low power consumption.

    If the PV plant has a shelter which can provide an appropriate temperature range and ac power is available, the desk-top type should be selected as it is generally cheaper.

7) ac vs. dc power source: Another discriminator among the laptop (or portable) PCs is the power source. There are about two dozen suppliers but only a handful can provide laptops which operate on low-voltage dc power (Grid Plus, Toshiba, and Compaq are good examples).

8) For on-site DAS, the PC must have a slot card with A/D conversion. Some of the laptop PCs require an external box which houses this interface (e.g., Toshiba 1600).

Signal Conditioning Devices

The signal conditioning devices convert the sensor signal from one range to another, and dc to dc, or dc to digital signal. There are essentially two specific types of signal conditioning — the isolation amplifier which does both signal conversion and isolation of input from the output (i.e., "galvanic" isolation), and the voltage divider.

Table 3-20.   Comparison of Main Memory Storage Devices

Type	Advantages	Disadvantages
Floppy Disk (FD), 3.5- and 5.25-inch	- Most widely used - Best handling characteristics - Cost lower than IC card memories	- Lower reliability than solid state memories - Subject to breakage - Low operating temperature limits (5 to 45°C)
Hard Disk (HD)	- No memory limitation - Better packaged, more reliable than FD - Operates at lower temperatures than FD - Ten times faster than FD	- Not easily transportable - High power consumption
Bubble Memory	- Large memory (512 kB) - -20°C to +65°C - Good vibration shock tolerance	- Not compatible with PCs - Limited sources (Hitachi, Fujitsu) - Problems on spare parts and availability - Error-prone - More complex on control - Highest cost of all memory devices - Poor logistics
RAM (CMOS) Module	- More widely used - Lower Cost (1,000 DM vs. 2,000 DM for Bubble Memory) - Good performance record - NES unit: 32 kB (1,900 DM) - AEG unit: 256 kB - Convenient to use	- More susceptible to noise than Bubble Memory - Static charge can erase memory
IC (RAM) Card	- Much easier operator interface - Size and shape like a credit card - More rugged than FD - Easier to mail than FD	- New, very little performance reliability data

Two of the best sources of isolation amplifiers are Analog Devices and Hartmann & Braun. The Analog Devices 5B series takes ± 10 mV, ± 10 V, and 0-20 mA signal, and a separate unit handles the PT 100 (RTD) temperature sensors. It is very rugged and reliable. The Hartmann & Braun units are very reliable, easy to set up, and adjust to different input signal types (up to 500 Vdc input). The main difference lies in their cost — about DM 250 per unit (i.e., per data channel) for Analog Devices, as compared with about DM 800 for Hartmann & Braun. The amplifiers are packaged in card-pluggable modules (see Fig. 3-72 for a typical Analog Devices module).

In general, isolation amplifiers should be used for the measurement of:

- High dc bus voltages (even when voltage dividers are used)
- Battery current across a shunt (current)
- Other parameters that require signal conditioning because of the DAS input limitations.

Voltage dividers are often necessary before the isolation amplifiers, for example, stepping down from the 500-Vdc level PV array and battery voltages, to the 10-Vdc level. The voltage divider is simply a two-resistor network. The key precaution in designing such a network is to ensure that 1) the resistors are capable of sustaining the power dissipated, and 2) at least 0.1% resistors are selected for measurement accuracy.

Power, Ampere-hour, and Energy Meters

These devices have been used mainly for the monitoring of:

- Solar energy on the plane of the array and a horizontal surface
- PV energy output
- Inverter energy output

162

Power, ampere-hour, and energy meters are actually sensors because they give certain values of output in real time. The power sensor has analog voltage multipliers resulting in the V · I product which is power. Then, an electronic (analog) integrating circuit operating on power as the input yields energy over a given duration of time. The ampere-hour counting circuit functions in a similar manner, but it uses only the current as input. Power and current integrators available in the EC are shown in Figure 3-73. ENEL/Advel in Milan has made a special 4-channel design of current and power integrators which they have implemented at several PV plants [55].

Many of the data loggers and PC-based DAS described earlier have the capability to multiply two channels (V · I) to calculate power and then to integrate the power profile to obtain the energy. The computer can also calculate Ah by integrating the current. So, one may question the need for duplication. One answer is that, if such a DAS is on-line, the analog counters and meters serve as a backup, thus increasing the reliability of data collection at the site. Experience has shown that even the best DAS's encounter stoppages.

There is another very important benefit of counters with digital readout devices — they provide a real-time display of available solar energy and output energies from the array and inverter. Most of the DAS's either do not have the mathematical function or the capability to be programmed to display the energies in real time. The energies are usually calculated afterwards via a separate off-line PC. Quite often, these types of real-time displays are very valuable to the site operators, owners, and visitors.

The conventional ac energy counters operating on a standard 50 Hz line are very reliable, and they are available from a number of suppliers. For variable frequency output inverters such as for pumping, the sources of ac energy meters are few in number. In Germany, the two suppliers of this type of meter are Sun Power and RDE. These meters can also be used as dc power integrators. For all these meters, it is essential to acquire units with pulse outputs for automatic recording of energy data at predetermined time intervals.

Modems

Users often need to send information to other computers via telephone modems and to do it quickly. The rate of transferring data is important, especially when making long-distance calls. This rate used to be 300 bits per second (bps) which is about 30 characters per second. Now modems are available with a faster communication capability, like the Smartmodem 2400 (Hayes Microcomputer Products) which has 1,200- and 2,400-bps rates.

In Germany, the new regulation which started in January 1990 allows the use of modems that are not fully compatible with the "AT" Hayes standard. Earlier, only Ramlauf or Dr. Neuhaus was permitted. The key problem here is that many of the operating software packages from the USA such as Carbon Copy and Procomm are often difficult to use, or are incompatible with the German standard modems. A typical modem available in Germany is Fury 2400 TI/MNP from Dr. Neuhaus.

Miscellaneous Hardware

Other sensors used in PV plant monitoring are those for ambient and PV module temperatures and wind speed. These are standard off-the-shelf items with many sources. For temperature measurement, PT100 has become a standard item in many PV plants.

The other hardware items that can add significantly to the overall cost of the DAS are lightning overvoltage protection devices and a meteorological/solar sensor tower or mast. Lightning protection devices were discussed in Section 3.1.4.

Some of the projects decided to erect a mast or tower to mount solar and meteorological sensors such as:

- Horizontal and tilted solar sensors
- Ambient air temperature sensor
- Wind speed sensor (anemometer)

Sun Power Solar Energy Meter

Sun Power 2-channel Integrator

AEG kWh Meter

Delta-T Solar Energy Integrator

Figure 3-73.   Power/Energy Meters

Unless dictated by non-technical constraints, *masts should be avoided to reduce the cost.* The best location for the solar sensors is the array structures, provided there is no shadowing, and the sensors are easily accessible for cleaning. This will significantly reduce both hardware and installation labour needed for the solar sensors.

The wind speed sensor should be mounted at the highest part of the PV array to avoid obstruction of the wind flow to the sensor. One distinct advantage of fastening the tilted solar sensors onto the tilted array structures as compared with a separate mast is that the orientation of the sensor (azimuth and tilt angle) is automatically aligned with that of the PV panel whenever the array tilt angle is adjusted.

3.3.4   SOLAR IRRADIANCE SENSORS

Measurements of solar irradiance are essential in practically all fields of solar energy research and applications. Both instantaneous and daily energy values are needed to evaluate the performance of installed PV systems. It is important, therefore, that sensors used for continuous measurements of solar irradiance outdoors give accurate readings and be constructed to withstand the environments normally encountered. In short, they must be qualified for outdoor use in the same manner as PV modules.

Two basic types of solar irradiance sensors are the thermopile and Si-based devices. Table 3-21 lists the basic features of commonly used Si and thermopile pyranometers. The interest in Si cells as the transducers for solar intensity stems from their low cost, and for the PV application, the similarity in the response characteristics between sensor and the PV array. They are available on the commercial markets about 3 to 5 times cheaper than the thermopile devices. A brief description of these sensors and calibration methods is given in the following paragraphs.

Thermopile Pyranometers

Thermopile-based pyranometers have been widely used for solar irradiance monitoring in the meteorological stations and for solar thermal research and application work for several decades, and their characteristics are well known [56-59]. The popular ones now available are the 2π, 160° field-of-view instruments from Kipp & Zonen, Schenk, and Eppley Laboratories.

These sensors have been widely accepted throughout the world as the principal instruments for irradiance measurement, and they are traceable to the World Radiometric Reference (WRR). For these reasons, any of the above secondary thermopile sensors can be used as the "super" sensor for calibration of Si sensors.

Si-based Sensors: Excellent for Field Use But They Need Recalibration Initially at Time of Installation

Si-based sensors have been reported to be technically comparable to the thermopile sensors within the desirable accuracy [60-62]. But, if they are not calibrated initially at the time of installation against a good working thermopile reference, the result can be large errors in all key parameters which are based on this measurement, such as PV array power, PV energy, and both system and PV efficiencies. For example, during a routine visit and inspection in 1988 at several pilot plants, the Haenni sensors (used for the plane-of-array irradiance monitoring) were found to have substantially lower sensitivity than the reference sensor (Kipp & Zonen CM11). Differences in irradiance measurements of up to -17% were attributed to the incorrect calibration factor (CF) from the supplier and not degradation of the solar cell. From this and other experiences it was apparent that most commercial Si sensor suppliers do not have an adequate standardized procedure for running their calibration measurements.

Table 3-21.   Basic Features of Several Solar Irradiance Sensors

A.   Si Types

Sensor/Model	Shunt Rsh/ Output @ 1000 W/m^2	Cover	Sensor	Cost (DM)
Haenni Solar 130	325 Ω 100 mV	White, curved plastic (Perspex opal 040)	Cz mono-Si, 0.5 cm^2	800
Li-cor 200 SZ	No Rsh/68 μA	White flat acrylic	(Similar to above but uses smaller detector)	700
Matrix	0.4 Ω/38 mV	Dome and flat pyrex	Cz mono-Si, 4 cm^2	650-700
Mesa	1.0 Ω/70 mV	Flat, pyrex	Cz mono-Si, 4 cm^2	900
Ref. cell in module pkg. (Siemens)[1]	0.010 Ω/40 mV	Flat, textured glass	Cz mono-Si, 100 x 100 mm	1,200
Individual cell packages (AEG, Italsolar, Photowatt, etc.)[1]	10-20 mΩ/ 50-70 mV	Flat glass	Circular (100 mm dia) Si Square (100 x 100 mm) Si	400 to 1,000
Ref. cell, ASTM Construction (limited sources)	no Rsh/100 mA	Flat quartz	Cz mono-Si, 4 cm^2	2,000

B.   Thermopile Types

Sensor/Model	Output @ 1 Sun (mV))	Cosine Response at 10° Elev. (%)	Response Characteristics (in seconds)[2] Time Constant 1/e	95% of Final	Cost (DM)
Eppley Labs, PSP	8-10	±3	1	3	3,000
Kipp & Zonen, CM 11	4-5	< ±3	5	12	1,900
Philipps Schenk, T-8101	15	+5	3	10	2,200

[1]   Contains temperature sensor.
[2]   Time constant for Si cells is in the range of a fraction of ms.

Reference Si Sensors

The terrestrial photovoltaic application community always had two choices of sensors, either a Si-based unit or thermopile instruments. The heritage of the Si sensors is the space programme. For space PV engineers the need to accurately calibrate the satellite solar panels working outside the earth's atmosphere was the prime motivator to create "reference" Si sensors. These sensors were therefore calibrated close to AMO (Air Mass Zero) intensity levels by testing at very high altitudes, on airplanes, and on balloons for every new manufacturing technology and variations of the same cell material (e.g., different base resistivities, surface treatments, suppliers, etc.), even though they are all monocrystalline types. For the space programme, the cells actually flown and tested on airplanes and high-altitude balloons were called "primary standards," and all other units which were calibrated against them under a solar simulator became "secondary" standards. The balloon-flown calibration is still employed today, and these primary standards cost over $ 10,000 [63].

For terrestrial PV applications, calibration and construction criteria for the reference Si sensors were patterned after the space programme practices, but based on AM 1.5 intensity established by the U.S. ERDA PV Programme in 1976 [64]. Afterwards, a series of standardization efforts resulted in several interim documents concerning calibration methods and design of reference cells [65-67]. These reference cells were mainly intended to calibrate the performance of PV devices (at or near 1000 W/m^2 intensity), and not for continuous outdoor use. The sunlight calibration method established by the ASTM which specified conditions such as AM and air turbidity limits, turned out to be impractical, and thus, was never put into real practice. Now in mid-1991, many researchers and system designers consider these atmospheric parameters unnecessary for sunlight calibration purposes [60,62,68,69].

Simplified Outdoor Calibration Method

No simple, reliable field calibration method was available for plant operators and users until recently. The field calibration method developed and now being evaluated under a CEC-sponsored Concerted Action [1,59] is considered adequate. However, many workers maintain that indoor calibration combined with a spectral mismatch correction is the best way to accurately calibrate any Si reference device. Others, including CEC's Concerted Action task working group, argue that outdoor calibration is technically better than indoor testing because no solar simulator can match the natural sunlight spectrum, and, moreover, anyone without special equipment and skills can do it.

The principal criteria, established by the CEC's Concerted Action group for sunlight calibration and construction of Si-based solar irradiance sensors, are as follows:

Test Conditions:

- For high accuracy, orient both test and reference sensors in the sun-tracking position
- For calibrating a tilted sensor, reduce the "solar window" as defined below
- Sky condition: Relatively clear and no clouds in front of the sun at the time of data-pair measurements (i.e., clouds may pass by between data recordings).

Measurements:

- Number of days: 1 day
- Number of data pairs: 10 minimum
- For sensors oriented to the sun: acquire data at solar noon $\pm$ 2 hours
- For sensors on fixed-tilted surface: acquire data at solar noon ± 0.5 hour.
- Solar intensity stability: $\pm 2\%$ maximum during data-pair measurements
                                               as indicated by the reference sensors

Analysis:

- Calculate the average of slopes (i.e., CF) of the sensor output voltage vs reference sensor irradiance (Wm^{-2}), using the data for the solar window specified above.
- Use solar irradiance data $> 800\ Wm^{-2}$

Sensor Construction and Performance Specification:

- Use a precision shunt resistor (across cell terminals) of at least 0.1%, preferably made of manganin material, for short-circuit current measurement
- 50 mV maximum reading across the shunt resistor at 1,400 W/m^2 intensity (including reflected rays)
- Solar cell (detector) must have high internal shunt resistance and AR coating
- Cover glass (flat or dome), preferably quartz or pyrex must be free of optical defects and have low light reflection, especially at high incidence angles (Note: Because of large effects of reflectance, a good quality dome cover is highly recommended).
- Detector field-of-view angle of 160°

Figure 3-74 illustrates the basic measurements and procedure for determining the Calibration Factor (CF) of a test sensor. The slope of the Y-X line is the CF. The problem here is that there are several ways to arrive at this slope: 1) Use of linear regression analysis which is a "least-squares-fit" procedure, 2) ratio of the sum of all Y-data to the sum of X-data, and 3) the average of the ratio of each corresponding Y-X data-pair. The last method (item 3) which averages all discrete slopes is preferred because it ensures the intersection of the origin [60]. The linear regression analysis (item 1) often results in non-zero intercept which is incorrect.

Available Si-based Sensors

The Si-based sensors may be divided into two groups, *commercial* and *custom-designed* as shown in Figures 3-75 and 3-76, respectively. Matrix, Haenni, and Licor units are all acceptable, but special attention should be given to their cosine-response characteristics, especially at large incidence angles [60,62].

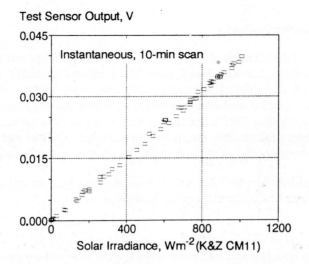

Figure 3-74. Derivation of Calibration Factor (CF) Using Simultaneous Measurements of Test and Reference Sensor Outputs.

Matrix

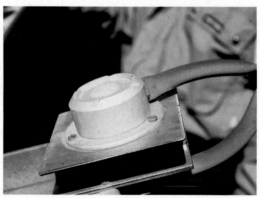

Haenni 130

Dodge Product SS100

Licor 200SZ

Figure 3-75. Commercially Available Si-based Sensors

Reference cell in ASTM package (Available from Univ. Politecnica, Madrid, ES)

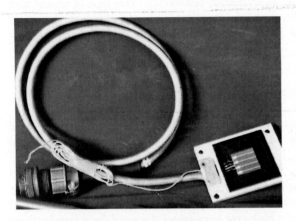

Mesa (Konstanz, FRG)

AEG (Wedel, FRG)

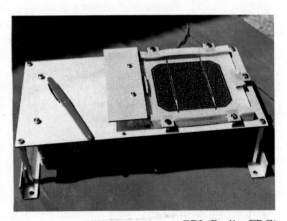

GBI (Berlin, FRG)

Siemens (Munich, FRG)

Italsolar (Rome, IT)

Figure 3-76. Custom-designed Reference Si Sensors

Custom-designed units such as the Siemens (in the multi-cell package) and other single-cell units have proved to be very stable and acceptable for long-term field use [60,70]. The main problem is that they (i.e., initial prototype versions) cost two to five times more than the best commercially available types. There is still a need for several manufacturers of low-cost reliable Si sensors in Europe.

Some of the Si sensors come equipped with voltage-to-frequency converters. This type of sensor with built-in signal processors should be avoided because of increased cost and the fact that most available data acquisition systems readily accept low-voltage signals directly from the solar sensor with a built-in shunt resistance.

Other Solar Irradiance Sensors and Related Equipment

Figure 3-77 shows the basic types of sensors used for monitoring the global irradiance on fixed horizontal and tilted surfaces, diffuse irradiance on the horizontal plane, and direct normal and global normal irradiance.

Both Si and thermopile sensors with light-collimation tubes are available for direct normal irradiance measurements, with Si types usually four to five times cheaper. However, continuous monitoring of direct normal and global normal irradiance incurs a high cost because of the need for a sun-tracking platform. As a partial solution to this problem, a novel method of monitoring the global normal irradiance (GNI) with a stationary sensor was recently developed by WIP under a CEC JOULE project [71,72]. Figure 3-78 shows a prototype unit. By orienting a 5-detector sensor assembly base plate south facing and tilted to the latitude, an accuracy within 3% of the GNI sensor on a sun tracker has been demonstrated. This device is potentially very useful for meteorological stations and PV sites with different array structures (fixed and sun-tracking, flat-plate and/or concentrating systems).

The cost of sun trackers, designed specifically for solar sensors, varies from about DM 4,500 for a single sensor holder (Eppley ST-1) to about DM 20,000 in Germany for a 90-day autonomous tracker which has a self-adjustment capability.

3.3.5   DATA COLLECTION RELIABILITY

A key problem in the past has been the unreliability of continuous operation of the DAS because of its many serial elements, more complex computer, and mechanical data recording devices. One way of improving the reliability of data collection is to use parallel or redundant data acquisiton devices for important parameters which you may not want to lose because of PC or data logger failure. For example, Ah and Wh counters can be installed to record solar and PV energies, respectively.

The use of energy counters (see Fig. 3-73) is not yet widely exploited in present monitoring systems for PV plants. By adding these devices, the overall reliability of the collection of key data can be significantly improved. Likewise, the capability for on-site performance assessment will be greatly enhanced. Another advantage of Ah and Wh meters is that they allow simple monitoring of the plant operation in real time by the local site operators. The problem, however, is that reliable and accurate integrators and meters for dc voltages and power are not yet widely available. This is not the case for standard ac kWh meters which have proven to be reliable and are widely available at a cost of about 1,000 DM each in Germany.

A more conventional approach is to install redundant monitoring hardware, such as 1) an identical set of DAS front end (e.g., isolation amplifier and IMP) and a PC, or 2) use a separate data logger in addition to the PC/DAS for only the critical parameters. The first solution is referred to as full redundancy and the second, partial redundancy.

The effect of full redundancy in one serial chain of electronics can be illustrated as follows. Given a combined failure rate of 250 per 10^6 hours for one channel, the reliability of a fully redundant system for one year is 0.75 as compared to 0.5 for the non-redundant case. For more details on reliability assessment and design approaches, see Appendix 29.

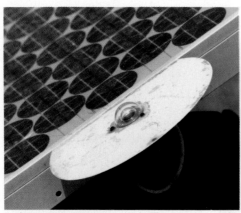

Plane-of-the array Solar Irradiance Sensor
(K & Z CM5/6 at KYTHNOS PV Plant).

Horizontal (left) and Plane-of-the array Sensors
(Matrix 1G at Ensheim's PV House)

Direct Normal (Eppley NIP) and
Total Normal (Matrix 1G and
Licor 200SZ) on a Sun Tracker

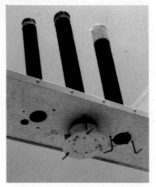

Direct Normal Sensor (50-cm
collimating Tube with Eppley
PSP) on a Sun Tracker

Diffuse Irradiance Sensor (K & Z CM11
with Shadow Band

Direct Normal Sensor (on Sun Tracker) and Diffuse
Sensors (at ZSW/Univ. of Stuttgart PV Test Facility)

Direct Normal (Eppley NIP and ST-1 Tracker)

Figure 3-77.   Monitoring of Global Horizontal, Diffuse, Direct Normal, and Global Normal Solar
Irradiances in the Field

Figure 3-78. A Novel Method of Monitoring Global Normal Irradiance (GNI) Was Recently Developed by WIP. A prototype unit on a completely stationary mount (tilted, south-facing) has demonstrated an accuracy within 3% of the sun-tracking reference GNI sensor over a period of four months.

3.3.6   COST OF DAS AND SENSOR DEVICES

Table 3-22 is a list of all main hardware and software required in the DCS in some combination determined by the system to be monitored. The overall cost of the DCS is best illustrated by giving specific data pertaining to the actual PV plants installed recently. Table 3-23 is a list showing the total cost of DAS installed in 1990. This shows that, based on recently available DAS equipment combined with improved monitoring methods that resulted from lessons learned, the installed cost of the DCS was about 2,000 to 3,000 DM per measurement parameter in 1990. This is considerably below the 4,000 to 5,000 DM per channel cost in the 1982 time period [73].

3.3.7   SIMPLIFIED METHOD FOR DETERMINING AVAILABLE PV POWER AND ENERGY

The power capability of the PV field or an array is the maximum power that it can supply under given conditions of solar irradiance and array temperature. In other words, it is the peak power point of the array I-V curve at the array dc bus at a given time. When this peak power is integrated over a given time duration (hour, day, month, etc.), the total energy capability of the PV field is obtained for that duration. This parameter is very useful for assessing how much of the plant's real energy production capability is being utilized, which we define as:

$$\text{Array Utilization Factor} = \frac{\text{Actual array output}}{\text{Array output capability}}$$

The array output can be expressed in either power or energy. The actual array output, the numerator in the equation above, is usually calculated from V,I measurements, and the array capability, the denominator, is calculated using actual irradiance and PV module temperature measurements. If a module temperature sensor is not available, the module temperature can be substituted by either ambient air temperature or the open circuit voltage of the entire PV array or any of its strings, with an appropriate factor.

Table 3-22.   Monitoring Equipment, Devices, Software and Their List Prices (1990) in Germany

Description	Unit Cost (DM)	Description	Unit Cost (DM)
Data Loggers/DAS		**Solar Irradiance Sensor**	
- Schlumberger		- Thermopile	
o IBM A, B, or C	3,600	o Kipp & Zonen CM11	1,960
o S-net Card	2,580	o Eppley, PSP	3,700
o IMPULSE software, basic	3,000	o Schenk	2,000
- AEG DAM 800	12,030	- Silicon	
- HP 3421A and peripherals	16,000	o Matrix 1G	600
- Campbell Scientific		o Licor 200 SZ	600
o 21X (with 40 kB RAM)	7,000	o ATSM housing	725
o CR10	6,000	o Haenni 118	800
- NES MODAS 12	17,800	o AEG Reference Cell	710
- RDE		o Siemens Multi-cell Reference	1,200
o Handler, NS-256	5,100		
o Options Box, MK 1108	1,250	**Meteorological Sensors/Mast**	
o Data Logger, MK256	4,800	- Temperature PT 100, JUMO	100
- Delta-T Logger	7,000	- Temperature, Thies	780
- Stella Solartechnik (12-ch), without PC	20,000	- Anemometer, Thies	500
- Australian Datataker	6,700	- Sensor carrier	1,200
- Lambrecht, ADLAS with readout unit	9,700	- Mast	800
Computer (PC)		**kWh Meters (Integrators)**	
- AT 286 with monochrome VDU	2,500	- Sun Power	
- AT 286 with colour VDU	4,800	o Solar irradiance/power integrator, 3-ch	4,500
- Toshiba 1000	2,500	o Solar irradiance integrator, 2-ch	1,950
- Toshiba 1200	6,000	o Current integrator	1,950
- Toshiba 1600, incl. External Box	10,800	o Power integrator, ac and dc	3,350
- Toshiba 3200 Laptop	12,000	- Power integrator, RDE, ac and dc	4,300
- Grid Case Plus	13,700	- Power integrator, Prokein & Koschik, 1-ch	1,200
- NEC Ultralite	12,500	- ac kWh meter-Helios	1,000
- Zenith Data LT 286, 20 MB	8,300	- ac kWh meter-Schlumberger	1,200
- Zenith Data LT 286, 40 MB	10,000	- 4-ch. amp-hr meter, Advel	6,000
- COMPAQ Desktop 286	9,000	- Microvolt integrator MV2, Delta-T	450
Software		**Modem**	
- MS-DOS 3.2	148	- Ramelow, Telemodem	1,600
- Word 5.0	1,490	- Dr. Neuhaus, Fury	1,150
- Winword	1,620		
- Quattro	400	**Miscellaneous Items**	
- Pagemaker	2,450	- Cabling, per project	2,500
- Carbon Copy	790	- 19" Rack	2,000
- Labtech Notebook	3,000	- Nuts, bolts, tubing, etc., per project	1,000
- IBM Postscript	2,000	- Printer	1,200
		- Electrical components, per project	1,000
Signal Conditioning			
- Analog Devices Isolation Amplifier	250		
- Hartmann & Braun Isolation Amplifer	800		
- Voltage Divider	560		

A simple method for determining the maximum available array power or energy is described in Appendix 27. The method is applicable to any PV array configuration and size. Both ZAMBELLI and VULCANO, among the pilot plants, and German National PV Projects have implemented this method in real-time and non-real-time (i.e., off-line calculations). Any of the PV projects which acquires solar irradiance (on plane of array) and back-of-the-module temperature measurements can make use of this method for on-site and/or off-site use.

Table 3-23.   Cost Summary for Actual Monitoring Systems Designed and Installed by WIP-Munich for Several National PV Projects in Germany (BMFT Programme)

| | | 1990 Cost (DM) | | |
Project	DAS	Equipment*	Design/ Installation	Total
1. Stand-alone with Diesel - 10 kW PV - 39 Measurements	Laptop/IMP/ kWh Meters	55,000	14,000	69,000
2. Stand-alone/Grid-connected - 10 kW PV - 20 Measurements	COMPAQ/IMP/ kWh Meters	48,000	14,000	62,000
3. Grid-connected - 15 kW PV - 10 Measurements	DAM 800/ kWh Meters	38,000	14,000	52,000
4. Grid-connected - 8 kW PV - 30 Measurements	Laptop/IMP/ kWh Meters	37,000	14,000	51,000

* Includes DAS, signal conditioning devices, sensors, and miscellaneous hardware

3.3.8   PERFORMANCE ANALYSIS AND PRESENTATION

For the medium- to high-power PV plants with a detailed monitoring requirement, a standard format was established for reporting monthly performance data based on Table 2-6 (in Chapter 2). This format is applicable to monitoring Types II, III, and IV described in Table 3-18. Eventually, only energy recordings (e.g., insolation on the plane of the array, actual and theoretical maximum array output, and inverter output), on a daily basis may suffice in most of the PV plants.

Table 2-6 gives a general list of all processed or calculated parameters for a day or one month for a generic PV plant. This table can be used to select parameters for Types III and IV type monitoring. Some of the terms which are not widely used are:

- Sunlight average value, based on the theoretical sun hours in a day

- Actual sun hours (the summation of total time in which the plane-of-the-array irradiance is greater than 100 W/m^2)

- Array utilization factor, the ratio of the actual and the available array energy (calculations made on an hourly basis and summed up for each day, month, and year)

- Array power and energy capabilities, calculated on the basis of several measured parameters for each hour, day, and month(see Appendix 27).

3.4   PROJECT MANAGEMENT

Most of the initial knowledge about PV plant design, installation, and operation of large PV plants in the European Community has come from the original pilot plants. These pilot plants were the proving grounds, not only for first generation PV system designs and components, but also for the training of future project managers, designers, operators, and users alike in project management methods and procedures. For example, a vast majority of the current PV projects, both CEC and national, are utilizing the pilot plant designers or their successors.

Project management encompasses the supervision and coordination of all phases of the project, including design, manufacturing, shipping, installation, and operation. In addition, customer coordination and satisfaction is a very important function. The key lessons learned in the project management category are as follows:

1) In any size project, be it a PV house or larger centralized PV plant, the financing organization(s) should require the following items to be part of the contractual requirements:

   - Information about the ownership of the plant for at least five years after plant installation, including the responsible person to be contacted.

   - A plan for the maintenance work and the system operational mode (i.e., grid-connected, stand-alone, etc.) for at least five years after plant installation, including the specific organization(s) which will be responsible for the actual maintenance work.

   - Manuals and documentation required for plant operation and maintenance (also see Section 3.1.7).

2) Adequate funds should be set aside for the plant designer to do the following:

   - Training of users in the operation and maintenance of the plant for at least one year after its installation. In the case of a fairly large complex plant, a good procedure is to do the training during the design and installation phases while the key designers are still on the job. Following the installation, designers tend to disappear very quickly.

   - Preparation and submittal of adequate documentation to do basic trouble-shooting and maintenance.

3.5   NON-TECHNICAL ASPECTS

3.5.1   SOCIAL BENEFITS ASSESSMENT

The Commission undertook a study on social benefits and effects in 1988-1989 as one of the Concerted Action projects. The aims of this study were to:

- Determine the social/economic benefits and effects of installed plants on users, owners, designers, and operators

- Formulate suggestions and actions regarding the social aspects of future projects, including plant operator/designer interfaces

- Identify and study non-technical issues

The approach taken in this study was essentially to gather information focusing on non-technical areas and to evaluate some of the social factors. Contributors were mostly members of the Concerted Action group identified in early 1988. It was evident that stand-alone systems present greater social impacts than the centralized PV plants. For this reason, a case study by a sociologist was made on a Spanish project involving over 100 PV houses [74]. By studying direct experiences related to the design and operation of the PV plants by the users, operators, and designers, techniques were formulated not only to improve the usefulness of the existing systems but also to suggest possible actions for the users and technical personnel in future projects.

The significant results of the above study are summarized as follows:

1) For the large number of widely distributed small PV systems:

   - Sociologists are very effective mediators between the user and system designers and installers. Unfortunately, most projects do not include such an activity, and therefore, no funds are available.

   - Users should pay the cost of PV energy to help them feel that they are an important part of the project demonstration.

   - Success in making the best use of a PV system is possible if the project personnel work with the users in a timely manner, especially in the first months after the PV system start-up. A periodic information pamphlet or newsletter for general distribution would be very helpful.

2) For the centralized PV systems like the pilot plants, social impacts and benefits can be substantial, and there are large potential pay-offs in promoting renewable energy sources. To this end, there should be a more concerted effort, and perhaps collaboration between the plant owners and the CEC, to find a better way of implementing a public information dissemination on operational PV plants. This should address both technical and non-technical issues and include brochures written in both local and English languages. PV plants in Europe have attracted visitors from many other countries; brochures in English have proved to be very helpful in many of the pilot plants situated in non-English speaking countries.

3) After the PV plant had been installed and some technical performance data had been submitted to the Commission as required by the contract, there was a lack of follow-up activities by the plant owner in the areas of performance assessment and public information dissemination which tend to go together.

4) Some of the PV plant owners should examine profit-making or money-raising activities to help defray the cost of plant maintenance and operation, and even public information dissemination activities. This should include organized tours in conjunction with the local and national Chambers of Commerce.

5) Plant designers should pay more attention to good human factors engineering, user needs, and social factors such as in the design of:

   - Billboards for large PV plants which are generally displayed at the entrance to the plant
   - Brochures about the plant for local and out-of-country visitors
   - On-site displays and devices specifically for the operators and site visitors
   - Simple diagnostic tools and procedures for the users

6) "Social" monitoring is as important as "technical" monitoring of the plant. Future projects must be more attentive to public relations, especially with users, operators, and visitors.

3.5.2   LEGAL AND INSTITUTIONAL ISSUES

In most EC countries, there is a lack of government legislation and policies towards the dispersed PV systems serving the public. The key legal and institutional issues are:

- Ownership of PV components (utility vs. houseowner)

- Sun-access rights

- The right to sell PV electricity to the utility

- Safety standards

- Insurance (theft and liability)

- Tax credit for the installation of renewable energy systems

- Equitable rates for the sale of PV energy to the utility

To facilitate the promotion of photovoltaics in the public sector and utilities, appropriate legislation is urgently needed.

3.6 PUBLIC INFORMATION DISSEMINATION

Plant Brochures

The brochures prepared and distributed at various plants provide excellent coverage of the PV plants but mostly from the technical viewpoint. Perhaps the less technical and more social-economic, environmental, and ecological aspects should be considered in the future to supplement the technical brochures. In on-going and future projects, plants should explore how to best inform the non-technical public, including possible future roles of the Commission and the plant owners.

Figure 3-79 shows parts of a 3-page foldout which gives a good description of the TERSCHELLING PV/Wind Project. This brochure was issued in response to the Commission's request at the start of the TERSCHELLING follow-on project in late 1987. The contractor, Ecofys, had undertaken plant improvement, maintenance, and R&D tasks which included a public information dissemination effort. The brochure text is printed in both Dutch and English, and is a good model for others to follow. The following pilot plant owners or managers have prepared excellent brochures available to site visitors:

- PELLWORM   - GIGLIO
- VULCANO   - FOTA

Figure 3-80 shows examples of billboards at the PELLWORM, VULCANO, and ZAMBELLI pilot plants. Since these billboards are usually at the entrance, they provide the first impression of the solar facility to site visitors. Thus, these billboards are very important and should be well designed and laid out.

3.7 REFERENCES

[1] G. Beer, G. Chimento, F.C. Treble, and J.A. Roger, "Concerted Actions: PV Modules, Array, and Solar Sensors," CEC Contract EN3S-140-1, Conphoebus Final Report, December 1989.

[2] S. McCarthy, M. Hill, and A. Kovach, "Concerted Actions: Battery Management," CEC Contract EN3S-0137-IRL, NMRC Final Report, December 1989.

[3] J. Schmid and R. v. Dincklage, "Concerted Actions: Power Conditioning and Control," CEC Contract EN3S-0141-D, Fraunhofer Institute Final Report, December 1989.

[4] M.S. Imamura and E. Ehlers, "Concerted Action Project: Data/Plant Monitoring," WIP 89-7, CEC Contract E3S-0142-D, November 1989.

[5] A. Haenel, M.S. Imamura, and P. Helm, "Coordination and Control of PV Concerted Action Projects and Work on Specific Tasks: Data and Plant Monitoring, Array Structures, and Social Effects Study," Vol. III - ARRAY STRUCTURES. Final Report Work Performed Under CEC Contract EN3S-0142-D, November 1989.

[6] M.S. Imamura and P. Helm, "Concerted Action Project: Monitoring, Array Structures, and Social Effects," WIP-89-9, CEC Contract EN3S-0142-D, November 1989.

[7] Private communication with K. Krämer of BEWAG, November 1991.

[8] L. Barra, S. Catalanotti, F. Fontana, and F. Lavorante, "An Analytical Method to Determine the Optimal Size of a Photovoltaic Plant," Sol. Energy 33 (1984) 509.

[9] Sharp Brochure, "PV Systems," 1985.

[10] H. Saha, "Design of a Photovoltaic Electric Power System for an Indian Village," Int. J. Solar Energy, Vol. 27, No. 2, pp. 103-107, 1981.

[11] M.S. Imamura, R. Hulstrom, and C. Cookson, "Definition and Design of Residential Photovoltaic System," Martin Marietta Corp., Final Report MCR 76-394, Contract NAS 3-19768, 1976.

[12] M.S. Imamura and B. Khoshaim," A New Approach to Optimum Sizing and In-orbit Utilization of Spacecraft Photovoltaic Power System," Acta Astronautica, Vol. 15, No. 12, pp. 1019-1028, 1987.

Het Terschelling zonne- en windenergie project
The Terschelling solar and wind energy project

Novem

overzichtsfoto Terschelling-systeem
overview of the Terschelling system

Het zonne- en windenergieproject op het waddeneiland Terschelling is als Europees energie-demonstratieproject in 1983 opgezet. Dit voor Nederland unieke zonnecelsysteem wordt gebruikt voor energie-onderzoek en wordt financieel door NOVEM en de EG ondersteund.

The solar and wind energy project on the Frisian island of Terschelling was started in 1983 as an European energy pilot project. This solar cell system is unique in the Netherlands and, with financial support from Novem and the CEC, it is used for energy research.

foto windturbine
photo of wind turbine

foto accu-ruimte
photo of battery room

Een ander probleem bij Nederlandse PV-systemen met opslag in accu's is dat het accusysteem vaak zeer groot moet zijn om de elektriciteitsvraag zowel 's zomers als 's winters te dekken.

Een interessant aspect van het Terschelling-systeem is dat de benodigde accu-capaciteit veel kleiner is dan bij gebruik van enkel of een zonnegenerator of een windturbine. Reden hiervoor is dat het aanbod van zonne- en windenergie en daarmee de opbrengsten van de zonnegenerator en van de windturbine elkaar redelijk aanvullen.

De huidige windturbine is van Lagerwey en heeft een maximaal vermogen van 75 kW. De ashoogte van de molen is 30 m, zodat de molen voldoende boven de nabijgelegen bebouwing uitkomt. De molen heeft twee rotorbladen (wieken) met een spanwijdte van 15,6 m. De generator is van het zogenaamde asynchrone type.

Als de molen meer energie levert dan er op de zeevaartschool gebruikt wordt, kan dit overschot via een gelijkrichter in de accu's opgeslagen worden.

Systeemonderzoek

In een complex system als het Terschelling-systeem worden de energiestromen gestuurd met een regelcomputer. Daarnaast is er veel apparatuur aanwezig voor energiemetingen aan het systeem en voor meteorologische metingen. De meetgegevens worden elke minuut en elk uur opgeslagen. Met name dit uitgebreide meetsysteem maakt dat het Terschelling-systeem goede mogelijkheden voor onderzoek biedt.

Advies- en onderzoeksbureau ECO-FYS te Utrecht voert in samenwerking met andere onderzoeksinstellingen, zoals de Rijksuniversiteit Utrecht, het onderzoek aan het Terschelling zon- en windsysteem uit. Het onderzoek wordt gefinancierd door NOVEM Nederlandse Maatschappij voor Energie en Milieu bv te Utrecht en de Commissie der Europese Gemeenschappen (DG XII) te Brussel.

Er worden onder andere experimenten gedaan om de regelstrategie tijdens netgekoppeld bedrijf en tijdens autonoom bedrijf te testen en te optimaliseren.

Bij normaal bedrijf is het zon- en windsysteem aan het openbare elektriciteitsnet gekoppeld (net-bedrijf). Als er meer elektriciteit wordt geproduceerd

only a solar generator or only a wind turbine was used.

The present wind turbine is a Lagerwey and has a nominal power of 75 kW. The shaft of the windmill is 30 m above ground level, so that the mill rises sufficiently above the nearby buildings. The windmill has two rotor blades with a span of 15.6 m. The generator is of the so-called asynchronous type.

If the turbine produces more energy than is used on the nautical college, this surplus can be stored in the batteries through a rectifier.

System research

In a complex system like the Terschelling system, the energy flows are controlled by means of a control computer. In addition, the system itself has considerable measurement equipment both for energy measurements etc. and for meteorological measurements. The measurement data are stored each minute and each hour. This extensive measurement system in particular is one of the reasons why the Terschelling system offers good possibilities for research.

ECOFYS Research and Consultancy in Utrecht conducts the research into the Terschelling solar/wind system, in cooperation with other research institutes, such as State University of Utrecht. The research is financed by NOVEM Netherlands Agency for Energy and the Environment in Utrecht and the Commission of the European Communities (DG XII) in Brussels.

Experiments are conducted in order to test and optimize the control strategy in grid connected operation as well as in autonomous operation.

During normal operation the system operates grid connected. If the electricity produced exceeds the electricity needed by the college, the surplus can be

Figure 3-79. Example of a Pilot Plant Information Brochure (by NOVEM, Netherlands Agency for Energy and the Environment, and Ecofys in Utrecht, NL). The text is written in Dutch (left column) and English.

ZAMBELLI

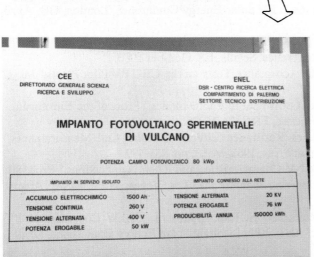

VULCANO

PELLWORM

Figure 3-80. Examples Showing PV Plant Billboards at the Plant Entrance

[13] M. Buresch, "Photovoltaic Energy Systems; Design and Installation," McGraw-Hill, New York, NY, USA, 1983.

[14] PRC Energy Analysis Compana, Solar Photovoltaic Applications Seminar: "Design, Installation, and Operation of Small Stand-alone Photovoltaic Power Systems," U.S. DOE/CS/32522-T1, 1980 (Dept. of Energy, Washington, DC, USA).

[15] R.N. Chapman and J.P. Fernandez, "User's Manual for SIZEPV: A Simulation Program for Stand-alone PV Systems," SAND 89-0616, March 1989.

[16] D.F. Menicucci and J.P. Fernandez, "User's Manual for PVFORM: A Photovoltaic System Simulation Program for Stand-alone and Grid-Interactive Applications," Sandia National Laboratories, SAND 85-0376, 1988.

[17] D.L. Evans and F.T.C. Bartels, Battery Sizing Criteria for Stand-alone Photovoltaic Power Systems," AS/ISES, Philadelphia, PA, USA, 1981.

[18] P. Groumpos and G. Papageorgiou, "An Optimal Sizing Method for Stand-alone Photovoltaic Power Systems," Int. J. Solar Energy, Vol. 38, No. 5, pp. 341-351, 1987.

[19] H.L. Macomber, J.B. Ruzek, F.A. Costello and staff of Bird Engineering, "Photovoltaic Stand-alone Systems: Preliminary Engineering Design Handbook," NASA CR-165352, NASA Lewis Research Center, 1981.

[20] C. Soras and V. Makios, "A Novel Method for Determining the Optimum Size of Stand-alone Photovoltaic Systems," Solar Cells, 25, pp.127-142, 1988.

[21] K. Träder, "PV Pilot Plants in Europe - Status of Operation," Int. J. Solar Energy, Vol. 4. pp. 281-295, 1986.

[22] P. Helm, "Manual of Photovoltaic Pilot Projects, Part A - Status and Improvements," WIP Report, CEC Contract EN3S-0007-D(B), Munich, FRG, February 1986.

[23] K. Krebs and M. Starr, "First Session of European Working Group on Photovoltaics," Proc. of Meeting in Ispra, IT 14-16 September 1985

[24] K. Krebs and M. Starr, "Second Session of European Working Group on Photovoltaics," Proc. of Meeting at Sophia Antipolis, FR, 10-12 March 1986.

[25] K. Krebs and M. Starr, "Third Session of European Working Group on Photovoltaics," Proc. of Meeting in Madrid, ES, 22-24 October 1986.

[26] K. Krebs and M. Starr, "Fourth Session of European Working Group on Photovoltaic," Proc. of Meeting in Lugano, CH, 1-3 April 1986.

[27] K. Krebs and M. Starr, "Fifth Session of European Working Group on Photovoltaics," Proc. of Meeting in Cork, IRL, 30 September-2 October 1986.

[28] G. Blaesser, K. Krebs, H. Ossenbrink, and E. Rossi, "Acceptance Testing and Monitoring the CEC PV Plants," Proc. of the 6th European PV Solar Energy Conference, London, GB, April 1985.

[29] G. Blaesser and K. Krebs, "Summary of PV Pilot Plant Monitoring Data 1984-1985," Proc. of the 7th European PV Solar Energy Conference, Seville, ES, October 1986.

[30] F.C. Treble, "Lessons Learned from the Acceptance Tests on the CEC PV Pilot Plants," Int. J. Solar Energy, 1985, l. 3, pp. 109-122.

[31] K.H. Krebs, "Quality and Performance Assessment in Photovoltaics," Proc. of the Euroforum - New Energies Congress, Vol. 1, Saarbrücken, FRG, 24-26 October 1988.

[32] M. Wolf and H. Rauschenbach, "Series Resistance Effects on Solar Cell Measurements," Advanced Energy Conversion, Pergamon Press, 1963, Vol.3, pp. 455-479.

[33] M.S. Imamura and P. Brantzig, "Dark I-V Characteristics and Their Applications," Proc. 8th Intersociety Energy Conversion Engineering Conference, Las Vegas, NEV, USA, August 1972.

[34] F.S. Huraib, M.S. Imamura, N. Eugenio, and N. Rao, "Status of 350-kW Concentrator PV System After Five Years," Proc. of the 7th EC Photovoltaic Solar Energy Conference, Seville, ES, 27-31 October 1986.

[35] A.B. Maish and J.L. Chamberlain, "PV Concentrators Today and Tomorrow," Proc. of the 10th European PV Solar Energy Conference, Lisbon, Portugal, 6-12 April 1991.

[36] Unpublished data, USA-Saudi Arabia SOLERAS 350-kW Concentrator PV Project, Martin Marietta Corp., Denver, CO, USA, 1987.

[37] R.D. Miller and K. Zimmermann, "Low Cost Solar Array Project, Phase II. Final Report," The Boeing Company, U.S. DOE Contract No. NAS 7-100-954833, April 1981.

[38] A Critical Comparative Analysis of Support Structure Costs, WIP, March 1982.

[39] U. Beyer and R. Pottbrock, "Design, Construction, and Operation of a 340-kWp Photovoltaic Plant," Proc. of the 9th European PV Solar Energy Conference, Freiburg, FRG, September 1989.

[40] B. Mortensen and M. Jorgensen, "Experiments on Battery Storage," CEC Contract ESC-R-097-DK, Final Report, Jydsk Telephone, October 1989.

[41] Private communication with K. Krämer of BEWAG on their 792-cell battery string multi-MW peak-power plant, June 1991.

[42] M.S. Imamura and R. Donovan, "Cell Level/Discharge Protection System," Proc. of the 12th IECEC, Washington, D.C., USA, September 1977.

[43] M.S. Imamura, R.L. Donovan, L.A. Shelly, "Microprocessor-based Battery Protection System," Proc. of the 10th IECEC, Newark, DE, USA, August 1975.

[44] M.S. Imamura, "PV System Technology Development in the European Community," Proc. of the Euroforum - New Energies Congress, Vol. 1, Saarbrücken, FRG, 24-28 October 1988.

[45] Private Communication with G. Zamboni of AGSM (Verona, IT) and A. Sorokin of TEAM (Rome, IT) and on the battery R&D at the ZAMBELLI PV Pumping Station, July 1991.

[46] A.A Salim, "A Simplified Minimum Power Dissipation Approach to Regulate the Solar Array Output Power in a Satellite Power Subsystem," Proc. of the 11th Intersociety Energy Conversion Engineering Conference, Lake Tahoe, NEV, USA, Sept. 1976.

[47] A.A. Salim, "Regulation and Control of Solar Array Power — A Minimum Power Dissipation Approach," Proc. of the 1st International Telecommunication Energy Conference, Wash., DC, USA, 25-27 Oct. 1978.

[48] F. Huraib, M.S. Imamura, and A.A. Salim, "Design, Installation, and Initial Performance of 350-kW Concentrator PV Plant for Saudi Arabian Villages," Proc. of the 4th European PV Solar Energy Conference, Stresa, IT, 10-14 May 1982.

[49] A. Goetzberger and J. Schmid, "Recent Progress of PV Systems of Intermediate Power Range," Presented at the 3rd International Photovoltaic Science and Engineering Conference, Tokyo, JAP, 3-6 November 1987.

[50] U. Beyer, B. Dietrich, R. Hotopp, and R. Pottbrock, "Operating Results of the 340-kWp Plant, Kobern-Gondorf, and Design and Construction of the 330-kWp, Neurather See," Proc. of the 10th European PV Solar Energy Conference, Lisbon, Portugal, 6-10 April 1991.

[51] Private communication with Dr. M. Tortoreli and her unpublished data on harmonic distortions of inverters and harmonic propagation in the grid network, December 1991.

[52] A. O'Riordan, S. McCarthy, and G.T. Wrixon, "PV Systems Research at the 50 kWp PV Installation on Fota Island," Proc. of the CEC/DGXII Contractors Meeting on PV Systems, Munich, FRG, 28-29 October 1991.

[53] A.T. Veltmann, "Inverter Power Quality Measurements at ECN," Proc. of the PV/Utility Interface Working Group Meeting at Foggia, IT, on 4-5 December 1991.

[54] M. Imamura and P. Sprau, "Status of the PV Monitoring Project, MUD: German National PV-Programme," presented at the European Working Group Meeting on PV Monitoring, Berlin, FRG, 6-7 June 1991.

[55] C. Brambilla and A. Iliceto, "A Low-cost Measurement and Data Acquisition System for Small PV Power Plants," Proc. of the 8th European PV Solar Energy Conference, Florence, IT, May 1988.

[56] A.J. Drummond, et al: "Radiation Instruments and Measurements," Part VI, IGY Instruction Manual, Pergamon Press, New York, USA, 1958.

[57] J.R. Latimer, "Radiation Measurements," International Field Year for the Great Lakes, Technical Manual Series No. 2, Ottawa, Canada, 1972.

[58] J.E. Hay and D.I. Wardle, "An Assessment of the Uncertainty in Measurements of Solar Radiation," Vol. 29, No. 4, pp. 271-278, 1982.

[59] G.A. Zerlaut, "The Calibration of Pyrheliometers and Pyranometers for Testing Photovoltaic Devices," SOLAR CELLS, 7 (1982-1983).

[60] M. Grottke, M.S. Imamura, W. Bechteler, and O. Mayer, "Comparative Assessment of Silicon Cell Pyranometers," Proc. of the 10th European PV Solar Energy Conference, Lisbon, Portugal, April 1991.

[61] M.B. Mahfood and M.S. Imamura, "Long-term Characteristics of Si and Thermopile Devices for Solar Irradiance Measurements," Proc. of the 7th European PV Solar Energy Conference, Seville, ES, October 1986, p. 349.

[62] F. Chenlo, N. Vela, and J. Olivares, "Comparison Between Pyranometers and Encapsulated Solar Cells as Reference PV Sensors. Outdoor Measurements in Real Conditions," Proc. 21st IEEE PV Conf. Orlando, FL, USA, May 1990.

[63] Private Communication with R. Diamond and J. Albeck of Spectrolab (Sylmar, CA, USA), November 1990.

[64] "Terrestrial Photovoltaic Measurement Procedures," NASA Tech. Memo TM 73702, Lewis Research Center, NASA, Cleveland, OH, USA, 1977.

[65] "Standard Method for Calibration and Characterization of Non-concentrator Terrestrial Photovoltaic Reference Cells under Global Irradiation," ASTM Standard E 1039-85

[66] "Standard Specification for Physical Characteristics of Non-concentrator Terrestrial Photovoltaic Reference Cells," ASTM Standard E 1040-84, 1984.

[67] CEI/IEC Publication 904-2 (1989): "Photovoltaic Devices — Part 2:   Requirements for Reference Solar Cells."

[68] M.S. Imamura, M.B. Mahfood and M. Hussain, "Simplified Calibration Method for Silicon Solar Irradiance Sensors," Proc. of the 7th European PV Solar Energy Conference, Seville, ES (1986) p. 54.

[69] F.C. Treble, "The Calibration of Primary Terrestrial Flat-plate Reference Devices — A Proposed International Standard," Proc. of the 7th European PV Solar Energy Conference, Seville, ES, 1986, p. 89.

[70] Private communication with Dr. S. Guastella of Conphoebus (Catania, Sicily) in May 1992 on their unpublished data under CEC's Concerted Action on PV Array and Solar Irradiance Sensors.

[71] Commission of the European Communities, Catalog of JOULE Programme 1989-1992, Contract JOUR0013, Systems Development Concerted Actions, 1991.

[72] M.S. Imamura, P. Helm, G. Beer, S. Guastella, and G. Chimento, "A Novel Method of Monitoring Global Normal Irradiance with a Stationary Sensor," Int. J. Solar Energy, 1992, Vol. 11, pp. 211-217.

[73] F.C. Treble, J.M. Tücke, and K. Träder, "Recommendations for Data Monitoring and Processing," presented at the Final Design Review Contractor Meeting, Brussels, BE, 30 November-2 December 1981.

[74] M. Montero Bartolomé, "Social and Anthropological Aspects in the Implementation of PV Systems," CEC Concerted Action on Social Benefits and Effects, under contract to WIP, January 1990.

[75] G. Blaesser, K. Krebs, and W.J. Zahiman, "Power Measurements at the 80-kW PV Plant Vulcano," Technical Note No. I.88.54, CEC-JRC, Ispra, IT, May 1988.

[76] G. Blaesser, W.J. Zahiman, "Power Measurements at the 50-kW PV Plant Nice," Technical Note No. I.88.43, CEC-JRC, Ispra, IT, April 1988.

Chapter 4

PHOTOVOLTAIC PILOT PLANTS: OPERATING EXPERIENCES AND LESSONS LEARNED

The main intent of this chapter is to present a brief summary of operating experiences and lessons learned from the Pilot I and II PV projects. These projects were co-financed by the Commission of the European Communities within the framework of the non-nuclear R&D programmes from 1981 to 1989. After plant overview summary, the discussion covers the following topics, patterned after those of Chapter 3:

- Operation and performance

- Monitoring

- Maintenance and safety

4.1   PV PLANT CHARACTERISTICS

The three groups of PV plants co-financed by DG XII during the period 1981-1989 were the 16 original pilot plants, three PV-powered houses, and three new PV system applications. The first two groups are referred to as Pilot I, and the new system applications as Pilot II projects. The total number of projects, their installation dates, and total installed power are summarized in Table 4-1. These projects are briefly described in the appendices as indicated below:

Project	Date Installed	Installed PV Power (kW)	Appendices
- Pilot I plants	1981-1984	1,110	1-16
- PV residences	1983-1985	9.1	17-21
- Pilot II plants and subsystem development	1988-1989	23.6	22-25

Table 4-1.   Summary of Installed Plants in CEC Pilot I and II Programmes (1980-1989)

| Group | Number of Projects | Operational Modes | | | Installation Date | Total PV Power (kWp) |
		Both S-A and G-C	S-A Only	G-C Only		
Pilot I Programmes						
- Pilot plants	16	6	10	0	1982-1984	1,110
- PV residences	5	1	1	3	1984-1985	9.1
Pilot II Programmes						
- New system application*	3	2	1	0	1988-1989	23.6*
					Total:	1,133

*   Does not include the 50-kW PV for the combined thermal/electric plant to be installed by 1992 by Stadtwerke in
    Saarbrücken, FRG
S-A - Stand-alone
G-C - Grid-Connected

The different design hardware approaches used and the main attributes of the original 16 pilot plants are as follows:

- System type: 10 stand-alone and grid-connected; 6 stand-alone only; 2 with wind generators; 4 with Diesel or gas generators

- PV array: Up to 420 Vdc, all fixed-tilted flat plate, monocrystalline and polycrystalline Si cell modules

- Battery: 90 to 2,500 Ah cell size; 110 to 346 Vdc nominal discharge voltage; up to 173 cells in series; cell type: all Pb-acid, flooded, and vented with both flat pasted and tubular positive electrode construction

- Battery charge control: 7 with series regulators; 7 with array string shedding; 2 with simple on-off control of the array

- Boost regulator: Used in 2 plants (2- and 25-kW units)

- Inverter: in 15 plants (6 to 450 kVA; most plants have self-commutated inverters)

Figures 4-1 to 4-19 show the simplified block diagrams of various pilot plants, which are standardized to facilitate comparisons and for ready reference. Note that most of the figures reflect the original system configuration. The improved versions are included, e.g., AGHIA ROUMELI and PELLWORM, in Figures 4-2 and 4-13, respectively.

Tables 4-2 to 4-6 list key features of the PV plants in the Pilot I Programme. The description has been standardized in these tables in terms of the following parameters:

- Plant information including project cost   - Converter/charge controller
- Site characteristics, including solar   - State of charge calculator
  irradiance and meteorological data   - Rectifier
- System design characteristics   - UPS
- PV array   - Energy meters
- Battery   - Control unit
- Inverter

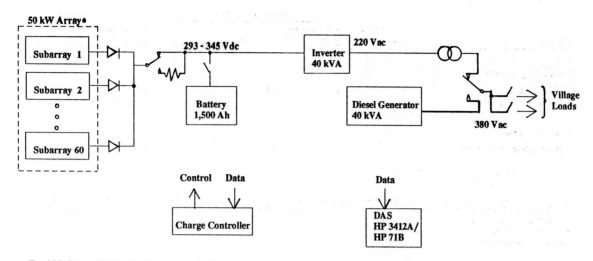

Figure 4-1.   Power Flow Diagram of the Original AGHIA ROUMELI PV Pilot Plant
(see Fig. 4-2 for the modified charger)

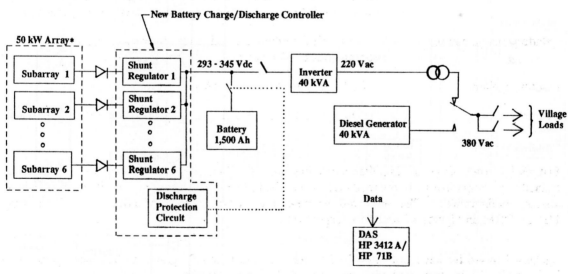

Figure 4-2.   Power Flow Diagram of the AGHIA ROUMELI PV Pilot Plant Showing
Improved Battery Charge/Discharge Controller Installed in May 1988

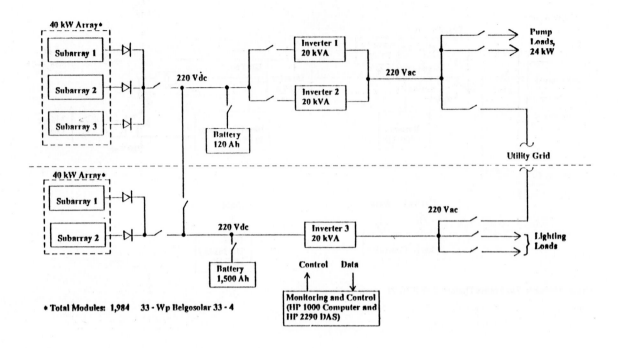

Figure 4-3. Power Flow Diagram of the CHEVETOGNE PV Pilot Plant

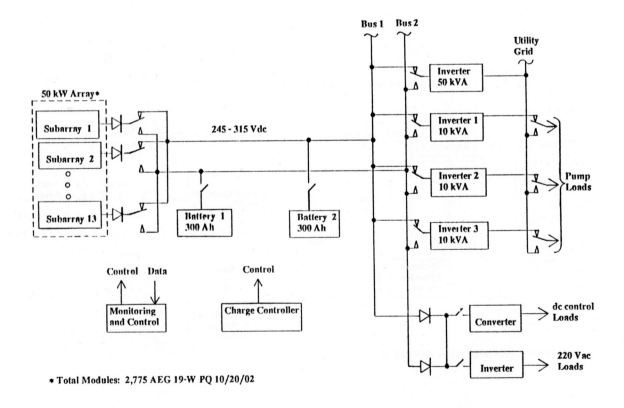

Figure 4-4. Power Flow Diagram of the Original FOTA PV Pilot Plant (see Appendix 3 for the Current Configuration which is grid-connected and rearranged into a single-bus system).

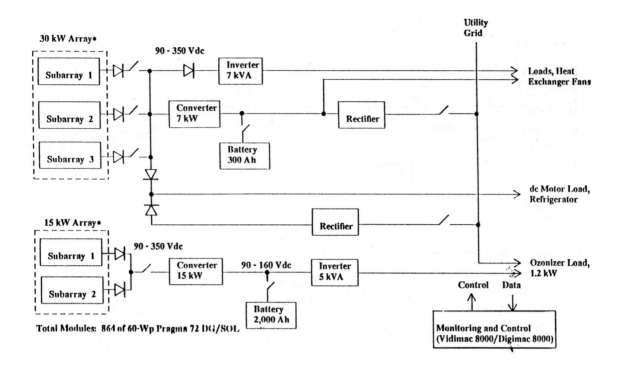

Figure 4-5.   Power Flow Diagram of the GIGLIO PV Pilot Plant

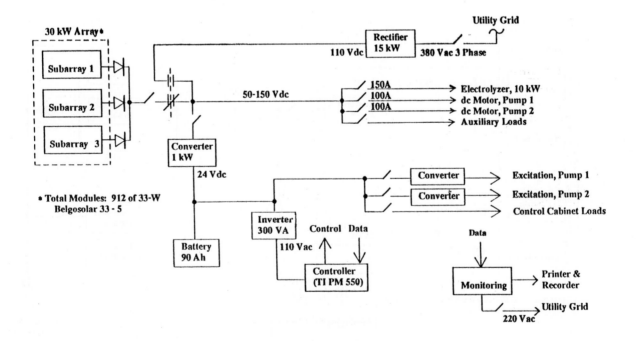

Figure 4-6.   Power Flow Diagram of the HOBOKEN PV Pilot Plant

187

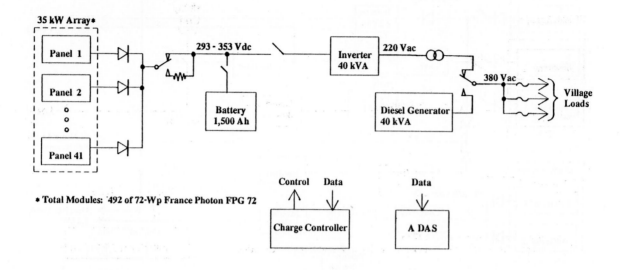

Figure 4-7. Power Flow Diagram of the KAW PV Pilot Plant

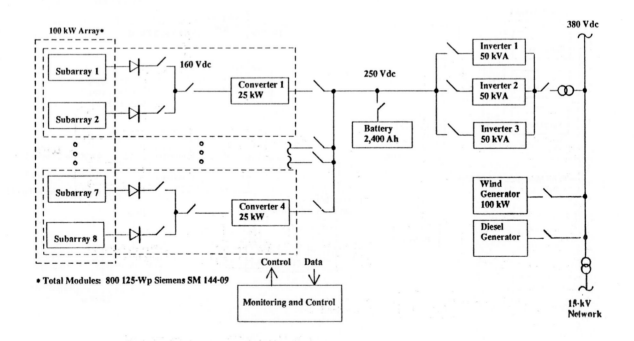

Figure 4-8. Power Flow Diagram of the KYTHNOS PV Pilot Plant

188

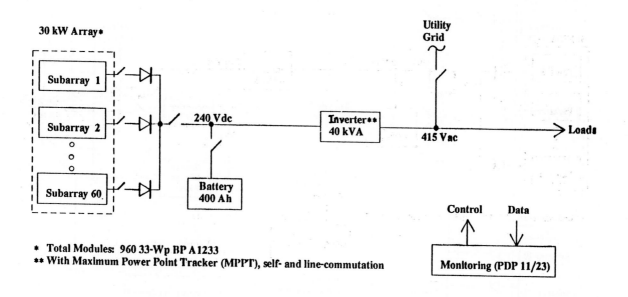

* Total Modules: 960 33-Wp BP A1233
** With Maximum Power Point Tracker (MPPT), self- and line-commutation

Figure 4-9.   Power Flow Diagram of the MARCHWOOD PV Pilot Plant

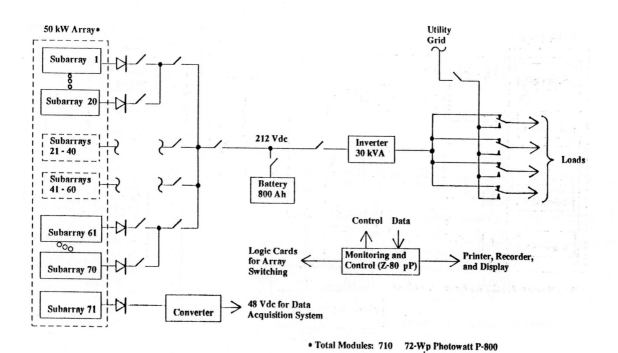

* Total Modules: 710   72-Wp Photowatt P-800

Figure 4-10.   Power Flow Diagram of the MONT BOUQUET PV Pilot Plant

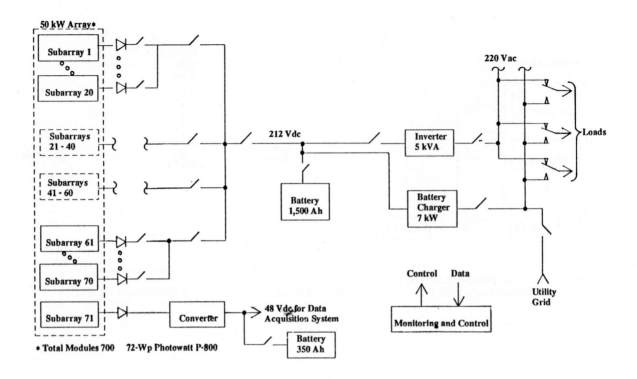

Figure 4-11. Power Flow Diagram of the NICE PV Pilot Plant

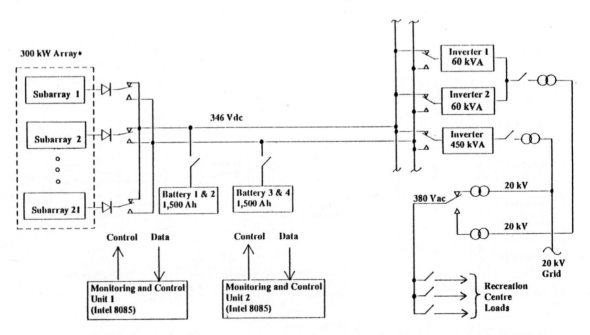

Figure 4-12. Power Flow Diagram of the Original PELLWORM PV Pilot Plant

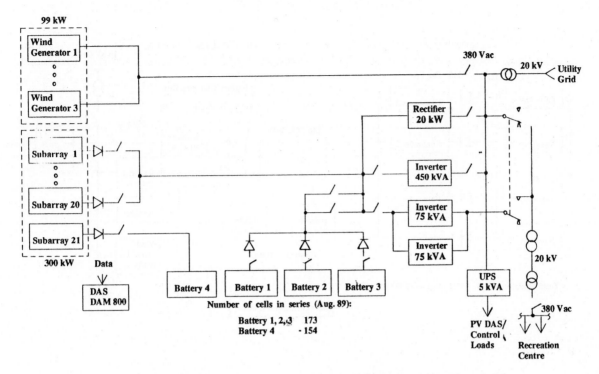

Figure 4-13. Power Flow Diagram of the Improved PELLWORM PV Pilot Plant (mid-1988). See Fig. 4-12 for the original 2-bus configuration.

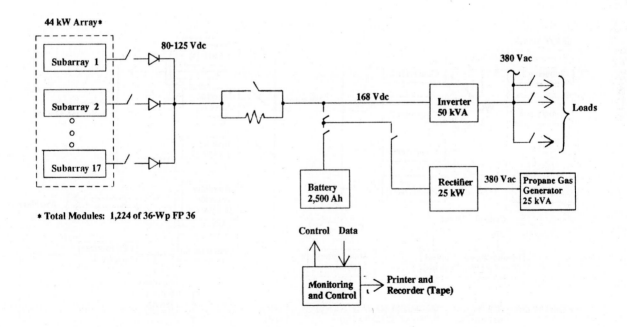

Figure 4-14. Power Flow Diagram of the Improved RONDULINU PV Pilot Plant

191

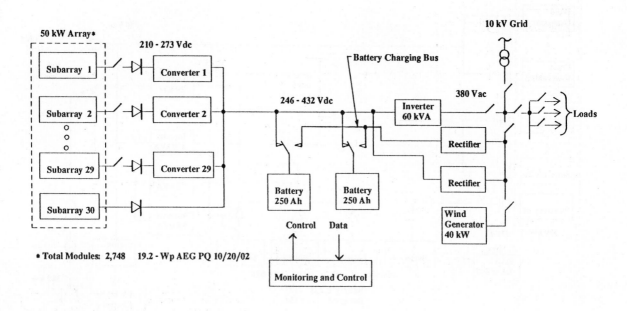

Figure 4-15. Power Flow Diagram of the TERSCHELLING PV Pilot Plant

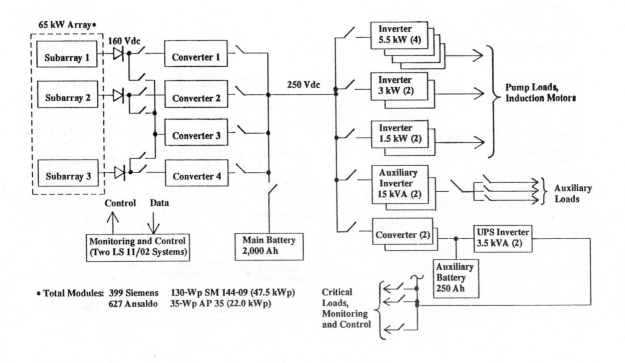

Figure 4-16. Power Flow Diagram of the TREMITI PV Pilot Plant

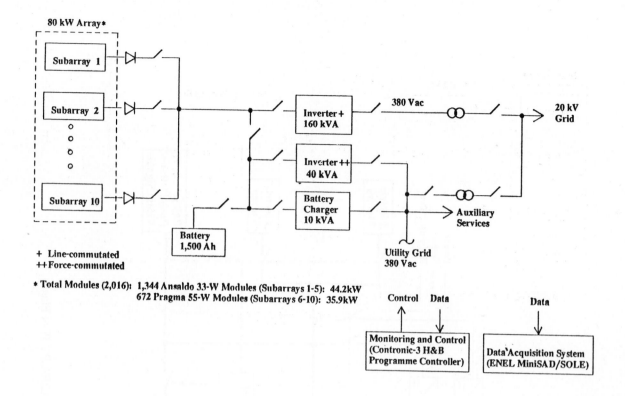

Figure 4-17. Power Flow Diagram of the VULCANO PV Pilot Plant

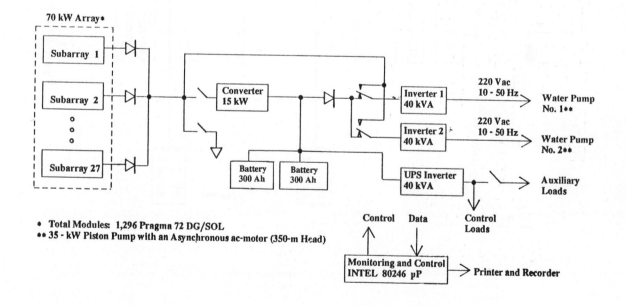

Figure 4-18. Power Flow Diagram of the ZAMBELLI PV Pilot Plant

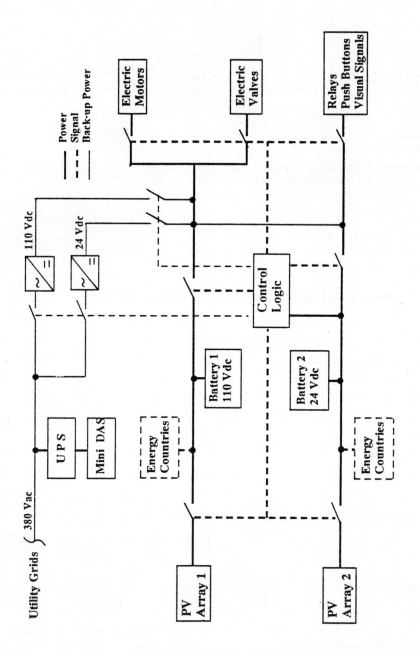

Figure 4-19. Power Flow Diagram of ANCIPA PV Pilot Plant

Table 4-2.   Principal Features of the 16 PV Pilot Plants Sponsored by DG XII

PV Plant (Country)	Array Power Rating (kW)	Primary Application	System Type*	Array Bus Voltage (Vdc)	Battery Voltage (Vdc)	Battery Capacity (Ah)	Total Inverter Rating (kVA)+	Rectifier Rating (kW)	Converter Rating (kW)	Other Power Sources
AGHIA ROUMELI (GR)	50	Village	S-A	292-353	292-353	1,500	40	---	---	Diesel (40 kVA)
CHEVETOGNE (BE)	40/23	Swimming pool pumps/lighting	S-A	198-264	198-264	1,500 120	20 x 2/20	---	---	Grid
FOTA (IRL)	50	Dairy Farm Pumps	Both	245-315	234-312	300 (x2)	50 (L-C) 10 x 3	---	2	Grid
GIGLIO (IT)	30/15	Refrigeration/ water treatment	S-A	90-360/ 90-350	228-350/ 90-160	2,000 300	7/ 5.7	7	7/ 15	Grid
HOBOKEN (BE)	30	Electrolyzer/pumps	S-A	50-150	22-29	90	---	14	1	Grid
KAW (FR)	35	Village	S-A	292-353	292-353	1,500	40	---	---	Diesel (40 kVA)
KYTHNOS (GR)	100	Village grid	Both	128-192	225-300	2,400	50 x 3	---	25 x 4	Diesel (1 MW) Wind (100 kW)
MARCHWOOD (GB)	30	Grid/DAS	Both	216-295	216-288	400	40 (L-C)	---	---	Grid
MONT BOUQUET (FR)	50	TV/FM transmitter	S-A	191-260	191-254	800	30	---	---	Grid
NICE (FR)	50	Air traffic control equipment	S-A	191-260	191-254	1,500	5	7	---	Grid
PELLWORM (FRG)	300	Recreation and grid center	Both	230-415	311-415	1,500 (x4)	75 x 2 450 (L-C)	20	---	Wind (3 x 30 kW)
RONDULINU (FR)	65	Village	S-A	80-125	151-202	2,500	50	25	---	Propane-gas (25 kVA)
TERSCHELLING (NL)	50	Merchant marine school	Both	210-273	346-432	250 (x2)	60	20 x 2	2 x 29	Wind (75 kW)
TREMITI (IT)	65	Desalination/pumps	S-A	128-192	225-300	2,000 250	61 **	---	20 x 4	---
VULCANO (IT)	80	Village grid	Both	240-420	240-310	1,500	140 (L-C) 50	10	---	Grid
ZAMBELLI (IT)	70	Water pumping	S-A	200-320	200-270	300 (x2)	40 x 2 ++	---	15	---

Note: Roof-mounted array at FOTA, HOBOKEN, MUNICH, and NICE only.

*   stand-alone (S-A), Grid-Connected (GC).
+   All inverters are self (or force) commutated, except those marked "(L-C)" which are line-commutated types.
** 4 - 5.5 kVA, 2 - 3 kVA, 2 - 1.5 kVA, 2 - 15 kVA
++ Variable frequency.

Table 4-3. Key Features of PV Pilot Plants, AGHIA ROUMELI to GIGLIO

PARAMETER	UNITS	AGHIA ROUMELI	CHEVETOGNE	FOTA	GIGLIO
Plant Information					
- Location, Country	---	Crete Is., GR	Rochefort, BE	Fota Is., IRL	Giglio Is., IT
- Operational Date	---	Oct '82	Nov '82	Jun '83	Jun '84
- Plant Designer	---	Seri Renault	IDE	NMRC	Pragma
- Plant Owner	---	PPC	Domaine Provincial	NMRC	ENEA
- Plant Operator	---	PPC	ACEC	NMRC	ENEA
- Cost	ECU	1.1 M ('83)	1.22 M ('83)	1.30 M ('83)	1.62 M ('84)
Site Characteristics					
- Latitude	Deg N	35.2	49.9	51.7	42.9
- Longitude	Deg E,W	33 E	2.1 E	8.8 W	12 E
- Altitude	m	sea level	sea level	sea level	300
- Horizontal Irradiance					
o Sensor/Type	---	K&Z CM11		Eppley PSP	K&Z CM11
o Annual Energy	kWh/m^2	1,700	1,000	1,100	1,500
o Daily Avg Energy	kWh/m^2	4.66	2.74	3.01	4.11
- Tilted Irradiance					
o Sensor/Type	---	Haenni 118		Eppley PSP	Haenni 118
o Annual Energy	kWh/m^2				1,787
o Daily Avg Energy	kWh/m^2			3.20	4.90
- Amb. Air Temp., max/min	°C	38/4	30/-18	28.9/-6.7	35/-1
- Wind speed, max	m/s	33	38	32	33
System Design Characteristics					
- Operational Mode	---	Stand-alone	Both	Both	Stand-alone
- Array Power, Rated	kW	50	63	50	45
- Total Inverter Power, Rated	kVA	40	60	80	12
- Total Battery Energy, Rated	kWh	450	356	160.8	312
- Array Bus Voltage Range	Vdc	290-390	220	245-315	90-350
- Battery Voltage Range	Vdc	292-360	198-264	268	90-160
- Array Field Area	m^2	3,800	2,100	717	3,000
- Array Tilt Angle	deg.	25	30	45	25
- No. Series Cells/String	---	816	576	720	576
- Total No. Solar Cells	---	48,960	71,424	55,500	62,208
PV Array					
- Module Manufacturer	---	France Photon	Belgosolar	AEG	Pragma
- Module Part No.	---	FPG 72	33-4	PQ 10/20	72 DG
- Module Power, Rated	W	68	33	19.2	57
- Module Area	m^2	0.824	0.474	0.258	0.855
- No. Modules	---	720	1,984	2,775	864
- Area of all Modules	m^2	593	940	715	738
- Solar Cell Material	---	mono	mono	poly	mono
- Solar Cell Dimension	mm	100 Dia	100 Dia	100 x 100	100 Dia
- No. Cells per Module	---	68	36	20	72

(Page 1 of 3)

Table 4-3.   (Cont.)

PARAMETER	UNITS	AGHIA ROUMELI	CHEVETOGNE	FOTA	GIGLIO
Battery					
- Cell Manufacturer	---	Varta	Oldham	Varta	Varta
- Cell Part No.	---	Vb 2415		Vb 2306	Vb 2420 : Vb 2306
- No. Series Cells/Battery	---	150	110 : 110	134	60 : 120
- Cell Capacity, Rated	Ah	1,500	120 : 1,500	300	2,000 : 300
- Total Energy, Rated	kWh	450	26.4 : 330	80.4	240 : 72
- No. Batteries	---	1	1 : 1	2	1 : 1
Inverter					
- Manufacturer	---	Jeumont Schneider	ETCA	AEG	Silectron
- Part/Model	---				
- Output Power, Rated	kVA	40	20	50 : 10	5 : 5
- Commutation Type	---	Self	Self	Line : Self	Self : Self
- Phase/Frequency	---/Hz	3/50	3/50	3/50 : 3/50	1/50 : 1/50
- Quantity	---	1	3	1 : 1	1 : 3
- Input Voltage Range	Vdc	293-353	198-264	245-315	200-340 : 120-330
- Peak Power Tracking	---	No	Yes	No	Yes : Yes
Converter				(none)	
- Manufacturer	---				Silectron
- Part No.	---				
- Quantity	---				1 : 1
- Output Power, Rated	kW dc				15 : 7
- Input Voltage	Vdc				90-350 : 150-370
- Output Voltage	Vdc				90-160 : 288
- Control Type (Primary)	---	Constant V	Constant V	Constant V	Constant V
- Charge Voltage Limit (CVL)	---	2.3 to 2.5 V per cell			
- CVL Adjustable	---	Yes		Yes	Yes
State of Charge Calculator			(none)		
- Manufacturer	---	Varta		Varta	Silectron
- Part/Model	---	Logistronic		Logistronic	
- Quantity	---	1		1	2
Rectifier		(none)	(none)	(none)	
- Manufacturer	---				
- Part No.	---				
- Quantity	---				1
- Output Power Rating	kW				10.35
- Input Voltage	Vac				220
- Output Voltage	Vac				220
UPS		(none)	(none)		(none)
- Manufacturer	---			AEG	
- Part/Model	---			Irva Inverter	
- Quantity	---			1	
- Output Power Rating	kVA			1	
- Input Voltage	Vdc			245-315	
- Output Voltage	Vac			220	

(Page 2 of 3)

Table 4-3. (Concl.)

PARAMETER	UNITS	AGHIA ROUMELI	CHEVETOGNE	FOTA	GIGLIO
Data Acquisition Unit					
- Manufacturer	---	HP	HP	Campbell Scientific	Digimac
- Part/Model	---	3052	2290	21X	8000
- No. Parameters Measured	---			18	
- Storage	---	FD		RAM/HD	
- Transmission	---	No	No	Yes	
Energy Meters					(none)
- Manufacturer	---	Schlumberger		Siemens	
- Quantity	---	1 (ac)		1 (ac)	
- Display Type	---	Mech. counter		Mech. counter	
Control Unit					
- Manufacturer	---	Seri-Renault		AEG	Digimac
- Part/Model	---			Geamic	Vidimac
- Computer	---	HP 1000			
Auxiliary Generator		Diesel	(none)	(none)	(none)
- Manufacturer	---				
- Part/Model	---				
- Quantity	---	1			
- Output Power Rating	kVA	40			

(Page 3 of 3)

Table 4-4. Key Features of PV Pilot Plants, HOBOKEN to MARCHWOOD

PARAMETER	UNITS	HOBOKEN	KAW	KYTHNOS	MARCHWOOD
Plant Information					
- Location, Country	---	Antwerp, BE	French Guyana, FR	Kythnos Is., GR	Leatherhead, GB
- Operational Date	---	Jun '83	Jan '83	Jun '83	Nov '83
- Plant Designer	---	ENI	Seri Renault	Siemens	BP Solar
- Plant Owner	---	Gensun	Commune di Regina	PPC	BP Solar
- Plant Operator	---,	ENI	EDF	PPC	BP Solar
- Cost	ECU	0.62 M ('83) BF 24.8 M	0.82 M ('82) FF 5.18 M	1.99 M ('83)	2.29 M ('83) £ 1.3 M
Site Characteristics					
- Latitude	Deg N	51.4	4.5	37.4	51
- Longitude	Deg E,W	3 E	52 W	24.4 E	1.4 W
- Altitude	m	sea level		134	sea level
- Horizontal Irradiance					
o Sensor/ Type	---	K&Z CM10	K&Z CM11	K&Z CM15	
o Annual Energy	kWh/m^2	950	1,500	1,700	
o Daily Avg Energy	kWh/m^2	2.6	4.11	4.66	
- Tilted Irradiance					
o Sensor/Type	---	Haenni 118	Haenni 118	K&Z CM5	
o Annual Energy	kWh/m^2				
o Daily Avg Energy	kWh/m^2				
- Amb. Air Temp., max/min	$^\circ$C	35/-15	35/20	40/-2	27/-5
- Wind speed, max	m/s	27	28	31	42
System Design Characteristics					
- Operational Mode	---	S-A*	S-A*	S-A & G-C*	S-A & G-C*
- Array Power, Rated	kW	30	35	100	30
- Total Inverter Power, Rated	kVA	None	40	150	40
- Total Battery Energy, Rated	kWh	2.16	450	600	96
- Array Bus Voltage Range	Vdc	110	292-353	160	240
- Battery Voltage Range	Vdc	22-29	292-353	225-300	240
- Array Field Area	m^2	415	1,567	7,500	3,000
- Array Tilt Angle	deg.	30	5	35	55
- No. Series Cells/String	---	304	816	360	576
- Total No. Solar Cells	---	34,656	33,456	123,840	34,560
PV Array					
- Module Manufacturer	---	Belgosolar	France Photon	Siemens	BP Solar
- Module Part No.	---	33-4	FPG 72	SM 144-09	A 1233
- Module Power, Rated	W	33	72	125	33
- Module Area	m^2	0.474	0.824	1.5	0.442
- No. Modules	---	912	492	860	960
- Area of all Modules	m^2	432	405	1,200	424
- Solar Cell Material	---	mono	mono	mono	mono
- Solar Cell Dimension	mm	100 Dia	100 Dia	100 Dia	100 Dia
- No. Cells per Module	---	38	68	144	36

* S-A: Stand-alone
   G-C: Grid-connected

(Page 1 of 3)

Table 4-4. (Cont.)

PARAMETER	UNITS	HOBOKEN	KAW	KYTHNOS	MARCHWOOD
Battery					
- Cell Manufacturer	---	Oldham	Varta	Varta	Lucas
- Cell Part No.	---		Vb 2415	Vb 2412	'P' Series
- No. Series Cells/Battery	---	12	150	125	110
- Cell Capacity, Rated	Ah	90	1,500	1,200	400
- Energy, Rated	kWh	1.16	450	600	96
- No. Batteries	---	1	1	1	1
Inverter					
- Manufacturer	---		Jeumont-Schneider	Siemens	AEG
- Part/Model	---				
- Output Power, Rated	kVA		40	50	40
- Commutation Type	---		Self	Self	Self/Grid
- Phase/Frequency	---/Hz		3/50	3/50	3/50
- Quantity	---		1	3	1
- Input Voltage Range	Vdc		293-353	250 + 20% - 10%	190-300
- Peak Power Tracking	---		No	Yes	Yes
Converter				(none)	
- Manufacturer	---	Polyamp	Seri-Renault	Siemens	BP Solar
- Part No.	---	50-150/375			
- Quantity	---	1		4	
- Output Power, Rated	kW dc	1		25	
- Input Voltage	Vdc	50-150		160	
- Output Voltage	Vdc	24 ± 5%		250 + 20% - 1%	
- Control Type (Primary)	---	Constant V	Constant V	Constant V	Constant V
- Charge Voltage Limit (CVL)	---	2.3 to 2.5 V per cell	2.3 to 2.4 V per cell	2.3 to 2.4 V per cell	
- CVL Adjustable	---	Yes	Yes	Yes	Yes
State of Charge Calculator		(none)			(none)
- Manufacturer	---		Varta	Varta	
- Part/Model	---		Logistronic	Logistronic	
- Quantity	---		1	1	
Rectifier			(none)		(none)
- Manufacturer	---	Van de Weygaerde			
- Part No.	---	1			
- Quantity	---	14			
- Output Power Rating	kW	380			
- Input Voltage	Vac				
- Output Voltage	Vdc				
UPS		(none)	(none)	(none)	(none)
- Manufacturer	---				
- Part/Model	---				
- Quantity	---				
- Output Power Rating	kW				
- Input Voltage	Vdc				
- Output Voltage	Vac				

(Page 2 of 3)

Table 4-4.   (Concl.)

PARAMETER	UNITS	HOBOKEN	KAW	KYTHNOS	MARCHWOOD
Data Acquisition Unit					
- Manufacturer	---	Thomson ENI	Elsyde	HP	Micro Consultants
- Part/Model	---		ADLAS	3021	IRIS/PDP II-25
- No. Parameters Measured	---				24
- Storage	---				2 x 5.25" FD
					2 x 12" HD
Energy Meters					(none)
- Manufacturer	---	Schlumberger	Schlumberger	Siemens	
- Quantity	---	1 (ac)	3 (ac)	1 (ac)	
- Display Type	---	Mech. counter	Mech. counter	Mech. counter	
Control Unit					
- Manufacturer	---	TI	Seri-Renault	Siemens	DEC
- Part/Model	---	PM 550		B8011	
- Computer	---				PDP 11/23
Auxiliary Generator		(none)	Diesel	Diesel	(none)
- Manufacturer	---				
- Part/Model	---				
- Quantity	---		1	4	
- Output Power Rating	kVA		40	250	

(Page 3 of 3)

Table 4-5.   Key Features of PV Pilot Plants, MONT BOUQUET to RONDULINU

PARAMETER	UNITS	MONT BOUQUET	NICE	PELLWORM	RONDULINU
Plant Information					
- Location, Country	---	Nimes, FR	Nice, FR	Pellworm, FRG	Paomia, Corsica, FR
- Operational Date	---	Apr '83	Mar '83	Jul '83	Jun '83
- Plant Designer	---	Photowatt	Photowatt	AEG-Telefunken	Leroy-Somer
- Plant Owner	---	TDF	Aéroport International	City of Pellworm	SER
- Plant Operator	---	Photowatt	Photowatt	AEG	Leroy-Somer
- Cost	ECU	1.05 M ('83)	1.02 M ('83)	4.65 M ('83)	0.95 M ('83)
Site Characteristics					
- Latitude	Deg N	44.2	43.7°N	54.44	42.2
- Longitude	Deg E,W	4.3 E	7.4 E	7.2 E	8.6 E
- Altitude	m	607			440
- Horizontal Irradiance					
o Sensor/Type	---			K&Z CM11	K&Z CM5
o Annual Energy	kWh/m^2		1,600	1,100	1,600
o Daily Avg Energy	kWh/m^2		4.38	3.01	4.38
- Tilted Irradiance					
o Sensor/Type	---			AEG 4 x 4 cm	K&Z CM5
o Annual Energy	kWh/m^2				1,650
o Daily Avg Energy	kWh/m^2				4.52
- Amb. Air Temp., max/min	°C	30/-15	26/6	20/-10	35/0
- Wind speed, max	m/s	42	43	42	48
System Design Characteristics					
- Operational Mode	---	S-A*	S-A*	S-A & C-G*	S-A*
- Array Power, Rated	kW	50	50	300	44
- Total Inverter Power, Rated	kVA	30	5	600	50
- Total Battery Energy, Rated	kWh	169.6	318	2,076	420
- Array Bus Voltage Range	Vdc	212	212	346	160-250
- Battery Voltage Range	Vdc	212	212	311-415	168
- Array Field Area	m^2	2,000	3,000	28,500	1,500
- Array Tilt Angle	deg.	60	45	40	60
- No. Series Cells/String	---	720	720	960	512
- Total No. Solar Cells	---	50,400	50,400	351,360	44,064
PV Array					
- Module Manufacturer	---	Photowatt	Photowatt	AEG	France Photon
- Module Part No.	---	PW P-800	PW P-800	PQ 10/20	FPG 36
- Module Power, Rated	W	72	72	19.2	36
- Module Area	m^2			0.26	0.408
- No. Modules	---	710	700	17,568	1,224
- Area of all Modules	m^2			4,568	505
- Solar Cell Material	---	mono	mono	poly	mono
- Solar Cell Dimension	mm	100 Dia	100 Dia	100 x 100	100 Dia
- No. Cells per Module	---	72	72	20	36

* S-A: Stand-alone
  G-C: Grid-connected

(Page 1 of 3)

Table 4-5.   (Cont.)

PARAMETER	UNITS	MONT BOUQUET	NICE	PELLWORM	RONDULINU
Battery					
- Cell Manufacturer	---	Oldham	Oldham	Varta	Oldham
- Cell Part No.	---	SYT 8	SZT 12	Vb 2415	SZT
- No. Series Cells/Battery	---	106	106	173	84
- Cell Capacity, Rated	Ah	800	1,500	1,500	2,500
- Capacity, Rated	Ah	800	1,500	6,000	2,500
- Energy, Rated	kWh	170	318	2,076	420
- No. Batteries	---	1	1	4	1
Inverter					
- Manufacturer	---	Aérospatiale	Aérospatiale	AEG	Aérospatiale
- Part/Model	---				
- Output Power, Rated	kVA	30	30	75 : 450	50
- Commutation Type	---	Self	Self	Self : Line	Self
- Phase/Frequency	---/Hz	3/50	3/50	3/50 : 3/50	3/50
- Quantity	---	1	1	2 : 1	1
- Input Voltage Range	Vdc	180-260	180-260	346 +20% -15%	120-200
- Peak Power Tracking	---	No	No	No	No
Converter				(none)	(charger)
- Manufacturer	---				
- Part No.	---				
- Quantity	---				1
- Output Power, Rated	kW dc				
- Input Voltage	Vdc				80-125
- Output Voltage	Vdc				168
- Control Type (Primary)	---	Constant V	Constant V	Constant V	Constant V
- Charge Voltage Limit (CVL)	---			2.3 to 2.4 V per cell	
- CVL Adjustable	---	Yes	Yes	Yes	
State of Charge Calculator		(none)	(none)		(none)
- Manufacturer	---			Varta	
- Part/Model	---			Logistronic	
- Quantity	---			1	
Rectifier		(none)			
- Manufacturer	---		Aérospatiale		Leroy-Somer
- Part No.	---				L.S.
- Quantity	---		1	1	1
- Output Power Rating	kW		7	20	18
- Input Voltage	Vac		220-380		380
- Output Voltage	Vdc		180 & 260		165-200
UPS		(none)	(none)		(none)
- Manufacturer	---			AEG	
- Part/Model	---			6360/23/25FVS	
- Quantity	---			1	
- Output Power Rating	kW			5	
- Input Voltage	Vdc				
- Output Voltage	Vac				

(Page 2 of 3)

Table 4-5.   (Concl.)

PARAMETER	UNITS	MONT BOUQUET	NICE	PELLWORM	RONDULINU
Data Acquisition Unit					
- Manufacturer	---			AEG	ASITEC
- Part/Model	---			DAM 800	
- No. Parameters Measured	---			20	16
- Storage	---			RAM/HD at Wedel (via Modem)	Cassette Tape
Energy Meters					
- Manufacturer	---			Siemens and Berlin	
- Quantity	---			1 (ac), 1 (dc)	
- Display Type	---			Mechanical	
Control Unit					
- Manufacturer	---	Photowatt	Photowatt	AEG	Leroy-Somer
- Part/Model	---				
- Computer	---	Z-80		Intel 8085	(none)
Auxiliary Generator		(none)	(none)	(none)	Propane-gas
- Manufacturer	---				Peugeot
- Part/Model	---				CEB 25
- Quantity	---				1
- Output Power Rating	kVA				25

(Page 3 of 3)

Table 4-6. Key Features of PV Pilot Plants, TERSCHELLING to ZAMBELLI

PARAMETER	UNITS	TERSCHELLING	TREMITI	VULCANO	ZAMBELLI
Plant Information					
- Location, Country	---	Terschelling Is., NL	Tremiti Is., IT	Vulcano Is., IT	Verona, IT
- Operational Date	---	Jun '83	Mar '84	Oct '84	Jun '84
- Plant Designer	---	Holecsol	Italenergie	ENEL	Pragma
- Plant Owner	---	Hogere Zeevart-school	CPIM	ENEL	AGSM
- Plant Operator	---	Ecofys	Rossetti	ENEL	AGSM
- Cost	ECU	1.76 M ('84)	2.67 M ('84)	2.5 M ('83)	1.97 M ('83)
Site Characteristics					
- Latitude	Deg N	54	42.1	38	45
- Longitude	Deg E,W	5.2 E	16.5 E	15 E	11 E
- Altitude	m	Sea Level	60	350	830
- Horizontal Irradiance					
o Sensor/Type	---	K&Z		K&Z CM 5/6	Eppley PSP
o Annual Energy	kWh/m^2			1,549	1,200
o Daily Avg Energy	kWh/m^2			4.24	3.29
- Tilted Irradiance					
o Sensor/Type	---	AEG 5 x 5 cm		K&Z CM 5/6	Haenni 118
o Annual Energy	kWh/m^2		1,500	1,765	1,300
o Daily Avg Energy	kWh/m^2		4.11	4.84	3.6
- Amb. Air Temp., max/min	°C	20/-10	40/-5	37/-2	35/-10
- Wind speed, max	m/s		36	33	33
System Design Characteristics					
- Operational Mode	---	G-C*	S-A*	G-C & S-A	S-A
- Array Power, Rated	kW	50	65	80	70
- Total Inverter Power, Rated	kVA	60	42	40/160	80
- Total Battery Energy, Rated	kWh	180	500	390	130
- Array Bus Voltage Range	Vdc	210-273	160	230-420	100-300
- Battery Voltage Range	Vdc	360	250	240-310	190-260
- Array Field Area	m^2	600	1,800	4,000 : 4,000	10,825
- Array Tilt Angle	deg.	40	22	35	45
- No. Series Cells/String	---	600	360	768 : 576	576
- Total No. Solar Cells	---	54,960	57,456 : 22,572	48,384 : 48,384	93,312
PV Array					
- Module Manufacturer	---	AEG	Siemens : Ansaldo	Ansaldo : Pragma	Pragma
- Module Part No.	---	PQ 10/20	SM 144-09 : AP 35	AP 33D : 72 DG	72 DG/SOL
- Module Power, Rated	W	19.2	130 : 35	33 : 55	55
- Module Area	m^2	0.26	1.5 : 0.45	0.44 : 0.86	0.86
- No. Modules	---	2,748	399 : 627	1,344 : 672	1,296
- Area of all Modules	m^2	714	599 : 282	594 : 580	1,115
- Solar Cell Material	---	poly	mono : poly	poly : mono	mono
- Solar Cell Dimension	mm	100 x 100	100 Dia : 100 x 100	100 x 100 : 100 Dia	100 Dia
- No. Cells per Module	---	20	144 : 36	36 : 72	72

* G-C: Grid-connected
S-A: Stand-alone

(Page 1 of 3)

Table 4-6.   (Cont.)

PARAMETER	UNITS	TERSCHELLING	TREMITI	VULCANO	ZAMBELLI
Battery					
- Cell Manufacturer	---	Varta	Varta	Varta	Tudor
- Cell Part No.	---	Vb 2305	Vb 2420	Vb 2415	4TF 230/HC
- No. Series Cells/Battery	---	173	125	130	108
- Cell Capacity, Rated	Ah	250	2,000	1,500	300
- Total Energy, Rated	kWh	90	500	390	130
- No. Batteries	---	2	1	1	2
Inverter					
- Manufacturer	---	Holec	CO.EL.	Borri : ENEL and MARELLI	Silectron
- Part/Model	---			GFS 150 : D14461-010	
- Output Power, Rated	kVA	60	7.5 : 3	40 : 80	40
- Commutation Type	---	Line	Self : Line	Self : Line	Variable freq.
- Phase/Frequency	---/Hz	3/50	3/50 : 3/50	3/50 : 3/50	3/10-50
- Quantity	---	1	4 : 3	1 : 1	2
- Input Voltage Range	Vdc	324-432	250 : 250	230-310 : 230-420	100-300
- Peak Power Tracking	---	Yes	Yes	No : Yes	Yes
Converter				(none)	
- Manufacturer	---	Holec	Siemens		Silectron
- Part No.	---				BO
- Quantity	---	29	4		1
- Output Power, Rated	kW dc	2	25		15
- Input Voltage	Vdc	210-273	160		100-300
- Output Voltage	Vdc	346-432	250		180-270
- Control Type (Primary)	---	Constant V	Constant V	Constant V with I limit	Constant V/2-step
- Charge Voltage Limit (CVL)	---			298-310	260/248
- CVL Adjustable	---	Yes	No	Yes	Yes
State of Charge Calculator					
- Manufacturer	---	Varta	Varta	ENEL	Silectron
- Part/Model	---	Logistronic	Logistronic	PLC Contronic 3	
- Quantity	---	1	1	1	1
Rectifier			(none)		(none)
- Manufacturer	---			BORRI	
- Part No.	---			SE015 B 008 A	
- Quantity	---	2		1	
- Output Power Rating	kW	20		10	
- Input Voltage	Vac	380		380 (3-phase)	
- Output Voltage	Vdc	430		260	
UPS		(none)	(none)		
- Manufacturer	---			CO.EL.	Silectron
- Part/Model	---			Compact	
- Quantity	---			2	1
- Output Power Rating	kW			3.5/15	3
- Input Voltage	Vdc			250 (for 15 kVA) 220 (for 3.5 kVA)	180-280
- Output Voltage	Vac			220	220

(Page 2 of 3)

Table 4-6.   (Concl.)

PARAMETER	UNITS	TERSCHELLING	TREMITI	VULCANO	ZAMBELLI
Data Acquisition Unit					
- Manufacturer	---	WIP-Munich		ENEL-Marconi	TEAM/SELTA
- Part/Model	---	PCIMP-4		Mini SAD/3	INTEL 80286
- No. Parameters Measured	---	51		48	30
- Storage	---	HD		Cassette Tape	HD 20 MB
Energy Meters					
- Manufacturer	---			Compagnia Elettromeccanica Meridionale	Sun Power
- Quantity	---	2 (ac)		5 (ac)	3
- Display Type	---	Mechanical		Mechanical	Electrical
Control Unit					
- Manufacturer	---	Satt Control		H&B	TEAM/ORSI
- Part/Model	---	Satt Con 31	LS 11/02	PLC-Contronic 3	
- Computer	---				PCM/ITER
Auxiliary Generator		(none)	(none)	(none)	(none)
- Manufacturer	---				
- Part/Model	---				
- Quantity	---				
- Output Power Rating	kVA				

(Page 3 of 3)

4.2   PLANT SITING AND OPERATIONAL STATUS

Site Layout and Landscaping

All pilot plants conformed to the general criterion, requested by the Commission, to safeguard the environment by designing the array field and the plant layout to harmonize with the surrounding landscape. In many of the pilot plants, an architect was employed to ensure that this requirement was fully complied with. The results are best illustrated in the individual pilot plant descriptions and photographs included in Appendices 1-24.

Operational Status

The major part of the 16 pilot plants were operational as of late 1990. Several plants have ceased operation for several reasons. An example is MONT BOUQUET whose PV array was badly burned by a local bush fire. There are plans to refurbish the plant and put it back into operation when a litigation concerning the insurance coverage has been resolved.

(If you wish to visit any one of the PV pilot plants, please inquire about the actual operation status with the plant operator.)

During the plant improvement phase (1987-1989), the monitoring systems in many plants were refitted with new, more reliable equipment (e.g., ZAMBELLI, TERSCHELLING, AGHIA ROUMELI, PELLWORM, and FOTA).   The new Pilot II plants such as BERLIN and POZOBLANCO installed new PC-based monitoring systems, all of which have demonstrated significantly improved data collection and processing capabilities.

4.3   PHOTOVOLTAIC ARRAYS

4.3.1   MODULES AND ARRAYS

Module Construction Features

The PV modules used in the pilot plants were supplied by the following manufacturers in Europe:

- AEG-Telefunken (FRG)
- Ansaldo (IT)
- BP Solar (GB)
- France-Photon (FR)
- Helios (IT)

- IDE (BE)
- Isofoton (ES)
- Photowatt (FR)
- Pragma (IT) (now Italsolar)
- Siemens Solar (FRG)

As of late 1990, all the above manufacturers were still in business, except for France-Photon and IDE. Details of the PV modules from these manufacturers may be found in Appendices 1-24.

The main types of materials and components used in the construction of the modules in addition to the frames are:

- Glass/polyvinylbutyral (PVB)/glass
- Glass/ethylene vinyl acetate (EVA)/glass
- Glass/PVB/Tedlar-coated aluminium foil
- Glass/EVA/Tedlar-coated aluminium foil
- Glass/transparent silicone/white silicone polymer
- Glass/transparent silicone/glass

All modules used tempered (toughened) glass covers. In most of the current second and third generation modules, the two most common materials among the above six types are glass/EVA/glass and glass/EVA/Tedlar foil.

Module Performance

During the initial acceptance test measurements, up to 5% power degradation was observed [1] on modules which used PVB. The manufacturer discontinued the use of this material on second generation modules.

The only other significant problem noted during the initial measurements was that in many cases, the actual module output was lower than the manufacturer's advertised rating by as much as 14% [2]. The estimated average power conversion efficiencies of the first generation modules and cells in these modules are:

	Polycrystalline	**Monocrystalline**
- cells:	8.4%	12.4%
- modules:	6.5%	8.5%

The AEG and Ansaldo modules used 100-mm square polycrystalline Si cells while all others utilized monocrystalline 100-mm diameter circular Si cells in the first generation modules (Pilot I plants). Note that most PV modules now employ square cells in order to achieve a high packing density and consequently a small increase in module efficiency (about 1 to 1.5%). This small increase in module efficiency, however, becomes quite significant to the overall cost of the PV plant as its PV array size increases. There has been a trend towards the use of polycrystalline cells, mainly because of their availability and lower cost.

Array Operational Experience

- In the 16 PV pilot projects, nearly 32,000 PV modules were produced and installed. The suppliers had replaced about 0.25% of these PV modules as of 1986 because of cracks in the cover glass. In most cases, however, the module performance was not affected. This problem mainly concerned the 20-W AEG modules used by FOTA, PELLWORM, and TERSCHELLING (see Fig. 4-20) due to a combination of poor mounting design and installation procedure. The problem was corrected by AEG on their second-generation modules by providing a mounting bracket on the module frame which reduced the stress on the cover glass.

- Less than 0.02% of the modules have failed due to defects in the electrical components (cells, connectors, diodes, cables). Damage to the cover glass from hailstorms was reported by GIGLIO [3], and vandalism was reported by TERSCHELLING and PELLWORM. Only 0.04% of the modules have been damaged by weather conditions.

- During an inspection by the Commission in 1985, signs of degradation were observed such as discoloration of cells and encapsulation material, and humidity inside the modules. However, no power degradation in the arrays was noticed.

- Manual cleaning of the module covers is apparently not necessary. Even at sites with high air pollution, no dirt and other particulate deposits on the module have been noticed.

- Over the total operating time of the pilot plants, the modules have proven to be one of the most reliable components in the system. However, in most plants, only qualitative information is presently available on the long-term stability of the PV modules. An exception is the VULCANO plant, where in early 1989, in an effort to investigate possible degradation of modules, ENEL and Conphoebus took out 200 modules (about 10% of the total), performed indoor I-V tests under controlled conditions, and compared the results with the original I-V data before the modules were installed. They concluded that no apparent power degradation is noticeable after four years of field operation [4].

Note: To determine the degradation of the array field accurately, it is essential to acquire the I-V characteristic curve of reference or control modules in the field at least once a year. Random selection and testing of modules for degradation determination purposes is useless because no precise information can be obtained. Accurate determination of module power degradation is possible only if a "fingerprint" I-V curve is acquired before the installation on each pre-selected module. For this purpose, a minimum of 10 modules is desirable.

- Following the field tests on selected plants in 1988 and 1989, JRC concluded that no measurable array degradation had occurred at VULCANO, NICE, TERSCHELLING, or AGHIA ROUMELI [5-8].

Table 4-7 lists the PV array outputs for the original pilot plants. In some cases, two sets of the rated power data are shown, based on actual measurements (e.g., the first set in 1982 or 1983 and the last one in 1988). The significance of the data shown in this table are as follows:

- Array-level data cannot be used to determine power degradation with a high degree of accuracy. For this purpose, it is best to use module-level data and string I-V measurements.

Figure 4-20.   Cracked Glasses on PV Modules at TERSCHELLING

Table 4-7.   Summary of PV Array Power Output Measurements and Rating at Pilot Plants

Plant	Date	Total Module Area (m^2)	Simulator Power (kW)**	Nominal Power (kW)+	Rated Power (kW)*	Ratio of Rated to Nominal Power (%)
Aghia Roumeli	Dec. '82	592.5	45.9	50.0	44.0	88
	Jun. '88	592.5	45.9	50.0	43.9	88
Mont Bouquet	Mar. '83	687.4	47.7	50.0	47.1	94
Nice	May '83	687.4	47.7	50.0	46.9	94
	Mar. '88	687.4	47.7	50.0	44.0	88
Fota	Jun. '83	712.4	45.7	50.0	45.9	92
Chevetogne	Jun. '83	861.0	63.5	63.0	58.1	92
Hoboken	Jun. '83	403.1	28.3	30.0	29.3	94
Terschelling	Jun. '83	706.2	45.3	50.0	45.9	92
	Jun. '88	706.2	45.3	50.0	43.4	87
Pellworm	Jun. '83	4,514.0	289.9	300.0	286.0	95
Kythnos	Jul '83	1,176.0	94.4	100.0	86.3	86
Rondulinu	Jul. '83	503.6	42.8	44.0	41.8	95
Marchwood	Nov. '83	426.2	31.5	30.0	29.4	98
Tremiti						
PV field A	May '84	280.9	20.8	20.0	21.0	105
PV field B	May '84	586.5	47.1	45.0	46.9	104
Giglio	Jul. '84	742.2	48.3	45.0	47.1	105
Vulcano						
PV field A	Sep. '84	602.1	41.9	45.0	44.2	98
	Apr. '88	602.1	41.9	45.0	43.9	98
PV field B	Sep. '84	581.3	37.6	35.0	35.9	103
	Apr. '88	581.3	37.6	35.0	32.9	94
Zambelli	Jun. '87	1,084.0	71.3	70.0	71.3	102

+   Design rating stated in the contracts
*   Calculated using actual string measurements on site, condsidering assumed cable and mismatch
** Estimated using average indoor data from 8 randomly selected modules

- The ratio of nominal to rated power ranged from 88 to 105%. Ratios above 100% indicate that the PV array was adequately sized to meet the contractually specified rating which is the nominal value. However, for proper establishment of the PV field power rating at the main dc bus, the ratio must adequately allow for module and string mismatch, power degradation for a defined time period, actual module output, and the statistical spread of all modules. Thus, projects with ratios above 104% did indeed size the array adequately.

4.3.2   ARRAY SUPPORT STRUCTURES AND FOUNDATIONS

Altogether 11 different support structures are used in the pilot plants, using wood, galvanized steel, or aluminium.

Irrespective of the materials used, all structures were found to be in good condition four to six years after installation. No indications of overload through wind forces or oscillations were found. These structures require little maintenance. On the wood structures used at PELLWORM, screws used for mounting the modules had to be tightened once a year due to the expansion and contraction of the wood beams.

Only screws, bolts, and braces made of stainless or galvanized steel were used for fastening structural parts. While stainless steel caused no problems, heavy corrosion is seen on the galvanized parts. In one case (HOBOKEN), corrosion led to the expensive replacement of most of the module fastening elements. Some of the galvanized attachment brackets and bolts on the KYTHNOS array are heavily rusted (see Fig. 4-21), although the structures are apparently holding up very well six years after installation. A simple lesson learned here is that stainless-steel bolts, nuts, and screws should be used to achieve long life, although they incur higher cost.

Figure 4-21. Rusted Nuts and Bolts on PV Panel Attachments at KYTHNOS.

No major problems have been experienced on the foundations for the support structures. At TREMITI, however, the foundation could be affected by erosion from rain water in the long term. At other sites located on sloping sites, this problem has been lessened by building terraces (e.g., at RONDULINU). At VULCANO, the water draining from the panels is fed into a storage tank.

The use of spaces between the rows of arrays for grazing to keep the grass low, has also proven satisfactory. Compared with other solutions, this alternative is particularly cost-effective for array fields located on steep slopes.

Operational experience so far indicates that the array support structures have not caused any problems. However, the combination of module, structure, foundation and cabling was, in many cases, not optimized relative to the material and installation cost. Array structures have generally been overdesigned and recent studies indicate that a significant cost reduction is possible [9].

Strain gauge measurements were performed in many of the PV pilot plants to study the effects of wind load on the array structures, but they did not produce satisfactory results. This is mainly due to the fact that the structures were overdesigned and that the location of the strain gauges or load cells in the array field was not optimized to reflect the total wind load with respect to the peak wind load. In many cases average hourly data were recorded, but average values are not useful for proper wind load analysis (peak values are needed). In most of the PV pilot plants, not even the recalibration of these load sensors was accomplished.

Strain gauge, load cell, and/or accelerometer measurements should be made only when ultra-light-weight structures are employed. These measurements can be used to determine the total wind force and identify critical structural areas where the highest stress is expected. For design purposes, measurements of wind speed and wind direction, at the reference height of 10 m and at the highest part of the panel, are needed to establish general rules for determining the wind load.

4.3.3   CABLING, CONNECTORS, AND TERMINATION BOXES

No damage to the array field or panel cables has occurred at any of the pilot plants. Humidity in the terminal boxes has been a frequent problem, causing ground isolation fault alarms. These types of defects, however, were corrected during the initial operation of the plant.

In most plants, the termination or junction boxes at the module and subarray or string connection points presented no major difficulties. Problems, such as water seepage into the junction boxes at VULCANO and ZAMBELLI, were quickly resolved by applying additional sealants. However, in general, it is better to drill a small drainage hole at the bottom of electrical termination boxes so that any moisture condensation can dry up rapidly during daylight and collected water can escape. it is important that the holes be drilled at the lowest gravitational point and be covered with porous material to prevent the entry of small insects.

The free-hanging wires interconnecting the modules have not caused major problems. However, cable breaks due to movements caused by the wind and birds resting on the wires can present problems. Fixing the wires to the modules or the structure, and if possible, protecting them from the sun, are the best solution from the technical and aesthetic points of view.

In restrospect, it was generally agreed by the pilot plant operators and designers that the following strategies could have reduced initial capital and installation costs at many plants:

- Eliminate the module termination boxes wherever possible
- Minimize the number of termination and intermediate boxes in the array field
- Use reliable quick-connect/quick-disconnect plugs for module interconnection
- Use large modules or pre-assembled panels

4.4   BATTERIES

Almost all the PV plants described in this handbook use lead-acid cells of the vented, flooded-electrolyte type. The PV plants at BRAMMING and VILLA GUIDINI use cells with *tubular* positive electrodes rather than the more conventional *flat-pasted* positive electrodes. Three suppliers, Tudor, Fiamm, and Lyac-Power, among the seven manufacturers of the battery cells, supplied the tubular electrode types.

Figure 4-22 shows installed batteries at several plants. Table 4-8 gives a summary of the key features along with the cell suppliers of batteries used in the Pilot and other selected PV plants in the EC. Included in this table are the number of cells in series in a battery and number of batteries used. For the plants listed in Table 4-8, if two or more batteries are listed, it means that they are strapped in parallel, often without diode isolation between the batteries and without measurement of the current flow in each battery string.

BERLIN (Prof. Hanitsch of Technical University of Berlin)

KYTHNOS (Dr. S. Pressas of University of Patras)

ZAMBELLI (late Dr. Braggion of AGSM)

TREMITI

NICE

GIGLIO (Dr. S. Li Causi of ENEA)

Figure 4-22.   Batteries at Several Pilot Plants

Table 4-8.   Principal Features of Batteries in PV Pilot Plants (DG XII) and Other Large Stand-alone PV Plants in the European Community

PV Plant	Cell Mftr	Number Batteries	Number Cells in Series	Battery Voltage	Battery Rated Capacity (Ah)	Battery Rated Energy (kWh)
AGHIA ROUMELI	Varta	1	150	292-353	1,500	450
CHEVETOGNE	Oldham	1 1	110	198-264	1,500 120	330 160.8
FOTA *	Varta	2	134	234-312	300 (x2)	160.8
GIGLIO	Varta	1 1	120 60	228-350/ 90-160	2,000 300	480 36
HOBOKEN	Oldham	1	12	22-29	90	2.2
KAW	Varta	1	150	292-353	1,500	450
KYTHNOS	Varta	1	125	225-300	2,400	300
MARCHWOOD	Lucas	1	110	216-288	400	88
MONT BOUQUET	Oldham	1 1	106 24	191-254 46-60	800 350	169.6 16.8
NICE	Oldham	1 1	106 24	191-254 46-60	1,500 350	2,076 16.8
PELLWORM *	Varta	4	173	311-415	1,500 (x4)	2,076
RONDULINU	Oldham	1	84	151-202	2,500	420
TERSCHELLING	Varta	2	180	346-432	250 (x2)	180
TREMITI	Varta	1 1	125 110	225-300 209-275	2,000 250	500 55
VULCANO	Varta	1	130	240-310	1,500	390
ZAMBELLI	Tudor	2	108	200-270	300 (x2)	129.6
ANCIPA		1 1	12 55	24 110	540 540	13 59.4
BERLIN	Hagen	1	140	266-350	504	100
BRAMMING	Lyac-Power	2	108	194-260	200	43.2
DELPHOS *	Tudor	1	250	450-600	2,400	1,200
MADRID	Tudor	1	170	324-408	420	142.8
POZOBLANCO	Tudor	1	12 54	24 106	1,120 1,600	26.9 172.8
VILLA GUIDINI	Fiamm	1	12	26-29	1,600	38.4

*   In 1989, FOTA and PELLWORM started to operate their plants in grid-connected mode, and their batteries are available (as of late 1990) for secondary mode of operation. Likewise, DELPHOS has operated in a similar manner.

4.4.1   PROBLEMS EXPERIENCED

The pilot plants were installed in the period from 1982 to 1984 . Actual failures encountered up to 1990 are summarized in Table 4-9 [10]. Cell failures can be classified as shorted, open-circuited, or low capacity. Regarding the relative frequency of types of failure, shorted cells have been reported more often than low-capacity or open-circuit types. This is perhaps because shorted cells are much easier to detect, combined with the fact that open-circuit failures rarely occur. The diagnostic method for low-capacity cells is not straightforward because of the need for monitoring of individual cell voltages and specific gravity, deep-discharging of the battery, and cell mismatch assessment.

Table 4-9.   Known Battery Cell Failures During the First Six Years of Plant Operation

PV Plant	Total Number			Failures	
	Batteries	Cells	Type	No. of Cells	Possible Causes/Remarks
AGHIA ROUMELI	1	150	Internal short	13	(Note 1)
DELPHOS	1	250	Cracked case	1	H_2 combustion resulting in high internal pressure build-up
FOTA	1	134	Internal short	40	(Note 1)
KAW	1	150	Internal short 2) Cracked cases	4 18	(Note 1) (Note 2)
NICE	1	106	1) Cracked cases/ damaged joints	22	Very high internal pressure buildup, 1 cell due to H_2 combustion; Severe overcharging under long float charge
PELLWORM	4	692	Internal short 2) Deformed cases	2 19	(Note 1) (Note 3)
VULCANO	1	130	Low capacity 2) Cracked case	11 8	(Note 1) (Note 2)
ZAMBELLI	2	216	Internal short	3	(Note 1)

Note:

1) Internal short is of two types, low and high resistance. The latter can also be categorized as a low-capacity failure mode. Possible causes are a combination of a) too many deep discharges, b) leaving the battery in partially-charged state too frequently, c) lack of timely recharging from insufficient PV energy and/or inadequate battery maintenance equipment (i.e., auxiliary power source and rectifier) and equalization procedures, and d) cell performance mismatch.

2) Cracks on the vertical sides of the case, usually along the middle area. Possible cause: plate growth and expansion.

3) Cases bulged out along the center of vertical walls. Most probable cause: severe overcharging via the 450-kVA inverter (reverse power flow), combined with resulting high cell temperature. This occured during the weekend in an attempt to perform recharging or equalization charge.

4.4.2   FAILURE MODES AND POSSIBLE CAUSES

The reported failures and abnormal events indicate that there are two dominant failure modes: 1) a direct internal short with no ability to recharge, 2) low capacity with a limited ability to recharge. The latter is often classified as "graceful degradation" because the capacity of the cell decreases until it is no longer capable of receiving and storing energy. The internal resistance of such cells may not indicate a shorted condition on charge and discharge.

Short-Circuit and Low Capacity

These two types of failures are difficult to distinguish since low-capacity cells are usually on the verge of becoming completely shorted. Many possible causes have been cited, including 1) excessive overcharging, 2) overdischarging, causing cell voltage reversals, 3) excessive DOD, 4) excessive cycling, 5) undercharging, 6) "poor" maintenance, 7) inadequate charge control, 8) loss of positive plate material, 9) operation in excessively high temperatures.

Precise determination of the causes of gradual capacity fading is perhaps futile without a knowledge of key operating data, especially cell-level measurements. Lack of adequate battery performance data from installed plants was mainly the result of poor monitoring equipment, and a failure to identify critical monitoring parameters early in the design phase.

Open-Circuit Mode

An open-circuited cell is indicated by a zero or random voltage reading across the cell terminals. This can happen only if there is a broken interconnection inside the cell or when the cell has lost all of its electrolyte and the plates have "dried up." Complete loss of electrolyte due to case cracking has occurred, although very rarely. Two causes of case cracking are:

- Internal pressure buildup resulting in sudden rupture of the plastic case
- Mechanical pressure exerted on the case walls due to positive plate growth, which is a slow process

Case cracking has occurred more frequently than expected (e.g., VULCANO, KYTHNOS, and KAW). The cause(s) are rather difficult to determine, but the most plausible explanation is that growth of the positive plates leads to pressure on the cell walls which eventually splits the case. This type of case crack is shown in Figure 4-23. Another cause is simply weakening of the plastic case from a combination of possibilities like crazing and exposure to wide temperature fluctuations.

Explosions

The failure of cells at NICE created a safety hazard. Explosions of several battery cells, including a fire in one cell, damaged about 20 cells (out of over 100 cells in series). Photographs in Figure 4-24 show various types of damages. Practically all damaged cells showed evidence that they were subjected to a tremendous amount of internal pressure, causing the prismatic cell cases (acrylic or styrene type) to burst open and/or uproot the vent plug and metal terminal studs at their mechanical bonding joints. One cell showed evidence of a "flash" burn inside the cell (i.e., sudden combustion of H_2-O_2 gases).

After an analysis of about 20 ruptured cells at NICE, performed by the Commission's representatives, it was concluded that the primary cause of these failures was blockage of the small holes in the vent plugs (see Fig. 4-24) which prevented the release of internal pressure from the H_2 and O_2 generated during the charging process [10]. The battery at NICE happened to be on continuous float charge using a rectifier for an undetermined number of days. The float charge is the contributing, rather than the primary cause. That is, if the vent plugs had been able to release the internal pressure, normal float charging would not have caused a destructive pressure buildup.

In another PV plant (DELPHOS), a single cell out of over 200 cells in a series exploded, followed by a complete spillage of its electrolyte. This cell was reported to show signs of a "flash" fire similar to that at NICE. For combustion to occur, there must be a source of an electric spark. The cell in question had a temperature sensor inside the cell, which required a bias voltage supply. Thus, a probable source of the spark which caused the explosion was the wiring of the temperature sensor shorting against the plates during the charging process. Even during normal charging, H_2 and O_2 can easily accumulate in the free space in the cell to a level which can ignite. A simple lesson learned from this case is not to install any electrically energized sensor inside a battery cell.

Figure 4-23.   Cracks in Cell Case (KAW Pilot Plant).
This has also occurred at VULCANO.

4.5   BATTERY CHARGE CONTROL AND DISCHARGE PROTECTION

Charge Control Types Used: Series, Shunt, and Array String Switching

The battery charge control methods used in the pilot plant are constant voltage limit type with the current tapering off when the battery voltage reaches the preset voltage limit. To accomplish this function, five plants (KYTHNOS, GIGLIO, TERSCHELLING, TREMITI, and ZAMBELLI) used a regulator, i.e., a dc-dc converter in series with the PV array. Furthermore, three plants, KYTHNOS, TREMITI, and TERSCHELLING, resorted to a "boost" converter, i.e., input voltage is boosted or stepped up to obtain a higher output voltage. GIGLIO and ZAMBELLI decided on a "buck" converter. Both types of series regulators have worked well for over six years. The efficiency characteristic of the GIGLIO converter, however, is slightly lower than the KYTHNOS one, even though the former operates at a slightly higher input voltage (see Appendices 4 and 7, respectively).

Seven of the pilot plants (CHEVETOGNE, FOTA, MARCHWOOD, MONT BOUQUET, NICE, PELLWORM and VULCANO) used an array shedding approach. All of these plants employed their main computer for array string/group on-off control, except MONT BOUQUET and NICE where the conventional analog (hard-wired) circuits were used for array string control.

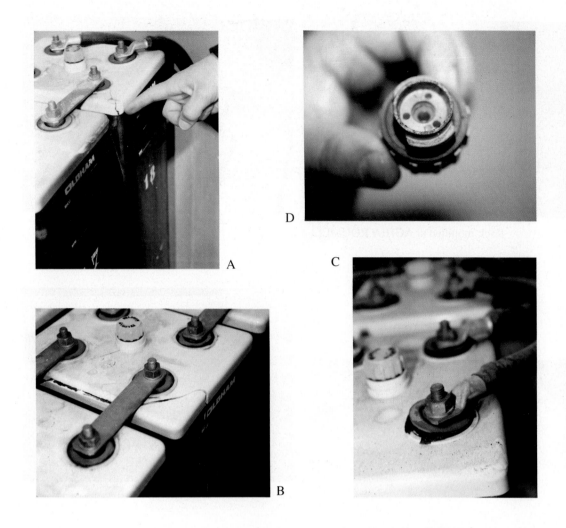

Figure 4-24.   Damaged Cells at NICE from Apparent High Internal Pressure Build-up.
A, B, C, show different damaged parts on the cell; faulty vent plug (photo D
shows inside view) and continued float charging are the probable cause.

The mini- and micro-computers implemented in the early 1980's were expensive and not very
reliable. Software mechanization was equally difficult. Nevertheless, the computer-based array
shedding for battery charge control worked very well following the initial adjustment and trimming
period for FOTA, PELLWORM, MARCHWOOD, and VULCANO.

On the other hand, the simple on-off control of the entire PV array at both AGHIA ROUMELI and
KAW was considered unsatisfactory, particularly in view of the large number of battery cell failures
encountered there.

SOC Calculator

Figures 4-25 and 4-26 show views and the basic functional diagram of the Varta Logistronic unit,
respectively. It was designed to provide both monitoring and control of battery operation. Eight
pilot plants each attempted to use Logistronic unit, but they proved to be unsatisfactory. Of the eight
plants, only this unit at KYTHNOS is working. (Note: The Logistronic unit at KYTHNOS is used
only to display the battery data: parameters like SOC and battery voltage are not utilized for control
purposes). A key observation is that an SOC with a 100% resettable capability is useful, but it should
only be used as a secondary charge control parameter — not as a primary control signal.

Rack-mounted at AGHIA ROUMELI

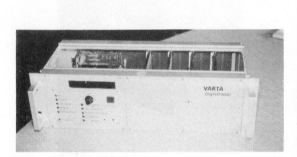

Figure 4-25.   Varta SOC Determination Equipment (Logistronic)

There are other SOC equipment in use. An example is the Technai unit shown in Figure 4-27 which has the same basic features as the Logistronic equipment.

4.6   POWER CONDITIONING

Inspection of the pilot plants in 1985 and 1986 and later discussions with the plant operators revealed that most of the operational problems encountered in the PV plants were related to inverters and equipment used for battery charge control purposes.

Inverters

Figures 4-28 and 4-29 show some of the inverters used in the pilot plants and a PV house. The main problems encountered on the inverters were the following:

- Electrical shorts due to corrosion in printed circuit cards
- Synchronization with the grid when it exhibits frequency
- Frequency control
- Inability to supply high starting currents (for induction motors)
- Sensitivity to high temperatures, dust, and humidity
- Lack of spare parts
- Unavailability of skilled maintenance and repair personnel

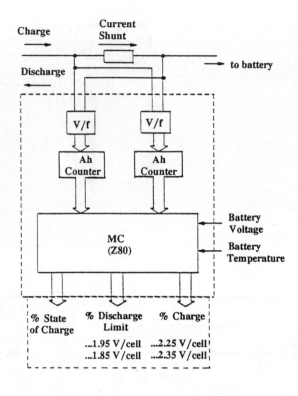

Figure 4-26. Functional Diagram of the
VARTA Logistronic Unit

The rated power efficiencies quoted by the manufacturers are based on full output power (90 to 94% range for most inverters used), but the operating efficiency of most PV plants has been significantly lower, because the average load has been much lower than full power.

One way to improve the overall energy efficiency of the inverter is to use multiple inverters and turn on only the required number of units based on the available input power from the PV array. This has been demonstrated to be practicable at plants like KYTHNOS which employed three inverters. Some of the plants remove the no-load demand by turning the inverters off when there is no load.

The 3-phase 50-kVA inverter used at AGHIA ROUMELI, made by Jeumont Schneider, has a square-wave output which is filtered by 3-phase power filter to achieve a sinusoidal waveform. Voltage regulation is accomplished through transformer tap changing. This was the cheapest way the ac output voltage regulation could be implemented, because the output waveform is not quasi-square wave but pure square wave without pulse width modulation.

Converters

Dedicated dc-dc converters with maximum power tracking (MPT) capability were used at TERSCHELLING, KYTHNOS, and ZAMBELLI. At ZAMBELLI, the inverter, in conjunction with the converter, regulates the speed of the pump motor to operate the array at its maximum power point. These converters also provide battery charge control as described earlier (see Section 4.5).

The series converters at KYTHNOS and ZAMBELLI operated well. But the 2-kW units (total quantity of 29) at TERSCHELLING continually failed due to defective transistors and high-temperature dissipation problems. Moreover, both the power conversion efficiency and the MTBF of the individual regulators have been found to be low.

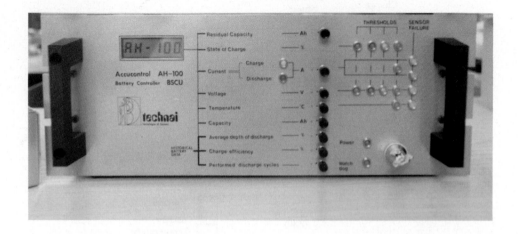

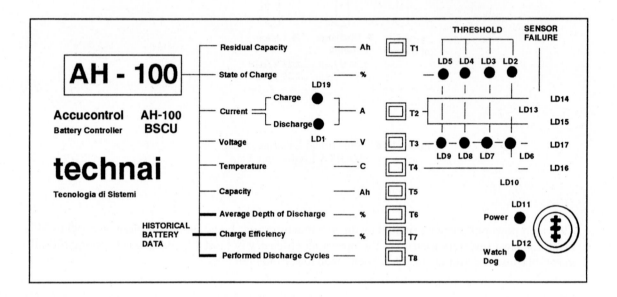

Figure 4-27. Technai SOC Equipment

Rectifiers

Rectifiers have been installed to recharge the batteries from auxiliary sources like wind and Diesel generators, or from the grid. In the latter case, the arrangement is similar to the conventional UPS (uninterruptible power system). The minimum requirement for the rectifier is to have adjustable voltage and current limits appropriate to the battery installed.

The use of a rectifier can extend the life of a battery by providing a full charge at a low charge rate. Most stand-alone systems, especially the dc systems without inverters, do not have auxiliary power sources and rectifiers to perform this function. The pilot plants listed below have included rectifiers, but have used them differently.

- BRAMMING   - GIGLIO
- HOBOKEN   - NICE
- RONDULINU   - TERSCHELLING
- PELLWORM   - VULCANO

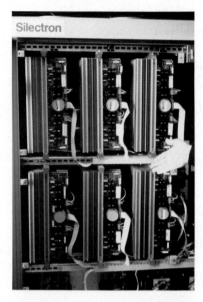

Control cards, inverter at ZAMBELLI

Inverter rack, ZAMBELLI

Transformer-less inverter at the Munich PV house

Synchronous Generator at TERSCHELLING
(Rotating Type)

Figure 4-28.   Inverters at ZAMBELLI, TERSCHELLING, and MUNICH

BRAMMING has made good use of the rectifier, as indicated by the performance and long lifetime of their battery. Among the large pilot plants, the RONDULINU plant design provided the best arrangement for maintenance recharging via the propane-generator/rectifier.

NICE used its grid-powered rectifier for battery charging as a standard procedure, but through a combination of failure of the vent plug to allow the internal gas to escape, too high a constant charging potential, and long periods of continuous float charging, they encountered battery damage (case cracks due to internal pressure buildup). This shows that care must be taken during extended trickle or float charging to monitor the individual cell voltages and to check the condition of the vent plugs. An adequate monitoring system could have prevented the cell explosions at NICE. In addition, the rectifier used for battery float charging must be designed with a capability to adjust the battery charge voltage limit (from about 2.23 to 2.45 V per cell) and the charge current limit.

None of the other plants appeared to have a well-defined standard procedure for the use of their rectifiers. This is, however, a very worthwhile area to explore as part of an attempt to extend the life of stand-alone batteries, and hence, the PV plant.

4.7   SYSTEM CONTROL AND POWER MANAGEMENT

The three types of power control and management strategies utilized at various plants are as follows:

- Single relay on-off switching of the array and battery via analog voltage comparators (used at AGHIA ROUMELI, KAW, RONDULINU, NICE, and MONT BOUQUET).

- Multiple-relay switching of array strings via a computer (to control the dc-bus voltage) and of batteries (FOTA and PELLWORM).

- On-off switching of the series converters and inverters via one analog signal (used at KYTHNOS, TREMITI, MARCHWOOD, HOBOKEN, and ZAMBELLI)

The main discriminator in the above strategies is analog vs. digital (i.e., computer) control. The computer control is very flexible, but it is more complex than the hard-wired analog approach and is thus less reliable. On the other hand, experience at AGHIA ROUMELI and KAW has demonstrated that the simplest control strategy (first item above) was not adequate for proper charge control and maximum battery life. In general, a more flexible charge control method was found to be mandatory in order to minimize premature battery failures.

The main problems associated with software changes were:

- High cost
- Unavailability of software personnel
- Use of non-standard programming language
- Computer capacity limits and computer faults

Software improvements were time-consuming and expensive because only personnel with special skills were able to implement and test the new software. In some projects (e.g., PELLWORM, VULCANO) the software improvements were made at the home facility rather than on site, which required frequent travelling. At PELLWORM major software improvements were implemented during the first year but only a minor effort was made in subsequent years.

In some projects, the microcomputers were programmable on site, e.g., TERSCHELLING, TREMITI, but, due to the unavailability of the original software personnel, appropriate changes could not be made. In the case of TREMITI, a very capable microcomputer was used but control software was not properly documented. At TERSCHELLING, the company which supplied the controller was not available for software improvements because of unfavourable market prospects for PV plants in the future.

Another problem associated with the microcomputer software was the programming language. At VULCANO, an HB TEU320 controller was used and only three people in Milan were capable of writing its software. In the FOTA project, a non-standard programming language was used, and this caused many difficulties in making software changes.

The hardware faults associated with microcomputers were mainly due to electronic card failures (e.g., in ZAMBELLI). The most serious problem was the time taken to repair the electronic cards. In the ZAMBELLI plant, the system was shut down for over a month because of a fault in the microcomputer. Critical spare parts should be defined during the design phase and be made available on site.

In HOBOKEN and ZAMBELLI the control strategy was to match the load to the available PV power by varying the speed of the pump motors. During high insolation with varying cloud cover, the system continuously adjusts the speed of the motors. The problems at HOBOKEN were pump overloading and frequent start/stop operations. At ZAMBELLI, the back EMF due to rapid reduction of the motor speed damaged the inverter on one occasion.

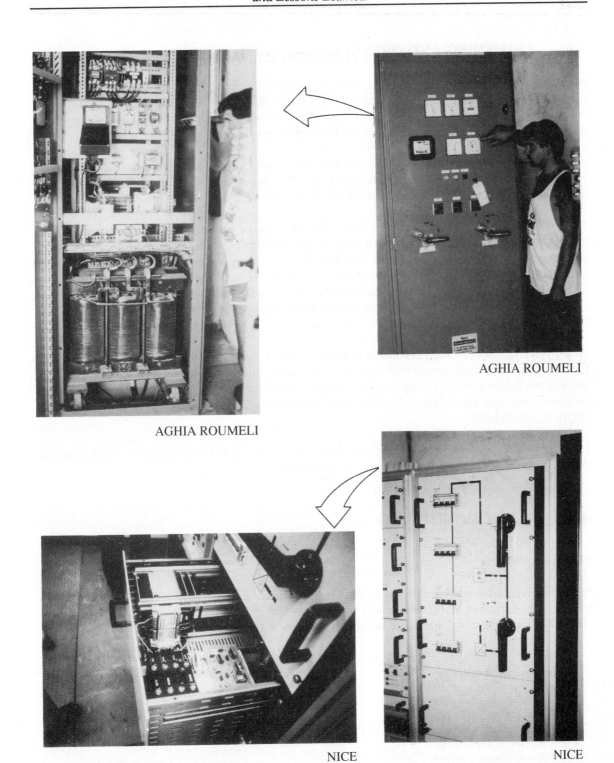

AGHIA ROUMELI

AGHIA ROUMELI

NICE

NICE

Figure 4-29. Inverters at AGHIA ROUMELI and NICE

4.8   LIGHTNING AND OVERVOLTAGE PROTECTION

Most plants did not experience direct lightning strikes, however, indirect lightning strikes have apparently caused most of the damage at KYTHNOS, MONT BOUQUET, NICE, and VULCANO. Nonetheless, the reported damages attributed to lightning are small in number, somewhat contradictory, and thus insufficient to draw solid conclusions [10,11].

At KYTHNOS, despite costly protection equipment which included lightning rods, the data monitoring system and two PV modules were allegedly damaged by lightning (see Fig. 4-30). A direct lightning strike at VULCANO completely destroyed one PV module.

Bad grounding of the dc and ac power lines at FOTA caused errors in both data acquisition and control systems, especially during thunderstorms with lightning flashes. The most commonly used protective devices for power and signal lines have been the surge arrestors, typically varistors, spark gaps, and zener diodes. However, care must be taken in the selection and use of these protection devices because NICE experienced a catastrophic failure of the varistors which caused a fire inside the control cabinet. These varistors were installed typically at the termination points of each power line from the PV array.

At KYTHNOS, Siemens Solar installed 9-m masts in the PV field, for lightning protection. However, the data acquisition system and two PV modules have allegedly been damaged by lightning. GIGLIO has had no lightning-related problems. An opposing view on lightning intercepting devices is that they are costly and are believed to attract lightning; however, it is difficult to technically prove or disprove their benefits.

In order to develop low-cost lightning protection criteria, more field experience is needed on the different protection strategies and devices currently in use.

4.9   PLANT MONITORING

Monitoring Criteria

Monitoring requirements were set up by the Commission in the latter part of the pilot plant design phase. The minimum set of parameters to be recorded on an hourly basis for "Analytical" and "Global" Monitoring were established as follows (also, see Table 2-5, Chapter 2):

Analytical Monitoring

- Irradiation, global horizontal
- Irradiation, array plane
- Ambient air temperature
- Array output voltage
- Array output current
- Converter output current
- Battery charge current
- Battery discharge current
- Dc line voltage (battery voltage)
- Current from dc generator or rectifier
- Current to all dc loads
- Current input to inverter
- Inverter energy input
- Inverter energy output
- Energy to ac loads
- Energy to (+) and from (-) grid
- Energy from ac generator

Global Monitoring

- Irradiation, global horizontal
- Array current
- Energy to all dc loads
- Energy to all ac loads
- Energy to (+) and from (-) grid
- Energy from auxiliary generator (Diesel, wind, etc)
- Hours load connected

Figure 4-30.   Modules Destroyed by Lightning at KYTHNOS

Until 1986, pilot plant data were required to be sent to CEC/JRC for further processing and analysis. The Analytical Monitoring was intended primarily for the pilot plants which were set up with R&D aims, and this required a reliable DAS.

The Global Monitoring criteria were established mainly for the DG XVII Demonstration plants and other small PV plants whose primary aim was to commercialize new energy technologies, and not to focus on R&D. Therefore, extensive monitoring was not required, and most demonstration plants were requested to use energy and current meters, rather than the more sophisticated DAS used in the R&D pilot plants.

Improvements in Monitoring Strategy

In 1988, a study of plant performance was done as a Concerted Action activity involving the pilot plant operators and designers of other large plants in Europe. The objectives were to review the performance and operating status of the plants and to determine whether changes were warranted in the monitoring requirements and methods for future PV plants. The conclusions of this Concerted Action study were as follows [10]:

- The monitoring systems had the highest failure rates in most of the PV plants.

- Some of the pilot plants which had operational DAS, collected sufficient "Analytical Monitoring" data, but in others, the off-site analysis efforts were hampered due to the lack of recorded plant data.

- There is a need for plant operators to be on site or at the home office, in order to perform "quick-look" analyses, especially at PV plants with batteries, because off-site analysis results are only available from a few weeks to months after data retrieval, and therefore cannot be used to prevent failures.

The solution approaches were to 1) establish improved monitoring criteria and diagnostic methods, 2) select and install more reliable monitoring equipment, and 3) incorporate a capability in the on-site DAS to calculate important diagnostic parameters in real-time. The resulting general list of parameters which were recorded on an hourly basis and a list of daily and monthly performance summary data from off-line analysis, are described in Chapter 2 (Section 2.12). These summary data are useful mainly for trend evaluation and not for "quick-look" diagnosis for any problems in the system.

Monitoring Equipment Performance

The DAS equipment selected and installed during the 1981 to 1984 time frame was costly, complex, and prone to stoppages [13]. Table 4-10 lists the DAS installed at different pilot plants and during 1982-1984. At that time, typical computer-controlled data acquisition systems were based on DEC PDP-11 or HP-1000 minicomputers together with front-end units for signal conditioning and A/D conversion. The standard output medium for these systems was the 8" diskette. These systems were reliable only when they were well protected against dust and temperature fluctuations.

Custom-made systems based on microprocessors, such as the Z80 and the Motorola 6800, were converted to the monitoring of PV plants. The new programmes had to be written in Assembly or non-standard languages used in industrial process control.

Table 4-10.   List of Original Monitoring Systems at Several Pilot Plants (Installed 1982-1984)

Project	Type	Storage
AGHIA ROUMELI+	Elsyde Data Logger	EPROM
CHEVETOGNE	HP-1000 + front end	8"-floppy
FOTA+	MC6800 custom design (AEG)	ECMA-46
GIGLIO	MC6800 custom design (Nuova Pignone)	8"-floppy
KAW	Elsyde Data Logger	EPROM
KYTHNOS+	μP-based custom design (Siemens)	ECMA-46
MARCHWOOD	PDP-11/34 + front end**	8"-floppy
PELLWORM+	μP-based custom design (AEG)	8"-floppy
RONDULINU+	MC6502-based custom design	ECMA-46
TERSCHELLING+	μP-based custom design (Holec)	ECMA-46
TREMITI+	PDP-11/23 + front end	8"-floppy
VULCANO	MiniSAD/3 custom design (ENEL)	ECMA-34*
ZAMBELLI+	Z80-based custom design	8"-floppy

+   These pilot plants have either replaced their original DAS equipment during the
    1988 to 1991 time period or planned to do so
*   Data were transmitted to Ispra on IBM-compatible 9-track 1,600 bpi tape
** "Front end" means signal conditioning and multiplexing electronics

The best recording medium in the early 1980's was the ECMA-46 tape cassette (e.g., Penny & Giles 7100) which caused problems in some installations but proved to be quite reliable in many others. A common problem was that the motor drive/tape transport unit stopped, resulting in loss of data. Two pilot plants (AGHIA ROUMELI and KAW) used the Elsyde data logger which had proven to be a reliable system when operated and maintained by the staff from the French meteorological service, but gave problems when untrained local personnel had to restart it after power failures or interruptions.

The above situation improved dramatically from 1987 to 1989 due to a new line of data acquisition equipment with floppy and hard disks, battery-powered RAM cassette modules, and RAM cards for data storage. The improvement efforts of several pilot plants and the new projects in the Pilot II programme have benefited from the rapid growth in laptops and μP-based data collection equipment. Table 4-11 lists current projects that have implemented recently available new DAS's.

Table 4-11.   Improved Monitoring Equipment Implemented in 1988-1989 at Several Pilot Plants

Project	Contractor	Installation Date	DAS
AGHIA ROUMELI	PPC (GR)	May '88	HP 3421A with HP 71B computer
BERLIN	EAB (FRG)	July '88	WIP design with PC-AT
FOTA	NMRC (IRL)	1989	NMRC design with Campbell Scientific 21X with PC-AT
PELLWORM	AEG (TST)(FRG)	Mid-'89	AEG DAM 800
POZOBLANCO	CIEMAT-IER (ES)	Mid-'89	Datataker (from Australia) with PC-XT
TERSCHELLING	Ecofys (NL)	Late '88 September '90	Ecofys' design with PC-XT WIP design with PC-AT
ZAMBELLI	AGSM (IT)	April '89	TEAM design with PC-AT

Significant results and observations concerning the DAS hardware are as follows:

1) Errors in the data, as reported by several PV plants, stem from the use of Si-based solar irradiance sensors that were not calibrated at the time of initial installation. As an example, the Haenni solar irradiance sensors, used in five pilot plants, were found 5 years after installation to be out of calibration by as much as 16%. This was not due to sensor degradation but simply because the manufacturer's original calibration had not been verified. This implies that the manufacturer's test procedure is the source of the calibration error. This is to be expected because the calibration method has yet to be standardized and is under discussion at the international level.

2) The practice of averaging all parameters based on 1-min scanned values, and then storing only 1-h averages is not appropriate for certain parameters. In many cases, instantaneous peak and minimum values are essential, e.g., wind speed, and lowest and highest battery voltages.

3) Only 3 out of 16 pilot plants provided data in sufficient quantity between 1984 and 1985 (FOTA, MARCHWOOD, and VULCANO). Several plants had no recorded data mainly due to failures of their monitoring equipment.

4) None of the monitoring systems were capable of long autonomous operation without human intervention or at least periodic surveillance.

5) The new monitoring equipment installed during 1988-1989 at several PV plants has largely resolved the basic reliability problem. Recently available monitoring equipment, such as the WIP design involving Analog Devices isolation amplifiers, Schlumberger IMP, and PC have already demonstrated a very high reliability in continuous operation and data storage at 12 PV plants (in the German national PV programme) [14].

Results of Recent Plant Monitoring Efforts

It is worth while to mention the significant results of monitoring activities during the period from 1988 to 1991. In the framework of the CEC's second R&D programme (1985-1989), several projects have carried out plant improvement and optimization actions.

Among the pilot plants, the best example of improved monitoring capability is ZAMBELLI. They have significantly enhanced their ability to conduct performance assessment both on-line (real time) and off-line. Tables 4-12, 4-13, and 4-14 list the daily summary data for the month of June for meteorological, array, battery, and system parameters, respectively. AGSM (Verona, IT) and TEAM (Rome, IT) had designed and put into operation an off-site computer programme which can process and print out these summary data one day after receiving the recorded data [15].

4.10   PLANT MAINTENANCE AND SAFETY

At least 21 PV plants discussed or referred to in this handbook operated at a voltage between 100 and 400 Vdc and one plant up to 800 Vdc. A large number of people have visited the PV sites during the period 1982 to 1990. Despite the potential dangers of high dc voltages in the array field and acid spills from the batteries, there have been no major incidents or personal injuries involving site operators and visitors.

Table 4-12. Daily Solar/Meteorological Data for ZAMBELLI PV Plant for the Month of June 1990

Day	POA Irradiance/Energy Energy (kWh/m²)	Average (W/m²)	Peak (W/m²)	Horizontal Irradiance/Energy Energy (kWh/m²)	Average (W/m²)	Peak (W/m²)	Sunlight Time Actual (h)	Theoretical (h)	Ambient Temp. Min (°C)	Average (°C)	Max (°C)	Wind Speed Average (m/s)	Peak (m/s)
01	6.92	456	949	7.99	526	984	13.13	15.19	11.1	15.6	18.3	8.3	14.7
02	5.15	339	958	6.14	403	1080	12.54	15.22	10.8	14.7	17.1	9.7	16.3
03	4.91	322	1159				13.11	15.24	10.6	14.2	17.3	8.9	18.2
04	1.01	66	290	1.19	78	381	8.97	15.26	11.3	12.3	13.5	3.5	10.0
05	1.87	122	847	2.25	147	985	9.44	15.28	11.7	13.6	15.8	4.7	10.9
06	1.23	81	310	1.48	96	384	9.40	15.30	12.9	14.1	16.0	4.7	10.4
07	2.43	159	1009	2.97	194	1035	12.90	15.32	12.2	14.2	16.9	5.2	11.6
08	2.00	130	788	2.81	183	1062	11.73	15.34	8.0	13.1	15.7		
09	3.74	243	1174	4.47	291	1209	11.18	15.36	9.6	12.8	16.1	9.8	17.9
10	5.26	342	1044	6.51	423	1212	12.67	15.38	7.9	11.7	15.6	10.2	25.6
11	7.17	466	960	8.76	569	999	12.81	15.39	9.6	15.4	18.5	4.5	8.7
12	1.00	65	153	1.18	76	177	13.00	15.43	8.7	10.7	14.6	5.5	8.9
13	2.99	194	485	3.92	253	624	16.00	15.45	7.6	13.2	14.9	3.0	6.4
14	4.16	270	948	5.04	327	1012	13.02	15.43	10.4	14.2	17.1	3.8	11.4
15	3.20	207	1085	3.82	247	1165	11.38	15.44	10.1	12.9	17.4	4.4	18.4
16	5.20	337	1104	6.26	405	1160	12.55	15.45	9.6	15.1	18.8	3.6	13.7
17	6.79	439	1019	7.82	506	1074	12.96	15.46	11.4	16.4	19.4	4.1	12.9
18	3.91	253	1072	4.56	295	1195	12.97	15.46	13.5	16.5	19.2	4.0	12.2
19	4.89	316	1016	5.81	376	1080	13.50	15.47	11.9	17.2	20.4	3.8	12.0
20	6.14	397	1095	7.06	456	1160	12.31	15.47	14.2	19.2	22.2	6.4	15.2
21	6.54	423	1056	7.74	500	1114	13.21	15.48	14.3	18.5	21.0	5.5	14.0
22	5.11	330	1044	5.90	381	1117	11.96	15.48	13.6	17.5	20.8	5.1	19.3
23	4.23	274	1129	4.94	319	1185	12.89	15.48	15.1	18.3	22.0	4.3	12.6
24	5.39	348	1027	6.33	409	1089	13.01	15.49	15.4	18.6	21.3	4.3	12.3
25	7.10	458	914	8.26	533	958	15.00	15.49	15.3	20.6	23.2	3.3	6.4
26	6.94	448	918	8.03	518	929	15.00	15.48	16.0	21.2	23.7	4.1	6.1
27	5.57	360	843	6.37	412	886	15.00	15.48	16.8	21.3	24.1	4.1	7.8
28	5.92	383	1015	6.53	422	1054	11.44	15.47	15.5	20.8	24.7	5.2	14.4
29	6.47	419	1013	7.37	477	1035	13.10	15.46	16.9	21.7	24.7	4.4	11.9
30	6.46	418	1004	7.41	479	1048	13.28	15.45	17.7	22.5	25.4	5.2	15.4
Monthly													
Lowest	1.	65	153	1.18	76	177	8.97	15.19	7.6	10.7	13.5	3.	6.1
Mean	4.66	302.17	914.27	5.48	355.21	979.07	12.65	15.4	12.32	16.27	19.19	5.3	12.95
Highest	7.17	466	1174	8.76	569	1212	16.	15.49	17.7	22.5	25.4	10.2	25.6
Total	139.7	---	---	158.92	---	---	379.46	462.1	---	---	---	---	---

Table 4-13.   Daily PV Array Performance Data for ZAMBELLI Pilot Plant for the Month of June 1990

Day	Actual and available Array Energy		Actual Array Power		Available Array Power		Array Efficiency		Array Temperature During Sunlight		Capacity Factor (%)	Array Utilization Factor (%)
	Actual (kWh)	Capability (kWh)	Average (kW)	Peak (kW)	Average (kW)	Peak (kW)	Average (%)	Peak (%)	Average (°C)	Peak (°C)		
01	310.7	465.3	20.46	38.68	30.63	62.18		6.60	25.8	40.8	18.50	66.78
02	274.1	352.6	18.02	56.25	23.17	73.94		6.59	21.6	38.7	16.32	77.74
03	265.7	335.6	17.44	65.22	22.02	79.08		8.92	21.5	41.1	15.82	79.20
04	48.1	73.9	3.15	17.33	4.84	20.95		6.45	10.1	20.6	2.86	65.02
05	100.9	133.1	6.60	46.32	8.71	58.28		6.87	13.8	29.8	6.00	75.78
06	58.7	89.5	3.84	17.40	5.85	22.03		5.50	11.6	24.1	3.50	65.59
07	117.5	172.5	7.67	50.56	11.26	69.76		7.02	18.0	32.1	7.00	68.13
08	80.4	143.6	5.24	24.05	9.36	57.51			14.5	28.7	4.78	55.97
09	179.7	262.0	11.70	63.11	17.06	85.59		7.12	16.2	36.5	10.70	68.59
10	245.8	364.8	15.98	38.36	23.73	69.15		6.00	17.8	37.8	14.63	67.37
11	408.5	479.7	26.54	55.18	31.17	62.68	0.01	10.62	25.9	41.4	24.31	85.15
12	27.9	74.0	2.15	5.61					8.5	11.8	1.66	37.68
13	137.3	211.2	8.58	24.12					17.8	28.7	8.17	64.99
14	245.2	288.8	15.89	57.77	18.72	64.11	5.26	8.85	23.4	37.5	14.59	84.91
15	175.2	222.6	11.35	60.55	14.42	75.49	4.89	6.81	20.4	38.9	10.43	78.70
16	184.0	370.6	11.91	62.58	23.99	85.05	3.16	7.12	28.1	44.6	10.95	49.65
17	301.1	471.4	19.48	58.45	30.50	77.05	3.96	7.24	28.5	42.3	17.92	63.87
18	207.1	269.4	13.39	51.88	17.42	73.59	4.73	7.14	25.1	39.0	12.32	76.87
19	192.6	343.7	12.45	58.91	22.22	66.35	3.52	8.55	28.8	42.5	11.47	56.05
20	303.8	408.1	19.63	59.25	26.37	71.21	4.42	7.76	32.9	47.6	18.08	74.45
21	312.6	435.1	20.20	60.03	28.11	75.70	4.27	8.83	31.4	44.6	18.61	71.85
22	246.3	345.0	15.91	57.27	22.29	70.70	4.30	7.13	29.6	45.3	14.66	71.40
23	94.2	310.6	6.08	61.90	20.07	77.56	1.99	6.56	24.4	42.1	5.60	30.31
24	88.5	395.4	5.72	50.83	25.54	79.79	1.47	6.53	29.5	42.3	5.27	22.39
25	372.1	469.6	24.81	52.14					31.3	44.9	22.15	79.24
26	384.5	459.0	25.64	51.84					31.3	45.2	22.89	83.78
27	241.6	369.4	16.11	45.09					29.2	46.2	14.38	65.40
28	326.4	387.4	21.11	57.48	25.05	65.04	4.92	6.87	34.3	48.1	19.43	84.26
29	307.3	429.4	19.88	55.30	27.77	65.15	4.24	8.51	36.7	48.0	18.29	71.58
30	329.7	423.5	21.34	57.97	27.41	65.04	4.56	9.97	35.6	47.4	19.62	77.85
Monthly												
Lowest	27.9	73.9	2.15	5.61	4.84	20.95	0.01	5.5	8.5	11.8	1.66	22.39
Mean	218.92	318.56	14.28	48.71	20.71	66.92	3.71	7.48	24.12	38.62	13.03	67.35
Highest	408.5	479.7	26.54	65.22	31.17	85.59	5.26		36.7	48.1	24.31	85.15
Total	6567.5	9556.8										

232

Table 4-14.   Daily Battery Performance Data for ZAMBELLI Pilot Plant for the Month of June 1990

Day	Voltage Max (Vdc)	Voltage Min (Vdc)	Cell Temp. (°C)	Energy Input (kWh)	Energy Output (kWh)	Capacity Charged (Ah)	Capacity Discharged (Ah)	Ah Charge Ratio (%)	Number Cycles	Daily DOD (%)	Min SOC (%)
01	254.0	223.0	15.9	18.00	15.32	72.31	68.73	105.2	1	11.45	83.1
02	254.1	222.7	17.0	20.51	16.88	82.77	75.49	109.6	1	12.60	83.3
03	254.0	222.7	17.4	24.62	19.51	99.85	87.49	114.1	1	14.59	83.5
04	243.3	217.7	17.2	19.06	27.06	79.78	122.28	65.2	2	20.36	80.0
05	241.7	221.5	16.4	27.14	24.38	112.29	109.34	102.7	1	18.22	83.0
06	248.0	222.4	16.1	24.63	17.27	101.11	76.87	131.5	1	12.82	82.1
07	251.4	217.9	16.3	32.30	30.72	131.08	138.23	94.8	1	23.04	81.9
08	256.6	217.7	16.5	38.20	36.98	154.83	167.62	92.4	3	27.93	75.4
09	254.5	221.6	16.0	28.88	24.85	115.66	111.49	103.7	1	18.59	82.9
10	255.0	222.8	15.2	28.35	20.15	114.69	90.40	126.9	1	15.07	82.9
11	255.0	221.6	15.3	18.80	17.50	74.95	78.75	95.2	1	13.12	83.0
12	240.2	216.5	17.0	17.43	16.86	72.50	75.13	96.5	1	10.08	81.7
13	256.5	222.4	15.0	29.98	21.21	119.75	95.04	126.0	1	12.42	83.2
14	261.3	220.5	15.9	22.83	19.76	90.38	88.77	101.8	1	14.80	85.0
15	261.1	223.0	16.7	29.38	17.82	46.11	62.40	73.9	1	13.25	82.7
16	261.9	223.6	16.4	16.54	15.08	64.64	66.96	96.5	1	11.16	84.8
17	261.9	220.6	17.6	21.00	18.58	82.46	83.52	98.7	1	13.92	85.2
18	260.3	222.8	18.9	27.95	17.55	112.52	78.38	143.6	1	13.06	83.4
19	260.3	223.6	19.2	17.02	13.84	67.46	61.58	109.6	1	10.27	84.2
20	260.0	222.9	19.8	19.55	17.71	76.78	79.50	96.6	2	13.25	83.7
21	260.9	220.7	20.7	23.08	18.94	91.89	85.00	108.1	1	14.16	82.6
22	247.6	222.9	21.0	22.88	18.42	93.10	82.32	113.1	1	13.72	83.4
23	247.7	223.4	20.9	20.48	16.82	83.50	74.65	111.9	1	12.44	82.7
24	253.5	222.7	20.5	23.38	20.62	93.23	91.95	101.4	1	15.32	82.6
25	259.5	223.2	21.5	20.13	16.34	80.28	73.21	109.7		10.42	84.4
26	250.1	223.5	22.2	18.83	13.19	75.70	58.79	128.8		8.98	83.7
27	259.0	223.8	22.9	19.08	15.81	76.00	70.41	107.9		9.48	83.9
28	247.3	222.9	22.7	17.36	19.26	70.70	86.21	82.0	2	14.37	82.2
29	259.0	223.2	22.9	21.21	11.78	85.01	52.36	162.4	1	8.72	83.4
30	259.6	223.9	23.4	15.30	14.88	60.78	66.49	91.4	1	11.08	86.1
Monthly											
Lowest	240.2	216.5	15.	15.3	11.78	46.11	52.36	65.2	1	8.72	75.4
Mean	254.51	221.92	18.48	22.8	19.17	89.4	85.31	106.71	1.19	13.96	83.
Highest	261.9	223.9	23.4	38.2	36.98	154.83	167.62	162.4	3	27.93	86.1
Total	---	---	---	683.9	575.09	2682.11	2559.36	---	32	---	---

4.11   REFERENCES

[1] F.C. Treble, "Lessons Learned from the Acceptance Tests on the CEC PV Pilot Plants," Int. J. Solar Energy, 1985, Vol. 3, pp. 109-22.

[2] G. Beer, G. Chimento, F.C. Treble, and J.A. Roger, "Concerted Actions: PV Modules, Array, and Solar Sensors," CEC Contract EN3S-140-1, Conphoebus Final Report, December 1989.

[3] S. Li Causi, S. Castello, G. Murzilli, "Photovoltaics for a Cold Store on Giglio Island, Sicily," Proc. of the European - New Energies Congress, Vol. 3, Saarbrücken, FRG, 24-28 October 1988.

[4] G. Belli and G. Chimento, "Determination of I-V Characteristics of PV Modules After Four-Year Service in Vulcano Pilot Plant," Proc. of the 9th European PV Solar Energy Conference, Freiburg, FRG, September 1989.

[5] G. Blaesser, K. Krebs, and W.J. Zahiman, "Power Measurements at the 80-kW PV Plant Vulcano," Technical Note No. I.88.54, CEC-JRC, Ispra, IT, May 1988.

[6] G. Blaesser, W.J. Zahiman, "Power Measurements at the 50-kW PV Plant Nice," Technical Note No. I.88.43, CEC-JRC, Ispra, IT, April 1988.

[7] G. Blaesser, W.J. Zahiman, "Power Measurements at the 50-kW PV Plant Terschelling," Technical Note No. I.88.77, CEC-JRC, Ispra, IT, July 1988.

[8] G. Blaesser, K. Krebs, and W. Zahiman, "Power Measurements at the 50-kW Plant Aghia Roumeli," Technical Note I.88.87, CEC-JRC, Ispra, IT, July 1988.

[9] A. Haenel, M.S. Imamura, and P. Helm, "Coordination and Control of PV Concerted Action Projects and Work on Specific Tasks: Data and Plant Monitoring, Array Structures, and Social Effects Study," Vol III - ARRAY STRUCTURES. Final Report Work Performed Under CEC Contract EN3S-0142-D, November 1989.

[10] M.S. Imamura and E. Ehlers, "Concerted Action Project: Data/Plant Monitoring," WIP 89-7, CEC Contract E3S-0142-D, November 1989.

[11] F.D. Martzloff, "Lightning and Surge Protection of PV Installations, Two Case Histories: VULCANO and KYTHNOS," National Inst. of Standards and Technology, USDOC NISTIR-89-4113, 1989.

[12] K. Krebs and M. Starr, "First Session of European Working Group on Photovoltaic," Proc. of Meeting in Ispra, IT 14-16 September 1985.

[13] G. Blaesser and K. Krebs, "Summary of PV Pilot Plant Monitoring Data 1984-1985," Proc. of the 7th European PV Solar Energy Conference, Seville, ES, October 1986.

[14] M. Imamura and P. Sprau, "Status of the PV Monitoring Project, MUD: German National PV-Programme," presented at the European Working Group Meeting on PV Monitoring, Berlin, FRG, 6-7 June 1991.

[15] A. Sorokin and G. Zamboni, "Zambelli PV Pumping Station (70 kWp) Operating Experiences and Results," Proc. of the 10th PV Solar Energy Conference, Lisbon, Portugal, 8-12 April 1991.

GLOSSARY

ACID - A chemical compound which yields hydrogen ions when diluted in water. In the lead-acid battery, the "acid" is the sulphuric acid (H_2SO_4) used in the electrolyte.

ACTIVE MATERIALS - The materials in a battery which react chemically to produce electrical energy. In a lead-acid battery the active materials are lead peroxide (in the positive plate) and sponge lead (in the negative plate).

AIR MASS (AM) - A measure of the length of a path through the atmosphere to sea level traversed by light rays from a celestial body. Expressed as a multiple of the path length for a light source at the zenith. AM1 intensity is 1,000 W/m^2. AM1.5 is a radiation spectrum with a power density of 832 W/m^2 (defined by the PV standardization committee). At any point, the air mass value is,

$$AM = \frac{1}{\cos \alpha}$$

where α = angle between the sun line and the normal to the horizontal surface

AMBIENT TEMPERATURE - The temperature of the surrounding medium; usually refers to room or air temperature.

AMORPHOUS - A non-crystalline solid that has neither definite form nor structure.

AMORTIZATION - The gradual reduction of a monetary amount over a period of time. A general term which includes various specific practices such as depreciation, depletion, write-off of intangibles, prepaid expenses, and deferred charges; or general reduction of loan principal.

AMPERE-HOUR CAPACITY - The number of ampere-hours which a cell or battery can deliver under specified conditions, i.e., charging conditions, the rate of discharge, temperature, and final voltage.

AMPERE-HOUR EFFICIENCY - The electrochemical efficiency of a battery expressed as the ratio of the ampere-hours discharged (output) to the ampere-hours recharged (input). It is usually expressed as a percentage.

ANALOG TO DIGITAL CONVERTER - An electrical circuit which changes the analog signal such as a voltage to a digital signal (a binary number).

ANGLE OF INCIDENCE - Angle between the normal to a surface and the direction of incident radiation; applies to the aperture plane of a solar collector.

ANTIMONIAL-LEAD - An alloy of antimony and lead which may contain small amounts of other metals as impurities or minor constituents. It is used in the manufacture of certain lead-acid battery components.

ANTIREFLECTIVE COATING - A thin layer of transparent material which decreases light reflection and increases light transmission through a glass or the surface of a solar cell.

APPARENT POWER - Made up of two components of ac power, a real (kW) and a reactive (kVAR) component; kVA is their right-angle or "vectorial" sum.

ARRAY - A mechanically integrated assembly of panels containing PV modules with support structures and foundations, cabling, and electrical junction boxes.

ARRAY FIELD - The aggregate of all PV arrays and components in the field generating power within a given system.

ARRAY RATED EFFICIENCY - For a flat-plate PV array, the peak available power at a solar irradiance of 1,000 W/m^2 divided by the product of 1,000 W/m^2 and the total module area in m^2.

ARRAY SUBFIELD - A group of PV arrays associated by a distinguishing feature such as field geometry and electrical interconnection.

AZIMUTH - Angle between the north direction and the projection of the surface normal into the horizontal plane; measured clockwise from north. As applied to the PV array, 180-degree azimuth means the array faces due south.

BACK SURFACE FIELD - A built-in electric field at the back of a solar cell to reflect photo-excited charge carriers back to the side of the cell where they can be collected. It helps increase the cell efficiency.

BALANCE OF SYSTEM (BOS) - Parts of a photovoltaic system other than the PV array subsystem. This term is intended to refer to all other subsystems related to the PV array (i.e., the power conditioning, energy storage, power distribution, control, and DAS). BOS is sometimes defined loosely as any hardware other than PV modules.

BASE LOAD - The minimum amount of electric power which a utility must supply in a 24-hour period.

BATTERY - Two or more cells electrically connected, generally in series to meet the required voltage and ampere-hour capacity. A battery sometimes consists of cells in parallel and series.

BATTERY CAPACITY - Generally, the total number of ampere-hours that can be withdrawn from a fully charged cell or battery at a given rate of discharge to a specified cut-off voltage.

BATTERY CYCLE LIFE - The total number of charge/discharge cycles that the battery can operate to deliver its required minimum depth of discharge (i.e., the last cycle in which the battery reached its discharge voltage limit before it supplied its allowable DOD).

BATTERY FINAL CUT-OFF VOLTAGE - Predefined end-of-discharge voltage for discharge termination purposes.

BATTERY SELF-DISCHARGE - Loss of energy while in open-circuit stand condition.

BRANCH CIRCUIT - A number of PV cells or modules connected in series to provide power at the system voltage.

BREAK-EVEN COST - The cost of a photovoltaic system at which the cost of produced electricity equals the price of electricity from a competing source.

BRIDGE - A circuit part or topology of an inverter or a rectifier, consisting of four semiconductor switches for a single-phase and six semiconductor switches for a three-phase power converter. These switches are arranged in a "bridge" shape.

BRIDGE, WHEATSTONE - A bridge circuit for determining the value of an unknown component by comparison to one of known value.

BROWN-OUT - A planned voltage reduction by a utility company to counter excessive demand on their generation and distribution system.

BYPASS DIODE - A diode connected in parallel with a number of cells in series to provide an alternate current path in case of shading, cell cracks, or cell circuit failures.

CALCIUM LEAD - An alloy of calcium and lead used in the manufacture of grids and other components for certain types of lead-acid batteries.

CAPACITY COSTS - Those cost reductions resulting from utility system capacity requirements because of an alternative source of electric power.

CAPACITY FACTOR - Actual PV output energy divided by nominal array rating times 24 hours. The term is used by the utilities for comparison and planning purposes on grid-connected systems.

CATHODIC PROTECTION - A method of slowing down the rate of deterioration from oxidation (i.e., rusting) in metal pipes, bridges, etc., resulting from a difference in the electrical potential of the surrounding ground and the metal. A small electric charge is applied between the metal to be protected and the ground to oppose the flow of electrons such that the applied voltage is greater than the oxidation voltage.

CELL (BATTERY OR SOLAR) - The smallest functional unit of a battery or PV array.

CHARGE RATE - The current applied to a cell or battery to restore its available capacity. The charge rate is in amperes but is commonly normalized with respect to the rated capacity of the battery. For example, given the rated ampere-hour capacity, C, and the actual charging current, Ic, the charging rate is expressed as:

C/Ic

Therefore, given a 500-ampere-hour battery being charged by a 50-ampere current, the charge rate becomes:

$500/50 = 10$ or $C/10$

(10-h or C/10 rate of charge current)

CIRCUIT BREAKER - A device used in the power distribution line, which opens its contacts when the current passing through it exceeds a certain value, thus protecting the power source from destructive over-current.

CLEARNESS INDEX - The ratio of direct normal solar irradiance at a given location to the extraterrestrial irradiance at a given time. This term can be applied also for a given time interval, e.g., for one day or month.

COMMON-MODE NOISE and VOLTAGE - The component of noise or voltage which is common to the dc output and return lines with respect to input neutral.

COMMUTATION - The action of transferring current from one switching device to another in a power conditioning unit. (Stand-alone inverters must be self-commutated but utility interactive inverters make use of line commutation where the circuit operation depends on the utility line connection. Utility interactive inverters may also be self-commutated. Other types of commutation have been used, such as natural and load commutation, but generally in special applications).

CONCENTRATOR - Devices that increase, or "concentrate" the sun's energy on a much smaller surface area. Types of concentrators are basically reflective (flat and parabolic) and refractive (Fresnel lens).

CONSTANT CURRENT DISCHARGE - Discharging of a cell or battery in which the discharge current is held constant throughout the entire discharge. It is commonly used for test purposes such as for voltage and capacity checks.

CONSTANT POTENTIAL CHARGE - A method of charging in which the voltage across the terminals of the cell or battery is held constant after it reaches the preset charge voltage limit.

CONSTANT VOLTAGE CHARGE - Same as CONSTANT POTENTIAL CHARGE.

CONVERTER (dc-dc) - An electrical equipment that changes the input dc voltage to another higher or lower dc voltage.

CYCLE - When referring to battery operation, a cycle comprises one discharge and one charge period occurring. As referred to an ac system, a cycle follows a sinusoidal waveform, with its value starting from zero, rising to a peak value in the positive direction, returning to zero, and then going in the negative direction in a similar manner.

CYCLE LIFE - (see BATTERY CYCLE LIFE).

CZOCHRALSKI (Cz) PROCESS - Method of growing a large-size high quality silicon crystal by slowly lifting a seed crystal from a molten bath of the material under careful cooling conditions. Most available monocrystalline solar cells are from Cz-grown ingots.

DATA ACQUISITION SYSTEM (DAS) - A key part of the DATA COLLECTION SYSTEM which has the function of collecting analog and digital signals, converting the analog signals to digital, multiplexing the signals, doing arithmetic and logical operations, and storing measured and calculated values. Depending on the specific manufacturer's model, the DAS may have its own display and printing devices built in (e.g., data loggers).

DATA COLLECTION SYSTEM - A complete system for collecting, analyzing, storing, and displaying measured and calculated data. It consists of sensors, signal conditioning, cabling, and the DAS.

DECLINATION - When referring to the sun, the solar declination angle is the angular position of the sun at its highest point in the sky with respect to the equatorial plane.

DENDRITE - A slender threadlike spike of material, metallic or crystalline structure.

DEPTH OF DISCHARGE (DOD) - Ampere-hours removed from a charged cell or battery in one discharge period. It is often expressed as a percentage of rated capacity.

DIFFERENTIAL CHANNEL or DIFFERENTIAL-MODE SIGNAL - As referred to electrical measurements, the signal has ground that is isolated from others. Also see SINGLE-ENDED CHANNEL.

DIFFUSE IRRADIANCE - Solar radiation component due to reflection and scattering by the atmosphere; in terms of three components of irradiance, the total (or global) normal minus the direct normal irradiance.

DIODE, RECTIFIER - Power semiconductor switch that conducts whenever its anode voltage is more positive than its cathode.

DIRECT CURRENT (DC) - Electric current in which electrons flow in one direction only; opposite of alternating current (ac).

DIRECT NORMAL (or BEAM) IRRADIANCE - Global normal irradiance minus the diffuse and reflected radiation.

DISCHARGE RATE - The current removed from a battery cell or battery. This current can be expressed in hours or fraction of rated capacity (see CHARGE RATE).

DISCOUNTED CASH FLOW - A cash flow occurring some time in the future which has been discounted by a given discount factor on a compounded basis; the present value of a future cash flow.

DISCOUNTED RATE OF RETURN - The effective periodic rate that would equate the present value of an investment with the accumulated present values of a stream of future cash flows, each appropriately discounted by the periodic rate.

DYNAMIC HEAD - The head or pressure loss in pipes due to friction inside the pipes.

EFFECTIVE VALUE - The value of a sine wave form of ac that is equivalent to a dc (0.707 x peak voltage or current).

ELECTROLYTE - An electrolyte is a conducting medium in the battery cell in which the flow of electric current takes place by the migration of ions.

ELECTROMAGNETIC INTERFERENCE - Undesirable energy usually emitted by switching power supplies such as inverters and converters; it can be conducted or radiated.

ELEVATION - Used in reference to the sun's position relative to the horizontal surface, specifically the angle between the horizontal plane and the direction of incident radiation.

END-OF-CHARGE VOLTAGE - The cell or battery voltage at which charging is normally terminated by the charging source or continued at constant voltage condition.

ENERGY PAYBACK TIME - The time required for any energy producing system or device to produce as much energy as was required in its manufacture.

EQUALIZATION - The process of restoring all cells in a battery to an equal state of charge.

EQUATION OF TIME - The difference between the standard time and the solar time (see also SOLAR NOON).

EQUINOX - The time when the sun in its apparent motion in the celestial sphere crosses the equator, and the solar declination is zero. March 21 is the vernal equinox and September 21 is the autumnal equinox in the northern hemisphere. The vernal equinox is more precisely defined as the point of intersection of the ecliptic plane and the equator on the celestial sphere.

ETHYLENE VINYL ACETATE (EVA) - An encapsulant used between the glass cover and the solar cells in PV modules. It is durable, transparent, resistant to corrosion, and flame retardant.

EXTRATERRESTRIAL IRRADIANCE - The solar flux outside the earth's atmosphere; see SOLAR CONSTANT.

FERRITES - Non-conductive ferromagnetic materials used for high-frequency transformer and inductance cores.

FILL FACTOR - The ratio of the maximum power output of a photovoltaic device and the product of its open-circuit voltage and short-circuit current.

FLAT-PLATE ARRAY - A PV array using non-concentrating modules with a flat surface for collecting solar energy.

FLOAT ZONE (Fz) PROCESS - Process used in the fabrication of single-crystal ingot. It uses induction heating at 1-3 MHz frequency and the liquid does not come in contact with any foreign material. This results in very pure crystal. Conventional Fz crystals contain two orders of magnitude less of carbon and oxygen impurities than Cz-grown crystals. Finished Fz cells exhibit typically 1-2% higher efficiency and are about 25% higher in cost than the Cz cells.

FORMATION - The process of activating the electrodes of a battery cell for normal cell operation. A cell or a complete battery upon installation is not necessarily fully formed and could require several cycles before complete formation. The battery will be able to give its usable capacity, but not its rate capacity.

FRESNEL LENS - A type of magnifying lens that focuses the direct normal component of sunlight onto one or more solar cells. The side of the lens facing the sun is flat and the side toward the solar cell has segmented "saw-tooth" construction to focus the light onto a circular or line pattern on a surface parallel to the lens surface. There are only two types used in PV applications, *point focus* and *line focus* fresnel lens. Typical devices are made of acrylic material which is either compression molded or cast.

FULL-WAVE RECTIFIER - A rectifier which produces a dc pulse output for each half cycle of applied ac voltage.

FUTURE VALUE - The amount yielded after compounding for a given number of periods at a stated rate of interest per period.

GALLIUM ARSENIDE (GaAs) - A black chemical compound that in its mono-crystal form has very good properties as a photovoltaic device.

GASSING - The evolution of gas from the electrodes in a cell. Gassing commonly results from the electrolysis of water in the electrolyte during charging.

GIGAWATT - One billion watts or one million kilowatts.

GLOBAL IRRADIANCE - Solar radiation flux (W/m^2) on a flat surface, which includes all solar radiation components (DIRECT NORMAL, DIFFUSE, and REFLECTED irradiances).

GRID - The network of ac power transmission lines and transformers used in central power systems. Grid power means utility power.

GROUND LOOP - An undesirable feedback condition caused by two or more circuits sharing a common electrical line.

HALF-WAVE RECTIFIER - Rectifier which permits half wave of an ac cycle to pass and reject reverse current of remaining half cycle. Its output is pulsating dc.

HARMONIC FREQUENCY - Frequency which is a multiple of fundamental frequency. For example, if the fundamental frequency is 50 Hz, second harmonic is 2 x 50 Hz or 100 Hz; third harmonic is 3 x 50 Hz or 150 Hz, and so on.

HEAT SINK - An aluminium chassis for absorbing and dissipating heat generated by power semiconductors, which assists in cooling the semiconductors.

HERTZ - A unit of frequency equal to one cycle per second.

HOT SPOT - An undesirable phenomenon of PV device operation whereby one or more cells within a PV module or array act as a resistive load, resulting in local overheating or melting of the cell(s).

HYBRID SYSTEM - Generally, a combination of two different power sources such as PV and wind, PV and thermal, or PV and Diesel. It often includes batteries.

HYDROMETER - Bulb-type instrument used to measure specific gravity of a liquid such as electrolyte in a battery cell.

INGOT - Conveniently shaped piece of metallic or other material, such as silicon, to be used for further processing, such as refining, cutting, or shaping; for example, Czochralski ingot from which monocrystalline Si-wafers are cut.

INSOLATION - Incoming solar radiation flux (W/m^2) or energy (Wh/m^2) received per unit area; also commonly referred to as irradiation.

INSULATED-GATE BIPOLAR TRANSISTOR (IGBT) - Power semiconductor switch made up of power MOSFETS and power bipolar transistors, used in high-frequency high-voltage inverters.

INVERTER - A power conversion device which changes dc voltage input to an ac voltage output.

I-V CURVE - The graphical representation of the current versus the voltage of a photovoltaic device as the load is increased from zero voltage to maximum voltage.

LANGLEY - Unit of solar irradiance; one Langley is one gram calorie per cm^2.

LEAD-ACID BATTERY - Electrochemical storage device that includes pure lead, lead-antimony, or lead-calcium types with sulphuric acid electrolyte. Two types are available, "sealed" and unsealed, often called the "maintenance-free" and flooded battery, respectively.

LIFE-CYCLE COSTING - A method of calculating the total cost or value of an item over its full life time, including interest, maintenance costs, fuel costs, replacement costs, etc.

LINEARITY - The absolute proportionality between input and output over the full output range; non-linearity is deviation from linearity.

LOAD - Any electrical device or appliance that uses power; often referred to as the "consumer".

MAINTENANCE-FREE BATTERY - Lead-acid battery in which the electrolyte is suspended in gel material; also called the VALUE-REGULATED battery.

MASTER-SLAVE OPERATION - A method of interconnecting two or more power conversion components so that one of them (the master) serves to control the others (the slaves). The outputs of the slave units remain equal to or proportional to the output of the master. For example, one master and other slave inverters can operate with all their outputs connected in parallel.

MAXIMUM POWER POINT - The operating point on a PV device I-V characteristic at which maximum power is delivered.

MAXIMUM POWER TRACKING - A control strategy whereby system operation is at or sufficiently close to the PV array maximum power point. Maximum power point tracking requires the presence of a load capable of accepting the full available power, such as a storage battery or a utility network. Many PV inverters, especially grid-connected types, have this maximum power tracking capability.

MICROMETRE ("Micron") - Equal to one-millionth of a metre (0.000001 m or μm).

MODEM - Modulator/demodulator. A device that converts signals from one form to a form compatible with another kind of equipment; commonly used for transmitting computer data over telephone lines.

MODULE - A number of PV cells connected together, sealed with an encapsulant, and having a standard size and output power; the smallest building block of the power generating part of a PV array.

MODULE AREA - Area of a PV module bounded by the outside edges of the module metallic frame.

MODULE PACKING EFFICIENCY or FACTOR - The ratio of the total active solar cell area to the total aperture area of a flat-plate module.

MODULE RATED EFFICIENCY - Peak available power of a PV module at 1000 W/m^2 irradiance and 25°C average cell temperature divided by the product of 1000 W/m^2 and the total module area in m^2.

MONOCRYSTALLINE SILICON - See SINGLE CRYSTAL SILICON.

MONOLITHIC (CONSTRUCTION) - Fabrication as a single structure or crystal.

MULTIPLE JUNCTION CELL - A photovoltaic cell with two or more cell junctions, each of which responds to a particular wavelength or colour of the solar spectrum to achieve greater cell efficiency.

NET PRESENT VALUE - The present value of all savings resulting from the PV system minus the present value of all costs associated with the PV system, including initial investment replacements, and operation and maintenance (less salvage value). Present value denotes the equivalent value at the present time of future cash flows; it is found by discounting future cash flows to the present to account for the time value (earnings potential) of money.

NOMINAL OPERATING CELL TEMPERATURE (NOCT) - The solar cell temperature at a reference environment defined as 800 W/m^2 irradiance, 20°C ambient air temperature, and 1 m/s wind speed with the cell or module in an electrically open circuit state.

ONE SUN - Solar radiation of 1,000 W/m^2, also referred to as Air Mass 1 intensity.

OPEN-CIRCUIT VOLTAGE - Terminal voltage of battery or PV device when no current is flowing in the external circuit.

PANEL - A collection of PV modules fastened together in the same plane, assembled, and wired.

PEAK WATT - Term used in performance rating of PV cells, modules, and arrays, usually at sunlight intensity of 1000 W/m^2 with an average cell temperature of 25°C.

PERIOD - Time duration (usually in seconds) of one cycle. It is equal to one divided by frequency ($T = 1/f$).

PHASE - Relationship between two vectors with respect to angular displacement.

PHOTON - A particle of light which acts as an indivisible unit of energy.

PHOTOVOLTAIC EFFECT - The generation of a voltage at the junction of two materials when exposed to light. A photovoltaic cell is a device that converts light directly into dc electricity.

PN JUNCTION - The line of separation between N-type and P-type semiconductor material.

POLARITY - Property of a device or a circuit which has poles such as north and south or positive and negative.

POLARIZATION - Defect in a battery cell caused by hydrogen bubbles surrounding the positive electrode, effectively insulating it from chemical reaction.

POLYCRYSTALLINE SILICON - Silicon which has solidified rapidly enough to produce many small crystals which are arbitrarily arranged. This type of cell has grain boundaries, randomly oriented, visible on the active surface.

POWER CONDITIONING (EQUIPMENT) - A class of equipment that regulates the output voltage and/or controls the quality of power; a collective term for voltage regulation circuits, battery charger, inverter, converter, transformer, and rectifier.

POWER FACTOR - The ratio of real power (watts) to apparent power (volt-amps) in an ac circuit. Displacement power factor is the ratio of fundamental watts to fundamental RMS volts times RMS amps.

POWER MOSFET - A transistor-based power semiconductor switch, easily driven and very fast.

PRESENT VALUE - The current equivalent value of cash available immediately for payment or a stream of payments to be received at various times in the future. The present value will vary with the discount interest factor applied to the future payments.

PRIMARY CELL - A battery cell that cannot be recharged.

PULSE-WIDTH MODULATION (PWM) - A method of voltage regulation used in power electronic converters whereby their output is controlled by varying the width, but not the amplitude, of a train of pulses which drive a power switch.

PYRANOMETER - An instrument for measuring total hemispherical solar irradiance on a flat surface, or "global" irradiance; thermopile sensors have been generally identified as pyranometers, however, silicon sensors are also referred to as pyranometers.

PYRHELIOMETER - An instrument that measures only the direct normal (beam) component of solar radiation using a collimated tube at normal incidence; it requires a 2-axis sun-tracker.

QUAD - A unit of energy in BTU. Quadrillion BTU = 10^{15} BTU = 1.7×10^8 barrels of oil = 2.93×10^{12} kWh

RATED CAPACITY - The rated capacity of a storage battery is usually the number of ampere-hours defined by the cell manufacturer, which the battery is capable of delivering when fully charged and discharged under specified conditions of temperature, current, and final voltage. Battery or cell rating is generally specified in ampere-hours. (Note: Rated capacity is available only when the battery is relatively new. The actual capacity available is a significant function of the discharge rate, and it gradually decreases with cycling.)

REACTIVE POWER - The reactive component of the ac power due to capacitance or inductance. The vectorial sum of real power (kW) and reactive power (kVAR) is the kVA.

REAL POWER - The non-reactive component of the ac power, or that resulting from the resistive element; see REACTIVE POWER.

RECTIFIER - A power conditioning equipment which converts ac power to dc; usually used for battery charging or as part of a UPS.

RELAY - An electromechanical switch activated by magnetic action through a coil.

REMOTE SITE - Site which is not connected to a utility grid.

RESPONSE TIME - Time period for the output voltage to change to 96 or 98% of the final value when a step change occurs in the input variable. See also TIME CONSTANT.

RETROFITTING - Placing solar energy devices on an existing building structure to provide part or all of the energy required by that structure.

REVERSE BIAS - Condition where the current-producing capability of a PV cell is significantly less than that of other cells in its series string. This can occur when a cell is shaded, cracked, or otherwise degraded or when it is electrically poorly matched with other cells in its string.

SEALED CELL or BATTERY - Cell or battery which has no provision for the addition of water or electrolyte, or for external measurement of electrolyte specific gravity. It has captive electrolyte and sealed vent caps; also called a valve-regulated or "maintenance-free" battery.

SECONDARY CELL - An electrochemical battery cell that can be recharged by passing a current through the cell.

SEMICONDUCTOR - A class of materials with electrical properties somewhere between those of metals (conductors) and insulators. In photovoltaic cells they absorb photons and emit electrons to produce electricity.

SHORT-CIRCUIT CURRENT - When exposed to a light source, the current produced by a PV device when its output voltage is nearly zero. It is determined by connecting a low resistance across the output terminals and measuring its voltage.

SIEMENS PROCESS - A commercial method of making purified silicon.

SILICON (Si) - A non-metallic element which constitutes more than one quarter of the earth's crust, and is found primarily in sand, but also in combination with other elements and in almost all rocks. In purified forms it acts as a semiconductor and has good properties for use as a photovoltaic device.

SINGLE CRYSTAL (or MONOCRYSTALLINE) SILICON - The perfect state of a solid in which all of the atoms are arranged in an ordered fashion. Single crystal silicon PV cells usually have a very even surface colour in contrast to the polycrystalline cells with grain boundaries.

SINGLE-ENDED CHANNEL - When referring to electrical measurements, the signal has ground common to others (i.e., not isolated).

SOLAR ALTITUDE ANGLE - The measure of the sun's angular distance from the horizon.

SOLAR CELL - Refers to a photovoltaic device which generates electricity when exposed to sunlight.

SOLAR CONSTANT - Defined and commonly accepted as 1,353 W/m^2 of solar flux outside the earth's atmosphere at one Astronomical Unit (i.e., mean distance of the earth from the sun, or about 150 million km).

SOLAR HOUR ANGLE - The unit of angular measurement of time. It is measured from solar noon and is positive before solar noon and negative after solar noon.

SOLAR NOON - The time at which the sun crosses the local meridian; usually the time at which the sun's altitude is at the highest point in the sky. Note that solar noon is different for locations at different longitudes.

SOLAR RADIATION - The radiation emitted by the sun; same as irradiance.

SOLAR TIME or SOLAR DAY - A measure of time from the moment the sun crosses the local meridian to the next time it crosses the same meridian.

SOLSTICE - The time when the sun in its apparent motion in the celestial sphere attains the maximum distance from the equator. In the northern hemisphere, June 21 is the summer solstice, and December 21 is the winter solstice, and the solar declination angle is maximum (northerly peak of + 23.45 degrees).

SPECIFIC GRAVITY - The specific gravity of an electrolyte in a lead-acid battery cell is the ratio of the weight of a given volume of electrolyte to the weight of an equal volume of water at a specified temperature.

STAND ALONE (SYSTEM) - A PV system that is not connected to an auxiliary power source or one that operates independently of the electric utility lines. It usually contains one or more batteries, but many pumping systems comprise only the PV array and the inverter.

STANDARD TEST CONDITIONS (STC) - A set of conditions at which many organizations test and rate PV modules:

- 1 kW/m^2 at defined spectral distribution at AM1.5
- Average cell temperature of 25 ± 3°C

STATE OF CHARGE - The available capacity in a cell or battery expressed as a percentage of rated capacity or in absolute value (Ah). For example, if 25 Ah have been removed from a fully charged new 100-Ah cell, the new state of charge is 75%. (Note: One cannot assume that the rated capacity is available throughout the battery life. The actual capacity of any battery decreases with cycling and ageing. Thus, extreme care should be taken in using this term and also in the calculation of state of charge.)

STRATIFICATION - A condition in a flooded lead-acid battery cell when the acid concentration of the electrolyte varies from top to bottom of the cell.

SUBSYSTEM - A subset of a system comprising one or more components which serve a given function of a system. A PV/battery system is usually grouped into PV array, battery, power conditioning, power distribution, control, and monitoring subsystems.

SUN PHOTOMETER - An instrument used to measure the turbidity in the atmosphere.

SURGE ARRESTOR - A protective device for limiting surge voltages by discharging or bypassing surge current; it also prevents the continued flow of fault current while remaining capable of repeating these functions. Arrestors are normally used in power and signal lines to protect the equipment from high current surges such as from overvoltage or lightning flashes.

THREE-PHASE ALTERNATING CURRENT - Combination of three alternating currents which have their voltages displaced by 120 degrees or one-third cycle

THIN FILM - A layer of semiconductor material; such as polycrystalline silicon or gallium arsenide, a few microns or less in thickness, used to make photovoltaic cells.

THYRISTOR (SILICON-CONTROLLED RECTIFIER) - A latching power semiconductor switch, turned on via its gate electrode. For dc applications, such as inverters, it needs an external commutation circuit to be turned off.

TILT ANGLE - The angle which the plane of a sensor or a PV module makes relative to the horizontal plane.

TIME CONSTANT - Time period required for the voltage of a capacitor in an RC (resistance-capacitance) circuit to increase to 63.2% (i.e., 1 - 1/e) of the final value, or reduce to 36.8% of final value. This term is often used to describe the responsiveness of a transducer, such as solar irradiance sensor, to step changes in the input quantity. Also see RESPONSE TIME.

TOTAL HARMONIC DISTORTION (THD) - The square root of the sum of the squares of the amplitudes of all the harmonics present in a wave function. As referred to the inverter, THD is expressed in percent of the amplitude of the fundamental component in the Fourier series [1] or:

$$THD = \frac{\left[\sum_{i=2}^{\infty} Y_i^2 \right]^{\frac{1}{2}}}{Y_1}$$

---

[1] The Fourier series is:

$$f(t) = (4/\pi) \cos(W_o t) - (1/3) \cos(3 W_o t)$$
$$+ (1/5) \cos(5 W_o t) - (1/7) \cos(7 W_o t)$$
$$+ ... + (1/2n-1) \cos([2n-1] W_o t)$$

The first term in this series is the fundamental component because W_o is the fundamental frequency.

where Y_i = the amplitude of the ith harmonic

Y_1 = the amplitude of the fundamental component

TRANSFORMER - Device which transfers energy from one circuit to another by electromagnetic induction.

TRANSISTOR - Semiconductor device derived from two words, transfer and resistor.

TRUE POWER - Actual power absorbed in a circuit.

TURBIDITY - A condition of the atmosphere which reduces its transparency to solar radiation due to scattering by atmospheric gases, dust, and aerosols.

UNINTERRUPTIBLE POWER SUPPLY (UPS) - Term used to define a reliable power source for special equipment such as computers and system controllers which must continue operating in the event of failure of the main power system. It usually contains a rectifier, battery, and inverter which serve as a back-up power supply to provide continuous power (i.e., on an uninterrupted basis).

VARISTOR - A voltage-dependent variable resistor, normally used to protect equipment from high voltages; e.g., from lightning strikes, by shunting the energy to ground.

VOLTAGE REGULATOR - An electrical device that controls its output voltage to predetermined levels.

WATT-LESS POWER - Power not consumed in an ac circuit due to reactance.

ZENITH - The point of the celestial sphere that is vertically above the observer. The solar zenith angle is the sun's angular distance from the zenith.

CONVERSION FACTORS

A. Engineering Units

To convert from	To	Multiply by
Solar Irradiance		
$cal \times cm^{-2} \times min^{-1}$	$J \times cm^{-2}$	4.1868
	$W \times m^{-2}$	697.8
$cal \times cm^{-2}$	$kWh \times m^{-2}$	0.01163
$J \times cm^{-2}$	$cal \times cm^{-2}$	0.23885
$J \times cm^{-2}$	$kWh \times m^{-2}$	0.0027778
$kWh \times m^{-2}$	$cal \times cm^{-2}$	85.985
$kWh \times m^{-2}$	$J \times cm^{-2}$	360
Langley	$cal \times cm^{-2}$	1.000
$W \times m^{-2}$	$cal \times cm^{-2} \times min^{-1}$	0.0014331
$W \times m^{-2}$	$mcal \times cm^{-2} \times s^{-1}$	0.023885
Power		
kilowatts (kW)	horsepower	1.341
watts (W)	Btu/hour	3.41
watts (W)	joules/sec	1
Energy		
calories (cal)	Btu	0.00397
joules (J)	calories	0.239
kilowatt-hours (kWh)	Btu	3410
Force		
dynes	newtons	0.00001
newtons (N)	pounds	0.225
Length		
centimetres (cm)	inches	0.394
kilometres (km)	miles	0.621
metres (m)	feet	3.28
millimetres (mm)	inches	0.0394
Area		
acres	hectares	0.405
acres	sq. metres	4047
hectares (ha)	acres	2.47
sq. centimetres (cm^2)	sq. inches	0.155
sq. metres (m^2)	sq. feet	10.76
Volume		
cubic centimetres (cm^3)	cubic inches	0.0610
cubic metres (m^3)	cubic feet	35.3
gallons, US, (gal)	litres	3.79
gallons, UK, (gal)	litres	4.55
litres (l)	quarts, U.S.	1.057
Mass		
kilograms (kg)	pounds	2.205
pounds (lb)	kilograms	0.454
tons (metric)	pounds	2,205
tons (metric)	kilograms	1,000
Pressure		
atmosphere	bars	1.013
atmosphere	grams/sq. cm	1,033
atmosphere	pounds/sq. in	14.7

B. Temperature Conversion

°C	°F	°C	°F	°C	°F
-50	-58	40	104	120	248
-40	-40	45	113	130	266
-30	-22	50	122	140	284
-25	-13	55	131	150	302
-20	-4	60	140	160	320
-15	5	65	149	170	338
-10	14	70	158	180	356
-5	23	75	167	190	374
0	32	80	176	200	392
5	41	85	185	225	437
10	50	90	194	250	482
15	59	95	203	275	527
20	68	100	212	300	572
25	77	105	221	325	617
30	86	110	230	350	662
35	95	115	239	400	752

°C = 0.5556 . (°F - 32)

C. Si Unit Prefixes

Factor	Prefix	Symbol
10^{18}	exa	E
10^{15}	peta	P
10^{12}	tera	T
10^{9}	giga	G
10^{6}	mega	M
10^{3}	kilo	k
10^{2}	hecto	h
10	deka	da
10^{-1}	deci	d
10^{-2}	centi	c
10^{-3}	milli	m
10^{-6}	micro	μ
10^{-9}	nano	n
10^{-12}	pico	p
10^{-15}	femto	f
10^{-18}	atto	a

D. One European Currency Unit (1.0 ECU) in June 1991

Country	Currency	
Belgium	42.3	BF
Denmark	7.89	DKr
France	6.98	FF
Germany	2.06	DM
Greece	225.3	Dra
Ireland	0.77	Irf£
Italy	1,530.00	Lit
Luxembourg	42.3	Flux
Portugal	179.5	Esc
Spain	127.40	Ptas
The Netherlands	2.32	HFl
United Kingdom	0.70	UK£
United States	1.14	US$

(Source: CEC/DG XVII, THERMIE, June1991)

Appendix 1

AGHIA ROUMELI PV PILOT PLANT

1. INTRODUCTION

This appendix describes the 50-kW AGHIA ROUMELI PV pilot plant and its status as of 1989. The plant has been in operation since November 1982. Figure 1 shows the array field and the control building.

The purpose of the plant is to demonstrate the technical, economic, and social suitability of photovoltaic generators to meet the energy requirements of villages far away from the utility grid. The plant was originally provided to meet the electrical power needs of 105 inhabitants in the village. Electricity is mainly required for water pumping, lighting, refrigeration, and other usual domestic appliances.

Information contained herein was obtained from ref. [1 to 15] and individual contacts with the plant designer and operator.

2. PLANT DESIGNER, OWNER, AND OPERATOR

SERI Renault Ingénierie was responsible for the design, installation, and initial operation of the plant. This plant has been handed over to the utility organisation, Public Power Corporation of Greece, which is now responsible for the plant. Their address is as follows:

Public Power Corporation
Attn: Renewable Energies Dept.
10, Navarinou St.
GR-10680 Athens
Greece

3. SITE INFORMATION

3.1 Location

Aghia Roumeli is a very isolated village on the southern coast of Crete Island (see Fig. 2). It can be reached either by a 1-hour boat ride from Sfakia or by a 5-hour walk from the village of Xiloskalo through the forest and the Samaria Gorge. Its coordinates are as follows:

- Latitude: 35.23°N
- Longitude: 23.97°E
- Altitude: 5 m

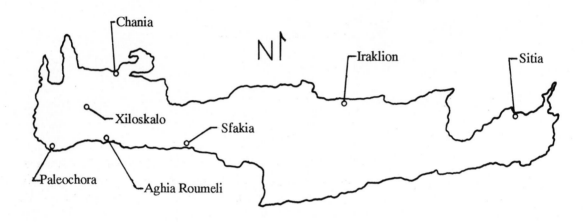

Figure 2. Location of the AGHIA ROUMELI PV Pilot Plant

3.2 Site Constraints

Public Power Corporation (PPC), the Greek National Utility, is not allowed by the government to lay power cables through the protected Samaria Park. Likewise, power connection along the coast with the utility serving the closest village was not planned at the onset of this project because it would require more than 15 km of cable under the sea; moreover, the low annual village load demand in 1986 did not justify extending the grid to Aghia Roumeli. The utility grid, however, was extended to the village in mid 1989.

3.3 Solar Radiation and Climatic Conditions

The following solar radiation and meteorological data were acquired from measurements at Paleochora and at Chania:

- Insolation, global horizontal:
  - o Annual: 1,700 kWh/m^2
  - o Daily average in July: 7.0 kWh/m^2
  - o Daily average in December: 2.5 kWh/m^2
- Ambient air temperature: 38°C maximum, 4°C minimum
- Wind speed: 120 km/h maximum at 2-m height

3.4 Load Demand

A typical energy demand of the villages in winter and summer periods is shown in Table 1, indicating the daily energy needs supplied by privately-owned Diesel generators.

Table 1. Daily Energy Demand of the Village [9]

Users	Winter (kWh)	Summer (kWh)
34 homes	17	34
7 hotels	35	105
8 shops	12	40
Total kWh:	64	179

4. SYSTEM DESCRIPTION AND PERFORMANCE

4.1 Plant Architecture and Layout

The general layout for the photovoltaic plant was drawn by a Greek architect under SERI Renault and PPC direction. The main considerations were environmental concern, protection against vandalism, and possible flooding of the site by the river. Figure 3 is a plan view of the PV array field layout and the building.

The building and the enclosure were constructed using local stones and pebbles available on site. Thus, the plant blended well with the site features.

4.2 PV System Configuration

Figure 4 is a simplified block diagram of the PV plant. Its major components and their key features are listed in Table 2. The array tilt angle of 25° (latitude minus 10 degrees) was selected to obtain a higher array output during summer months [10].

4.3 System Operation and Management

Old Configuration

Both the array and battery were connected to the Logistronic unit via one contactor. This contactor was operated by the controller. The array is capable of providing a nominal current of 166 A at the rated peak power (50 kWp/300 Vdc). The battery requires a current of about 150 A at a 10-hour charge rate so that normal battery charging can be performed by connecting the array directly across the battery. When the battery voltage reached an average of 2.35 Vdc per cell, charging was continued via a series resistor which dropped the charge current to a trickle (about C/40 rate).

Battery overcharging and overdischarging protection were a main concern. The Logistronic unit was intended to play a major part by calculating the battery state of charge (SOC). However, this function never worked properly due to either failure of its internal bias voltage power supply or its inability to count or keep track of the ampere-hour integral accurately. To prevent deep discharging, the inverter and/or the loads were manually disconnected when the battery dropped to 30% state of charge as indicated by the Logistronic unit. Whenever this happened, the operator was required to turn on the Diesel generator manually.

251

In 1989, the utility transmission network was extended to Aghia Roumeli. The PV system configuration, however, remained the same (i.e., manual connection of PV/inverter, Diesel generator, or the utility power source to the village grid) because the inverter was not designed for grid-connected operation.

The PV array field and the control building are at the lower left corner

24 Photon PV modules on one array panel

The control building is made of local stone

Figure 1. View of Array Field and Plant Control Building

252

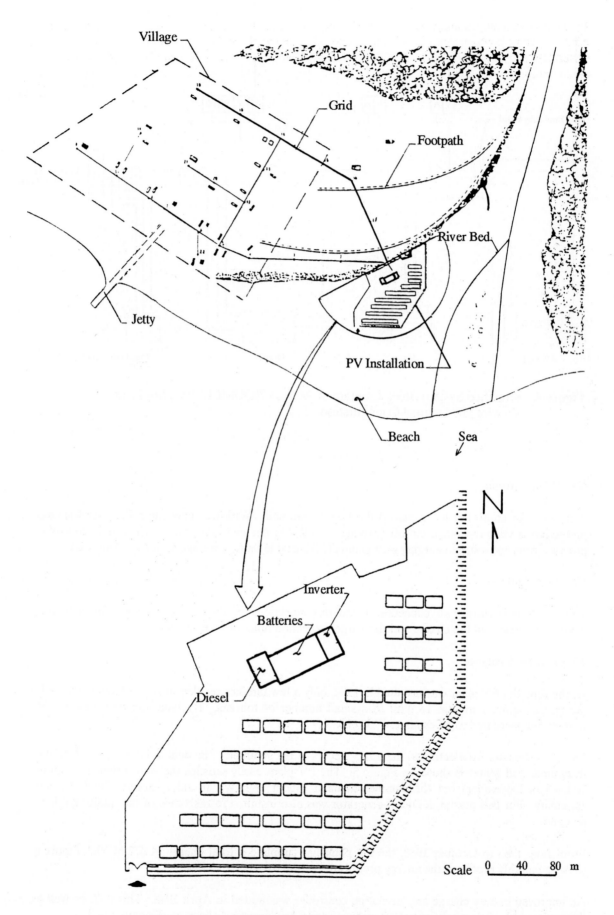

Figure 3. Layout of the AGHIA ROUMELI PV Pilot Plant

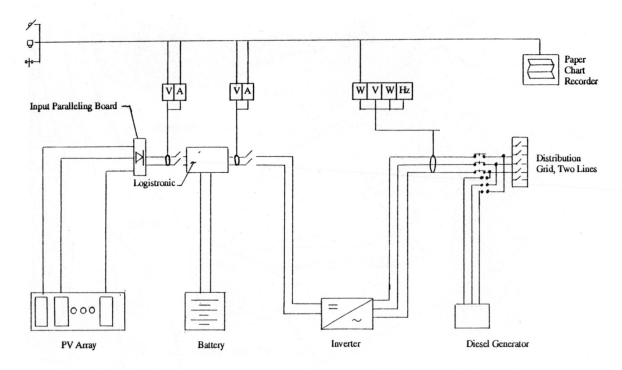

Figure 4. Simplified System Block Diagram of AGHIA ROUMELI PV Pilot Plant,
Showing the Original Configuration

New Configuration

A new charge controller has replaced the Logistronic unit. With this new controller overdischarge protection is via low-voltage cut-off (average of 1.9 Vdc per cell, adjustable). The new controller provides constant voltage charging with manually adjustable battery voltage limit (see Annex B).

4.4 Plant Control

The operator performs all connections and disconnections of the inverter. The plant's battery controller carries out the charge and discharge protection functions automatically.

4.5 Plant Performance

At the time the PV plant was installed in 1982, only a few families resided in Aghia Roumeli, mostly during the summer months to tend to a small number of tourists. By 1988, the total number of tourists had increased to an estimated 250,000.

The peak demand for electricity, therefore, occurs in the summer. The demand during a typical day in summer and winter is shown in Figure 5. The PV plant easily satisfies the winter demand (about 70 kWh in December) but the summer demand (about 180 kWh in July) exceeds the PV plant capability. For this reason, a Diesel generator was also installed to take care of the additional load demand.

From June 1983 to October 1988, the PV plant has supplied a total energy of 572 MWh. Figure 6 shows a monthly profile of the energy produced.

An improved battery charge and discharge controller was added in April 1988. This unit, as well as the battery with 12 new cells out of 150 cells in series, has apparently been working very well.

Table 2. Major Components and Their Features (AGHIA ROUMELI)

Solar Array
Rated Peak Power:	50 kW
Array Tilt Angle:	25°, South-facing
Tilt Angle Adjustment:	None
Nominal Bus Voltage:	300 Vdc
Number of Subarrays (Strings):	60
Total Number of Modules:	720
Number of Modules per String:	12
Module Manufacturer/Type:	France-Photon/FPG 72
Module Rating:	62.8 W
Number of Cells per Module:	68
Number of Bypass Diodes:	1 per 17 cells in series
Module Weight:	14.5 kg
Module Dimension:	64 cm x 128 cm
Module Area:	0.8192 m^2
Active Solar Cell Area/Module:	0.5325 m^2
Solar Cell Type/Size:	Monocrystalline/100 mm dia.
Cell Packing Factor:	65 %
Total Land Area:	3,000 m^2

Battery
Total Battery Energy Rating:	450 kWh
Cell Capacity Rating:	1,500 Ah
Voltage Range:	292-353 Vdc (300 V nominal)
Allowable Depth of Discharge:	80 %
Number of Batteries:	1
Number of Cells per Battery:	150
Cell Manufacturer/Part No.:	Varta Bloc/VB 2415
Cell Weight:	112 kg
Cell Dimension:	550 x 307 x 383 mm

Inverter
Rated Power:	40 kVA
Number of Inverters:	1
Type:	Self-commutated
Manufacturer:	Jeumont-Schneider
Input Voltage Range:	285-390 Vdc
Output Voltage/Phase:	380 Vac/3-phase
Frequency:	50 Hz
THD:	5 %
Efficiency, 100 % Full Load:	93 %
10 % Full Load:	85 %
Inverter Dimension:	230 x 67 x 88 cm
Inverter Weight:	785 kg

Controller (for Battery)
Manufacturer/Description:	Univ. of Patras (S. Pressas)
Model:	Charge/discharge controller
Features:	Constant charge voltage limit, 2.3 to 2.6 V per cell, adjustable; discharge limit, 1.8 to 2.0 V per cell, adjustable

Data Acquisition
Manufacturer/Description:	HP 3254 DAS and HP 9114 Computer (newly installed, 1988)

Auxiliary Power Source
Rated Power:	40 kVA
Manufacturer/Description:	Ansaldo Diesel Generator
Model:	H/Z Diesel, Type ASN 63, PF =0.8, Schoenebek motor, 92 BHP continuous

5. COMPONENT DESCRIPTION

5.1 Solar Array

5.1.1 Array and Cabling

The photovoltaic array is composed of 60 subarrays. Each subarray string has 12 modules connected in series. Each string is routed to the electrical terminations board in the building, which connects all of them in parallel.

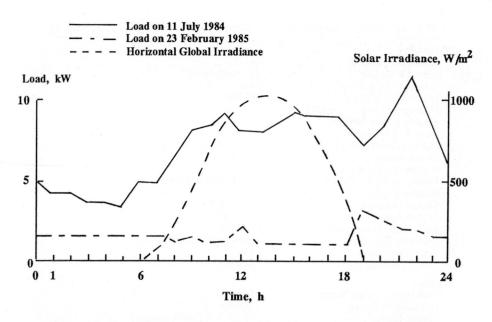

Figure 5. Typical Daily Profiles of Load and Horizontal Global Irradiance

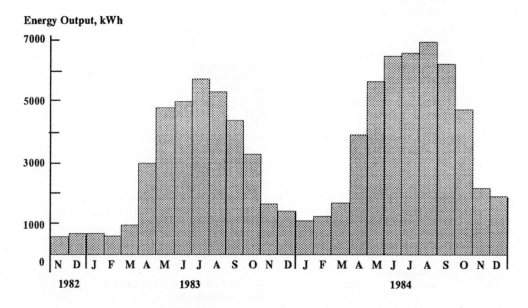

Figure 6. Monthly Energy Profile of PV Array

5.1.2 Modules

The module, manufactured by France-Photon (FP), contains 68 circular cells of 100-mm diameter monocrystalline silicon. Each module has one bypass diode across each 17 cells in series.

5.1.3 Array Support Structure and Foundations

The structures supporting the modules (see Fig. 7) were designed to minimize the manufacturing costs as much as possible. Each support structure holds 12 modules (representing approximately 860 Wp and 10 m^2) and comprises only the following elements:

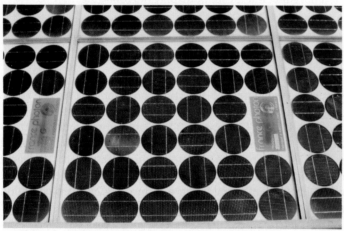

Lower Half of the Photon PV module (68 cells)

Figure 7. Array Support Structures and Foundations

- 2 footings, each one fixed on a concrete block by 2 threaded rods
- 1 central double beam and 2 bracketing sections fixing the modules onto the footings
- 2 threaded rods assembling the modules and the double metallic beam

Each structural subassembly weighs less than 25 kg. The panel is placed 1.5 m from the ground in order to ease access to the area under the modules. The panel tilt angle is 25°. A duct in the middle of the central double beam is provided for the module cabling.

The 12 modules in a panel are interconnected by a cable which leads to a junction box attached to one of the panel's supporting feet. Each panel is individually connected to the switchboard in the control cabinet.

5.2 Batteries and Charge Control Method

The single battery consists of 150 cells in series. Each cell has a capacity rating of 1,500 Ah (Varta VB 2415). The original battery control strategy used the state of charge (SOC) signal from the Logistronic unit. When the battery reaches its lowest SOC, the loads are connected to the Diesel generator, and the batteries are recharged by the array for two days. This Logistronic unit, however, did not function properly from the time of plant start-up. It was and is still capable of monitoring and displaying battery voltage, current, and temperature, but it cannot display the following:

- Ah charged
- Ah discharged
- SOC (%)

The battery charge control originally adopted consisted of inserting a power resistor in series whenever the predetermined charge voltage limit was reached (2.35 Vdc/cell). This effectively places the battery in a trickle-charge mode (about 30-40 A or C/40 rate). The plant improvement contract (1988-1989) which was supported by the CEC, PPC and the University of Patras (S. Pressas) called for the installation of a new system in April 1988 to provide a better charge control and discharge protection with high reliability. Annex B describes the new system in more detail. Since then, the battery operation has apparently improved in terms of maintaining a more fully charged condition and preventing complete battery discharge.

5.3 Power Conditioning

The power conditioning equipment comprises a single 40-kVA inverter manufactured by Jeumont-Schneider. It is self-commutated, using high power transistors and pulse-width modulation. The inverter input is connected to the dc bus using a small contactor at the positive and negative input terminals.

The other features of this inverter are as follows:

- Input voltage range:	285-390 Vdc
- Output voltage (stepped up from 220 Vac)	380 Vac, 3 phase
- Frequency:	50 Hz
- No load dissipation:	220 W (inverter), 700 W (transformer)
- Efficiency (rated):	93% at FL, 85% at 10% FL

Figure 8 shows the actual efficiency vs. load measurement made in August 1982.

5.4 Control System

The control system comprises a Varta Logistronic unit and a hardware circuit controller. The Logistronic unit was used solely for monitoring battery parameters shortly after installation because the SOC parameter it calculates never functioned properly.

5.5 Data Collection

An EPROM cartridge that complies with the specifications is used for data monitoring.

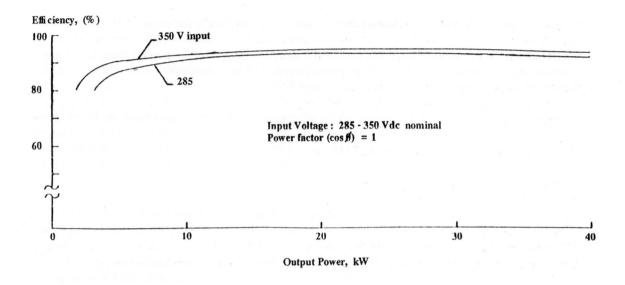

Figure 8. Power Efficiency vs. Output Power for Jeumont-Schneider Inverter

5.6 Lightning Protection

The lightning protection scheme was implemented after estimation of the probability of direct lightning strikes on the plant (see Annex A). Based on this assessment, the protection design consisted of a grounding wire netting using two 30-mm^2 copper wires over an area of 10 x 215 m.

5.7 Auxiliary Generators

The auxiliary generator used is a 40-kVA Diesel generator which was installed in July 1983.

5.8 Load Equipment

The PV plant feeds power into the village grid. The loads are usual domestic appliances.

6. MAINTENANCE

The plant is manually operated by two local inhabitants paid by PPC. The PV system normally supplies the day-time load and the Diesel generator, the night-time load. All connections and disconnections are made by the operator. Maintenance of the PV components is difficult as the maintenance personnel must come from Chania, a town on the northern coast of Crete, which is about a half-day journey considering the boat schedules.

7. SUMMARY OF PROBLEMS AND SOLUTION APPROACHES

The original plant used a relatively simple power management strategy. The main problem in the design was the poor battery charge control due to the simple charge termination method used. For example, when the battery reached a predetermined charge voltage limit, the array was either disconnected or the battery was placed in a trickle-charge mode via a power resistor. This often resulted in insufficient battery charging. There was also a need to protect the battery from excessive deep discharge. Should excessive discharge occur, the circuit would disconnect the load from the inverter and connect it to the Diesel generator. When the battery reached a sufficient state of charge, the load would be automatically reconnected to the inverter. Repeated deep discharges, operation at a partial state of charge, and a poor charge control method in high temperature conditions led to the shorting of 12 cells which had to be replaced in early 1988.

A key problem in operation is the manual mode of controlling the Diesel generator or the PV plant to supply power to the village grid. Even with the operator on site, a switch-over to the Diesel source takes 5 to 10 seconds, causing frequent "black outs" for the inhabitants. System automation would reduce switch-over time during power interruption and also avoid deep battery discharge which are detrimental to battery life.

One problem concerning the accuracy of performance data (i.e., PV power and energy efficiencies) was the use of the Si solar irradiance sensor. The sensor on a tilted surface was not calibrated (apart from that of the manufacturer) at the site or elsewhere against an acceptable thermopile pyranometer. A calibration check in 1988 [1] indicated that the sensor was reading about 17% lower than a working standard. This resulted in an optimistic (erroneous) prediction of the PV array power output and efficiency by 17%.

The original monitoring equipment was replaced by a more reliable set in order to provide continuous data.

8. KEY LESSONS LEARNED

A properly designed battery charge control and protection system is essential. Also, automation of the combined operation of Diesel generator and PV plant is very important to avoid many power outages, to protect the batteries, and to minimize manual intervention by the operator.

9. CONCLUSIONS

Since operation began, the plant has operated semi-automatically to supply power to the village on a 24-hour basis. The new charge/discharge controller has been working satisfactorily for about 1.5 years, and therefore outperformed its predecessor. In its present configuration, adequate manual monitoring and control is needed to prevent premature battery failures or degradation.

10. REFERENCES

[1] M.S. Imamura and E. Ehlers, "An Evaluation of Solar Irradiance Sensors at Aghia Roumeli," WIP Memorandum, Munich, FRG, August 1988.
[2] Unpublished report by S.A. Pressas, "New Battery Charge/ Discharge Controller for Aghia Roumeli," Univ. of Patras, GR, Laboratory of Electromagnetics, June 1988.
[3] J. Chadjivassiliadis, "Solar PV and Wind Power in Greece," IEE Proc. Vol. 134, Pt A, No. 5, May 1987.
[4] Aghia Roumeli Project Handout, Pilot Plant Optimization Meeting, Brussels, BE, 12-13 May 1986.
[5] P. Helm, "Manual of Photovoltaic Pilot Plants, Parts A and B," WIP report under CEC Contract EN 3 S-0007-D(B), Munich, FRG, February 1986.
[6] L.R. Selles, "Aghia Roumeli and Kaw Village PV Electrification," presented at IEE Meeting, Savoy Place, London, GB, 22 April 1985.
[7] J. Chadjivassiliadis, "Photovoltaic and Wind Power in Greece," presented at IEE Meeting, Savoy Place, London, GB, 22 April 1985.
[8] Aghia Roumeli Project Handout, Pilot Programme I Meeting, Tremiti Island, IT, 20-21 June 1984.

[9] A.N. Tombazis et al, "Kythnos Island and Aghia Roumeli Photovoltaic Pilot Projects - An Architectural Approach," Proc. of the 5th European PV Solar Energy Conference, Athens, GR, 17-21 October 1983.

[10] B. Aubert, "Aghia Roumeli Electricity Supply to an Isolated Village," Proc. of the EC Contractors' Meeting held in Hamburg and Pellworm, FRG, 12-13 July 1983.

[11] Aghia Roumeli Project Handout, Pilot Programme I Meeting, Mont Bouquet, FR, 12-14 April 1983.

[12] Aghia Roumeli Project Handout, Pilot Programme I Meeting, Crete, GR, 14-15 September 1982.

[13] Aghia Roumeli Project Handout, Pilot Programme I Meeting, Brussels, BE, 20-21 April 1982.

[14] B. Aubert, "Aghia Roumeli Electricity Supply to an Isolated Village," Proc. of the Final Design Review Meeting on EC Photovoltaic Pilot Projects held in Brussels, BE, 30 November - 2 December 1981.

[15] R. Buccianti, "Protection Against Lightning of Photovoltaic Generation Systems," CESI Report TS-200/2, prepared for the CEC, Milan, IT, September 1981.

ANNEX A. LIGHTNING PROTECTION ANALYSIS

The method used for estimating the probability of direct lightning strikes is specified in the CESI Study Report [15].

On the island of Crete, the Keraunic level (K) averages 30 days a year. Since the village of Aghia Roumeli lies on the seashore, an excess of this average value may be expected in comparison with the spot value for the site.

The earth-flash density (F) is defined by:

$$F = 0.023 \, K = 1.9 \text{ flashes/km}^2\text{-yr}$$

The equivalent exposed plant area is $A = 3,800 \text{ m}^2$.

The probability of a direct lightning strike on the plant is therefore:

$$P = F \times A = 1.9 \times 3,800 \times 10^{-6} \text{ flashes/yr}$$

$$= 0.0072 \text{ flash/yr}$$

And the average time between flashes, T, is:

$$T = \frac{1}{P} = 139 \text{ years/flash}$$

An important conclusion is that the average time between lightning strikes exceeds the expected lifetime of the plant by a factor of about seven. Based on this result, combined with the low probability of lightning occurrence, it may be assumed that the PV array is very unlikely to be subjected to direct lightning strikes. It was, therefore, decided that a careful grounding of all support structures of the array by two 30 mm^2 cross-section copper wires forming a netting measuring 10 x 215 m would be suitable and sufficient.

261

ANNEX B. THE NEW AGHIA ROUMELI BATTERY CHARGE REGULATOR AND DISCHARGE CONTROLLER [2]

A new battery charge regulator and discharge controller (BCDC) has been developed for the Aghia Roumeli PV plant. This unit has been incorporated into the system since early May 1988 and has replaced the previous controller. The previous method had caused a few problems concerning proper operation and protection of the battery. The new battery charge regulator is a pulse-width modulated shunt switch regulator (see Fig. B-1). It consists of six independent subregulators, each controlling the current produced by 10 PV strings. Each subset of 10 PV strings is paralleled through blocking diodes and produces a maximum short circuit current of 25 A and a maximum open circuit voltage of 480 Vdc.

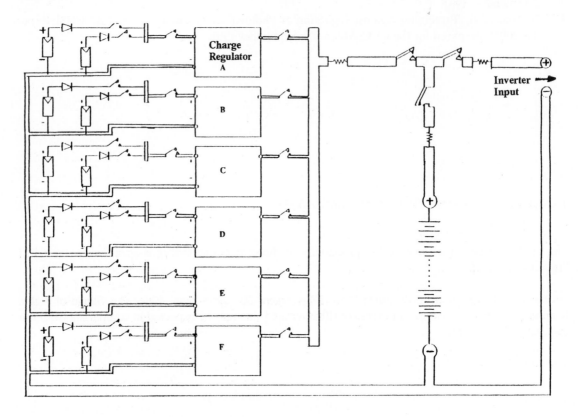

Figure B-1. General Schematic of New Charge Regulator and Discharge Controller

In the new system, the battery charging current is applied continuously. All array strings contribute to the battery charging current. The magnitude of the trickle charge current is controlled by the regulators (all 6 operating simultaneously) by narrowing or widening the battery charging current pulses at a frequency of 100 Hz. The I, V profile in Figure B-2 illustrates the typical performance of the new battery controller.

The grouping of PV strings and the number of independent shunt regulators were based on reliability considerations. If for some reason one out of the six subregulators fails, only 1/6 of the total PV power is lost, until the failure is corrected. However, each subregulator is oversized by a factor of two (i.e., designed for 50 A), so the PV power handled by the the failed subregulator can be managed by any one of the remaining five subregulators in operation. The new regulator incorporates temperature compensation of -6 mV/°C per cell. The temperature sensing devices (current sources) are mounted on the poles of the cells for monitoring internal cell temperature. The other key features of this battery charge/discharge unit are as follows:

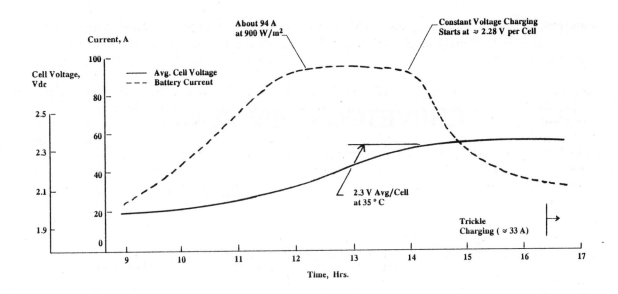

Figure B-2. Battery Charging Profile of the New Controller

- The maximum cell voltage is adjustable in the range of 2.2 to 2.5 volts per cell for normal (float) operation and in the range of 2.4 to 2.6 volts per cell for boost charging.

- The maximum voltage was set to 2.35 volts/cell for float operation and to 2.6 volts/cell for boost charging at 25°C. The change from floating operation to boost charging is done via a small switch on the PCB of each of the 6 subregulators.

- The use of cooling fans was avoided for reliability reasons and the high site temperature encountered. Cooling of the power semiconductors in the BCDC is done naturally, by mounting the power sections of each subregulator on a large heat sink with Rth = 0.2°C/W. The maximum heat sink temperature was estimated to be 53°C at 35°C ambient air temperature.

- The battery protection card inhibits the inverter's operation whenever the cell voltage is lower than the pre-set point to prevent deep discharges. The lowest allowable cell voltage is adjustable in the range of 1.7 to 2.1 volts/cell. It was set to 1.95 volts/cell due to the low discharge current measured during the night.

- A key objective in the design was to maximize the reliability of the new unit while minimizing its complexity because Aghia Roumeli is at a very isolated place and the PPC staff operating and maintaining the plant are not readily available.

Special care was taken to avoid corrosion of the printed circuits and their components due to the salty environment near the sea by the use of Vernice lacquer. Gold- or silver-plated contacts were used on all switches. Finally, every cable, including those of the data monitoring system is properly fused. The total power consumption of the new BCDC is 300 W and the efficiency is about 99%.

Appendix 2

CHEVETOGNE PV PLANT

1. INTRODUCTION

This appendix describes the 63-kW CHEVETOGNE PV pilot plant and its initial operation. The plant completion and start-up date was November 1982. Figure 1 shows the array field, the solar thermal collector panels, and the swimming pool. The thermal collectors have been in operation since 1981.

This plant supplies electric power for the following loads in the recreation facility:

- Pumps for a 2,100 m^2 solar thermal system for heating an outdoor swimming pool
- Lighting of the pool area in the evening

The information contained herein was obtained from ref. [1 to 8].

2. PLANT DESIGNER, OWNER, AND OPERATOR

The PV plant was established under the direction of IDE who also manufactured the PV module. This firm no longer exists. Plant maintenance and technical assistance to the owner is often supported by ACEC. The owner of the plant is the Domaine Provincial Chevetogne. Their address is as follows:

> Domaine Provincial Chevetogne
> B-5395 Chevetogne
> Belgium

The organization responsible for technical support is:

> ACEC
> P.O. Box 4
> B-6000 Charleroi
> Belgium

The 63-kW array supplies power to the pump motors

PV modules being mounted

Figure 1 The PV Array Field at CHEVETOGNE PV Pilot Plant

3. SITE INFORMATION

3.1 Location

This plant is installed in the Provincial Domaine of Chevetogne in the Province of Namur in southern Belgium (see Fig. 2). The Domaine has an area of 500 ha which includes camping places, sports centres, and an Olympic-size swimming pool. In 1985, about 70,000 people visited this facility.

CHEVETOGNE PV Plant (about
35 km southeast of Namur)

Figure 2. Location of the CHEVETOGNE PV Pilot Plant

3.2 Site Constraints

There were no major site constraints.

3.3 Solar Radiation and Climatic Conditions

The following solar radiation and meteorological data were acquired from measurements at Saint-Hubert (30 km SE of the site) and Uccle (150 km N of the site):

- Insolation, global horizontal: 1,000 kWh/m^2 annual
- Ambient air temperature: 30°C maximum, -18°C minimum
- Wind speed: 137 km/h maximum

3.4 Load Demand

The total power demand by the 14 pumps for water circulation in the solar heating system is about 24 kW. These loads are supplied by the 40-kW PV subsystem. The other consumers are the standard lamps along the road leading to the pool and around it. These loads are supplied by the 23-kW PV subsystem.

4. SYSTEM DESCRIPTION AND PERFORMANCE

4.1 Plant Architecture and Layout

The array field is installed on a small hill east of the swimming pool. The plant layout is shown in Figure 3.

The structures are designed so that the rows of PV panels follow the slope of the terrain, and the arrays blend into the landscape and do not disturb the skyline. Furthermore, the natural grazing area is preserved between the rows of the PV array.

4.2 PV System Configuration

The plant comprises two PV subsystems, a 40-kW unit for the pumping loads and one 23-kW unit for the lighting loads. Figure 4 is a simplified block diagram of the PV system. Its major components

View from northwest

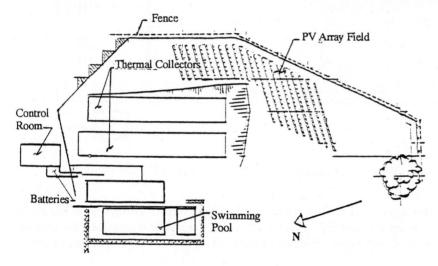

Figure 3. Layout of the CHEVETOGNE PV Plant

and their key features are listed in Table 1.

The 40-kW subfield is divided into three subarrays, and each subarray is connected to the dc bus via a dc contactor which is mainly used for battery charge control. This contactor permits the removal of each subarray individually from the dc bus. The 23-kW array is connected directly to another bus using a single dc contactor. Both the 40-kW and the 23-kW subsystems can be connected to the local grid via the ac bus.

The lead-acid batteries are charged by the dc-dc converter. The smallest subarray (13 kW output) is connected to the 120-Ah battery. Thus, during high insolation, one subarray can provide a charging current of 60 A.

4.3 System Operation and Management

The 40-kW subsystem is operated primarily in stand-alone mode for supplying pumping power to the outdoor swimming pool, and the 23-kW subsystem is operated with grid backup for providing power for lighting.

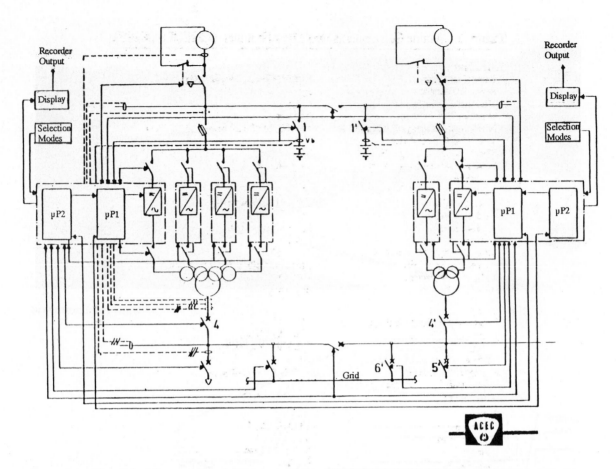

Figure 4. Simplified System Block Diagram of the CHEVETOGNE PV Pilot Plant

The two systems can be interconnected at their respective dc and ac buses. Furthermore, the excess energy is fed into the grid using the inverters. Both systems have separate microprocessor controls which operate independently of each other. Hardwired interlocking is provided to prevent undesired switching configurations.

4.4 Plant Control

The plant can be operated either manually, partially automatic, or fully automatic. There are ten different modes of operation which can be selected manually via push-button switches:

- Stand-alone with battery
- Stand-alone, battery and solar array
- Grid with battery charging
- Grid alone without battery charging
- Complete stop
- Complete stop except battery charging
- Grid connected, solar array, and inverter (in MPPT)
- Grid conected, solar array, and battery
- Grid connected with battery
- Manual operation

In the manual mode, the operator controls the system with the aid of the mimic diagram on the cabinets. Interlocking relay protection exists in this mode only. In the automatic mode, proper switching sequence of actuators and synchronization is performed automatically. In the stand-alone mode with array, battery, and inverter, the grid is not connected and MPPT is not used. In the grid-connected mode with the arrays and inverters, the inverters operate in MPPT, and they are synchronized with the grid.

Table 1. Major Components and Their Features (CHEVETOGNE)

Solar Array
Rated Peak Power:	63 kW (40- and 23-kW subfields)
Array Tilt Angle:	30°
Tilt Angle Adjustment:	None
Nominal Bus Voltage:	220 Vdc
Number of Subarrays (Strings):	124 (80 for 40-kW and 44 for 23-kW)
Total Number of Modules:	1,984
Number of Modules per String:	16
Module Manufacturer/Type:	Belgosolar (IDE)/33-4
Module Rating:	33 W
Number of Cells per Module:	36
Number of Bypass Diodes:	1 per 18 cells in series
Module Weight:	11 kg
Module Dimension:	1,028 x 412 x 50 mm
Module Area:	0.424 m^2
Active Solar Cell Area/Module:	0.298 m^2
Solar Cell Type/Size:	Monocrystalline silicon/100 mm dia.
Cell Packing Factor:	70.3 %
Total Land Area:	2,100 m^2

Battery
Total Battery Energy Rating:	356 kWh
Cell Capacity Rating:	120 Ah/1,500 Ah
Voltage Range:	220 Vdc (nominal)
Allowable Depth of Discharge:	80 %
Number of Batteries:	1 each of two ratings
Number of Cells per Battery:	110
Cell Manufacturer:	Oldham-France

Inverter
Rated Power:	60 kVA (20 kVA x 3)
Number of Inverters:	3
Type:	Self-commutated
Manufacturer:	ETCA
Input Voltage Range:	195 - 264 Vdc
Output Voltage/Phase:	220 Vac/3-phase
Frequency:	50 Hz
THD:	<3 %
Efficiency 100 % Full Load:	95 %
10 % Full Load:	92 %

Controller:	HP 1000 computer
Data Acquisition Manufacturer/Description:	HP 2250, ECMA-43 recorder

In the operational mode with the array, battery, and inverters in parallel with the grid, the battery voltage determines the array operating point, and consequently the MPPT is inhibited.

The pumps are actuated by the regulators in the heating system. As illustrated in Figure 5 the designer estimated that the number of pumps switched on is roughly proportional to the solar irradiance.

4.5 Plant Performance

The installation and checkout of the plant, with the exception of the inverters, was completed as of June 1983. The inverters were delivered in late 1983 and at that time the whole plant became operational.

The delivery of the inverter was late by half a year because it was especially designed with both self- and line-commutation capabilities, the latter with maximum array power tracking. Only about 50% of the array capability was used during the initial phases for two reasons: there were insufficient amounts of loads, and excess energy could not be fed into the grid.

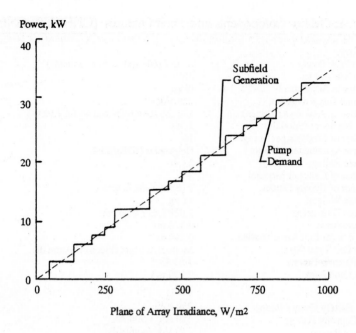

Figure 5. Predicted Relation Between Pump Load and
Solar Irradiance

The system sizing was initially based on the assumption of seasonal and daily matching of pumping energy need and insolation. In actual operation, the insolation level was often low with high diffuse irradiances. Even under these low solar intensity conditions, the PV array is capable of supplying electrical energy, but the solar thermal collector system was inoperative most of the time. Under combined operating conditions of low load demand, large surplus PV energy, and small battery capacity, the battery is quickly recharged, and the PV subarrays are switched off.

During the winter months, the electrical energy is used only by the illumination equipment, which represents a low fraction of the available array energy. According to a rough estimate, the PV array could supply 50% more energy than the actual demand.

5. COMPONENT DESCRIPTION

5.1 Solar Array

5.1.1 Array and Cabling

The array field has 124 parallel strings, 80 strings in the 40-kW subfield, and 44 strings in the 23-kW subfield. Each string consists of 16 PV modules connected in series. The designer grouped the modules via computer filing and analysis so that the modules in each string have the same value of maximum power point current. As a result, the module mismatch loss within one string was predicted to be no more than 1%. Each string is connected by a 5-mm^2 underground cable to the control room. The total wire and diode losses were estimated to be below 2%.

In the control room, a special panel is fitted with 124 3-position switches (one per string), which permit each string to be connected to the array bus, "short-circuited", or "open-circuited". In the open-circuited position, the string can be connected to a special variable load which allows measurement of the I-V curve of each string from the control room. This I-V load equipment gear is portable, and can be used for testing modules in the field.

5.1.2 Modules

The modules are Belgosolar 33-4 modules manufactured by IDE. Their features are as follows:

- Nominal output power of 33 W
- 6 monocrystalline silicon cells in series
- Twin glass, EVA-laminated encapsulation
- 2 shunt diodes

Two polarized plug-in connectors, especially developed for this plant, are used for module connection. These connectors reduce the installation time for wiring the modules on site to about half a minute per module.

Using the I-V characteristics of each module, which were stored in the computer file, the designer classified the modules according to their maximum power point current.

5.1.3 Array Support Structure and Foundations

A special type of support structure was designed for this plant (see Fig. 6). It consists of a 2-pole structure fitted with a beam supporting the modules whose frame provides part of the mechanical rigidity. The beams are connected to the poles in such a way that the rows can follow the slope of the terrain. Thus, it was not necessary to level the ground. The support structures are 1.1 m high and their span 3.8 m. The structures finally selected were of hot galvanized steel instead of painted steel as originally planned. The reason was mainly the lower cost of galvanized steel. The foundation is a concrete single pedestal footing.

5.2 Batteries and Charge Control Method

The plant has two lead-acid batteries, a 1,500-Ah battery connected to the 23-kW subfield dc bus, and a 120-Ah battery tied to the 40-kW subfield dc bus. Each battery consists of 110 cells (from Oldham manufacturer) connected in series.

The maximum charge rate recommended by the manufacturer was C/10 or 150 A for the 1,500-Ah battery. At a nominal dc bus voltage of 220 V, the expected peak array current for the 23-kW subfield is 105 A (23,000/220). Thus, the system designer assumed that this subfield can be connected directly to that battery without incurring problems. For the 120-Ah battery, the 40-kW subfield was divided into three subarrays, each of which could be turned on or off in sequence to control the charge current.

5.3 Power Conditioning

Three 20-kVA self-commutated inverters are used in this plant. The inverters utilize MOSFET transistors and pulse-width modulation techniques for waveform synthesis. In the 40-kWp subsystem, two inverters operate in parallel. All inverters are capable of operating in MPPT and with external commands.

The efficiency, including the filters, but not with the transformer, is 95% at full load and above 92% at 10% of full load. With the transformer the efficiency is more than 88% at 10% load. The total harmonic distortion is less than 3%.

Each unit is controlled by its respective microprocessor which assures commutation, matching with the grid, master/slave control of parallel units, and maximum power point tracking.

5.4 Control System

The HP 1000 microprocessor controller provides the automatic control and monitoring of the plant.

5.5 Data Collection

The HP 2250 data acquisition system performs the basic collection of performance data, and the ECMA-43 recorder is used for data storage.

271

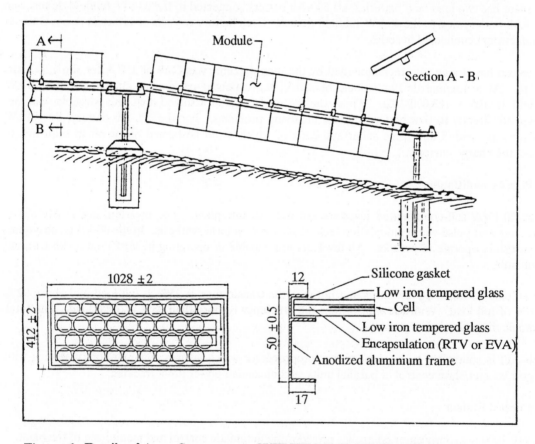

Figure 6. Details of Array Structures at CHEVETOGNE PV Pilot Plant

5.6 Lightning Protection

A statistical analysis during the design phase led to the conclusion that protection against direct lightning strikes is not warranted. Therefore no special protection is implemented against direct lightning strikes. A partial protection, however, is provided by the 10-metre high pole (for wind-measurement purposes) which is properly connected to the earth ground. Also, a gas-filled surge arrester against indirect strikes and high voltages is connected on each polarity of the dc bus.

A buried grounding network made up of 150 mm^2 of copper cables is connected to the array field structures. All PV module frames are grounded to the array structures through their fastening bolts.

5.7 Auxiliary Generators

The system has no auxiliary generators.

5.8 Load Equipment

The load equipment consists of ac motors, one for each of the 14 water circulation pumps, and ac lamps for the illumination of the pool area and along the road leading to it.

6. MAINTENANCE

The plant is fully automatic. The correct functioning of all components is checked out daily. An electrician of the Domaine who has been instructed on the system performs maintenance and repair work on non-electronic parts of the plant.

The inverter and control hardware and software have been maintained for some time at no charge by ACEC technicians. The inverters used in CHEVETOGNE have been especially manufactured for this plant. Therefore, only a few specialists can repair this equipment. Qualified personnel are not always available on short notice; therefore, to remedy even small problems takes some time.

7. SUMMARY OF PROBLEMS AND SOLUTION APPROACHES

The main problems encountered are summarized as follows:

- The total power degradation of the array is estimated to be 5%, but exact measurements have never been made. This degradation is believed to be at the PV module level, specifically from transmissivity loss at the glass cover to cell interface. Since IDE has stopped production of the PV modules, no follow-up effort was made to improve the module design.

- The inverter was made especially for the project and its delivery was 6 months late. After installation, the inverter was defective, and even though the problem was not severe, repairs took six weeks due to a lack of trained personnel.

- Modification of the control software was not possible because computer capacity was full.

- The plant control was susceptible to stoppages from electrical transients and disturbances.

- Excess PV energy was not fed into the grid because local power authorities were not willing to pay for the incoming energy; the inverter was designed for grid-connected operation.

- In the winter months, the system supplies only the lighting load, resulting in low utilization of the available PV energy. This problem is due to not supplying power to the grid.

- Detailed plant performance analysis has not been done after most of the optimization effort was completed.

- No documentation has been made, apparently because the plant was continually undergoing development and optimization. The users and equipment suppliers, however, do recognize the need for operation and maintenance documentation for use by non-trained personnel.

8. KEY LESSONS LEARNED

The PV system is under-utilized; the key lesson learned here is that both stand-alone and grid-connected operation should be better assessed early in the conceptual and detail design phase.

The designers believe that the experience gained from the design and operation of the system is insignificant because the plant is not considered to be typical for future PV systems with regard to its design, power output, and application.

The decision to design a new inverter with an integrated microprocessor controller caused both schedule and budget problems. In retrospect, a better approach at that time would have been to use existing inverters and implement only minimum modifications. Furthermore, one or two line-commutated inverters for grid-connected operation and separate self-commutated inverters would have been more effective.

The designer implemented too many operating modes that unnecessarily complicated the control subsystem. Some of the significant results and observations are:

- A 20-kW inverter with both self- and line-commutated capabilities was designed. This design should be made available for future systems that require such special features.

- Although some of the ten PV system operating modes are not essential for the basic application at this plant, this PV plant can be an effective test bed for R&D uses, if the owner can arrange a long-term collaboration with local universities. Moreover, the local universities can assist in the maintenance of the plant as well as in performing monitoring and analysis.

9. CONCLUSIONS

The CHEVETOGNE facility has both solar electric (PV) and solar thermal systems. The PV plant capacity is presently under-utilized. Nevertheless, it has attractive features that can serve both the day-to-day operation of the recreation facility and the research objectives of local universities interested in renewable energy work. As in some of the other pilot PV plants, a long-term operation and implementation plan should have been assessed carefully during the project formulation and conceptual design phases.

10. REFERENCES

[1] Chevetogne Project Handout, Pilot Plant Optimization Meeting, Brussels, BE, 12-13 May 1986.
[2] P. Helm, "Manual of Photovoltaic Pilot Plants, Parts A and B," WIP report under CEC Contract EN 3 S-0007-D(B), Munich, FRG, February 1986.
[3] Chevetogne Project Handout, Pilot Programme I Meeting, Tremiti Island, IT, June 20-21 1984.
[4] M. Van Gysel, "Powering of a Solar Heated Swimming Pool", Proc. of the EC Contractors' Meeting held in Hamburg and Pellworm, FRG, 12-13 July 1983.
[5] Chevetogne Project Handout, Pilot Programme I Meeting, Mont Bouquet, FR, 12-14 April 1983.
[6] Chevetogne Project Handout, Pilot Programme I Meeting, Crete, GR, 14-15 September 1982.
[7] Chevetogne Project Handout, Pilot Programme I Meeting, Brussels, BE, 20-21 April 1982.
[8] M. Van Gysel, "Powering of a Solar Heated Swimming Pool," Proc. of the Final Design Review Meeting on EC Photovoltaic Pilot Projects held in Brussels, BE, 30 November - 2 December 1981.

Appendix 3

FOTA PV PILOT PLANT

1. INTRODUCTION

This appendix describes the 50-kW FOTA PV pilot plant at the time of installation and its initial performance. The plant supplies power to operate a dairy farm whose seasonal power need is matched to the available insolation. Consequently, experience gathered with a PV system applied to some aspect of this dairy industry (cheese, butter factories, etc.) is directly applicable to a realistic potential use of PV power in northern Europe. Figure 1 shows the roof-mounted PV array.

The information contained herein was obtained from ref. [1 to 11] and individual contacts with the plant designer and operator.

2. PLANT DESIGNER, OWNER, AND OPERATOR

This plant was established under the administration of the National Microelectronics Research Centre (NMRC) of the University of Cork. It is now operated by this institute. Their address is:

National Microelectronics Research Centre
University College Cork
Lee Maltings
Prospect Row
IRL-Cork
Ireland

PV modules being installed on the roof

The largest roof-mounted PV array in Europe (50kW)

Figure 1. View of Completed Array at the FOTA PV Pilot Plant

3. SITE INFORMATION

3.1 Location

The island of Fota is situated in Cork harbour which is accessible by road via causeways and is 19 km distant from Cork. The railway between Cork and Cobh also crosses the island and stops at Fota.

The island has an area of 300 ha. A map showing the plant location is in Figure 2. The coordinates of the plant are:

- Latitude: 51.7°N
- Longitude: 8.8°W
- Altitude: 25 m

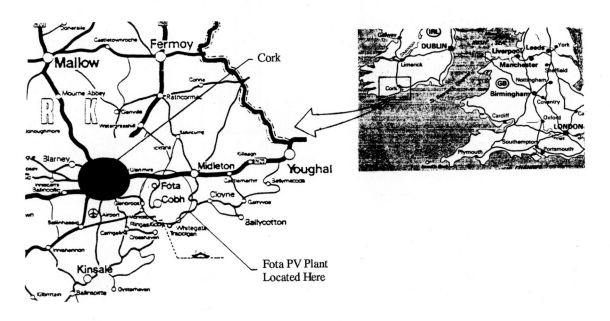

Figure 2. Location of the FOTA PV Pilot Plant near Cork (Southern Coast of Ireland)

3.2 Site Constraints

There were no major site constraints, other than the use of the farmhouse roof top for the arrays.

3.3 Solar Radiation and Climatic Conditions

The meteorological data listed below were obtained from the Roches Point Weather Station (6 km from Fota) and the solar radiation values from Valentia and Kilkenny meteorological stations (both about 160 km from Fota):

- Insolation, global horizontal:
  - o Annual: 1,100 kWh/m^2
  - o Daily average in June: 5.17 kWh/m^2
  - o Daily average in December: 0.52 kWh/m^2
- Ambient air temperature: 28.9°C maximum, 6.7°C minimum
- Wind speed: 116.6 km/h maximum

3.4 Load Demand

Figure 3 shows the load demand of the farm as measured during the period from October 1980 to September 1981, and indicates a 4.5 to 1 ratio between the maximum demand in July and the minimum demand in January. Figure 3 also shows a typical daily load profile of the farm, the main features being the morning milking between 8:00 a.m. and 10:00 a.m. and that in the evening between 4:30 p.m. and 6:30 p.m. At other times, the load alternates mainly between the milk coolers and machine sterilizers.

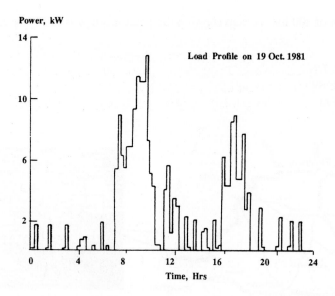

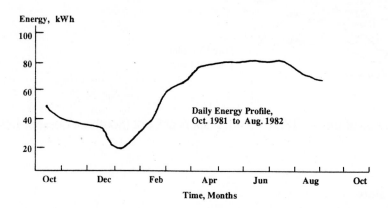

Figure 3. Typical Load Demand of Fota Island Dairy Farm

4. SYSTEM DESCRIPTION AND PERFORMANCE

4.1 Plant Architecture and Layout

The building on which the PV modules are roof mounted is 54 m long, 15 m high and 15 m wide. It is built using 20,000 high density 10 newton blocks. The building is divided into 11 bays comprising 10 internal walls used to support a galvanized steel structure on which the modules are bolted, thus forming an even plane. Figure 4 shows the layout of the main PV system components.

The roof has a 45° horizontal inclination and the building faces south. The middle bay is used as a control and battery room, the end bays as storage places, and the remaining eight bays as calving units for the dairy cows.

4.2 PV System Configuration

Figure 5 is a simplified block diagram of the original PV system. In 1988, NMRC modified this original configuration to parallel the two batteries into one (see Fig. 6). Major components of the system and their key features are listed in Table 1.

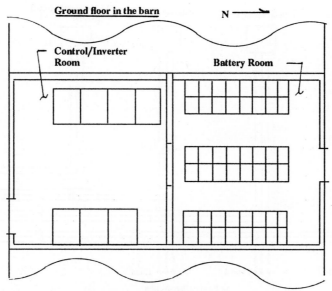

Figure 4. Layout of the Control/Inverter Room and the Battery Room

4.3 System Operation and Management

The 50-kWp system supplies the energy requirements of this dairy farm. The loads are induction motors used during the milking process, i.e., two milking machine motors (2.2 kW each), a cooler motor (3.5 kW), and four small motors (0.75 kW each) used to pump and stir milk and supply feed for the cows. In the previous system configuration these loads were supplied by 10-kVA inverters from one battery bus while the other battery in the second bus was charged by the solar array. All switching functions were under control of a micro-computer.

4.4 Plant Control

The key functions of the control system are as follows:

- Supply energy to the loads independent of the utility grid
- Feed excess energy into the utility grid
- Operate automatically

The following input/output signals are interfaced from the PV system to the control unit:

- 60 analog input signals (voltage, current, temperature, etc.)
- 39 digital input signals indicating status of switches (on/off) and status of hardware
- 1 analog output controlling the power flow through a line-commutated inverter
- 33 digital output signals operating ac and dc contactors, etc. 24K of control software is stored on EPROMs in the controller

4.5 Plant Performance

Figure 3 indicates that although the load is well-matched to the available solar radiation on a seasonal basis, it is not so on a daily basis, since most of the available insolation occurs at times other than those of the maximum power requirement, i.e., between 10:00 a.m. and 4:00 p.m. Therefore, in order to operate in a stand-alone state, a battery storage system was necessary.

Peak load demand is 90 kWh/d during the summer months, and it decreases to less than 5 kWh/d in mid-winter. Solar array output during the summer months exceeds 200 kWh/d and decreases to 30 kWh/d in mid-winter.

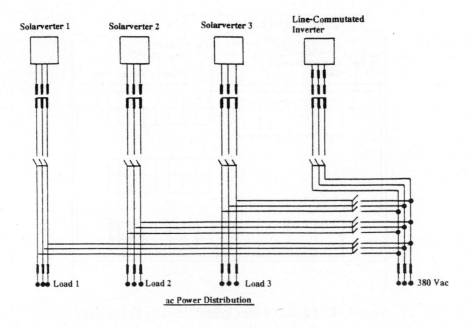

ac Power Distribution

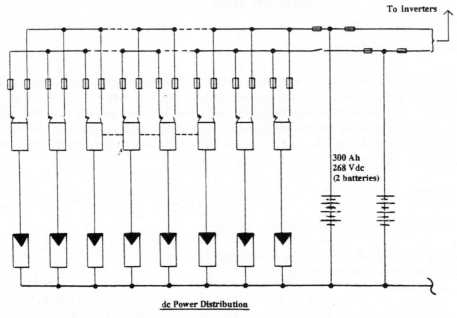

dc Power Distribution

Figure 5. Simplified Electrical Schematic of the Original FOTA PV Pilot Plant

5. COMPONENT DESCRIPTION

5.1 Solar Array

5.1.1 Array and Cabling

The solar array consists of 2,775 modules mounted on a steel structure. Each module is fixed by two bolts. A torque of 7.5 Nm was applied to each bolt. The solar array is divided into thirteen

280

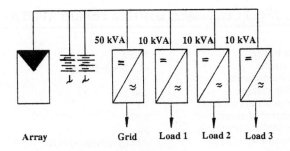

Final System Layout for Stand-alone Operation

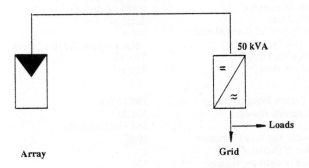

Final System Layout for Grid-connected Operation (Without Batteries)

Figure 6. Original and New System Layout,
FOTA PV Pilot Plant

subarrays. Each subarray contains six parallel strings whereby each string comprises 36 modules connected in series. Each subarray has an open-circuit voltage of 403 V and a short-circuit current of 14.4 A.

Each subarray has a junction box mounted on the inner wall of the building. A 2.5-mm² cable for each string is connected to the related junction box. From each of the 13 junction boxes, a three-core 10-mm² cable is tied to the array bus in the control room. When a failure occurs in the array string, the microcomputer detects a decrease in that subarray current and prints a message indicating a problem on that subarray. The defective string is then located by testing the strings in the subarray junction boxes. Three modules have built-in PT-100 temperature sensors connected to the microcomputer interface for monitoring.

5.1.2 Modules

The solar modules, type PQ 10/20/01, are supplied by AEG-Telefunken. Each module consists of 20 polycrystalline 100-mm square solar cells connected in series. The cells are embedded between two glass sheets using PVB material and bordered by a stainless steel frame. Two wires, attached to the series-connected cells through watertight connections, are provided as output terminals. A shunt diode is connected between the output terminals in each module. Each module is rated at 19.2 Wp with an 8.6-V effective voltage.

5.1.3 Array Support Structure and Foundations

In the higher site latitudes, a multiple-row array panel configuration requires large spaces between the rows in order to avoid shadowing. Thus the usual way of installing an array field using rows of

Table 1. Major Components and Their Features (FOTA)

Solar Array	
Rated Peak Power:	50 kW
Array Tilt Angle:	45°, South-facing
Tilt Angle Adjustment:	None
Nominal Bus Voltage:	268 Vdc
Number of Subarrays (Strings):	13
Total Number of Modules:	2,775
Number of Modules per String:	36
Module Manufacturer/Type:	AEG-Telefunken/PQ10/20/02
Module Rating:	19.2 W
Number of Cells per Module:	20
Number of Bypass Diodes:	1 per module
Module Weight:	3.85 kg
Module Dimension:	0.459 x 0.563 m
Module Area:	0.258 m^2
Active Solar Cell Area/Module:	0.2 m^2
Solar Cell Type/Size:	Polycrystalline Si/10 x 10 cm
Cell Packing Factor:	77.5 %
Total Roof Area:	810 m^2
Battery	
Total Battery Energy Rating:	160.8 kWh
Cell Capacity Rating:	300 Ah
Voltage Range:	268 Vdc (nominal)
Allowable Depth of Discharge:	80 %
Number of Batteries:	2
Number of Cells per Battery:	134
Cell Manufacturer/Part No.:	Varta Bloc/VB 2306
Cell Weight:	31.2 kg
Cell Dimension:	131 x 275 x 440 mm
Inverter	
Rated Power:	50 kVA (1 unit)/10 kVA (3 units)
Number of Inverters:	4
Type, 50 kVA:	Line-commutated
10 kVA:	Self-commutated
Manufacturer/Part No.:	AEG "MiniSemi"/AEG Solarverter
Input Voltage Range:	245-315 Vdc (both types)
Output Voltage/Phase:	380 Vac/3-phase (both types)
Frequency:	50 Hz (both types)
THD:	5 % (both types)
Efficiency, 100 % Full Load:	88 % (50 kVA), 95 % (10 kVA)
10 % Full Load:	75 % (50 kVA), 86 % (10 kVA)
Inverter Dimension:	0.754 x 0.480 x 0.337 (Solarverters)
Controller (for Battery)	
Manufacturer/Model:	Motorola 6800 microprocessor system
Features:	24 K of Control Software Memory
Data Acquisition	
Manufacturer/Model:	
- Original DAS:	The Controller is used for this function plus Penny & Giles Cassette Tape Recorder
- New DAS (installed in 1989):	Campbell Scientific 21X with modem at the PV plant

panels could not be chosen for this plant. Since a sizable area was available on the roofs of the farm building, it was decided to mount the PV modules there. The building is divided into 11 bays and the 10 separation walls are used to support a galvanized steel structure to which the modules are attached. Figure 7 shows the module mounting structure configuration and the PV modules as seen from below.

The modules are all mounted on one plane facing south, inclined at 45°. The features of this structure were determined by choosing the distance between the internal supporting block walls so that the total cost of the walls and their connecting steel framework for supporting the modules is minimized.

Roof construction   Array wiring underneath the PV modules

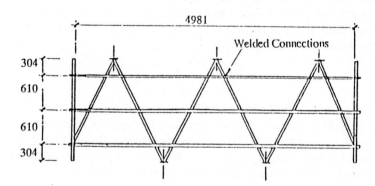

Figure 7. Details of PV Array Structure and Wiring

5.2 Batteries and Charge Control Method

The 600-Ah capacity is made up of two 300-Ah battery strings. Each string has 134 cells in series. The 2-battery configuration was initially selected to ensure a higher charge/discharge cycle life performance by both batteries than is achievable by a single large battery.

Since changes may occur in the array output due to clouds, the subarrays are connected to or disconnected from the batteries to maintain the battery voltage at 315 Vdc (avg. 2.35 V per cell) or lower. This limit was also used to minimize water loss in the battery cells. This voltage level, where internal gassing is expected to occur, is kept constant by successive switchings of the subarrays by the control computer.

When the batteries are deep-discharged during periods of prolonged low insolation, the line-commutated inverter is used as a rectifier to recharge the batteries from the utility grid. The control of the battery charging is based on current and voltage measurements and SOC calculation by the Varta Logistronic unit. However, the Logistronic unit did not function properly, so the battery charge control and discharge protection were implemented by the main control computer.

5.3 Power Conditioning

Power conditioning is performed using two types of inverters: a self-commutated type for supplying 3-phase ac loads and a line-commutated type for feeding excess energy from the system into the utility grid.

5.3.1 Self-commutated Inverter

Three 10-kVA self-commutated inverters known as "Solarverters" accept dc inputs of 245 to 315 Vdc and provide an ac output of 380 V, 3-phase. Switching transistors are used for providing the output waveform, thus eliminating the quenching circuits required for thyristor-operated inverters. The Solarverters operate at 95% efficiency under full-load and are protected by an automatic shut-down capability when a short circuit or an overload condition occurs.

The three Solarverters supply three independent loads on the dairy farm. By matching the loads to the inverter ratings in this manner, the inverter efficiency is increased. The inverters operate on the same bus on the dc side, however they are independent on the output side (ac).

In the original separate-battery configuration, each self-commutated inverter was connected either to battery 1 or 2. The dc contactors connecting the inverters to the batteries are controlled by the control unit. On the ac side, the three loads can be connected either to the inverters or the utility grid; furthermore, these ac contactors are controlled by the control unit as well. If a defect in the inverter occurs, the load is disconnected from it and reconnected to the utility grid. The load is not reconnected to the PV system until the defect has been remedied.

5.3.2 Line-Commutated Inverter

The 50-kVA line-commutated inverter known as AEG Mini-semi is a thyristor-operated unit. It requires the 50-Hz signal from the utility grid to operate. The current flow through the Mini-semi is regulated using a 0-10 Vdc signal from the microcomputer. The direction of current flow (inverter or rectifier operation) is also controlled by the micro-computer. This inverter is automatically disconnected from the dc bus in the event of certain alarms sensed by the control computer.

The inverter feeds excess energy into the utility grid, recharges the battery during periods of low insolation and can supply the loads.

The inverter supplies the loads by delivering power through the utility grid. This is not a standard operation mode, however it can be used when the self-commutated inverters malfunction.

In the original configuration the controller performed the following functions:

- connect the inverter to battery 1 or 2 using dc contactors
- determine the operation mode either as an inverter or a rectifier
- control the amount of power to or from the utility grid

In case of excess energy the controller works it out and signals it to the inverter. During emergency charging of the batteries a digital signal selects rectifier operation and an analog signal regulates the charging current.

5.4 Control System

The system is controlled by a Motorola 6800 microprocessor-based system designed by NMRC. The control software requires a total of 24 kBytes of memory.

5.5 Data Collection

The Motorola 6800 microprocessor also monitors data on a computer terminal and records them on a magnetic tape.

5.6 Lightning Protection

Lightning protection is implemented for the building, steelwork, and electrical equipment.

The building and steelwork are protected by installing seven galvanized steel lightning rods at the top of the building. The rods are connected to the earth grounding system by an 8-mm^2 solid galvanized steel conductor. The steelwork is bonded continuously at the top and the bottom by a 16-mm^2 stranded copper cable (PVC/PVC). The steelwork is grounded at four locations to minimize the resistance to earth. All metal ducting on the walls is grounded in the same manner.

The electrical equipment is protected by a varistor and spark gap combination in each of the 13 subarray junction boxes. When the dc busbar in the junction box exceeds 600 Vdc, the varistor and spark gap act as a short circuit to earth, thereby protecting the equipment in the control room from large surge currents. The varistors and spark gaps are positioned at the top of the junction box.

5.7 Auxiliary Generators

The system has no auxiliary generators.

5.8 Load Equipment

The load consists of seven induction motors used by the dairy farm during the milking process.

6. MAINTENANCE

The software was initially designed by AEG. It was later modified and is now maintained exclusively by NRMC. Power management and control hardware are maintained and repaired by AEG. Work on the conventional parts of the system is performed by local electricians.

7. SUMMARY OF PROBLEMS AND SOLUTION APPROACHES

At the start of operation, some PV modules were mechanically damaged due to over-tightened screws. This resulted in cracking of their cover glasses. Modification of the attachment procedure solved the problem.

The large starting currents of the induction motors could be supplied by the self-commutated inverter only to a limited extent. Several modifications were made to solve this problem. For the ac loads, they included re-wiring to star-delta configuration for starting and power factor correction capacitors. The inverter control has been changed to keep the torque of the motor constant by controlling its output voltage and frequency proportionally.

For economic reasons, the performance of the grid-commutated inverter has been limited to 40 kW. However, its output is insufficient to feed the maximum power of the PV array into the grid. Under low-load conditions and with a fully charged battery, parts of the PV array are switched off. This state of operation occurs usually at noon and under good insolation conditions. In order to prevent the energy loss resulting therefrom, the evaluation of insolation conditions expected during noontime has been added to the control software. This forecast is based on a comparison of the hourly maximum and the average insolation. Should both exceed predetermined limits, the control programme assumes an acceptable insolation condition and provides for the partial discharge of the batteries by feeding into the grid. Thus, during the hours of high insolation, the battery SOC is low enough to accept high charging currents.

The arrangement of two separate buses and two batteries was found not to be an optimum solution. Because the batteries are often not fully charged, an undervoltage condition occurs at inverter input when switching on large loads. As a result, the PV plant switches over to grid-connected operation. This problem is typical in the winter months, when both battery blocks are seldom fully charged. The load capacity of the battery was improved by forming one 600-Ah battery from the two 300-Ah batteries. This also simplified the power management by reducing switching, etc. Rather than sending power to the utility grid, the dairy farm could make use of the excess solar power at midday by introducing an ice-making facility (about 5-kW capacity). The ice would be generated around midday and be used for milk cooling during the afternoon, night-time, and early morning. The load demand and consequently the battery discharge would be reduced during the milking periods. This would also improve the load capacity of the batteries.

The problems associated with this project can be solved by further improving the power management. This will involve detailed analysis of the recorded data and the development of new control software.

The key problems encountered, some of which are discussed above, are:

- Cover glass cracking occurred on sixteen PV modules (0.04% of total). This was due to over-torquing the fastening screws. The non-planarity of mounting surfaces also contributed to the cracking of cover glasses.

- IC failures have occurred apparently from indirect lightning strikes.

Instrumentation noise and faulty measurements were encountered because one common grounding point was used for all dc and ac networks. The solution was to use two separate groundings, one for dc and the data monitoring system, another for ac devices.

The inverter could not supply the large initial current of induction motors. The solution was to use star-delta starting and power factor correction and a change in inverter control to keep motor torque constant by controlling its output voltage and frequency proportionally.

The inverter size (40 kW) was too small to handle the highest available array power. The solution was to switch off parts of the array and to effect proper control via software changes. In lieu of Logistronic state of charge (Ah) monitoring and control, the charge control method was modified to battery voltage limit via array string shedding.

The two-battery configuration was not a good design approach as it necessitated a complex power management software. The system was modified to a single-bus, single-battery arrangement in 1988, so that the two batteries are now strapped in parallel.

The PV array is rated at 50 kWp but the maximum power delivered to the grid is 34 kW. The batteries are too small to supply the load demand during night-time and early morning. The maximum PV array current is greater than the maximum allowed battery charging current. Consequently, the charge current must be limited by shedding array strings.

8. KEY LESSONS LEARNED

Splitting the dc bus into two and having two separate batteries to control caused a large complexity in the control software. NMRC estimated that the 25 kB of software could be reduced to 15 kB by going to a single-bus, single-battery configuration. Another lesson learned is that the load capacity of a double-battery configuration is lower than a single large battery (47% of annual load with a single battery and 29% for the double battery).

In theory, the performance criteria for the state of charge calculation in the Logistronic unit were defined properly. In practice, the unit was found to be unreliable. Follow-up and re-design at the time of installation were not accomplished because of other priorities in the project.

In this plant the undersized inverter was compensated by a complex software control. Future PV systems with complex controls should have:

- Skilled personnel in hardware performance and software programming
- A controller with high level language
- Facilities for on-site software improvement

9. CONCLUSIONS

The plant has now operated for over six years. After about one year of software optimization, the reliability of the plant had improved significantly. The primary improvement effort was to change the system from a 2-battery to 1-battery configuration.

Most of the problems have been solved by improving the control software. These improvements were based on the system experience and the continuous analysis of the recorded data.

10. REFERENCES

[1] S. McCarthy, G.T. Wrixon, and L. Keating, "Specific Actions on the Fotavoltaic Project," Proc. of the 8th European PV Solar Energy Conference, Florence, IT, 9-13 May 1988.
[2] S. McCarthy and G.T. Wrixon, "Optimization and Analysis of PV Systems Using a Computer Model," Proc. of the 7th European PV Solar Energy Conference, Sevilla, ES, 27-31 October 1986.
[3] Fota Project Handout, Pilot Plant Optimization Meeting, Brussels, BE, 12-13 May 1986.
[4] P. Helm, "Manual of Photovoltaic Pilot Plants, Parts A and B," WIP report under CEC Contract EN 3 S-0007-D(B), Munich, FRG, February 1986.
[5] G.T. Wrixon and S. McCarthy, "Field Experience of a 50-kWp Photovoltaic Array at Fota Island," Proc. of the 6th European PV Solar Energy Conference, London, GB, 15-19 April 1985.
[6] Fota Project Handout, Pilot Programme I Meeting, Tremiti Island, IT, 20-21 June 1984.
[7] G.T. Wrixon, "The Fotavoltaic Project: A 50-kW Photovoltaic System to Power a Dairy Farm on Fota Island, Cork, Ireland," Proc. of the EC Contractors' Meeting held in Hamburg and Pellworm, FRG, 12-13 July 1983.
[8] Fota Project Handout, Pilot Programme I Meeting, Mont Bouquet, FR, 12-14 April 1983.
[9] Fota Project Handout, Pilot Programme I Meeting, Crete, GR, 14-15 September 1982.
[10] Fota Project Handout, Pilot Programme I Meeting, Brussels, BE, 20-21 April 1982.
[11] G.T. Wrixon, "The Fotavoltaic Project - A 50-kW Photovoltaic System to Power a Dairy Farm on Fota Island, Cork, Ireland," Proc. of the Final Design Review Meeting on EC Photovoltaic Pilot Projects held in Brussels, BE, 30 November - 2 December 1981.

Appendix 4

GIGLIO PV PLANT

1. INTRODUCTION

This appendix describes the 45-kW GIGLIO PV pilot plant at the time of installation and its initial operation. This plant became operational in July 1984. Subsystem tests and adjustments were conducted for about a year, and in June 1985, the plant was turned over to the owners (ENEA and the local utility). Figure 1 shows the PV array field and the building which houses the refrigeration and ozonizer equipment, power conditioning, battery, and control equipment.

The plant supplies power to a large food refrigeration system (300 m³ capacity), and an ozonizer for water purification purposes (6 m³/h capacity). The main purpose of the project was to demonstrate the technical and economic feasibility of these two applications.

The information contained herein was obtained from ref. [1 to 9] and individual contacts with the plant owner and designer.

2. PLANT DESIGNER, OWNER, AND OPERATOR

The plant was designed by a consortium of companies under the leadership of Pragma. The owner is ENEA. Their address is as follows:

ENEA
Attn: PV Department
Via Anguillarese, 301
I-00060 S. Maria di Galeria, Rome
Italy

3. SITE INFORMATION

3.1 Location

The plant is located on the island of Giglio in Italy which is about 29 km from the west coast near the city of Orbetello. Orbetello is about 140 km northwest of Rome. A car ferry boat links the island to Porto San Stefano on the west coast several times a week during the summer and a limited number of times a week during the winter months.

The actual site of the plant is at an altitude of 300 m in Fosso di Valle Dell'Ortana below the resort beach city Campese, in the northwest part of Giglio Island (see Fig. 2).

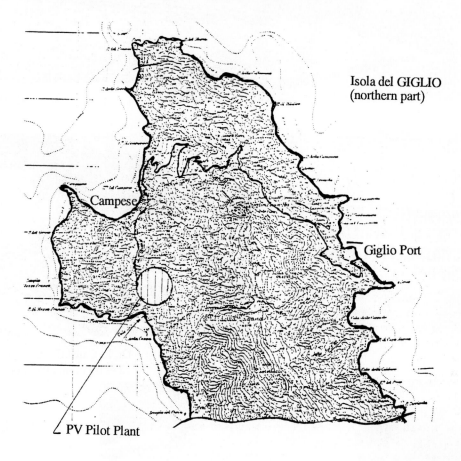

Figure 2. Plant Location on Giglio Island

3.2 Site Constraints

The plant site has been planned by the local government as an industrial area. The basic criterion used by the design team was to blend the array structures with the surrounding area such as limiting the height of the Mediterranean shrubs on the hillside.

3.3 Solar Radiation and Climatic Conditions

The solar radiation and meteorological data listed below were acquired from measurements on the island of Pianosa, 74 km from the site. They represent average figures calculated over a period of 23 years (1951-1973).

45-kW PV field

Dr. Li Causi (ENEA) inspecting the
72-cell Pragma PV modules used

Control building (cold store, PV electronics, battery)

Figure 1. Views of the GIGLIO PV Pilot Plant

- Insolation, global horizontal:	1,500 kWh/m^2 annual
- Ambient air temperature:	35°C maximum, 1°C minimum
- Wind speed:	120 km/h maximum

3.4 Load Demand

The PV system supplies a water processing unit (ozonizer) and a refrigeration unit. The nominal power consumptions of the two subsystems are:

- Ozonizer:	1.2 - 2.0 kW
- Cold storage:	14.0 - 24.0 kW

The refrigerator loads are the condenser fan, motor stator coil, inverter, and compressor motor.

4. SYSTEM DESCRIPTION AND PERFORMANCE

4.1 Plant Architecture and Layout

The PV array is set up on the slopes of a hill in the Fosso di Valle dell'Ortana district. Since a future industrial estate was planned by the local authorities in this area, it was considered an ideal site for a solar energy plant. In order to minimize disruption to the natural vegetation on the hill, the array structures were designed to match the surroundings. Figure 3 shows the layout of the plant.

The area prepared for the panel supporting structures slopes at the same angle as the average slope of the hill, so that the panels do not protrude above the Mediterranean shrubs growing on the hillside. Immediately below the array field is a building with a 240 m^2 area for the refrigerator and ozonizer systems, the battery, power conditioning, and control consoles. A vegetation "screen" is provided to hide the outline of the prefabricated building and provide some shade.

A gravel road links the plant to the nearby village on a resort beach. A servicing road to the PV field was installed in such a way as to minimize disturbance of the local landscape.

4.2 System Configuration

The system is divided into two array-battery subsystems, each with a dc-dc converter mainly for battery charging. A water disinfection system is powered by the 15-kW array, and the refrigerator is powered by the 30-kW array. Each system operates fully independently. Figure 4 is a simplified block diagram of the PV system. Its major components and their key features are listed in Table 1.

4.3 System Operation and Management

The plant control functions are as follows:

- Control of the dc-dc converter operation
- Control of battery charging and discharging
- Control of start-up and shut-down sequences

The converter with MPPT was designed to provide proper battery control, considering the sizes of the array and the battery. The control of the system was optimized also by trial and error, and by analyzing the performance data, such as:

- Battery charging and discharging profile
- Load matching with available array power
- System operation when the array output is in steady-state and in a transient condition (due to changes in solar irradiance levels)

291

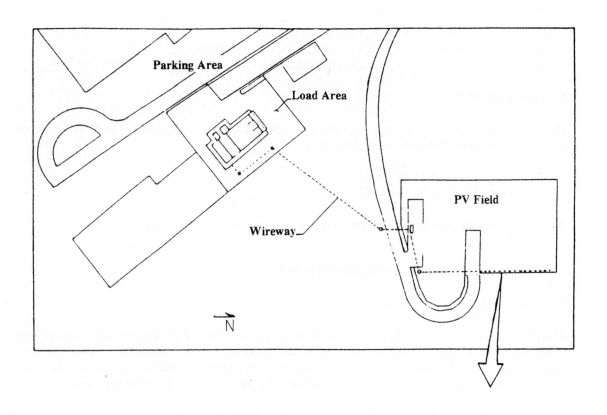

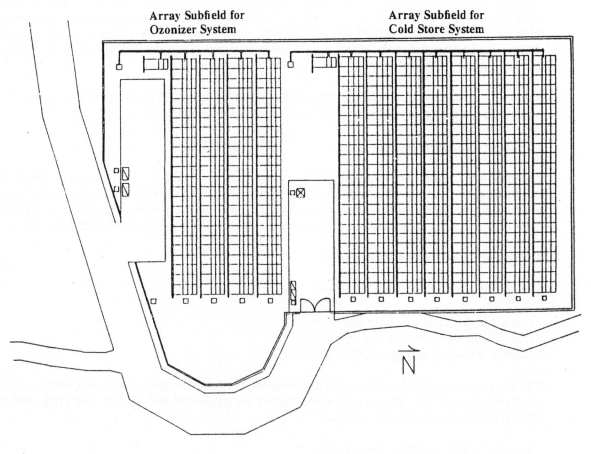

Figure 3. Plant Layout

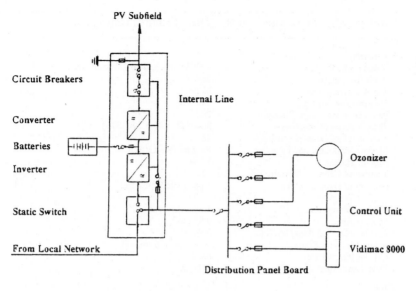

PV Subfield

Circuit Breakers

Internal Line

Converter

Batteries

Inverter

Static Switch

From Local Network

Ozonizer

Control Unit

Vidimac 8000

Distribution Panel Board

Diagram of Ozonizer Power Supply

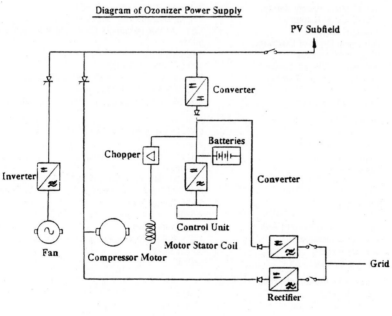

PV Subfield

Converter

Chopper

Batteries

Inverter

Converter

Fan

Control Unit

Compressor Motor

Motor Stator Coil

Grid

Rectifier

Diagram of Cold Store Power Supply

Figure 4. Simplified Block Diagrams of Ozonizer and Cold Store Systems

4.3.1 Ozonizer (Water Disinfection) System

The ozonizer operates up to 10 hours a day at full capacity to ensure proper purification of the water. The 15-kWp array subfield and the 2,000-Ah battery are provided to guarantee continuous operation and to supply required power for up to two days without sun. The maximum load demand is 1.9 kW. This load may also be supplied by the local utility grid.

The 15-kWp array subfield is wired as a single power source and may be connected to or disconnected from the dc bus via a single dc contactor. Array MPPT is done by the converter. The central controller acquires signals from the reference modules in the array field. These signals

Table 1. Major Components and Their Features (GIGLIO)

Solar Array

Rated Peak Power:	45 kW: (15- and 30-kW subfields)
Array Tilt Angle:	25
Tilt Angle Adjustment:	None
Nominal Bus Voltage:	240 Vdc
Number of Subarrays (Strings):	18
Total Number of Modules*:	864 (OS: 288, CSS: 576)
Number of Modules per String:	48
Module Manufacturer/Type:	Pragma 72 DG/SOL
Module Rating:	57 W
Number of Cells per Module:	72
Number of Bypass Diodes:	1 per 18 cells in series
Module Weight:	16 kg
Module Dimension:	1300 x 658 cm
Module Area:	0.855 m^2
Active Solar Cell Area/Module:	0.56 m^2
Solar Cell Type/Size:	Mono-Si, 100 mm dia.
Cell Packing Factor:	66 %
Total Land Area:	3000 m^2

Battery

Total Battery Energy Rating:	240 kWh (OS), 72 kWh (CSS)
Cell Capacity Rating:	OS: 2,000 Ah, CSS: 300 Ah
Voltage Range:	OS: 120 Vdc (nominal), CSS: 240 Vdc (nominal)
Allowable Depth of Discharge:	50 %
Number of Batteries:	1 of each type
Number of Cells per Battery:	120 for CSS, 60 for OS
Cell Manufacturer/Part No.:	Varta Bloc, OS: VB 2420, CSS: VB 2306
Cell Weight:	OS: 199.4 kg, CSS: 31.2 kg
Module Dimension:	OS: 393 x 392 x 550 mm, CSS: 275 x 134 x 440 mm

Inverter

Rated Power:	OS: 5 kVA, CSS: 7 kVA
Number of Inverters:	2
Type:	Self-commutated (both systems)
Manufacturer:	Silectron
Input Voltage Range:	OS: 220-340 Vdc, CSS: 120-330 Vdc
Output Voltage/Phase:	OS: 220 Vac/single phase, CSS: 220 Vac/3-phase
Frequency:	50 Hz (both systems)
THD:	<5 %
Efficiency, 100 % Full Load:	95 % (both types)
Inverter Dimension:	400 x 400 x 250 mm/400 x 400 x 400 mm
Inverter Weight:	30 kg/35 kg
Special Features:	MPPT

Converter for Ozonizer System

Power Rating:	15 kW
Input Voltage Range:	90-160 Vdc
Max. Input Current:	100 A +/- 10 A
Efficiency:	95 %

Converter for Cold Storage System

Power Rating:	7 kW
Input Voltage Range:	150-370 Vdc
Output Voltage Range:	210 Vdc
Max. Input Current:	33 A
Efficiency:	95 %
Power Consumption (max):	7 kW

Controller

Manufacturer/Model:	OS: Vidimac 8000, CSS: Independent unit
Features:	OS: Data Control, OS and CSS: Data Acquisition

Data Acquisition: Vidimac 8000 microprocessor

Rectifier

Power Rating:	10.35 kW
Input Voltage Range:	220 Vac, single-phase
Output Voltage Range:	220 Vdc
Efficiency:	85 %
Power consumption (max):	8.8 kW

* 4 additional modules installed as reference

(short-circuit current, open-circuit voltage and cell temperature) are processed to determine the MPP of the entire array. The central controller sends an analog signal to the converter to operate it at the MPP.

The maximum available current of the subfield is about 60 A (15 kW at 240 V) and the recommended maximum charging current of the lead acid battery 160 A (1600 Ah/10h). During normal operation the array is connected directly to the battery and maximum power is delivered to the battery.

When the battery voltage reaches its limit during charging and the array power exceeds the power demand, the converter inhibits the MPPT operation and switches to constant voltage charging. In this phase, the array operating point shifts to a higher voltage and lower current (i.e., to lower operating power).

4.3.2 Cold-store System

The load for the cold store is mainly the dc motor driving a 5-cylinder compressor. The control system varies the motor speed and the number of pistons in operation to match the available PV power.

The remaining available power from the 30-kW subfield is used to charge the battery through a converter and supply the motor excitation coil and the auxiliary ac loads (fan, oil pumps) via an inverter. A chopper amplifier operated by the central controller is used to vary the excitation current of the motor.

The control unit emits signals to adjust the loads for operation at MPP conditions. This is achieved by adjusting the current to the motor stator coil or operating cylinders in the compressor. As for the ozonizer system, MPPT is controlled by the central unit using signals from reference modules in the array.

4.4 Plant Performance

Immediately after installation, the average output of one array string was measured and found to be nominally 240 Vdc and 10.9 A, resulting in 2,616 W at standard reporting conditions. The total array output was thus estimated to be 47,088 W. The two PV subfields rated as 30 and 15 kW produced 31,392 W and 15,696 W, respectively, under standard conditions.

Table 2 gives monthly performance data during the period from July 1985 to June 1986. The low system efficiencies are due to not feeding the PV power into the grid. This is also indicated by the low PV array utilization factor.

5. COMPONENT DESCRIPTION

5.1 Solar Array

5.1.1 Array and Cabling

The photovoltaic array has a total of 864 modules, divided into 2 subarrays comprising 576 modules for the cold store subsystem and 288 modules for the ozonizer subsystem.

The array has 18 strings in parallel with 48 modules in series per string. The matching of modules was done by close grouping in the maximum power point current (0.2 ampere separation per string). This procedure theoretically made it possible to keep series mismatching losses extremely low. The mismatch losses for the entire array field were estimated to be between 2% and 3% based on actual measurements (0.2-A spread in MPP current per array string).

Table 2. PV Plant Performance Data Summary Sheet

PV Pilot Plant:	GIGLIO	Date of Report:	10-7-87	Site Information:	
Site Manager:		Tel.:	39-6-3048 3304	Longitude (deg N):	42.9
Proj. Lead:	Dr. Saverio Li Causi	Telefax:	39-6-3048 4110	Longitude (deg E or W):	12 E
				Altitude (m):	300

Year: 1985 / 1986

Item	Parameter	Units	Jan '86	Apr '86	May '86	Jun'86	Jul '85	Oct '85	Nov '85	Annual Total
1	Insolation Energy, Monthly									
	a. Plane of array	kWh/m^2	88.8	164	221	220	223	126	73.9	1775
	b. Horizontal Surface	kWh/m^2	59.8	150	213	217	215	93	52	1576
2	Insolation Power Daily Avg.									
	a. Plane of Array	kW/m^2-d	2.8	5.3	7	7.1	7.2	4	2.3	4.86
	b. Horizontal Surface	kW/m^2-d	1.9	4.8	6.8	7	7	3	1.6	4.31
3	Ambient Air Temperature									
	a. Daily Average	°C	10	16.6	21	21.9	26.7	18	14	17.9
	b. Daily Minimum/Maximum	°C	3/18	13/24	13/33	12/35	21/34	-1/31	7/23	-1/35
4	Wind Speed Daily Avg.	m/s	0.2	0	0.06	0	1.2	0.5	0.6	0.48
5A	Plant Energy Output									
	a. Array Field	kWh	1307	1124	2104	2287	1815	1002	653	17113
	b. Inverter(s)	kWh	1072	528	1350	1514	1253	608	393	11030
	c. Total Loads	kWh	1341	529	1351	1514	1496	654	1309	14275
6A	Plant Power Output									
	a. Array Field, Daily Avg.	kW	1.7	1.5	2.82	3.07	2.43	1.34	0.87	1.95
	b. Array Field, Peak	kW	11.7	7.2	13.8	13.9	11.7	10.9	12.2	13.9
	c. Inverter(s), Daily Avg.	kW	1.4	0.7	1.81	2	1.7	0.8	0.53	1.26
7A	Plant Operation Efficiency									
	a. Array Field Energy	%	6.0	2.7	3.8	4.2	3.3	3.2	3.6	3.9
	b. Inverter Energy	%	82	47	64	66	69	60	60	64
8A	Performance Indices									
	a. Array Field Energy Capability	kWh	1132	2469	3323	3313	3347	1897	1108	26634
	b. Array Field Utilization Factor	%	0.98	0.45	0.63	0.69	0.54	0.53	0.59	0.64
	c. System Efficiency	%	5	1.3	2.5	2.8	2.3	1.9	2.2	2.5
5B	Plant Energy Output									
	a. Array Field	kWh	365	1007	757	970	2120	783	486	10078
	b. Inverter(s)	kWh	350	749	574	768	1940	739	475	9000
	c. Total Loads	kWh	624	803	644	884	2337	952	495	11011
6B	Plant Energy Output									
	a. Array Field, Daily Avg.	kW	0.5	1.35	1	1.3	2.85	1.05	0.65	1.15
	b. Array Field, Peak	kW	21.7	19.5	18.2	22	20.7	20.05	20.7	22
	c. Inverter(s), Daily Avg.	kW	0.47	1	0.77	1	2.6	0.99	0.63	1.03
7B	Plant Operating Efficiency									
	a. Array Field Energy	%	0.8	1.2	0.7	0.87	1.9	1.3	1.3	1.15
	b. Inverter Energy	%	96	76	76	79	51	54	98	89
8B	Performance Indices									
	a. Array Field Energy Capability (Maximum Achievable)	kWh	2748	4939	6647	6764	6694	3795	2218	53457
	b. Array Field Utilization Factor	%	13	20	11	14	31	20	22	18.8
	c. System Efficiency	%	7.8	0.92	0.53	0.69	1.7	1.2	1.3	1.0

6A - 8A: PV Subsystem for Ozonizer System
5B - 8B: PV Subsystem for Cold Store

The layout of the wiring harness was designed to minimize cable loss. One bypass diode shunts across three modules in parallel. This was intended to protect the PV cells from "hot-spot" failures.

5.1.2 Modules

The modules produced by Pragma contain 72 circular monocrystalline silicon cells of 10-cm diameter. These cells are electrically connected as shown in Figure 5, including the bypass diode across 24 cells in series.

5.1.3 Array Support Structure and Foundations

Figure 5 shows the PV module mounting structures and foundations. The panel frame is made of hot galvanized steel. The foundations made of reinforced concrete are secured by means of expanding plugs which simplify the assembly of array structures. The module frame is attached to the array structure by a specifically designed bolting system. The entire structure is made of Fe 37 hot-dip galvanized steel and stainless steel bolts. To minimize labour costs, standard steel components were used as much as possible.

5.2 Batteries and Charge Control Method

There are two Varta lead-acid batteries, one for the ozonizer and the other for the cold store system. Their main characteristics are as follows:

	Ozonizer	Cold-store
- Capacity rating:	2,000 Ah	300 Ah
- No. cells in series:	60	120

The charge control technique used is essentially constant voltage with current input limited by the array and the load on the dc bus.

5.3 Power Conditioning

The power conditioning devices include dc-dc converters, dc-ac inverters, and ac-dc rectifiers. Two 15-kW converters produced by Silectron are used to provide battery charge control and array MPPT. These converters are of the "buck" type, i.e., their input voltage (array) must be higher at all times than their output voltage (battery). If the array voltage at MPP is lower than the battery voltage, the MPPT is abandoned and the array is operated at a higher voltage. The MPPT is achieved by adjusting the array voltage and current until the maximum power is reached. This is done by varying the apparent load impedance of the primary circuit of the converter. The converter is thus a variable ratio transformer in which the voltage ratio can be continuously changed using an analog signal from the central controller.

A similar dc-dc converter is used in the refrigeration system but its configuration varies with the PV array output. This converter supplies power to the central control unit, the batteries, and the motor stator coil via a chopper amplifier.

Two rectifiers (ac-dc) are supplied for operating the cold-store dc system from the single-phase utility grid. When the battery in the cold-store system reaches a low SOC it is recharged from the utility grid via the rectifier. The load is reactivated when the batteries have reached a sufficient charge as calculated by the SOC determination unit. The battery in the water disinfection system is recharged only by the PV array.

The system has one 5-kVA and one 7-kVA self-commutated inverter, each of which has an output voltage of 220 V.

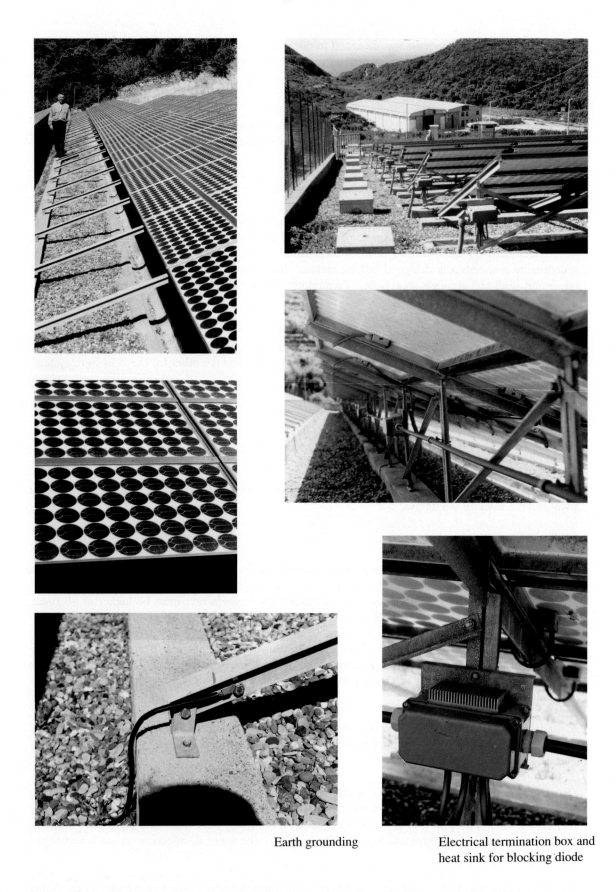

Earth grounding

Electrical termination box and
heat sink for blocking diode

Figure 5. Details of Array Support Structures and Wiring

5.4 Control System

Central system control is performed by Vidimac 8000 (a control and supervision system) combined with a Digimac 8000 control system. Its main features are:

Memory

- EPROM: 128 K
- RAM: 256 K
- Disk: 242 K

Inputs

- Analog: 46
- Digital: 45
- Storage Medium: Floppy Disk

5.5 Data Collection

Data monitoring is also performed by the Vidimac 8000 system.

5.6 Lightning Protection

The lightning protection for the arrays is implemented by the protective cables suspended over the array field, which serve as the "air terminals" (see Fig. 1) and varistors on each of the dc power lines. A "Faraday cage" is provided for the data monitoring equipment in the pre-fabricated building. The latter is made up of grounded copper and galvanized steel conductors at each corner of the building.

5.7 Auxiliary Generators

The system has no auxiliary generators.

5.8 Load Equipment

The load consists of a cold store and ozonizer which were described in Section 3.4.

6. MAINTENANCE

One local technician carries out periodic inspection and maintenance of the plant under subcontract to ENEA.

7. SUMMARY OF PROBLEMS AND SOLUTION APPROACHES

The local grid is supplied by a 1-MW Diesel generator. It is considered a "weak" grid and its line frequency is often unstable. Frequencies between 40 and 60 Hz have been measured, and the line-commutated PV inverter is often not able to synchronize with the grid frequencies. When this happens, uncontrolled commutation occurs and damages the inverter. Furthermore, these events result in a shutdown of the control system with the loss of all plant data. Restart of the plant is possible only manually which is a time-consuming effort.

There are two possible solutions to this problem. One is to improve the frequency response of the inverter to react promptly to the unstable line frequency. The other is to operate the system as a purely stand-alone plant (i.e., independent of the local grid). Because of the ease of implementation and the unreliability of the grid-connected operation, ENEA had planned to adopt the latter solution as part of plant improvement.

The maximum output voltage of the charge regulator is less than the gassing voltage of the battery. Consequently, the battery often may not reach full charge.

The processed well water has proven to be slightly acidic, which damages certain stainless steel components used in the ozonizer system.

8. KEY LESSONS LEARNED

The contractors have indicated that the most important experience gained from the operation of the plant is that the layout and the plant control system are too sophisticated. On the positive side, individual components such as the PV modules, the dc-dc converter and the cold store unit have shown good performance and proved to be acceptable.

To avoid loss of key plant data, the DAS must be separated from the main control unit that has provided both control and plant monitoring functions. An uninterruptible power supply is also highly desirable because of the unreliable utility grid.

The Haenni solar irradiance sensor was used for several years without initially calibrating it. This resulted in an error in array output prediction up to about 7%. A key lesson: although Si-based sensors are appropriate devices, they must be calibrated at the time of installation against a reference unit (such as the Eppley PSP or the Kipp & Zonen CM11).

9. CONCLUSIONS

The main elements of the PV plant have operated well. More careful attention is needed in an application like this in the areas of 1) weak and/or unstable grid characteristics and interfaces, and 2) characteristics of the water to be purified.

10. REFERENCES

[1] S. Li Causi et al, "Photovoltaics for a Cold Store on Giglio Island, Italy," Proc. Euroforum -- New Energies Congress, Vol. 3, Saarbrücken, FRG, 24-28 October 1988.
[2] Giglio Project Handout, Pilot Plant Optimization Meeting, Brussels, BE, 12-13 May 1986.
[3] P. Helm, "Manual of Photovoltaic Pilot Plants, Parts A and B," WIP report under CEC Contract EN 3 S-0007-D(B), Munich, FRG, February 1986.
[4] Giglio Project Handout, Pilot Programme I Meeting, Tremiti Island, IT, 20-21 June 1984.
[5] M. Tomassini, "Water Disinfection System and Cold Store," Proc. of the EC Contractors' Meeting held in Hamburg and Pellworm, FRG, 12-13 July 1983.
[6] Giglio Project Handout, Pilot Programme I Meeting, Mont Bouquet, FR, 12-14 April 1983.
[7] Giglio Project Handout, Pilot Programme I Meeting, Crete, GR, 14-15 September 1982.
[8] Giglio Project Handout, Pilot Programme I Meeting, Brussels, BE, 20-21 April 1982.
[9] G. Germano, "Water Disinfection and Cold Store," Proc. of the Final Design Review Meeting on EC Photovoltaic Pilot Projects held in Brussels, BE, 30 November - 2 December 1981.

ANNEX A. 1988 PROGRESS REPORT ON GIGLIO PV PLANT (Mostly from ref. [1])

The 45-kWp pilot plant "Isola del Giglio" was installed in 1984 to demonstrate the technical and economic feasibility of two important PV applications, i.e., food preservation in a cold store and water sterilization by an ozonizer. The performance of the overall plant has been investigated: experimental data analysis and the evaluation of both operation and maintenance activities in 1987 indicated that it would be advisable to disconnect the ozonizer and consequently reduce the energy demand from the local grid. The most important results obtained so far together with main maintenance activities carried out are discussed. The operation data for the refrigerator section are presented and the performance of both PV array and power conditioning subsystems are investigated. Finally, in order to improve the plant performance and to better meet local food preservation requirements, a new design of the whole PV plant has been established. The proposed new plant configuration and main improvements to be carried out are presented.

1. INTRODUCTION

The pilot plant "Isola del Giglio" originally comprised two main parts having the same Data Acquisition System (DAS) and two different control units and Power Conditioning Units (PCUs, see Fig. A-1). These two parts were the Cold Storage System (CSS) and the Ozonizer System (OS). The CSS has been working without any serious problem since June 1985 providing a significant contribution to local needs for food preservation. In fact, a few on-site shopkeepers, licensed by the Giglio Island authority, have used the CSS to preserve their food (typically vegetables and fruit) in the 300 m³ cold cell. On the contrary, the OS never reached steady state conditions of operation, due to failures of hydraulic and electric components. Moreover, the water had to be sterilized (because of chemical contamination of the well water).

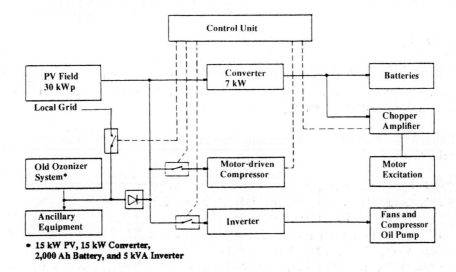

Figure A-1. Original GIGLIO PV Plant Configuration

In May 1987 the ozonizer was permanently disconnected and dismantled. The related subsystems (i.e., the 15-kWp PV array, the PCUs, and the electrical storage) are now being utilized to feed some plant ancillary equipment such as lamps, air conditioner, the DAS, and a few back-up devices for the CSS (see Fig. A-2). As a consequence, the percentage of energy absorbed from the local grid during the period June 1987-May 1988 was cut down. Moreover, the reliability of the CSS, i.e., the ratio between total working days and the number of days in the evaluation period has increased; finally, a decrease of the whole plant operation and maintenance (O&M) costs has been noted.

2. INITIAL EXPERIMENTAL DATA AND PERFORMANCE ANALYSIS

Since June 1985, much operational data have been collected in order to analyze the PV plant performance. The most important technical information concerns the working time and the energy balance of the CSS (see Table A-1). As previously mentioned, the use of available PV power (due to

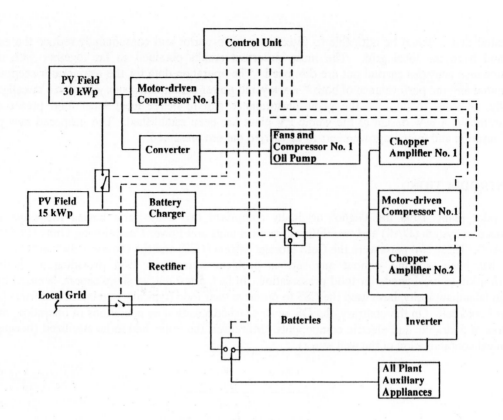

Figure A-2. Improved Plant Configuration

Table A-1. Performance of the Cold Storage System (June '85 - May '88)

Items	June 85 - May 87	June 86 - May 87	June 87 - May 88
Working Time (days)	332	360	362
Down Time (days)	33	5	3
Reliability (%)*	90.0	98.6	99.1
Time (%) in MPPT Mode**	22.5	19.0	24.0
Average PCU	95.1	94.3	93.2
Average System in MPPT	5.1	4.8	4.6
Energy (%) from Local Grid	13.2	3.7	0.4

* Ratio between working time and evaluation period
** MPPT mode occurs when motor-driven compressor is working

ozonizer disconnection in May 1987) is the major reason for the very low value of energy from the local grid (0.4%) obtained during the evaluation period June 1987-May 1988. On the other hand, the high initial value (13.2%) was basically due to an unsatisfactory setting up of plant components in the period June 1985-May 1986.

With regard to the O&M costs (see Table A-2), the first two-year operation period (June 1985-May 1987) was very expensive in comparison with the second one (June 1987-May 1988). As can be seen, in most cases a marked decrease in annual maintenance costs for single items occurred in the last one-year evaluation period. In fact, the only significant exception to this costs behaviour concerns the cooling system: the main reason for this exception was an extraordinary maintenance action in November 1987, when many deteriorated gaskets were replaced.

Table A-2. Operation and Maintenance Annual Costs (10^3 US$)

Items	June 85 - May 87 (Mean Value)	June 87 - May 88	June 85 - May 88 (Mean Value)
PV Array	10.1	5.1	8.4
Electrical Storage	5.3	4.1	4.7
Battery Management System	1.9	0.0	1.2
Power Conditioning Units	4.0	5.5	4.5
Cooling System	24.5	29.2	26.0
Data Acquisition System	55.7	34.2	48.5
On-site Operator	15.0	12.0	14.0
Total Cost:	116.5	90.1	107.3

The plant optimization achieved the following main results:

- The plant availability rose to 99.1% in the last evaluation period (the initial value was 90.0%).
- Due to the very low amount of energy taken from the local grid, the plant can now operate automatically as a stand-alone system.
- The whole plant O&M costs are still decreasing. The last two results are also related to the removal of the ozonizer.

Finally, a breakdown of the most important maintenance actions carried out, together with some related information, is presented in Table A-3. The most critical items of plant equipment, apart from their related O&M costs, have been the DAS, the OS inverter, and the Battery Management System (BMS). The DAS was replaced in March 1988, and the OS inverter has been working properly since May 1987, when it was disconnected from the local grid. The BMS, finally, has been improved by a more reliable algorithm.

3. PLANT IMPROVEMENT ACTIONS

The investigation of the overall plant performance and the analysis of both plant failures and normal maintenance operations suggested a few possible improvements to be carried out. Then, in order to increase the reliability of the CSS and to better meet local food preservation requirements, a new design of the whole plant was set up. Moreover, the proposed new plant configuration and the related operation and management logic will provide a higher system availability and lower O&M costs. The improvement effort was therefore carried out in 1988 and 1989.

Table A-3. Main Maintenance Actions (June 85 - May 88)

Items	Troubles	Number of Occurrences	Cause	Solution
DAS	Out of order and/or System shutdown	95	EMI and/or grid instability	Component replacements, Manual reset and DAS replacement
OS Inverter	Automatic Disconnection	77	Grid instability	Manual reset
BMS	False SOC estimate	75	Rough algorithm	Algorithm replacement
PV Array	Module breaking	16	Abnormal hail	Replacement
	Wiring failure	2	Rodent	Partial replacement
Battery	Cell failure	1	Sulphation	Element replacement
	Container breaking	2	Not identified	Element replacement
Cooling System	Freon leakage	3	Gasket deterioration	Gasket and pipe replacement

A block diagram of the new plant configuration is given in Figure A-2. The key features are:

- Battery charger: new equipment, operating as a charging step limited current regulator.
- Batteries: the original OS battery was increased by 60 cells in series to reach the plant operating voltage.
- Motor-driven compressors: one more compressor was added with the same characteristics as the existing one.
- New cooling system: the system was equipped with heat exchangers and forced air ventilation; this enabled a faster heat exchange in the refrigerator.

The key improvements made are the following:

- A second dc motor-driven compressor was added, with a new cooling system and heat exchangers with forced ventilation into the cold cell. The new compressor, which can be generally used as a back-up system, will be employed to cool the cell faster, or used when the insolation is not sufficient.
- The new cooling system and the old one wer interconnected to permit the interchangeability of the two compressors and/or the two heat exchangers, if required.
- The electrical storage subsystem was increased (to reach the proper operating voltage for both dc motor-driven compressors), and the battery charger was replaced. As a consequence, the dc compressor motor is able to operate for ten hours with the array at its maximum power point. On the basis of the experimental yearly mean value of the CSS energy demand, the autonomy of the refrigerator will reach about 11 days (taking into account the two-day thermal storage provided by eutectic plates already installed in the cold cell). Consequently, the local grid will be used only to charge the batteries in emergency conditions.
- The two PV arrays (30 kWp and 15 kWp) were interconnected to better utilize the total PV power. The choice among the four different available combinations will depend on the actual requirements to be satisfied; the reference parameters being cold cell temperature, insolation level, and battery state of charge (SOC).
- A new control unit was installed to automatically manage the whole plant, with a capability for remote control, and the data collection equipment was updated according to the new PV plant design.

4. CONCLUSIONS

In the initial operational phase, the pilot plant experienced several problems. Strong changes in the chemical characteristics of the water to be treated, combined with irregular inflow, caused many failures in the ozonizer's hydraulic circuit. This system often needed grid power which was often unstable both in frequency and voltage. This instability was the main cause of both DAS system shutdown and frequent OS inverter automatic disconnection.

The use of local grid energy increased the overall operational costs in the period June 1985 to May 1987. The disconnection of both the OS and the CSS from the local grid, as well as other optimization actions eliminated the major plant problems, and the PV plant has been working properly since May 1987.

The O&M annual costs have been cut by about 30%, the availability has increased from 90% to 99%, and the energy demand from the local grid has been sharply reduced from 13% to 0.4%. With further improvement of the PV plant, we believe that the O&M costs can be reduced to about $40,000/yr.

Appendix 5

HOBOKEN PV PILOT PLANT

1. INTRODUCTION

This appendix describes the 30-kW HOBOKEN PV pilot plant and its initial operation. This plant produced hydrogen by water electrolysis using PV power. Figure 1 shows views of this plant.

The basic loads are the 10-kW electrolyzer unit and two centrifugal pumps driven by 7.5-kW direct current motors which are also directly connected to the array bus without power conditioning. The hydrogen is to be used in the industrial processes of a non-ferrous metallurgic plant.

The information contained herein was obtained from ref. [1 to 8] and individual contacts with the plant designer and operator.

2. PLANT DESIGNER, OWNER, AND OPERATOR

The plant is installed on the factory site of Métallurgie Hoboken Overpelt (MHO). The development and maintenance of the system was carried out by Electrische-Nijverheids Installaties (ENI). Their address is as follows:

ENI
P.O. Box 389
B-2000 Antwerpen
Belgium

The owner of the plant is the GENSUN C50 consortium; MHO and ENI were part of this consortium.

Prefabricated 6-module panel being installed on rooftop

Foreground: Container for the H_2 Gas

Figure 1. Views of the PV Arrays and a Storage Tank for the H_2 Gas at the HOBOKEN PV Pilot Plant

3. SITE INFORMATION

3.1 Location

The system is installed in Hoboken, Belgium, about 15 km south of Antwerp at a latitude of 51.23°N (see Fig. 2).

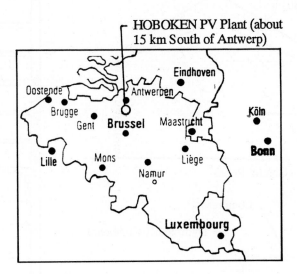

Figure 2. Location of HOBOKEN PV Plant

3.2 Site Constraints

There were no major site constraints other than to make use of the available roof area for the PV array.

3.3 Solar Radiation and Climatic Conditions

The following solar radiation and meteorological data were obtained from Uccle (near Brussels) and Deurne (near Hoboken). They represent monthly averages computed from 1963 to 1972.

- Insolation, global horizontal: 950 kWh/m^2 annual
- Ambient air temperature: 30°C maximum
- Wind speed: 80 km/h maximum

4. SYSTEM DESCRIPTION AND PERFORMANCE

4.1 Plant Architecture and Layout

The PV plant is located inside the MHO factory at Hoboken. The PV array is roof-mounted on top of a warehouse. The sloping roof of the warehouse offered a suitable location for the PV modules. The roof side is well exposed to the south with a 23° tilt angle and is not hindered by shadows from the surrounding buildings. Figure 3 shows the layout of the PV plant.

4.2 PV System Configuration

Figure 4 is a simplified system block diagram of the PV system and application loads. Its major components and their key features are listed in Table 1.

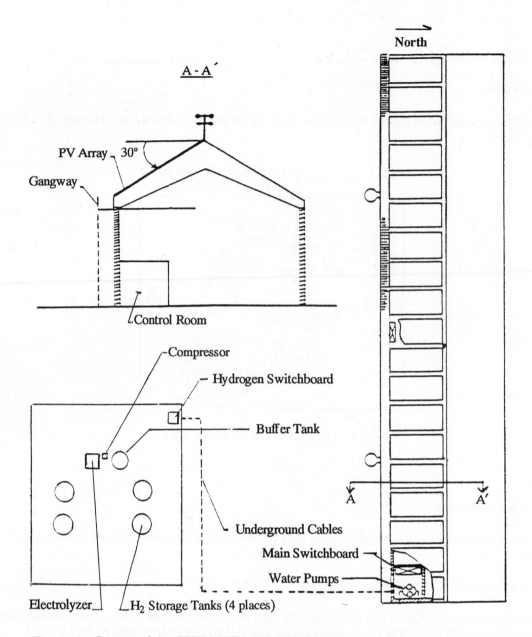

Figure 3. Layout of the HOBOKEN PV Pilot Plant

- A 30-kWp photovoltaic array is installed on the roof of an existing warehouse.

- A small control room is located inside the warehouse under the array. The main switchboard together with the load management system, two centrifugal pumps, each rated at 7.5 kW, and the data monitoring desk are located in this control room.

- The hydrogen production unit with a capacity of up to 12 kW stands in the open air about 50 m south of the warehouse. This unit is treated as an independent load with its own auxiliary control and alarm functions. There is also a separate electrical cubicle inside the hydrogen plant.

4.3 System Operation and Management

The plant implemented a relatively simple power management scheme, involving the electrolyzer, pumps, and available solar irradiance. In the power management algorithm, the electrolyzer has priority up to 12 kW. Above this power, the two 7.5-kW centrifugal pumps are switched on. The controller regulates their speed and input current as a function of solar irradiance.

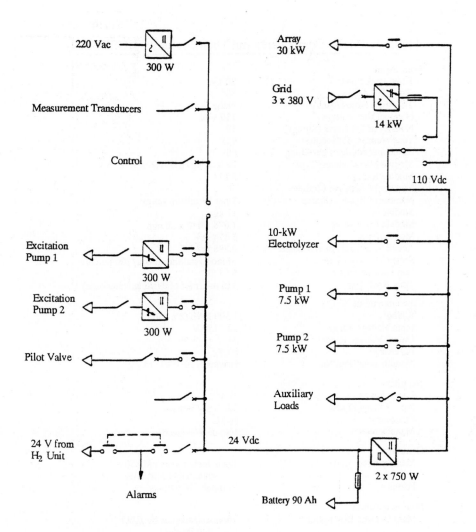

Figure 4. Simplified System Block Diagram of the HOBOKEN PV Pilot Plant

4.4 Plant Control

The plant controller and the monitoring system are independent of each other. The controller, powered by the 24-Vdc converter and battery which are connected to the PV array, performs the load management functions. The monitoring system is separately powered by the grid and collects hourly average data on an ECMA tape cassette.

4.5 Plant Performance

Key performance results at plant start-up are as follows:

- For the PV array, the manufacturer's module rating was found to be too optimistic. A large number of modules delivered did not meet the 33-W rating advertised. The module output based on manufacturer's data ranged from 26.2 to 39.9 W with an average of 34.7 W. Figure 5 is a histogram illustrating the spread in measured maximum power of PV modules. The net output of the PV array, however, met the 30-kW specified. Based on test data, mismatch and diode/wiring power losses were estimated to be 1.5 and 1.6%, respectively.

Table 1. Major Components and Their Features (HOBOKEN)

Solar Array
Rated Peak Power:	30 kW
Array Tilt Angle:	30°
Tilt Angle Adjustment:	None
Nominal Bus Voltage:	110 Vdc
Number of Subarrays (Strings):	19
Total Number of Modules:	912
Number of Modules per String:	48
Module Manufacturer/Type:	Belgosolar/IDE-Fabricable 33-4
Module Rating:	33 W
Number of Cells per Module:	38
Number of Bypass Diodes:	1 per 19 cells in series
Module Weight:	11 kg
Module Dimension:	1,078 x 440 x 50 mm
Module Area:	0.474 m²
Active Solar Cell Area/Module:	0.298 m²
Solar Cell Type/Size:	Monocrystalline silicon/100 mm dia.
Cell Packing Factor:	67.4 %
Total Roof Area:	415 m² (roof of existing warehouse)

Converter, dc-dc
Rating:	1,500 kWh (2 x 750)
Input Voltage Range:	50 - 150 V
Output Voltage Range:	24 V +/- 5 %
Efficiency:	83 %
Manufacturer/Part No.:	Polyamp/50-150/375

Rectifier
Power Rating:	14 kW
Input Voltage/Phase:	380 Vac/3-phase
Frequency:	50 Hz
Manufacturer:	Van de Weygaerde

Controller
Manufacturer/Model:	Texas Instruments PM 550 programmable controller
Features:	16 inputs/16 outputs

Data Acquisition
Manufacturer/Description:	Thomson/adapted by E.N.I. T.I. 840 Printer
Features:	ECMA 46 Magnetic Tape Recorder (Penny & Giles 7100)

Electrolyzer: A new type of high performance water electrolyzer and its direct coupling to the PV array were demonstrated. However, the electrolyzer proved to be the main source of problems for the plant. The key ones were the availability of membrane material for the electrolyzer stacks from a second source overseas, detection of H_2 in the O_2 outlet, the inability of the low pressure tubing H_2 to withstand temperatures higher than 70°C (it was replaced by high-temperature resistant material) before storage.

5. COMPONENT DESCRIPTION

5.1 Solar Array

5.1.1 Array and Cabling

The array is composed of 19 identical subarrays. Each subarray consists of 8 series-connected groups of 6 parallel connected modules. In order to minimize the mismatch losses, the parallel connected modules were selected on the basis of the same voltage output near the maximum power point, and the subarrays were formed by selecting groups with the most similar current outputs.

310

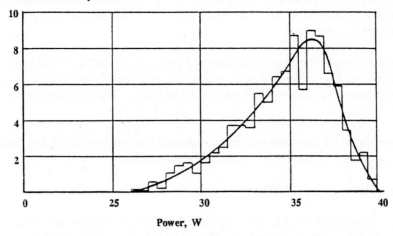

Figure 5. Spread in Actual Maximum Power for All PV Modules at HOBOKEN Pilot Plant

AMP two-wire twin lead Solarlok connectors are used to connect the modules in parallel. The wiring is carried out with a VOB (Belgian Standard) 4-mm^2 stranded wire which is the largest possible cable for the Solarlok connector. Wiring losses are minimized by connecting the module groups at both sides in a closed loop.

An LED is placed across the output of each module group in the nearest junction box. These LEDs are intended to help detect defective modules. Lighted LEDs can be seen through the transparent covers on the junction boxes.

The subarrays are connected with 10-mm^2 VOB wires to a terminal cubicle on the roof. Inside the cubicle is a three-position switch on each of the 19 subarrays, allowing a choice between open circuit, short circuit, or connection to the main switchboard.

5.1.2 Modules

The 912 modules were manufactured by IDE. Each module consists of 38 series-connected monocrystalline circular silicon cells. The terminals and the centre branch are wired to a watertight box containing two bypass diodes, one on each group of 19 cells. The box is fitted with a bushing connector. The cells are encapsulated in EVA between two tempered glass plates. The frame in anodized aluminium was especially shaped to fit the support structure.

The module features are as follows:

- cell diameter: 100 mm
- nominal peak power: 33 W
- nominal open circuit voltage: 21 V
- nominal short circuit current: 2.1 A
- dimensions: 1,078 x 440 x 50 mm
- weight: 11 kg

5.1.3 Array Support Structure and Foundations

The array was initially installed on the south side of a roof inclined at 23°. The roof covering made of corrugated asbestos sheets was not suitable to support a structure. This problem was solved by

311

bolting an independent galvanized steel structure to the steel columns of the building. At that time, the array tilt angle was changed to 30°. The south side of the roof measures approximately 60 x 8 m and is entirely covered by the array.

Access to the array is eased by a gangway along the edge of the roof and also by the cable ladders between the subarrays. The 152 groups were assembled and wired in the factory on galvanized steel frames which fit between the trusses of the supporting structure. The 20 trusses are connected to the ground.

Rubber shock absorbers are fitted between the galvanized steel structure and the aluminium module frames in order to minimize possible deformation of the modules.

5.2 Batteries and Charge Control Method

A standard dc-dc converter is used to recharge the 24-Vdc 90-Ah battery as well as to supply power to the system controller and pump excitation. These components are classified as a secondary power source and are not major elements of the PV system.

5.3 Power Conditioning

The primary load is the 10-kW electrolyzer which requires a wide range of medium dc voltage (105-115 Vdc) and high current (100A). Thus, no voltage regulation device is used. Figure 6 shows the electrolyzer and the peripheral components.

5.4 Control System

The control subsystem uses a TI process controller, PM 550. The inputs to the unit for management purposes are solar irradiance, electrolyzer current, pump 1 and 2 currents, and array voltage. The outputs are digital and analog. The digital signals serve to switch on/off the electrolyzer and pumps 1 and 2, and the analog outputs to control the speed of both pumps.

5.5 Data Collection

The monitoring system hardware includes a Thomson microprocessor, 40-channel analog to digital (A/D) conversion cards, TI 840 printer, and a Penny & Giles 7100 magnetic tape recorder. The recorded data and special sensors are as follows:

- Horizontal irradiance (with Kipp & Zonen CM 10 pyranometer)
- Plane of array irradiance (with Haenni Solar 118)
- Wind speed and direction
- Ambient air temperature (with PT100)
- Back of module temperature (with Ni-Cr-Ni)
- PV array voltage and current

5.6 Lightning Protection

Because of the low Keraunic level in the area, combined with the tall chimneys and masts surrounding the PV array, no protection devices against direct lightning strikes were fitted. For protection against induced transients, "TRANSORB" voltage suppressors were installed. Two of these devices were found to be burned out during the initial operational periods, reportedly due to overvoltages in the PV system.

5.7 Auxiliary Generators

The system has no auxiliary generators.

5.8 Load Equipment

The load consists of an electrolyzer (see Fig. 6) and two dc water pumps. Their typical design and operating features are as follows:

Electrolyzer:

- Input voltage: 105 to 115 Vdc
- Input current: 130 to 90A
- Stack temperature: 40 to 75°C
- 58 cells in series
- Alkaline water electrolysis

Pump/motor:

- Speed control via potentiometer on the main switchboard or by the system controller from no load (1,900 rpm) to full load (3,200 rpm)
- 7.5-kW 110-Vdc motor, 24 Vdc excitation
- All Weiler centrifugal pumps

6. MAINTENANCE

Development and maintenance of the HOBOKEN PV system have been carried out by Electrische Nijverheids-Installaties.

The plant is fully automatic. If it is shut down due to a failure, an alarm is initiated in the MHO operation centre. In case the MHO personnel are not able to restart the PV system, a message is sent to ENI. This procedure has proven satisfactory for general plant maintenance.

Since the PV system has been integrated into a large metallurgical factory, qualified maintenance personnel are available on short notice. Documents on the plant have been prepared, therefore, personnel not trained initially on this system were also able to carry out maintenance and repairs.

Problems of personnel availability occur periodically when development and maintenance of the control and monitoring software is needed.

7. SUMMARY OF PROBLEMS AND SOLUTION APPROACHES

Difficulties were encountered with the module attachment elements. Elastic vibration mounts were installed to avoid mechanical stress in the modules due to deformation of the structure under load. They were very susceptible to corrosion and frequently broke. They were replaced by galvanized bolts, although not an ideal solution since they also corrode.

Grounding problems in the form of leakage currents in the electrolyzer caused incorrect operation of the plant controls. These problems were eventually resolved but it was apparently difficult to locate the faults.

The dc-dc converters supplying regulated voltage for plant control have been damaged several times by an overvoltage from the PV array (181 V at low temperatures), although the maximum voltage of the array indicated by the supplier was 150 V. This problem was resolved by placing a resistor between the array and the converter which switches on at a certain array voltage.

The main operational problems were in the control of the two water pumps working in parallel and in the electrolyzer. Water pressure oscillations occurred during the initial parallel operation of the two

Electrolyzer   Buffer Tank

Electrolyzer (left)

H_2 Containers

Figure 6. The Electrolyzer and Peripheral Components

314

pumps. Excessive currents drawn by the pump motors because of these oscillations sometimes caused the plant to switch off. The change from current to power control of the motors improved the motor excitation, and it proved to be an adequate solution. However, software changes were difficult to implement due to lack of programmers.

The problems encountered with the diaphragms in the electrolyzer stack required the procurement of a special membrane from the U.S.A. High temperature-resistant material was needed to replace the ABS plastic tubing (70°C limit). Detection of H_2 in the O_2 circuit as well as O_2 in the H_2 circuit proved to be costly.

8. KEY LESSONS LEARNED

Key lessons learned are as follows:

- The direct coupling of the electrolyzer and PV array appears feasible, but an in-depth analysis is needed to evaluate the effects of all the operating variables with the PV source.

- A more extensive development of the electrolyzer and necessary peripheral equipment (like the H_2 detector) is essential before commercial application.

- Speed adjustment of the dc pumps proved to be a difficult task. Several changes in their field regulators were needed before satisfactory results were achieved.

- The detection of a faulty array condition by monitoring the subarray currents is feasible, but for the Hoboken array, it was found not to be possible.

9. CONCLUSIONS

The key objectives of the project were achieved. It was obvious that a longer period of performance assessment is needed to draw general conclusions for this type of application. Problems experienced were mainly in the application loads (i.e., the electrolyzer) and not in the PV system.

10. REFERENCES

[1] Hoboken Project Handout, Pilot Plant Optimization Meeting, Brussels, BE, 12-13 May 1986.
[2] P. Helm, "Manual of Photovoltaic Pilot Plants, Parts A and B," WIP report under CEC Contract EN 3 S-0007-D(B), Munich, FRG, February 1986.
[3] Hoboken Project Handout, Pilot Programme I Meeting, Tremiti, IT, 20-21 June 1984.
[4] C. Van Weyenbergh, "Photovoltaic Power Plant for Hydrogen Production and Water Pumping," Proc. of the EC Contractors' Meeting held in Hamburg and Pellworm, FRG, 12-13 July 1983.
[5] Hoboken Project Handout, Pilot Programme I Meeting, Mont Bouquet, FR, 12-14 April 1983.
[6] Hoboken Project Handout, Pilot Programme I Meeting, Crete, GR, 14-15 September 1982.
[7] Hoboken Project Handout, Pilot Programme I Meeting, Brussels, BE, 20-21 April 1982.
[8] G. Bertels, "Photovoltaic Power Plant for Hydrogen Production and Water Pumping," Proc. of the Final Design Review Meeting on EC Photovoltaic Pilot Projects held in Brussels, BE, 30 November - 2 December 1981.

Appendix 6

KAW PV PILOT PLANT

1. INTRODUCTION

This appendix describes the 35-kW KAW PV pilot plant and its performance as of 1988. The plant has been in operation since November 1982. Figure 1 shows the array field and the control building.

The purpose of the plant is to demonstrate the technical, economic, and social suitability of photovoltaic generators to meet the energy requirements of a remote community in French Guyana. The PV plant replaced two existing Diesel generators which supplied all the power needs of the community.

Information contained in this appendix was obtained from ref. [1 to 8] and individual contacts with the plant designer and operator.

2. PLANT DESIGNER, OWNER, AND OPERATOR

SERI Renault Ingénierie was responsible for the design, installation, and initial operation of the plant. This plant has been handed over to the local utility, Electricité de France, which is now responsible for the plant. For information about this plant, the following organization should be contacted:

> Renault Automation Seri
> Attn: Renewable Energies Dept.
> BP 66
> F-78184 St. Quentin Yvelines Cedex
> France

View of the 35-kW PV array looking north

Figure 1. Views of KAW Array Field and Control Building

3. SITE INFORMATION

3.1 Location

The plant is located at the village of Kaw in French Guyana near the equator. Its coordinates are:

- Latitude: 4.3°N
- Longitude: 52°W

Kaw is a village with 70 inhabitants. It can be reached from the main town of Cayenne after a two-hour drive (83 km) on a well-maintained trail but only during the dry season.

3.2 Site Constraints

Connection of the village to the Cayenne grid is not foreseen even in the distant future. The main reasons are the remoteness of Kaw and the low density of the population. Thus, the plant has to survive in a tropical climate and operate reliably.

3.3 Solar Radiation and Climatic Conditions

- Insolation, global horizontal:
  - o Annual: $1,500 \ kWh/m^2$
  - o Daily in July: $4.74 \ kWh/m^2$
  - o Daily in December: $3.32 \ kWh/m^2$
- Ambient air temperature: 38°C maximum, 4°C minimum
- Wind speed: 100 km/h maximum

3.4 Load Demand

The loads consist of normal household appliances, lighting and radios. In the past, electricity was produced by the two Diesel generators (30 kVA and 19 kVA) for public and private lighting, refrigerators, and cold stores. A typical daily load profile is shown in Figure 2.

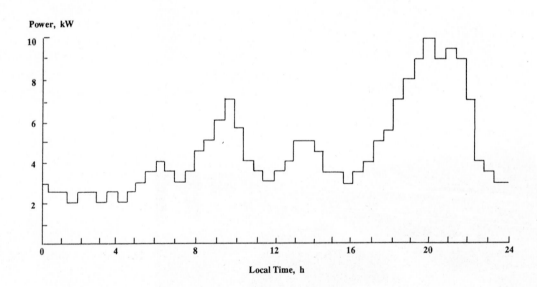

Figure 2. Typical Load Profile in June 1983

4. SYSTEM DESCRIPTION AND PERFORMANCE

4.1 Plant Architecture and Layout

The layout of the plant as shown in Figure 3 was designed after a visit to the site. The land close to the village was donated by the Village Community Council. The site is in a wide flat clearing hidden by a group of trees and bushes. This helped to avoid a possible undesirable contrast between the

PV array field and control building

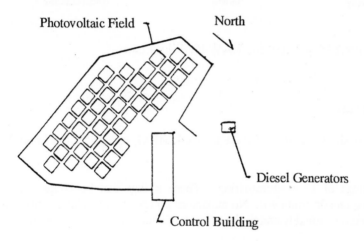

Figure 3. Layout of the KAW PV Pilot Plant

array field and the local tropical homes. A 1.8-m high wire fence has been erected around the PV plant.

The area for the PV plant was cleared of trees bordering the village, bearing in mind the possible extension of Kaw and future access roads.

The building which houses the storage battery and electronics was set up in the same style as the village. The design of the inside of the building as well as the vents in the walls permit a natural ventilation of the battery room.

4.2 System Configuration

Figure 4 is a simplified block diagram of the PV plant (Note: the arrangement and basic components are similar to the original AGHIA ROUMELI PV Pilot Plant which was also designed by SERI Renault). The array tilt angle was set at 5°, south-facing. Table 1 lists the major components and their features.

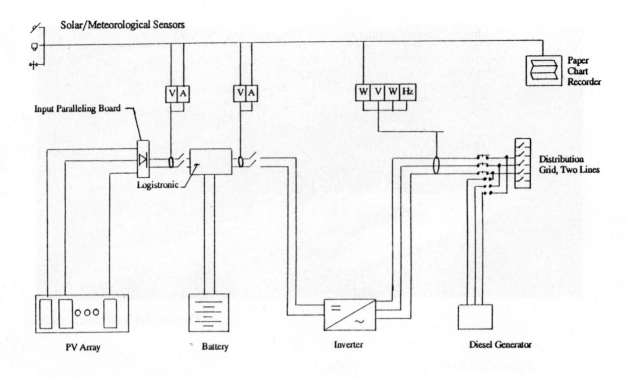

Figure 4. Simplified System Block Diagram of the KAW PV Pilot Plant

4.3 System Operation and Management

The operation of this system is very similar to that of AGHIA ROUMELI, except for the lower PV array size used at KAW.

The PV array consists of 41 strings of 12 modules/string. These strings can be connected or disconnected from the battery using one dc contactor. No maximum power point tracking is used but the nominal battery voltage is selected to roughly match the array maximum power point voltage.

The maximum current of the array is 116 A (35 kWp/300 V) and the nominal 10-hour charge current of the battery is 150 A. Thus, the designer assumed that no excessive overcharging can occur. When the battery drops to 70% SOC the inverter and loads are disconnected to minimize over-discharging following plant start-up.

4.4 Plant Control

Plant control and monitoring is accomplished by a hardwired controller installed as a back-up to the Logistronic unit which did not function properly following plant start-up.

4.5 Plant Performance

The following plant performance data were reported by SERI Renault in 1985 [2]:

	1983	1984
- Total operating time (h):	7,000	6,821
- Total energy produced (MWh):	29.2	39.2
- Average power delivered (kW):	4.1	4.5
- Mean energy demand (kWh/d):		101
- Mean energy supplied (kWh/d):		81

320

Table 1. Major Components and Their Features (KAW)

Solar Array
Rated Peak Power:	35 kW
Array Tilt Angle:	5°, South-facing
Tilt Angle Adjustment:	None
Nominal Bus Voltage:	330 Vdc
Number of Subarrays (Strings):	41
Total Number of Modules:	492
Number of Modules per String:	12
Module Manufacturer/Type:	France-Photon/FPG 72
Module Rating:	72 W
Number of Cells per Module:	68
Number of Bypass Diodes:	1 per 17 cells in series
Module Weight:	13 kg
Module Dimension:	1,284 x 0.642 x 0.0425 m
Module Area:	0.842 m^2
Active Solar Cell Area/Module:	0.534 m^2
Solar Cell Type/Size:	Monocrystalline silicon/100 mm dia.
Cell Packing Factor:	65 %
Total Land Area Used:	1,567 m^2

Battery
Total Battery Energy Rating:	450 kWh
Cell Capacity Rating:	1,500 Ah
Voltage Range:	292-353 Vdc
Allowable Depth of Discharge:	70 %
Number of Batteries:	1
Number of Cells per Battery:	150
Cell Manufacturer/Part No.:	Varta Bloc/VB 2415
Cell Weight:	152 kg
Cell Dimension:	550 x 307 x 383 mm

Inverter
Rated Power:	40 kVA
Number of Inverters:	1
Type:	Self-commutated, pulse-width modulatlion
Manufacturer/Part No.:	Jeumont-Schneider/61DG864
Input Voltage Range:	292-363 Vdc
Output Voltage/Phase:	380 Vac/3-phase
Frequency:	50 Hz
THD:	6 %
Efficiency, 100 % Full Load:	94 %
10 % Full Load:	85 %
Inverter Dimension:	1.5 x 0.84 x 2.3 m
Inverter Weight:	1,000 kg

Controller (for Battery)
Manufacturer:	Seri Renault
Features:	Charge current termination at a pre-set battery voltage limit

Data Acquisition
Manufacturer/Description:	ELSYDE/ADAS with EPROM Cartridge, 13 channels (original DAS)

Auxiliary Power Source
Rating:	40 kVA

The plant operator reported that the total energy delivered by the PV plant as of March 1986 was 83.5 MWh during a total operating time of 17,100 h.

The basic status of the system as of 1988 is as follows:

- User load demand has been continually increasing
- All PV modules are functional
- No corrosion or mould were encountered in the structures
- No significant power loss has occurred
- There has been a significant deterioration in the capacity of the battery

5. COMPONENT DESCRIPTION

5.1 Solar Array

5.1.1 Array and Cabling

The PV array comprises 41 panels, each with 12 series-connected PV modules. The panels are connected in parallel. The nominal array dc bus voltage is 330 Vdc.

The 12 modules are interconnected via Amphenol plugs. The interconnection cables lead to a connecting box attached to one of the panel's supporting legs. The panels are individually linked to the paralleling board in the control cabinet. The series blocking diodes are also situated here. Thus, the voltage of each panel can be individually checked at the control cabinet, which allows detection of possible cabling or module failure.

In the same way, the intra-panel cabling enables each module to be easily disconnected and checked. Also, the installation of the cable and connection of the 12 modules can be carried out in less than 10 minutes by two unskilled workers.

Total power losses due to module mismatch, field wiring, and diodes were estimated to be less than two percent.

5.1.2 Modules

The modules, manufactured by France-Photon, were especially developed for the R & D programme of the European Commission. Each module has 68 circular 100-mm monocrystalline silicon cells in series and two bypass diodes in the termination box behind the module.

5.1.3 Array Support Structures and Foundations

The mounting structures for the modules were designed to minimize the manufacturing costs as much as possible. Figure 5 shows the support structures and foundations. The structure for one panel of 12 modules has the following elements:

- Two footings, each fixed on a concrete block by two threaded rods
- One central double beam and two bracketing sections fixing the modules onto the footings
- Two threaded rods assembling the modules and the double metallic beam

The panel is placed 1.5 m from the ground in order to provide access to the area under the panel. A duct in the middle of the central double beam is provided for the cabling of modules.

5.2 Batteries and Charge Control Method

The battery consists of 150 Varta 1,500-Ah lead-acid battery cells in series. The size was selected to permit low-rate discharges. The battery charge control method adopted consists of inserting a power resistor in series with the battery whenever a predetermined voltage limit was reached (2.35 Vdc/cell). This effectively places the battery in a trickle-charge mode (about C/40 rate).

5.3 Power Conditioning

The power conditioning consists of a 40-kVA self-commutated inverter which uses transistors and PWM technology. This inverter was especially designed by Jeumont-Schneider for use in stand-alone PV plants. Highest possible efficiency was a design goal from low to full load condition. To meet this requirement the inverter is linked to an output transformer to minimize losses. In addition, to keep the total harmonic distortion (THD) below 6%, it was necessary to add a filter, which led to a slight lowering of overall efficiency and a small increase in no-load losses.

322

Panels tilted at 5° (site latitude = 4.5°)

Mr. P. Helm (WIP-Munich) inspecting the PV panels

Figure 5. Array Structures and Foundations

The inverter maintains its output voltage at 380 Vac with the input voltage of 292 to 363 Vdc. It was thus helpful to have a battery which minimized the variation of its input voltage between charge and discharge.

The measured efficiency of the inverter with its step-up transformer is 93% at full load and 85% at 10% of full load (see Fig. 6). The inverter is connected to the dc bus using two dc contactors (+ and -). These contactors are controlled by the control unit. To avoid damage to the inverter, the output of the inverter is interlocked with the output of the Diesel generator.

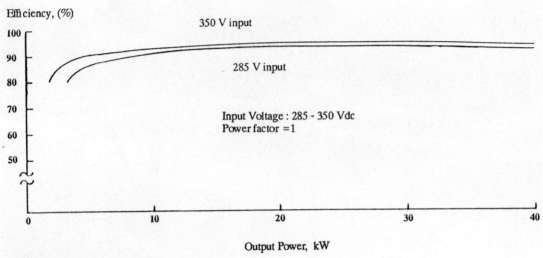

Figure 6. Inventer Efficiency vs. Output Power

5.4 Control System

The system was originally designed to be controlled by the Varta Logistronic unit. The Logistronic unit was intended to provide a battery state of charge signal to avoid both excessive overcharge and discharges greater than 70% DOD. The Logistronic unit did not function properly so its primary function, that of battery charge/discharge protection, was replaced by an analog circuit.

5.5 Data Collection

An automatic data acquisition system from ELSYDE, ADAS, was used for plant monitoring. The following measurements are recorded (13 channels total):

- Solar irradiance on horizontal and tilted (5°) surfaces and solar energy
- Ambient air temperature
- Wind speed
- PV, battery, and inverter output voltages and currents
- ac energy output
- Strain gauge on array support structure

5.6 Lightning Protection

The maximum Keraunic level was estimated to be 30 days per year. Calculations carried out according to a method recommended by the CEC have shown that for the KAW plant, the average interval between two lightning strikes hitting the installation is in the range of 300 years. As the estimated life span of the solar plant is 20 years, protection against lightning was not considered to be of critical importance.

The only protection method implemented is a careful grounding of the 41 separate array structures via two 30 mm² cross-section copper wires linking the whole array field.

5.7 Auxiliary Generators

A Diesel generator (40 kVA) is started automatically whenever PV power is not available.

5.8 Load Equipment

The consumer loads are conventional lights, refrigerators, fans, etc. used by private homes, school, and church.

6. MAINTENANCE

The plant is being maintained by EDF, the local utility organisation, with help from two local inhabitants. The main maintenance activities are:

- Cleaning of PV modules
- Mowing of grass under the panels
- Collection of data in the log book, including solar and inverter output energies, corrosion of structures, etc.

7. SUMMARY OF PROBLEMS AND SOLUTION APPROACHES

The main problems encountered are the increase in load demand and inadequate battery charge and discharge management due to Logistronic failure. A simplified back-up system was installed because of this failure, and it has worked adequately. Several battery cell cases have cracked, resulting in electrolyte spillage. An apparent decay in battery capacity was also noted. The plan was to replace all battery cells, and to add an improved charge controller.

8. KEY LESSONS LEARNED

Key lessons learned deal mainly with battery charge control and battery maintenance. They are similar to those of the AGHIA ROUMELI PV plant (see Appendix 1).

9. CONCLUSIONS

The plant has operated reasonably well after replacement of the Logistronic unit. Battery charge and discharge control is best done by proper monitoring of battery voltage. In particular, battery charge control needs sufficient flexibility and sophistication to ensure the following:

- Adequate overcharging every few days to achieve fully-charged conditions on a periodic basis
- Avoid long periods (days) in a partially charged state
- Prevent complete cell discharge, especially at very low rates

10. REFERENCES

[1] Kaw Project Handout, Pilot Plant Optimization Meeting, Brussels, BE, 12-13 May 1986.
[2] Letter from B. Aubert of SERI Renault, "Results After Two Years of Kaw PV Plant Operation by EDF," dated 23 October 1985.
[3] Kaw Project Handout, Pilot Programme I Meeting, Tremiti Island, IT, 20-21 June 1984.
[4] B. Aubert, "Rural Electrification in French Guyana," Proc. of the EC Contractors' Meeting held in Hamburg and Pellworm, FRG, 12-13 July 1983.
[5] Kaw Project Handout, Pilot Programme I Meeting, Mont Bouquet, FR, 12-14 April 1983.
[6] Kaw Project Handout, Pilot Programme I Meeting, Crete, GR 14-15 September 1982.
[7] Kaw Project Handout, Pilot Programme I Meeting, Brussels, BE, 20-21 April 1982.
[8] B. Aubert, "Rural Electrification in French Guyana," Proc. of the Final Design Meeting on EC Photovoltaic Projects held in Brussels, BE, 30 November - 2 December 1981.

Appendix 7

KYTHNOS PV PILOT PLANT

1. INTRODUCTION

This document describes the 100-kW KYTHNOS PV pilot plant and its initial performance. It was designed to verify the benefit of photovoltaics as a fuel-saver for 1,700-kW Diesel generators in a village grid on the island of Kythnos. This PV plant also operates in parallel with the 100-kW wind generators. The plant has been in operation since June 1983. Figure 1 shows the array field and the control building.

The information contained herein was obtained from ref. [1 to 9] and individual contacts with the plant designer and operator.

2. PLANT DESIGNER, OWNER, AND OPERATOR

SERI Renault Ingénierie was responsible for the design, installation, and initial operation of the plant. This plant has been handed over to the utility organisation, Public Power Corporation of Greece, which is now responsible for the plant. Their address is as follows:

Public Power Corporation
Attn: Renewable Energies Dept.
10, Navarinou St.
GR-10680 Athens
Greece

3. SITE INFORMATION

3.1 Location

Kythnos is an island in the Cyclades and is located approximately 60 km south of Athens. It can be reached by a 4-hour boat ride from Athens. The PV plant is about 4 km (20 minutes by car) from the

Array field after installation

PV array field blends into the landscape
View from the southwest

Battery housing

Figure 1. View of the Array Field and the Containers Used to House the Power Electronics and Batteries

boat harbour of Kythnos. The topography of the island is typically Aegean: gentle slopes, absence of flat fields, low mountains, and very sparse vegetation. Its coordinates are:

- Latitude: 37.4°N
- Longitude: 24.4°E
- Altitude: 134 m

3.2 Site Constraints

A key constraint was to blend the arrays into the environment. The array field was therefore located in four terraces, and indigenous rocks were used for some of the enclosures.

327

3.3 Solar Radiation and Climatic Conditions

The following solar radiation and meteorological data were based on data obtained at Syros for daily values and at Kythnos PV plant for hourly values:

- Insolation, global horizontal:
  - o Annual: 1,700 kWh/m^2
  - o Daily average: 4.66 kWh/m^2
- Ambient air temperature: 40°C maximum, -2°C minimum
- Wind speed: 112 km/h maximum

3.4 Load Demand

In the summer months of 1988 the village load had reached 8,000 kWh per day with ac power up to 600 kW. The daily energy needs decrease to about 3,000 kWh during the winter months. The typical monthly energy output of the three power sources (diesel, wind, PV) is shown in Figure 2. The transmission network consists of 15-kV 3-phase lines. Its total length is 32 km with 14 substations for the whole island.

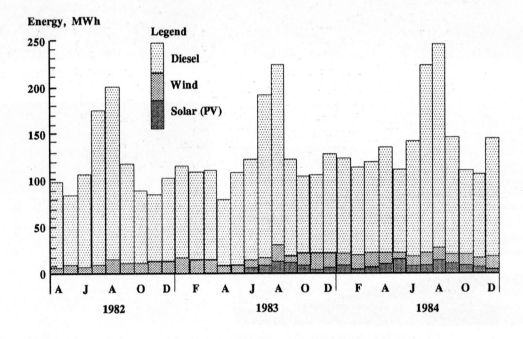

Figure 2. Typical Monthly Energy Output of Three Power Sources at KYTHNOS

4. SYSTEM DESCRIPTION AND PERFORMANCE

4.1 Plant Architecture and Layout

The site is located near the Diesel power station on a slope oriented to the south. Figure 3 shows the layout of the plant.

The architect and the contractors designed and installed the plant in such a way as to blend with the surrounding scenery. Traditional materials and techniques were applied. For the enclosure around the PV plant, dry walls were built of local stone slabs so as to match the rectangular houses scattered on the hillsides.

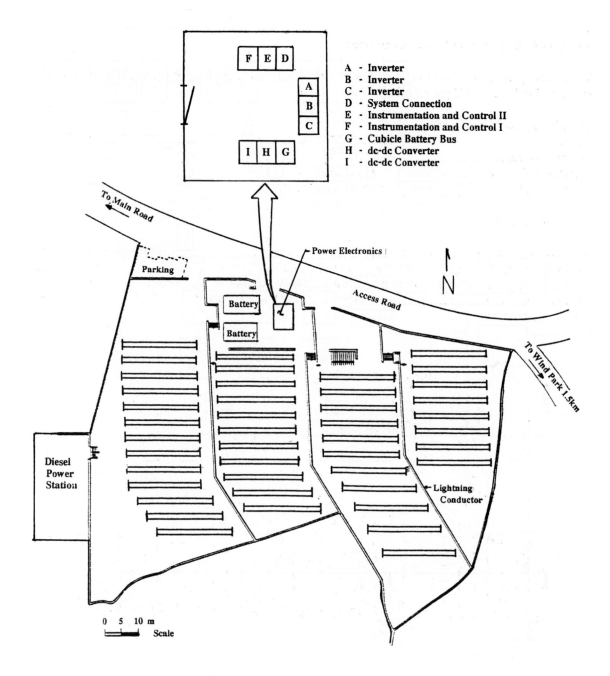

A - Inverter
B - Inverter
C - Inverter
D - System Connection
E - Instrumentation and Control II
F - Instrumentation and Control I
G - Cubicle Battery Bus
H - dc-dc Converter
I - dc-dc Converter

Figure 3. Layout of the KYTHNOS PV Pilot Plant

The four terraces for the PV field were slightly inclined so as to keep the drop at the end of each terrace to a height of one metre maximum. Dry stone walls were built to retain the soil and border the area.

Three containers housing the system equipment and the batteries were painted white and positioned at the north side of the plant area where they cannot shadow the solar modules and are easily accessible from the road.

4.2 System Configuration

Figure 4 is a simplified block diagram of the PV plant. The plant basically consists of the array field, one battery housed in two containers, and in the third container the power electronics (four

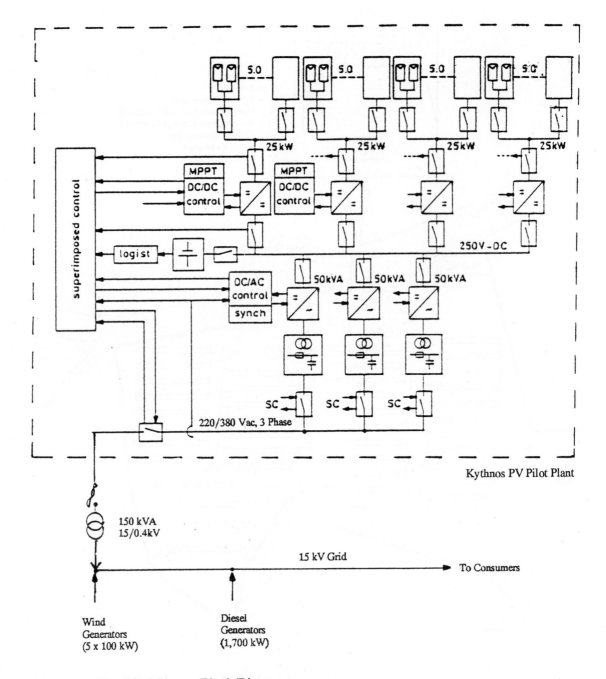

Figure 4. Simplified System Block Diagram

converters and three inverters) and system controllers. Its major components and key features are listed in Table 1.

4.3 System Operation and Management

The photovoltaic power plant supplies power to the 15-kV grid network in parallel with the existing Diesel power station and wind generators (see Fig. 4). The operation of the plant is performed automatically by the "superimposed controller". Operator interaction is minimal. A simple logic is implemented for power management and inverter control in order to optimize inverter efficiency (see Subsection 5.3.2).

Table 1. Major Components and Their Features (KYTHNOS)

Solar Array
Rated Peak Power:	100 kW
Array Tilt Angle:	35°
Tilt Angle Adjustment:	Adjustable, 60 to 15°
Nominal Bus Voltage:	160 Vdc
Number of Subarrays (Strings):	40
Total Number of Modules:	800
Number of Modules per String:	20
Module Manufacturer/Type:	Siemens/ SM 144-09
Module Rating:	125 W
Number of Cells per Module:	144 (36 in series, 4 in parallel)
Number of Bypass Diodes:	1 per 18 cells in series
Module Weight:	27 kg
Module Dimension:	1.5 x 1.0 x 0.08 m
Module Area:	1.5 m^2
Active Solar Cell Area/Module:	1.1304 m^2
Solar Cell Type/Size:	Monocrystalline silicon/100 mm dia.
Cell Packing Factor:	75 %
Total Land Area Used:	7,500 m^2

Battery
Total Battery Energy Rating:	600 kWh
Cell Capacity Rating:	2,400 Ah
Voltage Range:	250 Vdc (nominal)
Allowable Depth of Discharge:	30 %
Number of Batteries:	1
Number of Cells per Battery:	125
Cell Manufacturer/Part No.:	Varta Bloc VB 2424
Cell Weight:	134 kg (with acid)
Cell Dimension:	307 x 283 x 525 mm

Converter, dc-dc
Rated Power:	25 kW x 4
Manufacturer:	Siemens
Input Voltage Range:	160 Vdc
Output Voltage Range:	250 Vdc +20 %, -10 %
Efficiency:	94-96 % (10-100 % FL)

Inverter
Rated Power:	150 kVA (50 kVA x 3)
Number of Inverters:	3
Type:	Self-commutated (each one)
Manufacturer:	Siemens
Input Voltage Range:	250 Vdc +20 %, -10 %
Output Voltage/Phase:	380 Vac/3-phase (each one) +/- 10 %
Frequency:	50 Hz (each one)
THD:	5 % (each one)
Efficiency, 100 % Full Load:	94 %
10 % Full Load:	82 %

Controller
Manufacturer/Model:	Varta Logistronic Unit (original device) analog circuit (voltage comparator)

Data Acquisition
Manufacturer/Features:	Siemens with ECMA

Other Power Sources:	1,700 kVA Diesel generator, 8 units 100 kW wind generators, 5 units

4.4 Plant Control

The plant is controlled automatically without manual intervention. Manual control and adjustments can be made on the following:

- The charge voltage limit on each dc-dc converter (2.3 to 2.4 V per cell)
- On/off control of each converter and inverter
- Inverter power output to match the load demand

Parameters measured and recorded on an hourly basis are:

- Solar irradiance, global horizontal and global tilt (on array plane)
- Ambient air temperature
- Wind speed
- Converter, battery, and inverter voltages and currents
- Grid voltage and frequency

4.5 Plant Performance

The plant has operated very well for almost five years (as of 1988). From June 1983 to December 1988, the plant supplied 572,000 kWh to the grid. Figure 5 shows the monthly energy produced over several years as recorded by the ac energy meter. During the plant acceptance test by JRC-Ispra, the array field output at the standard test condition of 1,000 W/m^2 and 25°C average cell temperature was estimated to be 86.3 kW.

Energy, MWh

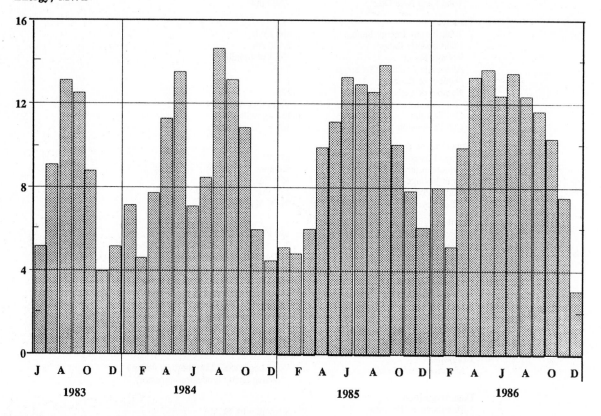

Figure 5. Monthly Energy Production Profile from 1983 to 1986

The system has been operated such that the battery SOC could fluctuate between 60 and 100% even though the allowable DOD is 30%. The battery state of charge signal from the Logistronic unit was intended to help provide battery protection and management. The calculated signal was found to be unreliable over a period of time so the basic charge control and deep discharge prevention was done by the dc-dc converter and the system controller, respectively. Other key problems encountered are listed in Section 7.

In 1979, the energy supplied by the PV plant was 170,000 kWh, which is about 18% of the total grid demand. The expected saving of fuel oil, however, exceeds 18% because unnecessary Diesel generators are switched off rather than being run idly. In fact, the PV plant often allowed the Diesel generator operation to be optimized.

Characteristics of the originally installed Diesel generators are as follows:

- Annual energy generation: 932,000 kWh
- Maximum daily energy generation: 5,100 kWh
- Minimum daily energy generation: 1,700 kWh

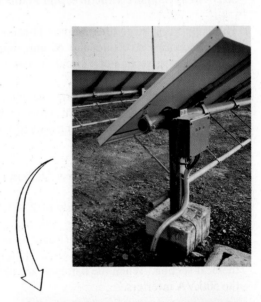

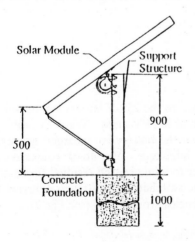

Electrical Connection Box

Figure 6. Details of Array Structures, Foundations, and Electrical Connection
Boxes at the KYTHNOS PV Pilot Plant

Because of the increasing power demand each year, the total capacity of the Diesel station was increased to 1,700 kW in 1987.

5. COMPONENT DESCRIPTION

5.1 Solar Array

5.1.1 Array and Cabling

The array contains a total of 800 PV modules, grouped into 40 strings, each consisting of 20 modules in series. Careful design of the field cabling and electrical termination boxes ensured low cable losses and easy trouble-shooting of wiring problems, as well as easy maintenance. Stranded single-core wires are used for power cabling.

5.1.2 Modules

Siemens module SM 144-09 rated at 125 W is the standard size used in the PV array. Each module contains 144 circular monocrystalline cells of 100-mm diameter. 144 cells are connected in four parallel strings each of 36 cells.

5.1.3 Array Support Structures and Foundations

The supporting structures employ H-section beam IPBL 120 and 114-mm diameter tube for the upper mounting structure and 60-mm diameter tube for the lower mounting structure. Figure 6 shows the cross section of a foundation and supporting structure. The PV panels are inclined at 35°.

The distance between the arrays was chosen to prevent shadowing of modules during the winter. Good stability of the support structure is achieved by balancing the module immediately above the horizontal support tube. During embedding in the concrete foundations, a jig was used to fix the H-section beams.

5.2 Batteries and Charge Control Method

The battery has a rated capacity of 2,400 Ah at a 250-Vdc nominal voltage. The batteries (Varta Bloc) have a low self-discharge and are provided with recombinators to minimize water loss. A microprocessor controlled unit (Logistronic) was intended to measure the state of battery charge and provide the control system with data. Battery charging is performed using the MPPTs as described in the next section. When the batteries are deep-discharged, they can be recharged from the grid using the 50-kVA inverters.

5.3 Power Conditioning

5.3.1 Converter

Four 25-kW dc-dc converters transform the 160-V input voltage to the 250-V battery system voltage. These units perform MPPT by adjusting the operating voltage of the PV array. When the battery is fully charged, the PV array operating point is adjusted so that the battery is recharged at constant voltage (2.35 V/cell) condition. In this case, the MPPT is inhibited. The dc-dc converters are controlled by the central controller with shifted pulses in order to minimize the residual ripple of the battery and charging current. At night-time the converters are switched off and reverse current flow to the PV array is prevented. The converter operates at efficiencies around 97% (see Fig. 7).

5.3.2 Inverter

Three 50-kVA self-commutated inverters were especially designed for this plant. These inverters use high power transistors and pulse-width modulation for voltage regulation.

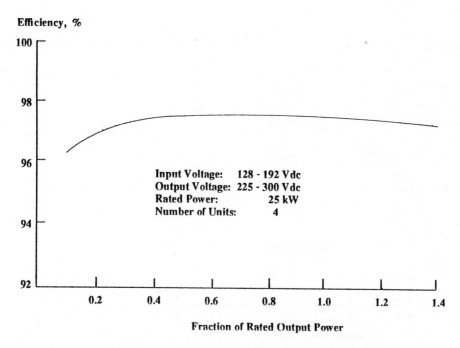

Figure 7. Efficiency Characteristics of Siemens 25-kW Converter

Each inverter may be switched on or off depending on the power requirement. This control method minimizes losses during partial load. These inverters may operate in a grid-connected or stand-alone mode. Several inverters can also be operated in parallel. The active and reactive power is uniformly distributed during parallel operation. The inverters are designed for high impulse load capacity and 4-quadrant operations. These inverters can also be used as rectifiers (in a reverse power flow mode) to recharge the batteries.

The three 50-kVA inverters are designed for parallel operation (ac side). Although the inverters are self-commutated, their frequency and power factor are controlled to match those of the local grid with which the Diesel power station and the wind generators are operating in parallel.

The system controller interprets a frequency drop in the grid voltage below the reference value as a demand for power. If the battery is sufficiently charged, the controller sends a command for the next inverter to turn on. Conversely, one inverter switches off when the frequency rises above the threshold. This strategy allows a higher overall efficiency at low-load conditions. The efficiency of the three inverters is illustrated in Figure 8.

When an inverter receives a signal to start, it automatically begins to function, synchronizes, hooks up to the line, and takes over its share of the load.

5.4 Control System

The system control consists of the "Superimposed Control" and the Logistronic unit. The controller Model B 8011 was designed by Siemens. The Varta Logistronic unit was intended to provide the required signal (state of charge) for overcharge and overdischarge protection. This signal as well as other essential signals for the automatic operation of the inverters are acquired by the superimposed controller.

5.5 Data Collection

The data acquisition system from Siemens consists of on-site monitoring and display devices and data recording equipment. The on-site display devices show the following parameters on continuous chart recorders in colour in real-time:

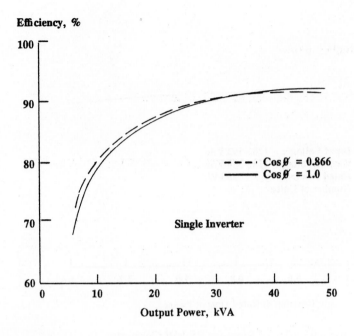

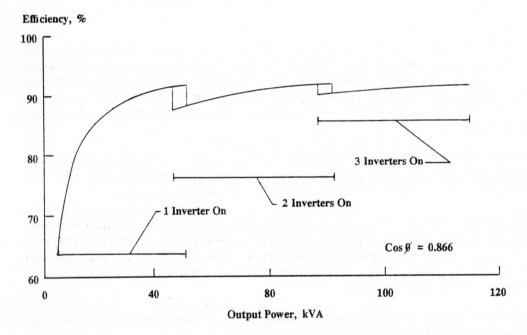

Figure 8. Efficiency Characteristics of Siemens 50-kVA Inverter

- Converter output current (4)
- Battery voltage, current, and state of charge
- Total active power supplied
- Solar irradiance
- Ambient air temperature

In addition, an ac kWh meter (Schlumberger) provides a continuous display of ac energy supplied by the PV plant to the grid. The hourly recording of data has been done on a Penny & Giles cassette; the instantaneously scanned measurements can be seen on the video terminal.

336

5.6 Lightning Protection

The lightning protection included the installation of:

- 10-m lightning rods within the array field area
- Overvoltage arrestors at both ends of the cabling between the solar generators
- Varistors on the series blocking diodes in the array strings
- Arrestors in the ac output of the plant

The lightning arrestors were chosen with an ignition voltage limiting the peak voltage to the voltage the insulation must withstand according to VDE standards.

5.7 Other Power Sources

Operating in parallel with the PV plant are a Diesel power station consisting of several generators (1,700 kW total) and five wind generators (100 kW each).

5.8 Load Equipment

The village 15-kV grid network serves as the load for the PV plant.

6. MAINTENANCE

The PV plant has run with little or no maintenance. Normal maintenance activities consisted in replacing the paper rolls in the continuous line recorders, extracting the recorded cassettes, and checking battery cell electrolyte levels and specific gravity periodically.

7. SUMMARY OF PROBLEMS AND SOLUTION APPROACHES

Key problems encountered and solution approaches are as follows:

- After initial installation, the Logistronic unit sustained damage attributed to lightning strikes. Since varistors were added, this unit has operated without problems. One problem with the key function of the Logistronic is the state of charge (SOC) signal which it calculates. This was found to become inaccurate as the period of calculation increases. A periodic reset of the SOC signal to 100% is essential for this signal to be useful.

- Damage also occurred from an apparent lightning strike to the following:

  o Two PV modules and their termination boxes
  o One converter
  o Data monitoring system
  o Varistors in the monitoring signal lines

- The ventilation fan in the converter cubicle has failed which causes overheating of the converter electronics.

- Failures in the inverter printed circuits due to corrosion caused inverter shut-downs.

- Unreliable operation of the system controller and one inverter was experienced in late 1987 and 1988. This was attributed to the salty humid atmosphere which oxidized the connection pins and flat metal conductors on the printed circuit boards.

8. KEY LESSONS LEARNED

Key lessons learned are as follows:

- Electronics must be designed for long-term operation in a hot, salty environment. Corrosion protection and passive cooling of electronic components are essential in coastal sites like Kythnos Island.

- Adequate lightning protection methods are still not well understood and further work is required. Some specialists on lightning effects believe that the 10-m poles dispersed inside the array field are not necessary.

- Battery ageing characteristics are largely unknown and an accurate state of charge calculation model has not yet been developed.

9. CONCLUSIONS

The PV plant has operated well in stand-alone and grid-connected modes for about four years. The malfunctioning of the electronics experienced in 1987 and 1988 is primarily attributed to corrosion by the hot, salty atmosphere. This is perhaps the only major weakness in the plant which was overlooked or not seriously considered during the component design and manufacturing phase.

10. REFERENCES

[1] J. Chadjivassiliadis, "Solar Photovoltaic and Wind in Greece," IEE Proc., Vol. 134, Pt A, No. 5, May 1987.
[2] Kythnos Project Handout, Pilot Plant Optimization Meeting, Brussels, BE, 12-13 May 1986.
[3] Kythnos Project Handout, Pilot Programme I Meeting, Tremiti Island, IT, 20-21 June 1984.
[4] A.N. Tombazis et al, "Kythnos Island and Aghia Roumeli Photovoltaic Pilot Projects - An Architectural Approach," Proc. of the 5th European PV Solar Energy Conference, Athens, GR, 17-21 October 1983.
[5] Consortium Siemens, "Kythnos Photovoltaic Power Plant," Proc. of the EC Contractors' Meeting held in Hamburg and Pellworm, FRG, 12-13 July 1983.
[6] Kythnos Project Handout, Pilot Programme I Meeting, Mont Bouquet, FR, 12-14 April 1983.
[7] Kythnos Project Handout, Pilot Programme I Meeting, Crete, GR, 14-15 September 1982.
[8] Kythnos Project Handout, Pilot Programme I Meeting, Brussels, BE, 20-21 April 1982.
[9] H. Ritter, "Kythnos Photovoltaic Power Plant," Proc. of the Final Design Review Meeting on EC Photovoltaic Projects held in Brussels, BE, 30 November - 2 December 1981.

Appendix 8

MARCHWOOD PV PILOT PLANT

1. INTRODUCTION

This appendix describes the 30-kW MARCHWOOD PV pilot plant and its initial operation. The plant completion and start-up date was November 1983. Figure 1 shows the array field. This plant is designed to operate in a stand-alone or grid-connected mode. The load ranges from 10 to 30 kW.

The information contained herein was obtained from ref. [1 to 9] and individual contacts with the plant designer and owner.

2. PLANT DESIGNER, OWNER, AND OPERATOR

The designer and owner of the plant is BP Solar. Their address is as follows:

    BP Solar International Ltd.
    Solar House
    Bridge Street
    GB-Leatherhead Surrey KT22 8BZ
    United Kingdom

3. SITE INFORMATION

3.1 Location

The plant is installed on the CEGB Power Station site at Marchwood. Marchwood is located on the southern coast of the United Kingdom near Southampton. It is situated at sea level. Its coordinates are:

- Latitude: 51°N
- Longitude: 1.4°W

3.2 Site Constraints

There were no major site constraints.

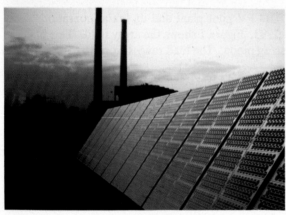

At sunset

Figure 1. View of the Array Field at the MARCHWOOD PV Plant

3.3 Solar Radiation and Climatic Conditions

The following solar radiation and meteorological data were acquired from the UK Meteorological Office at Dungeness (1931-1975):

- Insolation, global horizontal:
  - o Daily maximum in summer months: 4.9 kWh/m^2
  - o Daily average in winter months: 0.5 to 1.5 kWh/m^2
- Ambient air temperature: 27°C maximum, -5°C minimum
- Wind speed: 151 km/h maximum

3.4 Load Demand

The output of this plant is either directly coupled to the existing Marchwood site distribution network at 415 Vac, 3-phase, 4-wire, 50 Hz, or arranged to supply segregated loads of 10 to 30 kW.

4. SYSTEM DESCRIPTION AND PERFORMANCE

4.1 Plant Architecture and Layout

Figure 2 shows the layout of the plant. The boilers in the original coal-fired power station were converted to burn oil instead of coal before the station was commissioned. The coal storage area on a concrete surface was therefore never used. This site was attractive for the PV array because little or no civil engineering work was required, and the PV array structures could be easily installed on site and bolted down to the existing concrete pad. The site installation time was therefore minimal. Another consideration was the possible transfer of the PV plant to an alternative site later on.

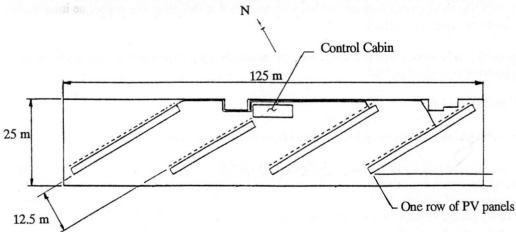

Figure 2. Layout of the MARCHWOOD PV Pilot Plant

4.2 PV System Configuration

The plant consists of a 30-kW PV array, 400-Ah battery, 40-kVA inverter, and a data acquisition system. Figure 3 shows a simplified block diagram of the PV system. Its major components and their key features are listed in Table 1.

4.3 System Operation and Management

The plant is capable of the following modes of operation:

- Grid-connected, with or without the battery
- Stand-alone, with or without the battery

The PV plant was designed for the following modes of operation:

- PV array/inverter feeding into the grid; inverter in MPPT
- PV array/inverter feeding the selected loads; inverter in MPPT
- PV array/battery/inverter feeding into the grid
- PV array/battery/inverter feeding the selected loads

MPPT is accomplished using a microprocessor. This controller also provides battery charge control. The processor goes through the following sequence to maintain the peak power point operation:

- Determine power input to the inverter by multiplying the measured voltage and current input.

- On the basis of the input power thus calculated, adjust the inverter output to feed more or less power depending on the difference between the values of two successive power calculations (i.e., if the new power value minus the previous power value is greater than zero, the inverter feeds more power). The processor repeats this procedure to stay at peak power point within one percent.

In the grid-connected mode with no battery, the processor controls the operation as follows:

- If the array output is below the no-load losses of the inverter, the inverter uses the grid power to supply the no-load losses for about two minutes.

- If this condition is the same after another two minutes, the array is shorted and the inverter is switched off.

- If the array short-circuit current reaches a value such that the array can supply the inverter no-load losses, the inverter is restarted.

The MPPT operation can also be implemented manually via a switch and a potentiometer located at the inverter dc control panel.

4.4 Plant Control

The following functions can be implemented from the control console:

- Operational mode and PV system configuration control
- Array string on/off control
- Battery on/off control
- Battery charge voltage limit adjustment
- Inverter on/off control
- MPPT (manual)

In designing the monitoring system, the following factors were considered:

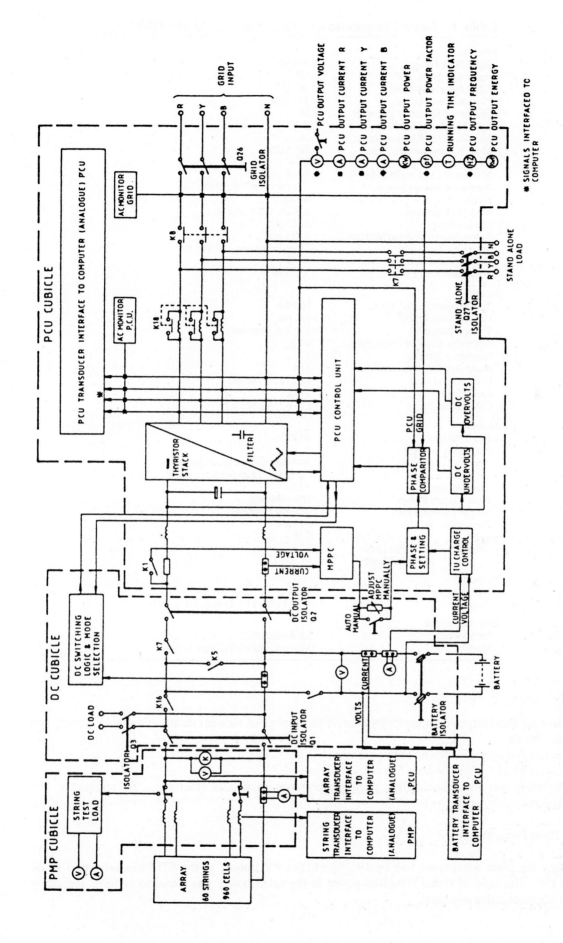

Figure 3. Simplified Schematic of the MARCHWOOD PV Pilot Plant

343

Table 1. Major Components and Their Features (MARCHWOOD)

Solar Array
Rated Peak Power:	30 kW
Array Tilt Angle:	55°
Tilt Angle Adjustment:	+/- 15°
Nominal Bus Voltage:	240 Vdc
Number of Subarrays (Strings):	60
Total Number of Modules:	960
Number of Modules per String:	16
Module Manufacturer/Type:	BP A1233
Module Rating:	33 W
Number of Cells per Module:	36 (in series)
Number of Bypass Diodes:	1 per string of 18 cells
Module Weight:	6.5 kg
Module Dimension:	1,073 x 412 x 38.5 mm
Module Area:	0.424 m^2
Active Solar Cell Area/Module:	0.292 m^2
Solar Cell Type/Size:	Monocrystalline, 100 mm dia.
Cell Packing Factor:	66 %
Total Land Area Used:	3,125 m^2

Battery
Total Battery Energy Rating:	96 kWh
Cell Capacity Rating:	400 Ah
Voltage Range:	220 Vdc (nominal)
Allowable Depth of Discharge:	20 %
Number of Batteries:	1
Number of Cells per Battery:	110
Cell Manufacturer:	LUCAS 'P' Series
Cell Weight:	25 kg
Cell Dimension:	173 x 158 x 505 mm

Inverter
Rated Power:	40 kVA
Number of Inverters:	1
Type:	Self-commutated (but capable of grid synchronization)
Manufacturer:	AEG
Input Voltage Range:	190 - 300 Vdc
Output Voltage/Phase:	415 Vac/3-phase
Frequency:	50 Hz
THD:	5 %
Efficiency, 100 % Full Load:	92 %
10 % Full Load:	75 %

Controller
Manufacturer:	DEC
Model:	PDP 11/23 PLUS computer

Data Acquisition
Manufacturer:	Motorola/DEC
Model:	Motorola IRIS/DEC PDP 11/25

- Need for flexibility and expansibility relative to number of measurements, computing power, and data storage
- Unattended operation for up to two months
- Reusability of equipment after five years
- Marketing role in demonstrating plant performance to visitors (therefore, the equipment and displays must be attractive)

4.5 Plant Performance

During the plant acceptance test by JRC-Ispra, the PV output at the STC was estimated to be 29.4 kW. The ratio of actual measured power to the rated power was found to be 98%, indicating that adequate sizing was done.

344

5. COMPONENT DESCRIPTION

5.1 Solar Array

5.1.1 Array and Cabling

The PV array consists of 960 modules which are electrically grouped into 60 strings or subarrays. Each string has 16 modules in series, resulting in the array bus voltage of 240 Vdc nominal.

5.1.2 Modules

The PV module provides a rated peak output of about 33 watts at a nominal terminal voltage of 16 Vdc.

5.1.3 Array Support Structures and Foundations

The array structures are designed such that one panel holds 16 modules (see Fig. 4). Galvanized steel is used. The legs permit vertical adjustment for unevenness in the existing concrete foundation. The position of the front legs also permits lining up of adjacent subarrays so that each row appears geometrically aligned. One front attachment position is used for two adjacent subarray structures, but an additional brace is incorporated in every fourth subarray structure as shown in Figure 4.

The panels can be inclined to three different tilt angles, 70°, 55°, and 40°. The tilt angle can be adjusted simply by choosing the appropriate bolt position on the lower support member. The structural design also allows the panel to be lowered to a good working height for maintenance

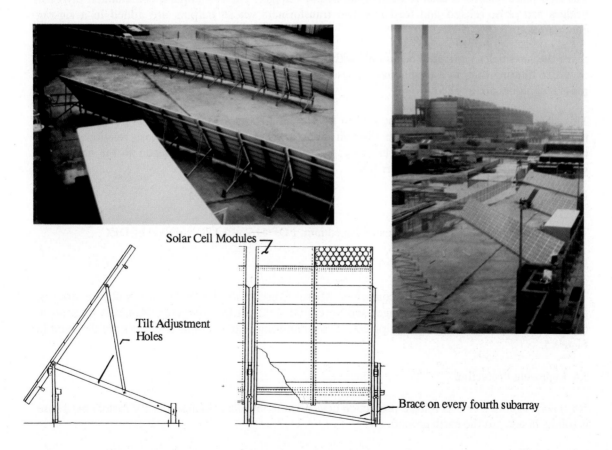

Figure 4. Array Support Structures and Foundations at the MARCHWOOD PV Pilot Plant

345

purposes. The distance between the rows of arrays was selected to minimize shadowing of the panels.

5.2 Batteries and Charge Control Method

The battery consists of 110 400-Ah Lucas lead acid cells in series. The special cell features as reported by the supplier are as follows:

- Positive plate contains lead alloy with some impurity, developed for strength and cycling ability
- Plates are wrapped with a sheet which minimizes stratification
- Cell case is made of polypropylene material
- High efficiency and low water loss

The cells are mounted on an epoxy-coated metal stand, and placed in a separate ventilated room in the control cabin.

The system controller provides constant voltage charging whenever the battery charge voltage reaches its predetermined limit. Voltage regulation is done by shedding or adding the array strings as necessary to maintain the bus voltage constant.

5.3 Power Conditioning

The power conditioning equipment consists of an inverter. This inverter contains an input filter, run up current limiter, impulse-commutated thyristor bridges, output filter, control unit and maximum peak power controller.

The step-pulse method is used to control the output voltage. The two impulse-commutated thyristor bridges are phase-related and feed into two transformers whose outputs are added in a zig-zag configuration.

Since the harmonics generated begin only with the 11th harmonic, this method offers the advantage of easier filtering than in a traditional 3-phase design in which only zig-zag addition and not step-pulse techniques are applied.

This inverter can supply power to the grid or operate in a stand-alone mode. The flow of power to the grid is achieved by connecting it through inductors and controlling its phase relationship with respect to the grid: If it is leading in phase, then power flows from the inverter to the grid. The greater the lead, the greater the power flow.

5.4 Control System

The system controller is a microprocessor-based unit, PDP 11/23 PLUS, supplied by DEC.

5.5 Data Collection

The data acquisition equipment comprises Micro Consultants (IRIS Level 1 software system, programmed in a BASIC-like language) and DEC PDP 11/23 PLUS computer. Annex A presents a more detailed description of this equipment. The monitoring system and its interfaces are depicted in Figure 5.

5.6 Lightning Protection

No special lightning or overvoltage protection devices were installed because the PV system hardware is solidly bonded to the earth ground.

5.7 Auxiliary Generators

The plant has no auxiliary generators.

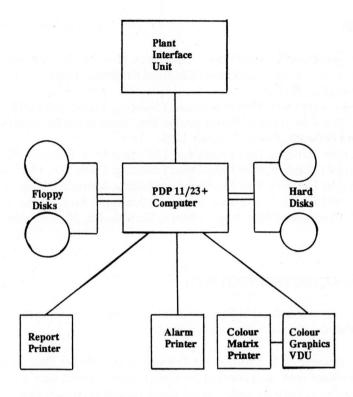

Figure 5. Monitoring System and Its Interfaces

5.8 Load Equipment

The loads are the monitoring equipment and the utility network.

6. MAINTENANCE

The maintenance activities consist of periodic inspections of the battery cell electrolyte level and specific gravity.

7. SUMMARY OF PROBLEMS AND SOLUTION APPROACHES

No major technical problems were apparently encountered. A programmatic problem is that there was no concerted effort to assess the plant performance and disseminate the results.

8. KEY LESSONS LEARNED

A sufficient amount of engineering time and effort for plant performance analysis and subsequent timely dissemination of the results should be made a part of the contractual requirement.

9. CONCLUSIONS

The plant has operated as designed and has demonstrated the basic modes of operation. Based on the performance data submitted to CEC in 1984 and 1985, the plant worked very well, and most notably in the grid-connected mode.

10. REFERENCES

[1] Marchwood Project Handout, Pilot Plant Optimization Meeting, Brussels, BE, 12-13 May 1986.
[2] "Marchwood Project" Meeting, Institute of Electrical Engineers, London, GB, 22 April 1985.
[3] Marchwood Brochure, 1984.
[4] Marchwood Project Handout, Pilot Programme I Meeting, Tremiti Island, IT, 20-21 June 1984.
[5] R.D.W. Scott, "30-kW Marchwood Power Station Site," Proc. of the EC Contractors' Meeting held in Hamburg and Pellworm, FRG, 12-13 July 1983.
[6] Marchwood Project Handout, Pilot Programme I Meeting, Mont Bouquet, FR, 12-14 April 1983.
[7] Marchwood Project Handout, Pilot Programme I Meeting, Crete, GR, 14-15 September 1982.
[8] Marchwood Project Handout, Pilot Programme I Meeting, Brussels, BE, 20-21 April 1982.
[9] R.D.W. Scott, "30-Kilowatt Marchwood Power Station Site," Proc. of the Final Design Review Meeting on EC Photovoltaic Pilot Projects held in Brussels, BE, 30 November - 2 December 1981.

ANNEX A. DATA ACQUISITION SYSTEM [9]

To satisfy the basic criteria defined in Subsection 4.4, the monitoring equipment, Micro Consultants IRIS Level 1 illustrated in Figure 5, was implemented.

The system was programmmed in a BASIC-like language to provide the plant input scanning and first level conversion routines such as thermocouple linearization. In this unit a number of inputs are averaged over one-minute periods. After initial conversion, appropriate inputs are checked against alarm limits and any violation notified to the main processor, Digital Equipment PDP 11/23 PLUS computer. This computer provides data archiving, alarm annunciation, operator interface, graphic displays and report printer.

For the routine unattended data archiving, the larger 10 megabyte removable cartridge disks are used. The data files are both cyclic for the short-term hourly, daily, and monthly files, and linear for the long-term yearly and five-year files. It is possible to dump selected files of data to floppy disks for removal and subsequent processing on other computers. The system is configured for three different file dumps including one specifically formatted for data required by the CEC.

The colour visual terminal displays both alphanumeric reports and high resolution graphs and histograms. The facilities are available through a selection of operator commands which allow the operator to define the start and finish time of the data to be plotted and the time interval of the samples. All displays, both alphanumeric and graphic, can be hard copied to a colour matrix printer.

Two conventional printers are also provided, one for recording alarms and significant events and one for printing tabular reports. Alarms are also announced on the top line of the VDU and, because the control room is not normally manned, are relayed to a lamp alarm in an adjacent gatehouse.

The unmanned operation of the system has also imposed the very important requirements for security of data and system recovery. In order to overcome the potential problems of restart after utility power failure, a battery-powered real-time clock is incorporated.

After a power failure and automatic reload and restart of the system, the data archiving task is able to identify the affected files and fill the appropriate records with invalid data markers. This invalid data marker is used to highlight failed or suspect signals during normal scanning. If any of the subsequently calculated averages have insufficient valid results, as compared with a configurable limit, an alarm is actuated.

Hence, by selecting proven and reliable hardware with a flexible and mature operating system such as RSX-11M and writing the application software in a structured language such as PASCAL, it has been possible to configure a monitoring system which will meet all the original design objectives. Sufficient flexibility and expansion capability has been incorporated in order to allow the addition of suitable facilities such as modelling, optimization, and control.

Appendix 9

MONT BOUQUET PV PILOT PLANT

1. INTRODUCTION

This document describes the 50-kW MONT BOUQUET PV pilot plant and its initial performance. Figure 1 shows the PV array and other components located outdoors. The date of initial plant start-up and operation was April 1983.

The main purpose of this plant was to demonstrate and study the use of a PV power source for telecommunications equipment.

The information contained herein was obtained from ref. [1 to 8] and individual contacts with the plant designer and AFME.

2. PLANT DESIGNER, OWNER, AND OPERATOR

The plant design and installation at Mont Bouquet, France was supervised by Photowatt and handed over to the owner, Télédiffusion de France (TDF). AFME and TDF, along with the CEC co-financed this project. Information concerning this plant can be obtained from the following organization:

> AFME
> Attn: Renewable Energies Dept.
> Sophia Antipolis
> Avenue Emile Hugues
> F-06560 Valbonne Cedex
> France

3. SITE INFORMATION

3.1 Location

The plant is installed on the top of a hill called Mont Bouquet, located about 30 km from Ales in southeast France. Its coordinates are:

- Latitude: 44.2°N
- Longitude: 4.5°E
- Altitude: 607 m

50-kW PV array powers telecommunications equipment

Figure 1. Views of the PV Array Field at the MONT BOUQUET PV Pilot Plant

3.2 Site Constraints

The PV plant was to be located along the ridge of the hill near the radio and microwave repeater station.

3.3 Solar Radiation and Climatic Conditions

The following solar and meteorological data were recorded at the Carpentras station of the Météorologie Nationale which is 80 km from the site.

- Insolation, global horizontal:
  - o Annual: 1,400 kWh/m^2
  - o Daily average in July: 7.1 kWh/m^2
  - o Daily average in December: 1.3 kWh/m^2
- Ambient air temperature: 30°C maximum, -15°C minimum
- Wind speed: 152 km/h maximum
- Snow height: 50 cm maximum

3.4 Load Demand

The main station loads (34 to 40 kW total) consist of three radio and three television repeaters. The radio repeaters require a total of 7 kW (single phase 220 Vac), and the TV repeaters total 24 kW (three-phase 380 Vac). Their annual energy demand is about 256 MWh.

4. SYSTEM DESCRIPTION AND PERFORMANCE

4.1 Plant Architecture and Layout

Mont Bouquet is a popular tourist site whose previous parking lot is now occupied by the PV plant. This plant was set up on the highest part of the surrounding countryside. The PV field is installed in an area 15 m wide by 150 m long on the south-facing slope. It has a gradient of 6% in the upper part and 9% in the lower part so that the distance between rows is smaller than it would be if the arrays were on a horizontal surface.

4.2 PV System Configuration

Figure 2 is a simplified block diagram of the PV system. Its major components and their key features are listed in Table 1. The system consists of the following:

- 50-kW PV array (710 Photowatt 72-W modules)
- 212-Vdc battery (106 France-Oldham 800-Ah cells at 10-h rate)
- 48-Vdc battery (24 France-Oldham 350-Ah cells)
- 30-kVA three-phase inverter

The basic plant configuration is similar to that of the NICE PV Pilot Plant, particularly in the PV array, battery interconnection, dc bus voltage regulation, and battery charge control.

4.3 System Operation and Management

The plant supplies power to the TV and FM transmitters but cannot supply the load fully on its own and therefore operates jointly with the local utility grid. The maximum available power of about 40 kW is available only around noon. This is about the time of peak power demand of this station. The battery recharges first, and usually it would operate between 80 and 75% SOC.

In case of loss of grid power, the PV plant supplies the station's power demand which is reduced to about half during the grid power outage.

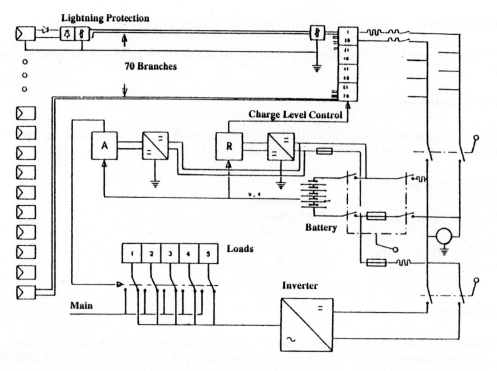

Figure 2. Simplified System Block Diagram of the MONT BOUQUET PV Pilot Plant

The array is divided into 71 array strings or branches and the number of these strings connected to the battery depends on battery voltage measured every 30 s. During periods of varying cloud cover, the strings can switch on/off continuously. The rectifier provides a source of charging power in the event of long periods of bad weather.

4.4 Plant Control

Manual intervention in the switching functions is possible. On-site displays of operating performance for various components are provided. The data monitoring system with its own PV power source records the solar, meteorological, and system performances (15 parameters).

4.5 Plant Performance

The PV array output measurements made by JRC-Ispra during the plant acceptance test in 1983 showed the peak power at STC to be 47.1 kW.

Following a forest fire that destroyed a part of the plant, the Université Claude-Bernard tested 634 modules. The resulting histogram of the module peak power at STC is shown in Figure 3 [1].

5. COMPONENT DESCRIPTION

5.1 Solar Array

5.1.1 Array and Cabling

The array comprises 71 strings. Each string has 10 PV modules in series. The cables for the 71 array strings are laid on the flat part of the galvanized steel troughs which are bolted onto the array support structures. The trough has a rectangular cross-section with perforations to allow water to escape. The return leg of the array is grounded to the earth at a central point.

Table 1. Major Components and Their Features (MONT BOUQUET)

Solar Array
Rated Peak Power:	50 kW
Array Tilt Angle:	60°
Bus Voltage:	212 Vdc nominal
Number of Strings:	71
Total Number of Modules:	710
Number of Modules per String:	10
Module Manufacturer/Type:	Photowatt /PW P-800
Module Rating:	72 W
Number of Cells per Module:	72
Number of Bypass Diodes:	1 per 18 cells in series
Module Weight:	16.7 kg
Module Dimension:	1.535 x 640 mm
Module Area:	0.9824 m^2
Active Solar Cell Area/Module:	0.5655 m^2
Solar Cell Type/Size:	Monocrystalline silicon/100 mm dia.
Cell Packing Factor:	57.6 %
Total Land Area Used:	2.000 m^2

Battery
Total Battery Energy Rating:	169.6 kWh
Cell Capacity Rating:	800 Ah/350 Ah
Voltage Range:	212 Vdc/48 Vdc (nominal)
Allowable Depth of Discharge:	30 %
Number of Batteries:	2 (1 x 212 Vdc, 1 x 48 Vdc)
Number of Cells per Battery:	106/24
Cell Manufacturer:	Oldham/SYT 8
Total Battery Weight:	7 tons

Inverter
Rated Power:	30 kVA
Number of Inverters:	1
Type:	Self-commutated
Manufacturer:	Aérospatiale
Input Voltage Range:	180 - 260 Vdc
Output Voltage/Phase:	380 Vac/3-phase
Frequency:	50 Hz
THD:	4 %
Efficiency, 100 % Full Load:	90 %

Controller
Manufacturer:	Photowatt

Data Acquisition
Manufacturer/Model:	DC 300 XL cassette,
Scan Rate:	1/min (faster possible)
Recording Rate:	hourly average values
Recording Device:	DC 300 XL cassette (ECMA 46)

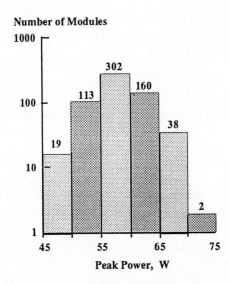

Figure 3. Histogram of the Peak Power Output for 634 Modules for the MONT BOUQUET PV Pilot Plant

5.1.2 Modules

The modules are 72-W size manufactured by Photowatt International. The cell assembly is laminated between glass/PVB and tedlar/aluminium/tedlar. Electrical connections are made with "quick-fastening" plugs. Six bypass diodes (one per 18 cells in series) are mounted in the termination box. Each module has 72 monocrystalline silicon cells, 100-mm diameter.

5.1.3 Array Support Structure and Foundations

A total of 142 panels of 5 modules each are arranged in 35 rows. The distance between rows is about 3.8 m. The panel tilt angle is 60°. Figure 4 shows the array structure and foundation configuration.

The concrete foundations were cast directly onto the rocky ground by using a mobile jig, thus ensuring a correct setting of the panel support tubes. Each panel has its own foundation block.

The module mounting structures were designed to allow a simple assembly and installation of modules and cabling without drilling mounting holes, but to withstand the structural load from wind speeds up to 180 km/h.

The rows of structures are mounted on pipes made of galvanized steel which are anchored into prefabricated reinforced concrete girders. The resulting configuration uses the panel as well as the module frame as part of the support structure itself. Capability is provided for some adjustment of height and tilt angle of the panel without needing precise alignment of the foundations.

5.2 Batteries and Charge Control Method

The main battery contains 106 France-Oldham cells of 800-Ah rating. The second battery rated at 350 Ah is used for the data monitoring system.

The charge control for the main battery is essentially a constant voltage charging method. This is accomplished by controlling the number of array strings connected to the battery bus via a logic card. The battery voltage and temperature are monitored every 30 s and the logic card selects a suitable number of array strings to be on or off in a pre-arranged sequence. It removes the array strings as required whenever the battery charges up to the charge voltage limit, so that this threshold is not exceeded.

5.3 Power Conditioning

The power conditioning equipment is the 30-kVA inverter. It is a self-commutated inverter manufactured by Aérospatiale. To improve its efficiency, its filter commutation starts at 3 kVA. Its nominal current may be exceeded by 25% for up to 100 s.

5.4 Control System

Analog logic cards are used for the control of array string switching and battery protection (i.e., digital computer is not used).

5.5 Data Collection

The data acquisition system scans the solar, meteorological, and PV system data every minute and records the average values of these data once every hour. The system was designed for autonomous operation up to three months.

5.6 Lightning Protection

Lightning and overvoltage protection consisted of providing proper earth grounding of all array structures and varistors (for overvoltage) in both ends of each array string cabling.

Ground Fault Detector

10 Photowatt modules use polycrystalline cells
(experimental); all others are monocrystalline

72-cell Photowatt PV module

Foundations

Figure 4. Details of Array Support Structures and Foundation at
the MONT BOUQUET PV Pilot Plant

5.7 Auxiliary Generators

The system has no auxiliary generators.

355

5.8 Load Equipment

The loads consist of radio and TV repeaters plus other peripheral equipment.

6. MAINTENANCE

Principal maintenance activities consisted of periodic checks of battery cell condition (electrolyte level, specific gravity) and replacement of the cassette tape (data monitoring system).

7. SUMMARY OF PROBLEMS AND SOLUTION APPROACHES

This plant was badly damaged by a local brush fire shortly after start-up, and consequently the designer/operator of the plant did not have sufficient opportunity to adequately assess its performance.

8. KEY LESSONS LEARNED

Although not much has been reported on this plant, lessons learned in the NICE pilot plant are applicable to MONT BOUQUET because of its similarity in the technical design.

9. CONCLUSIONS

A part of this PV plant was badly damaged by a local brush fire. The owner is planning to rebuild it after insurance claims have been settled.

10. REFERENCES

[1] M. Abou El Ela and J.A. Roger, "On Site Testing of Large PV Arrays Using a Portable I-V Curve Plotter," Proc. of the 8th European PV Solar Energy Conference, Florence, IT, 9-13 May 1988.
[2] Mont Bouquet Project Handout, Pilot Plant Optimization Meeting, Brussels, BE, 12-13 May 1986.
[3] Mont Bouquet Project Handout, Pilot Programme I Meeting, Tremiti Island, IT, 20-21 June 1984.
[4] P. Coureau, "Power Supply for TV and FM Emitters," Proc. of the EC Contractors' Meeting held in Hamburg, FRG, 12-13 July 1983.
[5] Mont Bouquet Project Handout, Pilot Programme I Meeting, Mont Bouquet, FR, 12-14 April 1983.
[6] Mont Bouquet Project Handout, Pilot Programme I Meeting, Crete, GR, 14-15 September 1982.
[7] Mont Bouquet Project Handout, Pilot Programme I Meeting, Brussels, BE, 20-21 April 1982.
[8] P. Coureau, "Power Supply for TV and FM Emitters," Proc. of the Final Design Review Meeting on EC Photovoltaic Pilot Projects held in Brussels, BE, 30 November - 2 December 1981.

Appendix 10

NICE PV PILOT PLANT

1. INTRODUCTION

This document describes the 50-kW NICE PV pilot plant and its initial performance. Figure 1 shows the PV array and other components located outdoors. This plant, which has been in operation since March 1983, produces solar energy to power selected loads at Nice Airport.

Sophisticated apparatus in use for ground monitoring and control of aircraft such as goniometer vectors, image number analysis, and sound measurement computers, require a reliable source of power which is relatively free of distortion and unacceptable power outages. A PV system was considered the best solution for this purpose because it can be set up independently of the utility grid.

The information contained herein was obtained from ref. [1 to 8] and individual contacts with the plant designer and AFME.

2. PLANT DESIGNER, OWNER, AND OPERATOR

The plant design and installation in Nice was supervised by Photowatt and handed over to the owner, Chambre de Commerce et d'Industrie of Nice and the International Airport Nice-Côte d'Azur, in November 1985. Their address is as follows:

> Chambre de Commerce et d'Industrie
> de Nice et des Alpes-Maritimes
> Aéroport International Nice-Côte d'Azur
> F-06056 Nice Cedex
> France

Information concerning this plant can be obtained from the following organization:

> AFME
> Attn: Renewable Energies Dept.
> Avenue Emile Hugues
> Sophia Antipolis
> F-06560 Valbonne Cedex
> France

PV panels are installed on top of the freight building near the airplane parking area

Mr. Plazy (AFME) inspecting the modules

Figure 1. PV Arrays on the Freight Building Roof at Nice Airport

3. SITE INFORMATION

3.1 Location

The plant is installed at the freight terminal airport building inside the Nice Airport complex. Its coordinates are:

- Latitude: 43.65°N
- Longitude: 7.40°E
- Altitude: 10 m

3.2 Site Constraints

The PV array structures were to be located on the roof of the building. Normal security regulations for international airports are imposed on plant visitors and external contractors.

3.3 Solar Radiation and Climatic Conditions

The following solar and meteorological data were recorded in 1951 and 1978 at the Météorologie Nationale at Nice Airport.

- Insolation, global horizontal:
  - o Annual: $1,600$ kWh/m^2
  - o Daily maximum (summer months): 6.0 kWh/m^2
  - o Daily minimum (winter months): 1.5 kWh/m^2
- Ambient air temperature: 26°C maximum, 6°C minimum
- Wind speed: 152 km/h maximum

3.4 Load Demand

The equipment connected to the PV system has a constant load of 6 kW all year long with an annual energy consumption of about 48,000 kWh.

4. SYSTEM DESCRIPTION AND PERFORMANCE

4.1 Plant Architecture and Layout

The PV arrays are installed on the roof of the freight airport building inside the Nice Airport complex. They are on the west side of the roof which is 250 m long and 24 m wide. The four rows of PV arrays are parallel with the side of the building facing south (see Fig. 1). In the layout and design of the array structures, careful consideration was given to the fact that the roof was not designed to carry a heavy weight. The four rows of PV panels are attached to the tubular legs embedded in the concrete foundations (see Section 5.1.3).

The remainder of the PV components are located at the south-west corner of the ground floor of this building(see Fig.2).

4.2 PV System Configuration

Figure 3 is a simplified block diagram of the PV system. Its major components and their key features are listed in Table 1. The system consists of the following:

- 50-kW PV array (710 Photowatt 72-W modules)
- 212-Vdc battery (106 France-Oldham 1,500-Ah cells
- 48-Vdc battery (24 France-Oldham 350-Ah cells)

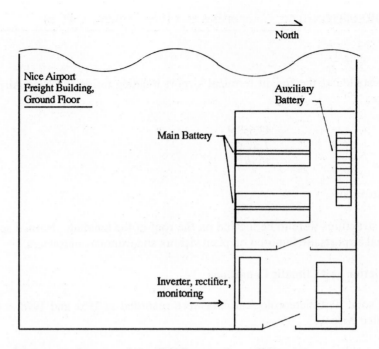

Figure 2. Layout of the PV Electronics and Battery

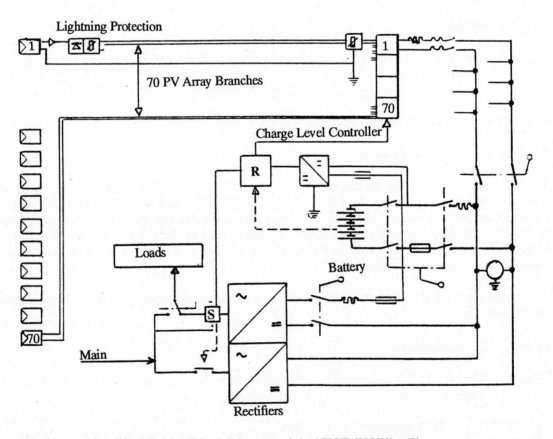

Figure 3. Simplified System Block Diagram of the NICE PV Pilot Plant

Table 1. Major Components and Their Features (NICE)

Solar Array	
Rated Peak Power:	50 kW
Array Tilt Angle:	45°
Bus Voltage:	212 Vdc nominal
Number of Strings:	71
Total Number of Modules:	710
Number of Modules per String:	10
Module Manufacturer/Type:	Photowatt /PW P-800
Module Rating:	72 W
Number of Cells per Module:	72
Number of Bypass Diodes:	1 per 18 cells in series
Module Weight:	16.7 kg
Module Dimension:	1.535 x 640 mm
Module Area:	0.9824 m^2
Active Solar Cell Area/Module:	0.5655 m^2
Solar Cell Type/Size:	Monocrystalline silicon/100 mm dia.
Cell Packing Factor:	57.6 %
Total Land Area Used:	3,000 m^2
Battery (2 batteries)	Main/DAS
Total Battery Energy Rating:	318 kWh/16.8 kWh
Cell Capacity Rating:	1,500 Ah/350 Ah
Voltage Range:	212 Vdc/48 Vdc (nominal)
Allowable Depth of Discharge:	30 %
Number of Batteries:	1 Main/1 DAS
Number of Cells per Battery:	106/24
Cell Manufacturer:	Oldham SZT12 (main battery)
Total Battery Weight:	10 tons (92.2 kg per cell) for main battery
Rectifier (battery charger)	
Rated Power:	7 kVA
Input Voltage/Phase:	220-380 Vac/3-phase
Output Voltage:	180 and 260 Vdc
Inverter	
Rated Power:	5 kVA
Number of Inverters:	1
Type:	Self-commutated
Manufacturer:	Aérospatiale
Input Voltage Range:	180-260 Vdc
Output Voltage/Phase:	220 Vac/single-phase
Frequency:	50 Hz
THD:	4 %
Efficiency, 100 % Full Load:	90 %
Controller	
Manufacturer:	Photowatt
Data Acquisition	
Manufacturer	DC 300 XL cassette,
Scan Rate:	1/min (faster possible)
Recording Rate:	hourly average values
Recording Device:	DC 300 XL cassette (ECMA 46)

- 5-kVA single-phase inverter
- 7-kVA battery charger (rectifier)

The basic plant configuration is similar to that of the MONT BOUQUET PV Pilot Plant, particularly in the PV array, battery interconnection, dc bus voltage regulation, and battery charge control.

4.3 System Operation and Management

This PV plant is required to provide uninterrupted power at all times. To meet this requirement, the charger which is powered by the grid is automatically switched onto the battery whenever necessary. Thus, the battery power is always available to the inverter. The system controller is hardwired with special logic cards for array string switching and for on/off control of the inverter, battery, and charger.

The array is divided into 71 array strings or branches and the number of these strings connected to the battery depends on battery voltage measurements made every 30 s. During periods of varying cloud cover, the strings can switch on/off continuously. The rectifier is used for charging in the event of long periods of bad weather.

4.4 Plant Control

Manual control of all switching functions is possible. On-site displays of operating performance for various components are provided. The data monitoring system with its own PV power source records the solar, meteorological, and system performances (about 15 parameters).

4.5 Plant Performance

The PV plant has been operating continuously except for the periods in which component failures were experienced. The two main operational interruptions were in the spring of 1983 due to a fire in the overvoltage protection cabinet and the failures encountered in the battery in early 1988.

The overvoltage protectors (varistors) were placed at both ends of the array string cables. Those in the cabinet located near the power electronics caused a fire which was attributed to defective and/or under-rated varistors. As a result, these overvoltage protection circuits were redesigned with new varistors from another supplier. Moreover, they are no longer placed in the electronic cabinet located under the dc switches, but in a separate cabinet on the roof of the building above the control room. Otherwise, the basic cabling has not been modified.

The battery damage in 1988 was apparently due to a combination of excessive overcharging and plugged-up vent caps on the battery cells. Plans are being made to replace the battery cells and to implement a safer battery charging procedure.

The PV array output measurements made by JRC-Ispra showed the following results in 1983 and in 1988:

	Acceptance Test in 1983	Test in Apr. 1988
Peak power at standard reporting conditions (STC):	46.9 kW	44.0 kW

5. COMPONENT DESCRIPTION

5.1 Solar Array

5.1.1 Array and Cabling

The array comprises 71 strings. Each string has 10 PV modules in series. The cables for the 71 array strings are laid on the flat part of the galvanized steel troughs which are bolted onto the array support structures. The trough has a rectangular cross section with perforations to allow water to escape. The return leg of the array is grounded to the earth at a central point.

5.1.2 Modules

The modules are 72-W size manufactured by Photowatt International. The cell assembly is laminated between glass/PVB and tedlar/aluminium/tedlar. Electrical connections are made with "quick-connect" plugs. Six bypass diodes (one per 18 cells in series) are mounted in the termination box. Each module has 72 monocrystalline silicon cells, 100-mm diameter.

5.1.3 Array Support Structure and Foundations

The existing freight terminal building was erected in 1953. It was determined that the arrays could be installed on the roof of this building using reinforced concrete foundations which are simply laid on the flat roof area. Their weight was calculated to withstand the wind force on the array panels. Each prefabricated concrete foundation block, weighing about 50 kg, is placed on a 3-cm thick isolation pad to ensure an even load distribution. To avoid water penetration problems, reinforcements were added to the rows of girders to prevent deterioration of roof-sealing parts at the time of installation and in the future. The 284 prefabricated blocks, weighing 140 tons, were lifted up to the terrace with a crane. Figure 4 shows some details of the array structures, foundations, and cable trays.

The module mounting structures were designed to allow a simple assembly and installation of modules and cabling without drilling mounting holes, but to withstand the structural load from wind speeds up to 180 km/h.

The rows of structures are mounted on tubular legs made of galvanized steel, which are anchored into the prefabricated reinforced concrete foundations. The resulting configuration uses the panel as well as the module frame as part of the support structure itself. Capability is provided for some adjustment of height and tilt angle of the panel to allow for imprecisely installed foundations.

5.2 Batteries and Charge Control Method

Two batteries were installed. The main battery contains 106 France-Oldham cells of 1,500 Ah rating. The second battery rated at 350 Ah is used for the data monitoring system.

The battery charge control is essentially a constant voltage charging method. This is accomplished by controlling the number of array strings connected to the battery bus via a logic card. The battery voltage and temperature are monitored every 30 s and the logic card selects a suitable number of array strings to be on or off in a pre-arranged sequence. It removes the array strings as required whenever the battery charges up to the charge voltage limit, so that this threshold is not exceeded.

Battery discharge below the low voltage limit is prevented by the automatic connection of a battery charger (rectifier) on line. This rectifier operates in parallel with the PV array, and thus provides uninterrupted power to the loads. The ac loads may be reconnected to the utility grid via the ac contactor.

The 5-kVA inverter is connected to the dc bus through the dc contactor and to the ac load via the ac contactor. This unit requires no external control.

5.3 Power Conditioning

The power conditioning equipment consists of the inverter, rectifier (charger), and the converter. One 5-kVA self-commutated inverter manufactured by Aérospatiale is used for dc-ac conversion. The 7-kVA rectifier, also a conventional product from Aérospatiale, provides battery charging in place of the PV array during inclement weather conditions.

5.4 Control System

Analog logic cards are used for the control of array string switching and battery protection. (Note: same as MONT BOUQUET; no computer used.)

5.5 Data Collection

The data acquisition system scans the solar, meteorological, and PV system data every minute and records the average values of these data once every hour. The system was designed for autonomous operation up to three months.

363

500-kg concrete blocks for foundation

Cables on perforated galvanized steel channel

Figure 4. Details of Array Support Structures, Foundations, and Cable Routing

5.6 Lightning Protection

Lightning and overvoltage protection consisted of providing proper earth grounding of all array structures and varistors (for overvoltage) in both ends of each array string cabling.

5.7 Auxiliary Generators

The system has no auxiliary generators.

5.8 Load Equipment

The loads consist of equipment typically used for the ground control of aircraft traffic such as goniometer vectors, image number, and analysis and sound measurement computers.

6. MAINTENANCE

The principal maintenance activities consisted in periodic checks of battery cell condition (electrolyte level, specific gravity) and cassette tape replacement (data monitoring system).

7. SUMMARY OF PROBLEMS AND SOLUTION APPROACHES

The main problems encountered were as follows:

- In early 1988 about 17 of the 106 cells in the main battery were found to have cracks in their plastic cases. Their electrolyte spilled all over the floor of the battery room. The problem was caused by the plugged-up vent caps on the cells combined with continuous overcharging which induced a high pressure build-up inside each cell.

- A fire in the cabinet shortly after plant start-up was attributed to the faulty varistors; new varistors were added after a careful re-sizing of their turn-on voltage.

- The original inverter supplied by Aérospatiale did not function properly so it was replaced by a conventional inverter from Photowatt.

8. KEY LESSONS LEARNED

The key lessons learned are:

- Overcharge control and protection of batteries must be carefully designed and appropriate battery maintenance procedures implemented. This includes a more frequent inspection of the cell vent caps and their replacement if necessary. A catastrophic failure of the type encountered is very hazardous to both surrounding equipment and on-site personnel.

- The hard-wired logic cards for array string switching have advantages over computer-based designs. However, it was found to be important to have a little more flexibility for improving the switching logic. For example, monitoring every 30 s and decision making at this interval have caused the switching relays to be activated on/off continuously. This also affects the battery charge/discharge protection.

- The electrical loads are so low that almost 80% of the PV array capability is frequently unused. Future modifications should consider additional options such as allowing grid-connected operation periodically, to make better use of the available energy.

9. CONCLUSIONS

The PV plant has operated satisfactorily for some time. The main problems encountered and valuable lessons learned were in areas of overvoltage protection devices, battery charge control, maintenance, and protection, and dc bus voltage regulation via non-computerized array switching.

10. REFERENCES

[1] G. Blaesser and W. Zaaiman, "Power Measurements at the 50-kW PV Plant Nice," JRC Technical Note No. I.88.43, Ispra, IT, April 1988.
[2] Nice Project Handout, Pilot Plant Optimization Meeting, Brussels, BE, 12-13 May 1986.
[3] Nice Project Handout, Pilot Programme I Meeting, Tremiti Island, IT, 20-21 June 1984.
[4] P. Coureau, "Nice Airport Survey and Control," Proc. of the EC Contractors' Meeting held in Hamburg and Pellworm, FRG, 12-13 July 1983.
[5] Nice Project Handout, Pilot Programme I Meeting, Mont Bouquet, FR, 12-14 April 1983.
[6] Nice Project Handout, Pilot Programme I Meeting, Crete, GR, 14-15 September 1982.
[7] Nice Project Handout, Pilot Programme I Meeting, Brussels, BE, 20-21 April 1982.
[8] P. Coureau, "Nice Airport Survey and Control System," Proc. of the Final Design Review Meeting on EC Photovoltaic Pilot Projects held in Brussels, BE, 30 November - 2 December 1981.

Appendix 11

PELLWORM PV PLANT

1. INTRODUCTION

This appendix describes the 300-kW PELLWORM PV pilot plant at the time of installation and its initial operation as well as principal modifications implemented up to 1990. This plant was installed during the winter of 1982, and it became operational in June 1983. Figure 1 shows the views of the array field and the control building.

The plant supplies power to the recreation centre for the local community and visitors. When the batteries are fully charged, the excess energy is fed into the grid. Because the peak energy demand by the recreation centre occurs during the summer months, a reasonable matching of energy production and consumption profile is achieved.

The information contained herein was obtained from ref. [1 to 11] and contacts with the plant designer and operator.

2. PLANT DESIGNER, OWNER, AND OPERATOR

Until early 1989, the municipality of Pellworm was the owner and responsible for the PV plant. In early 1989, the Pellworm authorities transferred the ownership of this plant to the utility company, Schleswag AG, which is responsible for the local network. Their address is as follows:

Schleswag AG
Attn: Renewable Energies Dept.
Kieler Strasse 19
D-2370 Rendsburg
F.R. Germany

The 300-kW PV field and a building which houses the PV electronics and batteries

Mr. H. Lowalt (TST)

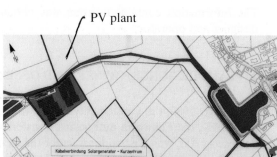

Kurzentrum (Recreation Centre)

Figure 1. The PELLWORM PV Pilot Plant

The recreation centre facility engineer takes care of routine plant operation and maintenance. On an on-call basis, maintenance and improvements of the plant are performed by the original designer, AEG-AG (name changed to TST in 1989), located in Wedel. Their address is as follows:

Telefunken System Technik (TST)
Industriestrasse 29
D-2000 Wedel
F.R. Germany

3. SITE INFORMATION

3.1 Location

The island of Pellworm is located in the North Sea, near Sylt, northwest of Hamburg FRG. It can only be reached at high tide by a ferry boat from the island Nordstrand west of Husum along the northwest coast of F.R. Germany (see Fig. 2). The coordinates of the PV site are as follows:

- Latitude: 54.4°N
- Longitude: 8.6°E

3.2 Site Constraints

No site constraints were identified.

3.3 Solar Radiation and Climatic Conditions

The following solar and meteorological data are based on recordings from 1972 to 1981 made by the weather station nearby in Sylt, F.R. Germany.

- Insolation, global horizontal: 1,100 kWh/m^2 annual
- Ambient air temperature: 20°C maximum, -10°C minimum
- Humidity: 80% average
- Wind speed: 25 km/h avg, 150 km/h maximum

3.4 Load Demand

Figure 3 shows the peak and off-season daily power demand profiles of the recreation centre supplied by the PV plant. Figure 4 shows the energy demand over the year at the time of initial plant operation in comparison with the expected PV plant power output, calculated using the solar insolation data from a meteorological station.

4. SYSTEM DESCRIPTION AND PERFORMANCE

4.1 Plant Architecture and Layout

The island of Pellworm is flat and parts of it lie below sea level. Its economy is based on farming, fishing, and tourism.

A key requirement for the PV plant was to minimize disturbance to the environment. The building was set up to blend with the other buildings on the island. Figure 5 shows the layout of the plant.

Sheep need about 80 cm free height to graze. The desired height for installing the modules at an angle of 40° with respect to the ground surface for easy connection of the individual modules was determined to be about 2.5 m for the top row and 1 m for the lowest row of modules. The structures made of different materials were separated from each other by about 30 m. From the main building, there is an unimpeded view of the landscape. In addition bushes have been planted around the field.

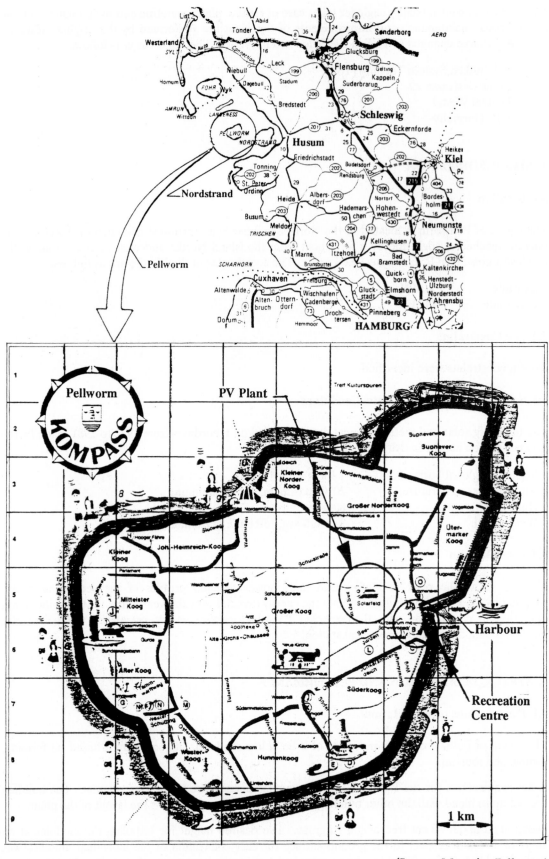

(Source: Magazine Pellworm)

Figure 2. Location of the PELLWORM PV Pilot Plant

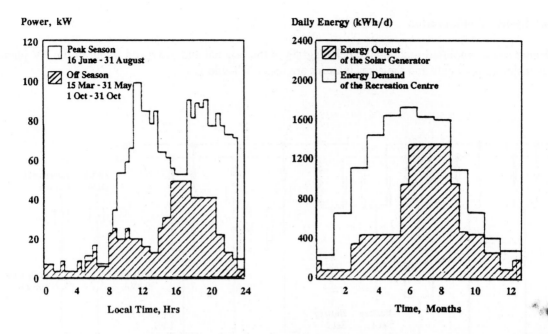

Power, kW

Daily Energy (kWh/d)

Figure 3. Daily Power Demand Profiles

Figure 4. Actual Daily Energy Demand and Expected PV Energy Output Over One Year

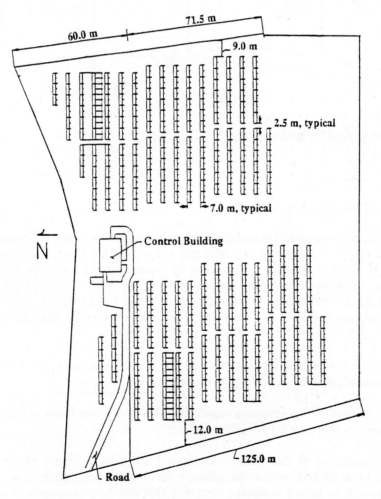

Figure 5. Layout of the PELLWORM PV Pilot Plant

371

4.2 System Configuration

Figure 6 is a simplified functional block diagram of the original and improved layout of the PV plant. Its major components and their key features are listed in Table 1.

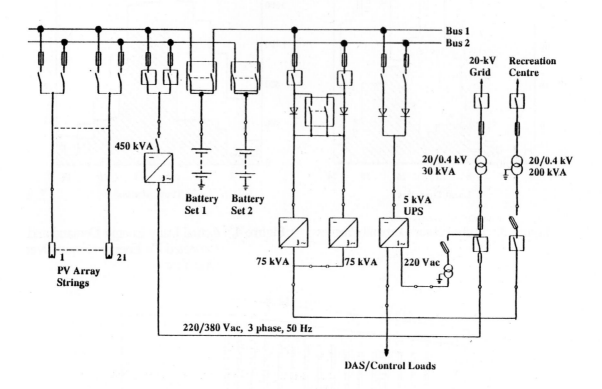

Figure 6. Simplified Schematic of the PELLWORM PV Pilot Plant

4.3 System Operation and Management

The 300-kWp plant was designed to supply energy to the recreation centre independently of the utility grid. Surplus energy is delivered to the utility grid using a 450 kVA line-commutated inverter. This unit can also be used to recharge the batteries by reversing the power flow from the grid during sunless days and periods of prolonged low insolation. Two 75-kVA self-commutated inverters are used to supply the recreation centre. The batteries are divided into two sets. One set acts as a buffer while the other operates in a stand-by mode, i.e., one set can be charging and the other set discharging.

The plant operates automatically and is controlled by a micro-processor-based control system. The same microprocessor system is used for data monitoring.

The maximum demand of the recreation centre is 100 kW at a power factor of 0.8, and the maximum daily consumption is about 1,800 kWh (1985 data).

The power management of this plant is performed using an elaborate switching configuration. Improvements were carried out on the software for about a year after initial start-up. It was therefore important to have a computer controller that can be reprogrammed on site. The two-battery-bus operation in the original design introduced a complex switching arrangement which resulted in extra software.

Table 1. Major Components and Their Features (PELLWORM)

Solar Array

Rated Peak Power:	300 kW
Array Tilt Angle:	40°
Tilt Angle Adjustment:	None
Nominal Bus Voltage:	346 Vdc
Number of Subarrays (Strings):	21 (366)
Total Number of Modules:	17,568
Number of Modules per String:	48
Module Manufacturer/Type:	AEG/PQ 10/20/01
Module Rating:	19.2 W
Number of Cells per Module:	20
Number of Bypass Diodes:	1 for 20 cells in series
Module Weight:	3.85 kg
Module Dimension:	563 cm x 459 x 11 mm
Module Area:	0.2584 m^2
Active Solar Cell Area/Module:	0.20 m^2
Solar Cell Type/Size:	Polycrystalline/10 x 10 cm
Cell Packing Factor:	0.774
Total Land Area Used:	16,500 m^2

Battery

Total Battery Energy Rating:	2.1 MWh
Cell Capacity Rating:	1,500 Ah (Battery: 6,000 Ah)
Voltage Range:	346 Vdc (nominal, +20 - 15 %)
Allowable Depth of Discharge:	80 %
Number of Batteries:	4
Number of Cells per Battery:	173
Cell Manufacturer/Part No.:	Varta Bloc/2415
Cell Weight:	152 kg, filled with acid
Cell Dimension:	307 x 383 x 500 mm

Inverter

Rated Power:	2 x 75 kVA; 1 x 450 kVA
Number of Inverters:	3
Type:	Self-commutated (75 kVA); line-commutated (450 kVA)
Manufacturer:	AEG
Input Voltage Range:	346 Vdc +20 % - 15 %
Output Voltage/Phase:	380 Vac/3-phase
Frequency:	50 Hz
THD:	3 % (75 kVA); 5 % (450 kVA)
Efficiency, 100 % Full Load:	91.4 % (75 kVA); 96 % (450 kVA), grid connected
10 % Full Load:	89.6 % (75 kVA); 94 % (450 kVA), grid connected
Inverter Dimension:	210 x 80 x 200 cm

Controller

Manufacturer:	AEG
Features:	µP 8085 board

Data Acquisition

Manufacturer/Model:	AEG, µP 8085 board, Penny & Giles 7100 cassette for storage
No. of Measurements:	69

Two computers, one each for hardware and operator control, were provided for the following reasons:

- To simplify the operator control and to facilitate improvements in the control software
- To link the controller to another computer via the standard RS232 interface
- To allow reprogramming of operator control functions in a high level language

In 1989, as part of the plant improvement effort, AEG implemented a single-battery-bus layout, in the new configuration (see Fig. 6) thus greatly simplifying the original control strategy. The primary operating mode now is grid-tied, with the three battery strings strapped in parallel at the single dc bus. One battery with 19 cells less than the others are simply connected to two array strings connected in parallel but separated from the rest of the PV array, and "float-charged" for maintenance purposes. In this operating mode, the batteries are not used during the time the PV energy is not available.

4.4 Plant Performance

This PV plant was turned on in July 1983. A subsequent optimization effort, mostly in the control software, was completed in early 1984.

From July 1983 to December 1984, the plant delivered 201,000 kWh. A more representative performance of the plant was from mid-March 1984 to the end of 1984. During this period the PV plant delivered 143,171 kWh to both the recreation centre and the grid. Only 17,296 kWh was supplied to the grid. Thus, the PV plant supplied 88% of the total energy required by the recreation centre.

Significant operational events and the plant performance history [4] are as follows:

- When batteries are completely discharged, as indicated by their low-voltage limits, the inverters shut down. This plant shutdown occurred several times in late 1983 and also in 1984. When this happens, the operator manually restarts the plant.

- A computer programme failure in 1984 resulted in low utilization of PV energy.

- A low-load condition resulted in high inverter losses which was expected.

- After one year of operation 40 PV modules had developed cracks in their glass covers, resulting in about 0.23% failure rate.

5. COMPONENT DESCRIPTION

5.1 Solar Array

5.1.1 Array and Cabling

The solar array is divided into 21 subarrays. The first 20 consist of 3 groups of 6 strings each. Subarray. 21 has only 1 group of 6 strings. Thus, there are 366 strings. With 48 modules in series per string, the total number of modules is 17,568.

The connector plugs on the modules facilitated the interconnection of modules. An additional securing sleeve prevents this type of plug from being opened by unauthorized persons. This securing sleeve is clamped to the module frame and thus the connecting cables are fixed additionally.

The six double-wire string cables of each subgroup are connected to a group junction box (6 x 48 modules in series are connected in parallel). Each string can be separated from the dc bus by removing its fuses. The six ground conductors are connected to the negative junction box. The six positive wires are connected to the array bus via two blocking diodes in parallel and in series with a fuse. Both buses and the earth ground are connected to the battery switchboard using a cable type NYY 3 x 16, NYY 3 x 25 or NYY 3 x 35 depending on their length.

The 61 array groups come together at the main switchboard. Three subgroup cables are connected at the switchboard terminal to form one group. The subgroup cables are embedded in trenches in the frost-free zone at approximately 80-cm depth.

5.1.2 Modules

The PV array uses AEG glass-encapsulated module, type PQ 10/20/01 with polycrystalline silicon cells. For more information, see Appendix 3, since the FOTA PV plant used the same module.

5.1.3 Array Support Structure and Foundations

The photovoltaic plant is installed in a total area of 28,000 m^2. Loadable stratum is located at a depth of about 10 m. The climatic conditions are very rough, stormy weather with salty, foggy air which may occur at any time of the year.

The total height of the support structure was minimized to reduce the wind load and to blend the array into the natural background. The ground area is maintained for permanent agriculture and sheep grazing. The lowest panel is 1 m above the ground while the highest point is 2.60 m. All panels are mounted at a fixed angle of 40° facing due south to obtain the maximum energy output over the whole year. Figure 7 shows the types of array structures and foundations.

Eighteen strings of modules are connected to one group. This group is mounted on 8 rigs. Each rig contains four rows of modules, each row consisting of 24 modules in series. Electrically, the rigs are divided into two parts: 24 modules in the 2 lower rows and the other 24 modules in the 2 upper rows.

A main goal in designing the structure was to minimize costs while considering the salty environment, high wind load, and bad soil carrying capacity. Tests were conducted to determine the minimum number of attachment points and mounting brackets. The results of these tests showed that only two fixing points per module in the middle of the long sides of the module were necessary. Further, only one girder for each row of modules is needed. Construction materials and methods were optimized to reduce costs. Material types such as painted, hot galvanized, stainless steel, aluminium, pressed pine wood, and bongossi (tropical hardwood) were tested. The hot galvanized steel and bongossi wood were chosen for the following reasons in addition to low cost:

- Galvanized steel is the most widely used metal in coastal regions.

- Its lifetime exceeds 15 years and it requires no additional protection. Manufacture of steel is conventional and its static values can be calculated.

- Bongossi wood has an extremely long lifespan, over 20 years and necessitates no special treatment.

- Costs for both materials are nearly the same.

The supporting structures of the four rows of girders were designed as triangles with a tilt angle of 40°. The calculations were based on the German standard DIN 1055, with a wind load of 120 km/h (i.e., 0.8 kN/m^2).

After the first calculation, a test rig was built to evaluate the natural frequency. The evaluation showed that a strut was necessary to stiffen the diagonal side of the triangle. The girders were connected by a U-bar every one third length between the trestles, and the diagonal wind braces were strengthened. The natural frequency of the rig was higher than the exciting frequency of the wind (maximum 10 Hz). For the higher impulsive moments there are no problems of vibration with the wood rigs. All mounting brackets (for the modules, for the steel as well as for the wood rigs) are made of stainless steel.

As stated, the soil conditions on site were bad. The site itself lies below sea level. The ground water level is at a depth of about 2 metres. The load-bearing capacity is only 8 kN/m^2 in the frost-free zone. The foundation had to be of very light construction. Thus prefabricated bars made of reinforced concrete were used for the foundations.

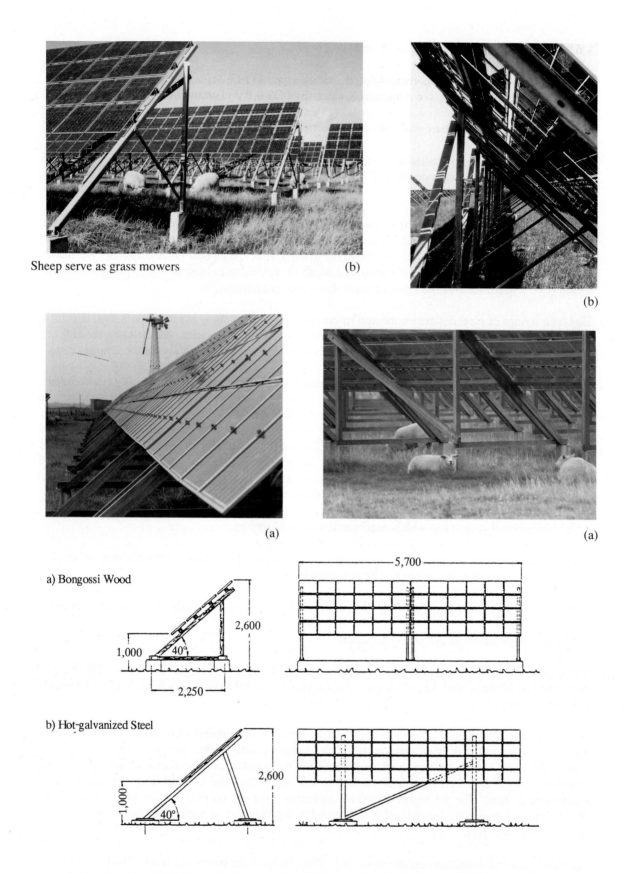

Sheep serve as grass mowers

(b)

(b)

(a)

(a)

a) Bongossi Wood

b) Hot-galvanized Steel

Figure 7. PV Array Structures at the PELLWORM PV Pilot Plant: a) Bongossi Wood and b) Hot-galvanized Steel

376

5.2 Batteries and Charge Control Method

The total battery rated capacity of 6,000 Ah is made up of four battery banks as illustrated in Figure 8. Banks A and B are connected in parallel to make up battery set 1, and similarly banks C and D make up battery set 2. The two battery sets were used independently in the original plant arrangement. Each bank has 173 cells in series with each cell rated at 1,500 Ah.

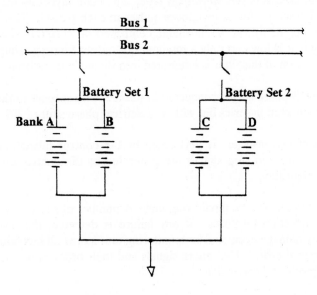

Figure 8. Connection of the Four Banks Made Up Two
Battery Sets (173 cells in each battery bank) in
the Original Plant Layout

The double-battery configuration was selected originally to ensure more complete charge/discharge cycles of both batteries than a single large battery would do. Both batteries operate alternately as buffer and stand-by batteries. The stand-by battery has no discharge current and consequently may be brought to full charge within one or two days. This mode of operation was selected to allow the batteries to be fully recharged at regular intervals.

During charging a number of PV subarrays are connected to a battery set. If the battery voltage is lower than the threshold, all subarrays are connected to it. To minimize water loss in the battery cells and prevent damage to cells due to overcharging, the charge voltage limit of 2.35 V per cell was used. A constant voltage is maintained by the computer by shedding or adding a required number of subarrays to the battery bus.

When the batteries become deep-discharged at times of prolonged low insolation, the line-commutated inverter is available for use as a rectifier to recharge the batteries from the utility grid (i.e., via reverse flow of power through the inverter).

The control of battery charging was originally based on the Varta Logistronic unit, but several problems were encountered with this device. Hence, the main computer controller was used for implementing the basic charge control and discharge protection.

5.3 Power Conditioning

5.3.1 Self-Commutated Inverters

The two 75-kVA self-commutated inverters which supply the recreation centre operate in parallel on the dc and ac sides. As commanded by the central controller, each inverter may be connected either to Bus 1 or Bus 2 via dc contactors. This was changed in 1988 to a single-bus configuration.

The commutation circuit used is of the McMurray type. The main thyristors are controlled in such a way that each inverter branch produces two power pulses in each period. They are combined in the main transformer. The output voltage is sinusoidal, composed of 30 square waves. In this case, the harmonic distortion content is low and only a minimum of output filtering (simple low-pass filter) is needed. The series induction of this filter is integrated into the main transformers.

These inverters have two 3-phase stages connected in parallel with respect to the dc side. On the ac side, these stages are connected in series but with an electrical phase shift of 30°.

The parallel operation of two inverters is performed by the control subsystem which switches the inverter on or off. The start and stop signals are generated by the control system according to an efficiency optimization algorithm.

Additionally each inverter has its own monitoring unit. A number of sensors are provided for status indication and failure detection purposes. If any failure is detected, the inverter is switched off automatically before any damage occurs. The monitoring unit stores all incoming failure signals and prints them in chronological order. The failure signals and their order of occurrence have been very useful for problem diagnosis and correction.

5.3.2 Line-commutated Inverter

The 450-kVA line-commutated inverter uses two 6-pulse thyristor bridges in parallel. This inverter operates at maximum 1,000 Adc at an output voltage of 380 Vac, 3-phase and is capable of operating in the reverse direction in order to recharge the battery. However, this method is used mostly in an emergency situation.

5.4 Control System

Two 8-bit microcomputer systems (INTEL 8085) are used to control the plant. Communication between these computers is via a RS232 link. One system includes a computer terminal for the operating personnel, a printer for permanent messages, and a Penny & Giles recorder for data storage.

One computer is used for accessing both analog and digital signals and the other for processing data and sending control commands to the first computer. These commands are then converted into analog and digital signals for control purposes.

Either the manual or automatic mode of control can be selected at the main control panel. Bar indicators on this panel show the position of the switching contactors and circuit breakers.

Since the control system was designed for automatic operation, special precautions were taken to ensure fail-safe operation. A combination of hardware and software measures were implemented in the system to reduce the risk of damaging the equipment or injuring the operator. With regard to hardware safety, each computer is controlled by a separate "watch dog" circuit. Whenever a microcomputer does not transmit a periodic control signal to the watch dog circuit, the computer assumes that a malfunction has occurred. Subsequently a cold start procedure (i.e., starting from the beginning of the programme) is initiated.

In the original plant configuration, the main computer provided the following control functions on the 450-kVA inverter:

- On-off switching
- Connection to busbar 1 or busbar 2 by using contactors
- Control of the amount of power flowing to or from the utility grid

When there is surplus energy, the central controller determines the surplus and sends a signal to the inverter. During emergency charging of the batteries, a digital signal begins rectifier operation and an analog signal regulates the charging current. If a defect occurs in this inverter, it is automatically disconnected from the dc bus (Note: when short-circuited, the batteries can deliver a few thousand kA). The controller also provides battery charge control and discharge protection. This function is described in section 5.2 above.

5.5 Data Collection

A total of 69 measurements are monitored via the control computer which requires a majority of these signals for control purposes.

The key features of the data acquisition unit and on-site display are as follows:

- 10-sec sampling rate for all control process data and one minute for meteorological and solar irradiance data

- Printing and tape recording is via the control computer

- 11 tables can be selected and displayed on the video terminal. These tables provide essential information pertaining to the plant operational status, power flow state, the battery, and other key components.

- Special events and alarms are printed automatically

- Recording of data every 10 minutes. (Note: the original configuration used the Penny & Giles magnetic tape cartridge. This tape had to be changed once a month. In 1989, this was replaced by the AEG DAM 800 monitoring system).

5.6 Lightning Protection

During the design phase special attention was given to protection from lightning strikes. Originally three methods of protection were considered:

- Construction of a number of lightning conductors distributed over the array

- Use of high-voltage surge arrestors in the main power lines such as metal-oxide varistors, air terminals, and high-current chokes

- Low-voltage protection of the measuring and signal circuits via surge arrestors, diodes, and isolation transformers

After a more detailed evaluation by L. Thione et al [11], it was decided that lightning conductors need not be installed in the array field because the probability of strikes in the field was calculated to be very low.

In the final design, it was decided that lightning protection would consist of the following:

- Use of air terminals at the control building

- Use of over-voltage surge arrestors in the PV array power cables

In the subarray junction box one metal-oxide varistor with an air gap protector is provided between the positive and negative conductors and earth ground. In the switchboards, protective devices are connected between each positive and negative conductor of the individual subarray cables and between the positive conductor and the protection earth. A lightning protection coil in each subarray line is intended to prevent the high current transients from entering into the dc bus.

Two components are provided for the string series diode in order to withstand positive voltage surges up to 2,000 V. Surge suppressors and isolation amplifiers are provided in the switchboard for low voltage protection of the data acquisition equipment.

5.7 Auxiliary Generators

The system has no auxiliary generators.

5.8 Load Equipment

The system provides electricity for the recreation centre, supplying lighting, heat pumps, etc.

6. MAINTENANCE

AEG (TST) still provides maintenance on behalf of Schleswag (utility). Operation of the plant is automatic. The operator checks the state of the plant and switches it on again if it has turned off due to overload or other fault conditions. Normal maintenance activities include checking the condition of the printer, the battery cells (electrolyte level, specific gravity, etc.), and the operating state of major components. This manual intervention has proven to be a good solution.

7. SUMMARY OF PROBLEMS AND SOLUTION APPROACHES

Key problems encountered and solution approaches implemented were as follows:

- About 25 modules developed cracks in their glass covers due to poor mounting design and installation procedure. No recurrence of this problem has been observed since the module mounting system was modified.

- Module mounting screws had to be re-tightened periodically because of the wooden structures.

- Breakages in the module interconnector cables have occurred. Thus free-hanging module connection cables were found not to be a good design practice.

- Array wiring grounding problems often occurred after rainfall. The solution was to put additional drainage holes in the bottom of junction boxes (for wire connections).

- Automatic control of energy management is not implemented, so manual load control was used originally to minimize the power fed to the grid. This has caused full battery discharge. Battery charge control problems became more complicated because of this special energy management.

- Having only one inverter on line has caused inverter shut-downs. Both inverters are required to provide the starting currents of two heat pump motors (20 kW each).

- The data recording unit failed frequently, especially during dry dusty conditions.

- Use of the 450-kVA inverter for battery recharging (i.e., use of the inverter as a rectifier via the utility grid) resulted in damaging 19 battery cells. This type of operation requires careful control and monitoring as it can result in extensive damage to the entire facility.

8. KEY LESSONS LEARNED

The primary lesson learned was to separate the control functions from the plant monitoring function (i.e., data collection) and to simplify the battery charging, protection, and operating strategies. For example, four banks of batteries should have been consolidated instead of splitting them into two sets operating independently. A single-bus systzem with the batteries paralled was implemented in 1988.

9. CONCLUSIONS

After about a year of control system improvement effort (mostly software changes), this PV plant has operated very well in both stand-alone and grid-connected modes of operation.

10. REFERENCES

[1] H.J. Lowalt, "PELLWORM PV Plant Improvements," Proc. of the 7th European PV Solar Energy Conference, Florence, IT, May 1988.

[2] Pellworm Project Handout, Pilot Plant Optimization Meeting Brussels, BE, 12-13 May 1986.

[3] P. Helm, "Manual of Photovoltaic Pilot Plants, Parts A and B," WIP Report Under CEC Contract EN3S-0007-D(B), Munich, FRG, February 1986.

[4] H.J. Lowalt and B. Proetel, "The 300-kW Pellworm Solar Power Station: Performance and Experience," Proc. of the 6th European PV Solar Energy Conference held in London, GB, 15-19 April 1985.

[5] Pellworm Project Handout, Pilot Programme I Meeting, Tremiti Island, IT, 20-21 June 1984.

[6] H.J. Lowalt, "300 kW Photovoltaic Plant Pellworm - Power Supply for the Recreation Centre," Proc. of the EC Contractors' Meeting held in Hamburg and Pellworm, FRG, 12-13 July 1983.

[7] Pellworm Project Handout, Pilot Programme I Meeting, Mont Bouquet, FR, 12-14 April 1983.

[8] Pellworm Projeçt Handout, Pilot Programme I Meeting, Crete, GR, 14-15 September 1982.

[9] Pellworm Project Handout, Pilot Programme I Meeting, Brussels, BE, 20-21 April 1982.

[10] H.J. Lowalt, "300 kW Photovoltaic Pilot Plant Pellworm," Proc. of the Final Design Review Meeting on EC Photovoltaic Pilot Projects held in Brussels, BE, 30 November - 2 December 1981.

[11] L. Thione, R. Buccianti, and L. Dellora, "Protection Against Lightning of PV Generation Systems," Presented at the Contractors' Meeting in Brussels, BE, 1981.

Appendix 12

RONDULINU PV PLANT

1. INTRODUCTION

This document describes the 44-kW RONDULINU PV pilot plant and its initial operation. The plant started operating in June 1983. It was designed as a stand-alone system to power the Rondulinu village in the Paomia area on Corsica Island, France. Figure 1 shows the PV field and surrounding areas.

Rondulinu is located on the spurs of the mountains at about 400 m elevation. No grid power is yet available at the PV plant, although the nearest grid network is only 4 km away. This network was planned to be extended to Rondulinu in 1989. The hamlet was used by the inhabitants of the mountain village of Vico as a wintering place for their cattle and ewes. This ancient habit has diminished over the years due to the lack of electricity on site. Only a few families live there in winter and a few more in the summer months.

The information contained herein was obtained from ref. [1 to 12] and direct contacts with the owner and the operator (Université de Corse/CRES-CNRS).

2. PLANT DESIGNER, OWNER, AND OPERATOR

The PV plant has been designed and installed by Leroy Somer. Information concerning this plant can be obtained from the following organization:

> Université de Corse/CNRS
> Lab. d'Hélioénergétique
> Route des Sanguinaires
> F-20000 Ajaccio
> France

44-kW array field from the Southwest

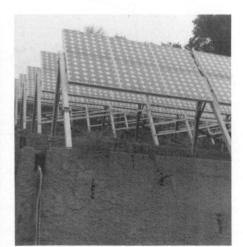

Control building utilized local natural materials

Figure 1. Views of the RONDULINU PV Pilot Plant

383

AFME in Sophia Antipolis, the co-financer of the pilot plant, is responsible for the plant's administrative supervision. Until 1987, AFME had delegated the monitoring of plant performance to the Université Claude-Bernard. In 1988, AFME turned over the plant improvement, operation, and maintenance activities to the University of Corsica and CNRS.

3. SITE INFORMATION

3.1 Location

The PV plant is located in Rondulinu which is about 6 km from Cargese, a city on the west coast of Corsica. Figure 2 shows its relative location. Cargese is situated about 35 km north of Ajaccio. The plant is visible from the road leading to the Paomia village and from the coast road. Its coordinates are:

- Latitude: 41.9°N
- Longitude: 8.6°E
- Altitude: 400 m

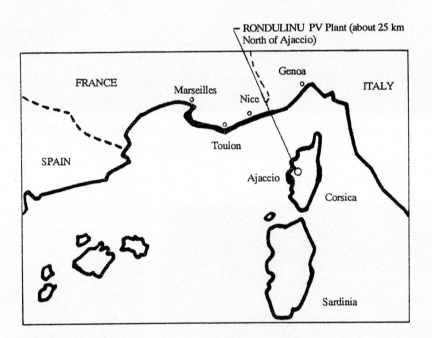

Figure 2. Location of the RONDULINU PV Pilot Plant

3.2 Site Constraints

Because the plant site is in an environmentally protected area, there were constraints to minimize disturbance to the natural landscape.

384

3.3 Solar Radiation and Climatic Conditions

The following solar and meteorological data, representing an average of 23 years, were obtained at Pianosa Island, 62 km from the PV site:

- Insolation, global horizontal:
  - o Annual: 1,600 kWh/m^2
  - o Daily average in July: 7.37 kWh/m^2
  - o Daily average in December: 1.85 kWh/m^2
- Ambient air temperature: 35°C maximum, 0°C minimum, 14.7°C average (daily)
- Wind speed: 173 km/h maximum

3.4 Load Demand

The plant is intended to supply power to private houses, milking systems, workshops, public lighting, and a water pump.

The estimated peak load demand used for system sizing purposes was 43 to 46 kW. Table 1 lists the various loads on a monthly basis.

Table 1. Peak Power and Energy Demand of Electrical Loads

Peak Power Demand (kW)

Load	J	F	M	A	M	J	J	A	S	O	N	D
Houses	16.8	16.8	18.9	18.9	18.0	18.0	19.2	19.2	18.0	18.0	18.9	16.8
Milking Systems	12	12	12	12	12	12	12	12	12	12	12	12
Pump	2	2	2	2	2	2	2	2	2	2	2	2
Workshop and Washing machines	12	12	12	12	12	12	12	12	12	12	12	12
Public lighting	0.8	0.8	0.8	0.8	0.8	0.8	0.8	0.8	0.8	0.8	0.8	0.8
Total	43.6	43.6	45.7	45.7	44.8	44.8	46	46	44.8	44.8	45.7	43.6

Daily Average Energy Demand (kWh/day)

Load	J	F	M	A	M	J	J	A	S	O	N	D	Average
Houses	31.5	31.5	40.5	40.5	45	54	72	72	54	45	40.5	31.5	46.5
Milking of ewes	6	6	8	8	12	12	12	12	12	8	6	6	9.2
Pump (Water)	4.1	4.1	5.3	5.3	5.9	7.1	9.5	9.5	7.1	5.9	5.3	4.1	6.1
Workshop and Washing machines	4	4	5	5	6	6	9	9	7	5	5	4	5.7
Public lighting	7.7	7.7	6.7	6.7	5.7	5.7	5.7	5.7	6.7	7.7	7.7	7.7	6.8
Total	53.3	53.3	65.5	65.5	74.6	84.8	108.2	108.2	86.8	71.6	66.5	53.3	74.3

4. SYSTEM DESCRIPTION AND PERFORMANCE

4.1 Plant Architecture and Layout

Because the plant site is in an environmentally protected area, there were difficulties in obtaining the construction agreement for the plant. Figure 3 is a diagram of the plant layout. The architectural

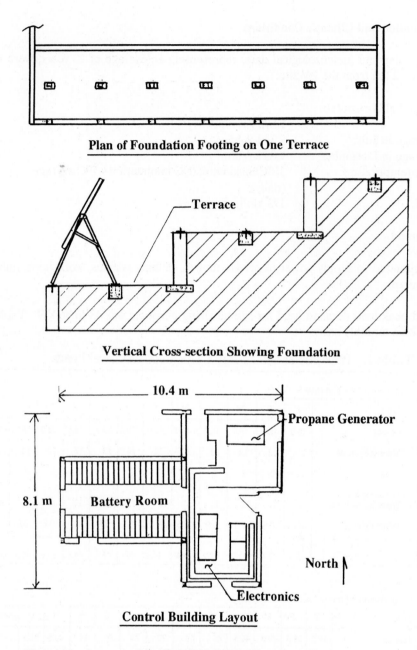

Plan of Foundation Footing on One Terrace

Vertical Cross-section Showing Foundation

Control Building Layout

Figure 3. Layout of the RONDULINU PV Pilot Plant

work was performed by a local architect considering the following basic criteria:

- PV array to be installed on the slope of the hill
- Use of locally available materials such as sandstone and pine wood
- Installation of a protective platform to avoid soil erosion from heavy rainfall
- PV array and building to harmonize with the landscape

4.2 System Configuration

The plant consists of a 44-kW PV array, a 50-kVA inverter, a 2,500-Ah battery, a 25-kVA petrol generator, a 25-kW rectifier, and a data monitoring system. Figure 4 shows the simplified block diagram of the plant. Major components and their key features are listed in Table 2.

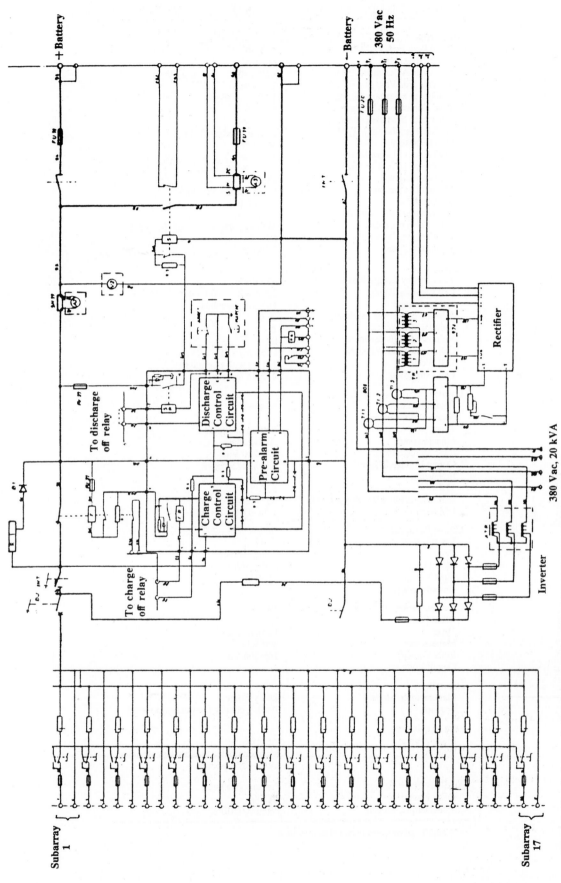

Figure 4. Schematic Diagram of the RONDULINU PV Pilot Plant

Table 2. Major Components and Their Features (RONDULINU)

Solar Array
Rated Peak Power:	44 kW
Array Tilt Angle:	60°
Tilt Angle Adjustment:	- 10° to 60°
Nominal Bus Voltage:	168 Vdc
Number of Subarrays (Strings):	102
Total Number of Modules*:	1,224
Number of Modules per String:	12
Module Manufacturer/Type:	France-Photon/FP 36
Module Rating:	36.72 W
Number of Cells per Module:	36
Number of Bypass Diodes:	1 per 18 cells in series
Module Weight:	14 kg
Module Dimension:	640.5 x 642 x 42.5 mm
Module Area:	0.4112 m^2
Active Solar Cell Area/Module:	0.2827 m^2
Solar Cell Type/Size:	Monocrystalline silicon/100 mm dia.
Cell Packing Factor:	68.7 %
Total Land Area:	5,000 m^2

Battery
Total Battery Energy Rating:	420 kWh
Cell Capacity Rating:	2,500 Ah (at 10-h rate)
Voltage Range:	168 Vdc (nominal)
Allowable Depth of Discharge:	80 %
Number of Batteries:	1
Number of Cells per Battery:	84
Cell Manufacturer/Part No.:	Oldham/SZT
Cell Weight:	191 kg with electrolyte
Cell Dimension:	212 x 576 x 820 mm

Inverter
Rated Power:	50 kVA
Number of Inverters:	1
Type:	Self-commutated
Manufacturer:	Aérospatiale
Input Voltage Range:	120-200 Vdc
Output Voltage/Phase:	380 Vac/3-phase
Frequency:	50 Hz
THD:	5 %
Efficiency, 100 %/10 % Full Load:	93 %/88 %
Dimension:	1.5 x 0.84 x 2.3 m

Charge Regulator
Rated Power:	50 kW
Type:	LSH
Manufacturer:	Leroy-Somer
Input Voltage Limit:	201.6 Vdc
Output Voltage:	Variable

Constant Current dc Generator
Rated Power:	18 kW
Type:	L.S.
Manufacturer:	Leroy-Somer
Input Voltage:	300 Vdc
Output Voltage:	165-200 Vdc

Controller
Manufacturer/Model:	Leroy-Somer
Features:	Analog circuits (no microprocessor)

Data Acquisition
Manufacturer:	ASITEC
Features:	Microcomputer, cassette tape recorder, video CRT, printer

Auxiliary Power Source:
	Propane generator
Rated Power:	25 kVA
Manufacturer/Model:	Peugeot/CEB 25

* 4 additional modules installed as reference

The criteria for sizing of the main PV components were as follows:

PV Array:

- Use of average daily insolation and average operating efficiencies

Battery:

- A 7-day storage capability, which was considered sufficient for a medium variation of solar energy in the winter
- A 3-day storage capability for July and August (load demand is greater in the summer)

Inverter:

- 50-kVA size, to allow for a maximum load of 62.5 kVA and in the worst case, to be able to start motors of about 8-kVA allowing for the initial starting current

Propane-Powered Generator and Rectifier:

- 25-kVA size, to be able to recharge the battery via the rectifier in no more than two days

4.3 System Operation and Management

In the primary mode of operation the control of battery charging and discharging is accomplished by the control circuit card. When the battery voltage reaches 201.6 Vdc (2.4 V/cell), it disconnects the battery from the array bus. When the battery voltage drops to 180.6 Vdc (2.15 V/cell), the control circuit reconnects the battery to the array bus. When the battery is supplying power to the inverter at night, the control circuit removes the inverter whenever the battery voltage decreases to 155.4 Vdc (1.85 V/cell).

During prolonged sunless periods, the propane-fuelled generator and the constant current dc generator in series are started up to recharge the battery. The generator starts automatically when the battery SOC is 20% or less (than its nominal capacity rating). If the battery is fully discharged, it takes an estimated two days to fully charge it (i.e., to reach 201.4 Vdc at a constant current of 100 A).

4.4 Plant Control

The basic control of the plant was implemented with analog circuits. Since a microprocessor is not involved, the reliability of the control system is inherently greater than a controller implemented with both analog and digital methods.

Manual control capability is possible at the control cabinet for the following:

- Inverter on/off
- Petrol generator on/off
- Array on/off
- Battery on/off
- Subarray on/off (for measuring the Voc and Isc of each of the 17 subarrays each containing six strings)

4.5 Plant Performance

A summary of key performance results and operational highlights is as follows:

- The spread of the actual power output for the PV modules at STC is (see Fig. 5):

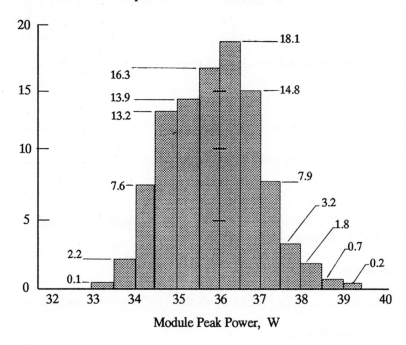

Figure 5. Histogram of Actual PV Module Output Power

o The range of power: 33 to 39.4 W
o Average: 36.0 W
o One standard deviation: 1.05 W

- Acceptance test results on the PV array in 1983 (by JRC-Ispra) were:

o Rated array power: 41.8 kW
o Ratio of rated to nominal power: 95%

- Based on the measurements obtained in 1987, J.A. Roger of the Université Claude-Bernard [1] reported an array field output of 36.8 kW with an estimated 3% mismatch loss. Furthermore, the 17 subarray outputs ranged from 1.9 to about 2.55 kW with an average of 2.2 kW.

- During the first few years of operation no significant operational data have been collected due to the lack of consumer loads on the PV system (see below).

- The 3-phase ac output of the PV plant is distributed as follows:

o Phase 1 for public lighting

o Phases 2 and 3 for about eight of fifteen houses in the village. Only four of these houses are inhabited the entire year. Thus, only a part of the PV power can be used, especially in winter. A typical house load composed of lighting, radio, TV, and refrigerator is about 1 kW.

- The connection of the consumer loads to the PV plant has been delayed. By 1984 only two houses had been connected because of financial, administrative, and partly sociological reasons. The older permanent inhabitants were somewhat reluctant to use electricity, perhaps even opposed to it.

36-cell Photon PV module

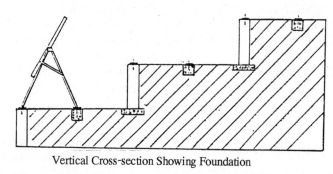

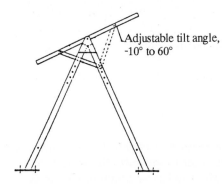

Adjustable tilt angle,
-10° to 60°

Vertical Cross-section Showing Foundation

Figure 6. Array Support Structures and Foundations

391

- In January 1986, the energy counters at the inverter console showed a consumption of about 10,000, 4,000, and 1,000 kWh for the three ac output phases, respectively. This indicated that the energy output of the plant from June 1983 to January 1986 was about 15 kWh per day.

- Because the battery was hardly used, the propane generator was seldom started up. For this reason, the generator was not run for three hours minimum per week as stipulated by its manufacturer as part of maintenance activity.

- Inverter failures (in one ac output phase) occurred in March 1984 and January 1985. Aérospatiale had no maintenance personnel for the inverter, so the head of the original inverter design team normally carried out the maintenance and repair work up to 1986 on a free-lance basis.

5. COMPONENT DESCRIPTION

5.1 Solar Array

5.1.1 Array and Cabling

The solar array comprises 1,224 modules which are mounted in 34 rows of six 6-module panels. To minimize field assembly cost and possible misconnection errors, the 6-module panels were pre-assembled, including the series connection of the six modules at the France-Photon factory.

A group of 12 modules (432 cells) are connected in series per string and a total of 102 strings in parallel are connected to the array bus in the control building. Furthermore, these 102 strings are divided into 17 subarrays (six strings each) for test purposes. The reason for this wiring arrangement was to permit trouble-shooting for defective PV modules or problems in the string circuit continuity via measurement of the string current.

The modules were classified by France-Photon on the basis of output current at rated power, and the modules for each string were selected on the basis of close matching of the current. This was done to minimize mismatch losses in each string. The mismatch losses were estimated to be less than 2%.

5.1.2 Modules

The France-Photon module used has the following features (see Table 2 for other details):

- Rated peak power: 36 W
- Bypass diode: One per 18 cells in series
- Nominal voltage: 13 V (21.5 V open circuit voltage)
- Encapsulation: Tempered low iron glass/Transparent silicone rubber/
                         White silicone rubber
- Connection: Waterproof plug-in connectors

5.1.3 Array Support Structures and Foundations

The support structures for the 6-module panels are triangular, with two identical posts made of galvanized steel. There are no additional parts, except bolts, in order to minimize the weight of the mechanical structure, but it was designed to withstand the worst wind and snow conditions at the site.

This structure permits a tilt adjustment capability of -10 to 60°. Figure 6 shows the structure and the foundation configuration.

The sizing of this structure was performed according to the French standard CM 66 for calculation of metal structures. The thermostress has been allowed for by the use of soft plastic washers.

5.2 Batteries and Charge Control Method

The battery consists of 84 lead-acid battery cells from France-Oldham. The cells are rated at 2,500 Ah at the 10-h rate. The electrodes contain a very low level of antimony-lead alloy. The battery discharge voltage ranges from 155.4 to about 180 Vdc with a nominal level of 168 Vdc.

The charging method involves simply terminating the charge when the battery voltage reaches a predefined limit of 201.6 Vdc (2.4 V/cell). The charging current can vary up to 250 amperes when the PV array is operating and a constant value up to 100 amperes when the petrol generator is providing the charging power.

Other battery features as defined by the cell manufacturer are as follows:

- Self-discharge rate: 4 % per month maximum at 20°C
- Storage capacity: 420 kWh at 250 A (10-h rate) 560 kWh at 80 A (30-h rate)

5.3 Power Conditioning

One 50-kVA self-commutated 3-phase inverter is employed. It uses power transistors and PWM techniques. The inverter consists of three star-connected single-phase inverters allowing a perfect phase equilibrium for any type of load. To avoid excessive no-load losses, each single phase inverter has a built-in filter commutation which switches in above 5% full load. A galvanic isolation between the three phases is achieved using an output transformer.

5.4 Control System

The control system was designed by Leroy-Somer. It implemented simple control strategies with analog circuits, and no microprocessor was used.

5.5 Data Collection

The monitoring system was manufactured by PME Informatique. It monitors the following:

- Global irradiance on horizontal surface and on plane of the array
- Sunshine duration
- Ambient air temperature
- Battery temperature
- Voltage and current of PV array, battery, inverter, and the generator
- Strain gauge data on array support structures

The solar and meteorological data are sampled once every five minutes, integrated and stored on an hourly basis in an ECMA 46 cassette tape. A video terminal and a printer are available (printout one page per day).

5.6 Lightning Protection

The Keraunic level of Ajaccio is 37 thunderstorm days/year and the earth flash density about 2.5 flashes/km^2-yr. Based on this flash density, the probability of direct lightning strikes was estimated to be 0.004 per year. For these reasons, no protection against direct lightning strikes was provided.

The following protection strategies were implemented for protection from indirect lightning strikes and overvoltages:

- Connection of the array support structures to the earth ground of the control building

- Connection of the negative terminals of all PV components in the control building to the main plant ground
- Full isolation of the two power circuit polarities from the earth ground
- Protection devices such as diodes, MOVs and coils were installed in the connection boxes in the field
- Installation of the following devices in the input and output power lines in the building and each electronic device:

  o Zener diodes
  o Surge suppressors with propane discharge
  o Surge suppressors and MOVs separated by small inductor coils

5.7 Auxiliary Power Source

The 25-kVA gasoline engine generator equipped with an alternator, is used to recharge the battery when PV array power is not available. The generator can also operate in parallel with the PV system to supply energy to the hamlet. It is regulated to give a constant current at any voltage under its nominal voltage (180 Vdc).

When the battery reaches its lowest level, 155.4 Vdc (1.85 V/cell), the generator is switched on and the battery is charged until it reaches its maximum voltage of 201.6 V. This may take about two days. The generator is then turned off, and the inverter is reconnected to the dc bus.

5.8 Load Equipment

The basic consumer loads are public lighting, typical house appliances, milking machines, washing machines, and a water pump.

6. MAINTENANCE

The maintenance activities consisted of the following periodic tasks:

- Check the electrolyte level and specific gravity of each battery cell
- Check the accuracy of the charge voltage limit
- Check the output of the inverter, generator, and the current of 17 array string groups

7. SUMMARY OF PROBLEMS AND SOLUTION APPROACHES

The key problems encountered were as follows:

- It took a long time to acquire a construction permit
- A malfunction of the inverter occurred (failure of one of three ac output phases)
- The connection of consumer loads was delayed by financial and administrative problems
- Failure of the monitoring system has occurred frequently due to malfunctions in its power supply

8. KEY LESSONS LEARNED

The main lessons learned are:

- A lack of electrical loads can cause unexpected problems for the PV plant, e.g., how to adequately maintain the battery and how to run the auxiliary generator under low-load conditions;

- The battery voltage regulation and charge control could be accomplished by array string control via the logic cards. This may provide better charging conditions for the battery because constant voltage charging results in higher capacity acceptance.

9. CONCLUSIONS

The design of the plant was simple enough to allow rapid installation by workers of average skill. The total installation time was 3.5 months.

10. REFERENCES

[1] M. Abou El Ela and J.A. Roger, "On-site Testing of Large PV Arrays Using a Portable I-V Curve Plotter," Proc. of the 8th European PV Solar Energy Conference, Florence, IT, 9-13 May 1988.

[2] J.A. Roger, S. Massad, and M. Abou El Ela, "Choice and Definition of the Best Figures of Merit for the Analysis of PV Plants - Application to the French Pilot Installation of Paomia," Proc. of the 7th European PV Solar Energy Conference, Seville, ES, 27-31 October 1986.

[3] Rondulinu Project Handout, Pilot Plant Optimization Meeting, Brussels, BE, 12-13 May 1986.

[4] P. Helm, "Manual of Photovoltaic Pilot Plants, Parts A and B," WIP report under CEC Contract EN 3 S-0007-D(B), Munich, FRG, February 1986.

[5] Rondulinu Project Handout, Pilot Programme I Meeting, Tremiti Island, IT, 20-21 June 1984.

[6] D. Mercier, "Solar Plant for a Remote Corsican Village," Proc. of the EC Contractors' Meeting held in Hamburg and Pellworm, FRG, 12-13 July 1983.

[7] Rondulinu Project Handout, Pilot Programme I Meeting, Mont Bouquet, FR, 12-14 April 1983.

[8] Rondulinu Project Handout, Pilot Project I Meeting, Crete, GR, 14-15 September 1982.

[9] Rondulinu Project Handout, Pilot Project I Meeting, Brussels, BE, 20-21 April 1982.

[10] D. Mercier, "Revitalisation d'un Village Corse," Proc. of the Final Design Review Meeting on EC Photovoltaic Pilot Projects held in Brussels, BE, 30 November - 2 December 1981.

[11] A. Louche, G. Simmonot, G. Notton, G. Péri, "Global analysis of the Rondulinu-Paomia PV Plant behaviour" - Proc. of the 8th European PV Solar Energy Conference, Florence, IT, 9-13 May 1988.

[12] A. Louche, G. Péri, G. Simmonot, G. Notton, "The Paomia-Rondulinu Photovoltaic Plant: a Stand alone Power Plant in a Rural Site - Characterization of the PV Array" - Proc. of the Euroforum New Energies Congress, Saarbrücken, FRG, October 1989.

Appendix 13

TERSCHELLING PV PLANT

1. INTRODUCTION

This appendix describes the 50-kW TERSCHELLING PV pilot plant. One aim of this project was to study the combination of solar and wind power systems designed to supply the energy demand of the navigation school. The load demand profile combined with low insolation levels in the winter months suggested the need for a large energy storage system. However, in lieu of a high capacity battery, a 40-kW wind generator was selected and installed originally. This generator was replaced by the 75-kW unit in 1989. Figure 1 shows the PV field in front of the school building.

The information contained herein was obtained from ref. [1 to 12] and individual contacts with the plant design improvement and maintenance subcontractor, Ecofys of Utrecht. Ecofys performed the improvement activities in 1987 and 1988.

2. PLANT DESIGNER, OWNER, AND OPERATOR

The PV system was designed and installed under the direction of Holecsol Systems. This private company (a subsidiary of the Holec Projects) has been liquidated in the meantime. The organisation responsible for the maintenance, operation and further experimentation of the plant (1987 to 1992) is:

> Ecofys Cooperatief Advies-en Onderzoeksbureau
> Attn: Renewable Energies Dept.
> Biltstraat 110
> NL-3572 BJ Utrecht
> The Netherlands

The plant is owned and operated by the Commissie Zonen Wind project of the municipality of Terschelling. Their address is as follows:

> Hogere Zeevaartschool Terschelling
> P.O. Box 26
> NL-8880 AA West-Terschelling
> The Netherlands

50-kW array field from the southwest

Electrical connection box

Figure 1. Views of the TERSCHELLING PV Pilot Plant

Ecofys has been subcontracted by the owner to provide operation and maintenance from late 1987 to 1992. After this period, maintenance will be done by the school and technicians with external support.

3. SITE INFORMATION

3.1 Location

The plant is located on the southwest part of Terschelling Island (see Fig. 2), which is the second largest Wadden Island in the northern part of The Netherlands. It is one of the most popular Dutch holiday resorts. The 'Hogere Zeevaartschool Willem Barentsz', the school for naval personnel is located at the south coast of the island. The plant is situated almost at sea level at the following coordinates:

- Latitude: 53.35ºN
- Longitude: 5.23ºE

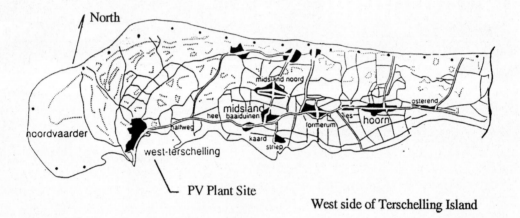

Figure 2. Location of the TERSCHELLING PV Pilot Plant

3.2 Site Constraints

The area available for the PV field was in front of the school building, facing the beach.

3.3 Solar Radiation and Climatic Conditions

The following data were obtained by the meteorological station nearby:

- Insolation, global horizontal:
  o Annual: 990 kWh/m^2
  o Daily average in June: 5.2 kWh/m^2
  o Daily average in December: 0.5 kWh/m^2
- Ambient Air Temperature: 9º C average
- Wind speed: 6.1 m/s daily average

3.4 Load Demand

The school's energy demand at the time of installation was estimated to be 35,000 kWh/year with a typical load pattern for a school. Meanwhile the energy demand has increased to 150,000 kWh/yr because the school kitchen load has also been connected to the system.

4. SYSTEM DESCRIPTION AND PERFORMANCE

4.1 Plant Architecture and Layout

The site area in front of the school has a slope of about 10° on the beach front, and faces due south. These circumstances offered an excellent opportunity to fit the PV field into the landscape. Taking into account the special status of the island and the site characteristics, local and provincial authorities had no objections to the installation of the solar and wind generators. Figure 3 shows the layout of the PV plant.

4.2 PV System Configuration

Figure 4 is a simplified block diagram of the system. Its major components and their key features are listed in Table 1.

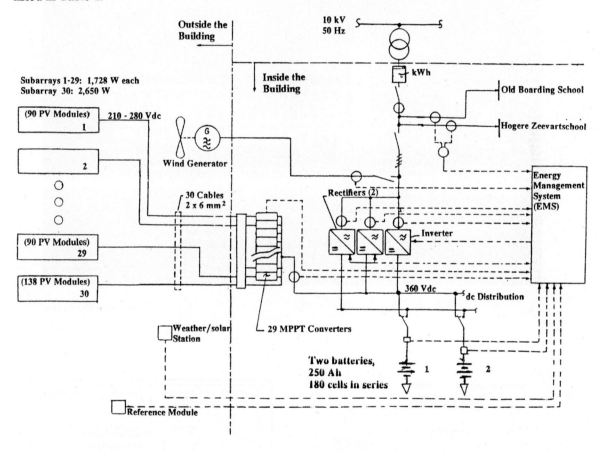

Figure 4. Simplified System Block Diagram of TERSCHELLING PV Pilot Plant

4.3 System Operation and Management

The combined operation of the 50-kW photovoltaic plant and the original 40-kW wind generator was predicted to supply about 95% of the energy needs of the naval school. Because of the load increase and a failure of the generator, Ecofys replaced the wind generator with a higher output unit (75 kW) in 1989. The plant was designed to be capable of both grid-connected and stand-alone operation. In both operating modes, a software Energy Management System (EMS) is involved.

4.4 Plant Control

The original plant control subsystem which involved the Logistronic unit was upgraded with an industrial process controller in 1988. The PLC also performs the monitoring functions. The Logistronic unit was intended to provide charge control and discharge protection signals to the PLC. The Logistronic unit, however, did not operate as intended at plant start-up, so the battery protection function was done entirely by the dc-dc converters.

4.5 Plant Performance

Figure 5 shows the estimated load profile along with PV and wind generator output. It can be seen that a combination of PV and wind generators is desirable for the prevailing local conditions. Table 2

View from the South

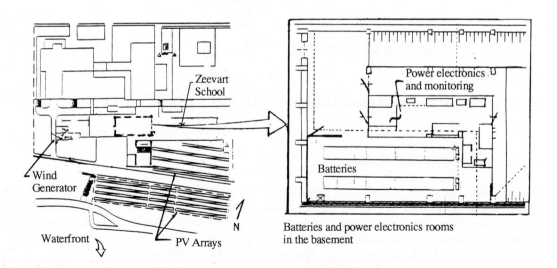

Figure 3. Layout of the TERSCHELLING PV Pilot Plant

Table 1. Major Components and Their Features (TERSCHELLING)

Solar Array
Rated Peak Power:	50 kW
Array Tilt Angle:	40°
Tilt Angle Adjustment:	None
Nominal Bus Voltage:	240 Vdc
Number of Subarrays (Strings):	30 (90)
Total Number of Modules:	2,748
Number of Modules per String:	30 (3 strings have 46)
Module Manufacturer/Type:	AEG/PQ10/20/01
Module Rating:	19.2 W
Number of Cells per Module:	20
Number of Bypass Diodes:	1 for 20 cells in series
Module Weight:	3.85 kg
Module Dimension:	563 x 459 x 11 mm
Module Area:	0.2584 m^2
Active Solar Cell Area/Module:	0.20 m^2
Solar Cell Type/Size:	Polycrystalline/100 x 100 mm
Cell Packing Factor:	0.774
Total Land Area Used:	600 m^2

Battery
Total Battery Energy Rating:	180 kWh
Cell Capacity Rating:	250 Ah
Voltage Range:	360 Vdc (nominal)
Allowable Depth of Discharge:	80 %
Number of Batteries:	2
Number of Cells per Battery:	180
Cell Manufacturer/Part No.:	Varta Bloc/2305
Cell Weight:	29.1 kg, filled with acid
Cell Dimension:	131 x 275 x 440 mm

Inverter
Rated Power:	60 kVA
Number of Inverters:	1
Type:	Line-commutated, dc-motor/synchronous generator
Manufacturer:	Holec
Input Voltage Range:	324-432 Vdc
Output Voltage/Phase:	380 Vac/3-phase
Frequency:	50 Hz
THD:	5 %
Efficiency, % FL:	
100 with synchronous machine:	87 %
without synchronous machine:	90 %
10 with synchronous machine:	65 %
without synchronous machine:	94 %
Inverter Dimension:	0.5 m x 1.5 m x 2 m

Converter
	MPPT
Rated Power:	2 kW
Number of Converters:	29
Input Voltage:	210-273
Output Voltage:	346-432
Special Feature:	MPPT, voltage boost

Controller
	Process Logic Controller (PLC), Midi/D08
Model:	Sattcontrol
Manufacturer:	

Data Acquisition:	(Provided by controller)
Auxiliary Power Source:	75-kW wind generator (replaced 50 kW unit on 12 May '89)

Rectifier (12 units)
Rated Current:	100 A and 40 A at 430 Vdc
Battery SOC Equipment:	Varta Logistronic Unit (not used)

lists the actual efficiency of the inverter as a function of load in the two operating modes. Table 3 lists the performance data from January 1988 to June 1988 in a standard format requested by CEC/DG XII.

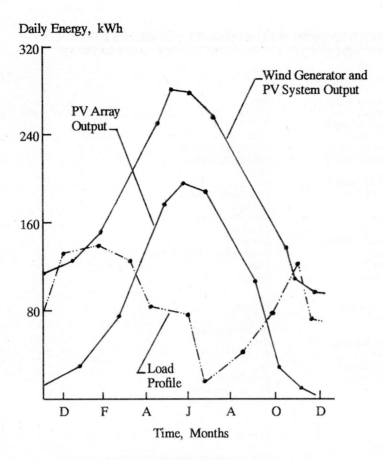

Figure 5. Load and PV/Wind Generator Output Profiles Over a One-year Period

Table 2. Inverter Efficiencies vs. Output Power

	Output Power, % of Full Load				
	10	20	50	80	100
Efficiency for grid-commutated inverter without synchronous motor, %	87	92	93-94	94	94
Efficiency for stand-alone inverter with synchronous motor, %	65	75	87	89	90

Note: Efficiencies include the output transformer and choke

The predicted annual output of the solar/wind generation system and school load were 45 MWh and 35 MWh, respectively. The excess energy available from the PV and wind generators is fed into the grid. Annex A presents some performance results on the main PV components.

5. COMPONENT DESCRIPTION

5.1 Solar Array

5.1.1 Array and Cabling

The array consists of 2,748 PV modules, divided into 29 subarrays of 90 modules each, and one subarray of 138 modules. The first 29 subarrays have three strings of 30 modules each, and the 30th subarray has three strings of 46 modules each.

402

Table 3. Monthly Performance Summary, January 1988 to June 1988

PV Pilot Plant:	TERSCHELLING	Date of Report:	31-8-88	Array tilt (deg):	40°
Site Manager:	W. Smit	Prepared by:	E.W. ter Horst	Latitude (°N):	53°35'
Project Manager:	T. van der Weiden	Organization:	31-20-732144	Longitude °E):	5° 23'
Organization:	Ecofys (Utrecht, NL)			Altitude (m):	5

		1988					
Parameter	Units	Jan	Feb	Mar	Apr	May	Jun
1. Solar/Temperature/Wind Information:							
a. Insolation, plane of array	kWh/m^2	.82	2.04	2.98	5.12	5.33	4.23
b. Insolation, horizontal surface	kWh/m^2	.50	1.3	2.19	4.41	5.33	4.41
c. Irradiance, plane of array, sunlight avg	W/m^2	.03	.09	.12	.21	.22	.18
d. Irradiance, plane of array, daily peak	W/m^2			.09	.18	.22	.18
e. Irradiance, horizontal, sunligh avg.	W/m^2	.02	.05				
f. Temperature, PV module	°C	6	5	5.1	7.8	13.1	13.9
g. Temperature, ambient air, peak	°C	---	---	---	---	---	---
h. Temperature, ambient air, daily avg	°C			---	---	---	---
i. Temperature, ambient air, lowest	°C						
j. Wind speed, daily avg.	m/s	7.5	9.5	7.5	6.5	6	5
k. Wind speed, daily peak	m/s						
2. Plant Output Information:							
a. Energy array field	kWh	533.33	1258.06	1687.5	4727.27	5382.35	4057.14
b. Energy, inverters	kWh	320	780	1080	3120	3660	2840
c. Power, array field, sunlight avg	kW	.72	1.81	2.27	6.57	7.23	5.63
d. Power, array field, daily peak	kW	---	---	---	---	---	---
e. Julian day of peak power occurrence	1-365						
3. Plant Performance Indices:							
a. PV array energy capability	kWh	1086.4	2545.13	3970.21	6599.12	7104.89	5461.34
b. Array utilization factor	%	.49	.49	.43	.72	.76	.74
c. Array energy efficiency	%	654.4	616.4	566.58	924.09	1009.82	958.32
d. Inverter energy efficiency	%	.6	.62	.64	.66	.68	.7
e. System energy efficiency	%	392.64	382.17	362.61	609.9	686.68	670.82
f. Plant availability	%	90	75	90	90	90	80
4. Battery Information:							
a. DOD, % rated Ah, daily avg	%	60	60	60	60	60	60
b. No. cycles*	--			--- not known ---			
c. Cell temperature, avg.	°C			--- not known ---			
d. Daily avg. recharge factor	--			--- not known ---			
e. Lowest state of charge reached	%			--- not known ---			
5. Operation & Maintenance Information:							
a. Total time DAS operational	%	0	0	0	0	0	50
b. Total inverter on-time: auxiliary	%						
pumping	%						
c. Number of key failures	--		0 windturbine	idem	windturbine	idem	idem
d. MTTR	h						
e. Sunlight hours:							
1) Daily avg, actual	h						
2) Daily avg., theoretical	h						
3) Total actual	h						
4) Total, theoretical	h						
6. Appliction Information:							
a. Consumer load energy	kWh	15380	14340	18220	11580	14260	1580
b. User energy supplied by PV plant	kWh	320	780	1080	3660	3660	2840
c. Total energy from grid to PV plant	kWh	15060	13560	17140	10600	10600	11740
d. Total water volume pumped (pumping head 37 bar)	m^3	---	---	---	---	---	---

* DOD >5% counts as one charge/discharge cycle

5.1.2 Modules

The PV modules used are AEG model PQ 10/20/01, each containing 20 series-connected polycrystalline cells. The module is identical to that of the FOTA PV plant (see Appendix 3).

5.1.3 Array Support Structures and Foundations

The mounting structures for the PV modules are made of prefabricated hot-dipped galvanized cold-rolled iron. The design was studied to permit simple erection on site. Figure 6 shows the array support structures and foundations. The modules are manufactured with short leads and special connecting plugs for easy and safe connection to a string. The cables are placed in vinyl wire channels with a rectangular cross-section.

Aerial view

Electrical termination box. Note the strain gauge above the electrical box

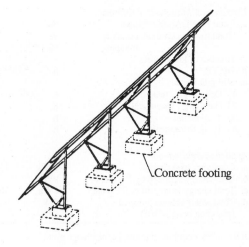

Concrete footing

Figure 6. Details of Support Structure, Foundation Footing and Electrical Termination Box at the TERSCHELLING PV Pilot Plant

404

5.2 Batteries and Charge Control Method

The energy storage system consists of two batteries, each containing 180 cells (Varta Bloc 2305) in series. The cells are rated at 250 Ah. Battery charge control is provided by the dc-dc converters described in Subsection 5.3.1. When the battery voltage reaches 432 Vdc (2.4 Vdc/cell average) the converter reduces the charge current by switching off the PV subarrays in groups of two. Deep-discharge is prevented by disconnecting the inverters from the dc bus whenever the battery voltage reaches a predetermined low cutoff value. To protect the batteries during periods of prolonged low insolation, they are recharged from the utility grid using a rectifier. The Logistronic unit was intended to monitor the SOC of the battery for charge/discharge protection purposes.

5.3 Power Conditioning

Power conditioning equipment consists of 29 dc-dc converters, one inverter, and two rectifiers.

5.3.1 Converter

This converter is a voltage boost-buck type with an MPPT capability. There are 29 converters, each tied to a PV source.

The design criteria for this MPPT is that the maximum power point voltage, Vm, remains under the battery voltage, Vb. Since neither Vm nor Vb are constant, the MPPT continuously bridges the gap between the maximum power point voltage and the actual battery voltage. The converter performs up- or down-conversion ("boost" or "buck") of input voltage by adjusting the gain of the dc-dc converter.

5.3.2 Inverter

The key elements of the inverter are the 60-kVA line-commutated current-source unit and a 50-kVA synchronous motor. The latter provides the emf needed for natural commutation of the inverter in stand-alone mode and compensates the reactive power in both stand-alone and grid-connected modes. The basic schematic of the inverter and its interfaces are shown in Figure 7. The inverter efficiency is generally higher in the grid-connected mode which does not use the synchronous motor as indicated in Table 4.

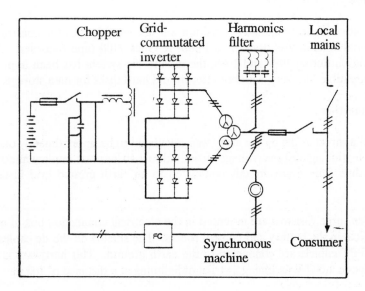

Figure 7. Schematic of the Inverter and Its Interfaces

405

Table 4. Inverter Efficiency Including Output Transformer and Choke

	Efficiency	
% Full Load	Grid-connected without Synchronous Motor	Grid-connected with Synchronous Motor
10	87.0	65
20	91.8	75
50	94.0	87
80	94.4	89.4
100	94.2	89.8

A 40-kVAR harmonics filter, consisting of three tuned branches, is provided to minimize voltage distortion, compensate for the lagging power factor, and minimize power loss in the inverter.

The chopper duty cycle controls the transfer of the active power to either the consumer or the grid. In stand-alone mode, the frequency is used for proper active power balance. The motor excitation determines the reactive power balance with the system. The inverter and the synchronous motor are able to support the starting surge currents of some of the consumers like the induction motors. In the grid-connected mode, if the motor is kept running, the synchronous motor may be used for power factor correction or as a UPS.

5.3.3 Rectifier

Two rectifiers are provided to permit battery charging from either the grid or the wind generator. Either one or both batteries can be connected to the charging bus for this purpose (see Fig. 4).

5.4 Control System

The plant is controlled by the Energy Management System which now uses Midi/D08 PLC (previously Satt microprocessor).

5.5 Data Collection

In the original plant configuration, data acquisition was accomplished by the control subsystem. Data storage in the initial version was done by a Penny & Giles 7100 tape recorder. As a result of an extensive modification during 1987 and 1988, the monitoring system has been improved significantly using an IBM compatible PC along with both floppy and hard disks for data storage.

5.6 Lightning Protection

The probability of a direct strike on the array was calculated to be approximately one every 200 years. Thus, normal protection against overvoltages from indirect strikes was implemented. All metal parts including the module frames are directly connected to an earth ground grid installed in the array field.

Twelve metal-oxide surge varistors are mounted in the electrical connection box of each subarray and in the MPPT cubicle in the battery and control rooms. The shields on the dc cables connecting each array with the MPPT cubicle are connected to the earth ground. This hardware is intended to limit the residual voltage to 1,000 V assuming a strike of lightning at a distance of 100 m.

5.7 Auxiliary Generators

One 40-kW wind generator was connected to the PV plant at the ac side of the inverter. This generator was replaced by a new 75-kW wind turbine in May 1989.

5.8 Load Equipment

The electrical loads consist of conventional ac devices and appliances at the school.

6. MAINTENANCE

The plant is operated by the technical personnel of the Hogere Zeevaartschool. They were instructed on the operation of the plant during the installation and start-up phase.

7. SUMMARY OF PROBLEMS AND SOLUTION APPROACHES

The inverter in Terschelling has operated reliably, with no-load losses of about 750 W in the grid-connected mode. The design allows switching off the inverter when it reaches a prescribed minimum load.

The reliability of the buck-boost converters with MPPT capability was found to be very low. The malfunction of one converter causes the corresponding PV array string to be no longer useful. The overheating of certain electronic parts has been the cause of frequent converter stoppages. This is due to the poor arrangement of heat-sensitive parts on the cards. Another reason is the very dense packing of the electronic boards in an electronic rack without forced ventilation.

The user and the supplier of the plant believed that one array string which is connected directly to the battery bus (bypassing the converter) supplies more energy than those with the MPPT converter in series. (Note: This was partly verified by tests conducted by Ecofys in 1989 as discussed in Annex A.)

The key hardware problems encountered are as follows:

- PV array: Cable cracks and ground leakages causing intermittent shorts
- Converters: Transistor failures
- Logistronic unit: Did not operate normally from the plant start-up
- Rectifiers: Defective from 1984 to late 1987
- EMS: Design errors in software
- Data monitoring: Tape recorder failures

The plant has been operated mostly in the grid-connected mode. In the stand-alone mode, its efficiency is low because of the losses in the synchronous motor. This mode of operation is therefore chosen only for demonstration purposes or when the grid is switched off. When the PV generator is operated in the stand-alone mode, the synchronous motor is first started using the line frequency (i.e., grid). Because of these operational characteristics, the PV system cannot provide real emergency backup power in case of loss of grid power.

The Logistronic unit could not provide proper battery charge regulation. After several attempts to correct this problem, the unit was turned off. The charge control now used is constant voltage mode (i.e., current tapering at voltage limit).

The battery cells have sustained a large water loss even though they were equipped with recombinator plugs. Thus, periodic addition of water is necessary. This is an indication of excessive overcharging and poor battery management.

The battery capacity is close to the lower limit. Consumers with high power needs can, in stand-alone operation, cause the inverter to switch off due to low battery voltage. Kitchen ovens are therefore connected to the plant in such a way that they can be switched on only in grid-connected mode or when the battery voltage is above the minimum selected value.

8. KEY LESSONS LEARNED

Due attention should be given to the proper layout, ducting and protection of array cables. Cable cracks have occurred frequently (Ecofys estimated 10 to 20 a year). Possible causes of this problem are faulty weather-proofing of the cable pipes, and inadequate space in the pipes to carry all the cables.

Cracking of glass covers in the PV modules is in most cases due to incorrect torquing of the fastening bolts as well as the poor design of the module attachment brackets.

Since charge control and discharge protection of the batteries are based on the total battery voltage level (i.e., 180 cells in series), normal mismatch in cell performance can cause some of the cells to be subjected to a higher overcharge during charging and polarity reversal during discharge. For example during one discharge period, Ecofys found one cell reversed at -2.4 V while the average cell voltage was 2.03 V per cell.

PV system efficiency is very low when operating in the stand-alone mode. Losses in the synchronous motor and the converter are contributing factors to this low efficiency.

9. CONCLUSIONS

Despite a number of defective parts, the PV plant has managed to supply power with a varying degree of success. For example, the average PV energy production in 1984 and 1985 was 1/3 of total possible capability, while in 1987 the production reached only 31,000 kWh. Maintenance turned out to be much more than expected. Significant improvements were made in 1987-1989.

10. REFERENCES

[1] E.W. ter Horst, "Repair and Optimization of TERSCHELLING PV Plant," Ecofys Semi-annual Report, November 1989.

[2] M. Imamura and P. Helm, "PV Pilot Plant Improvement and System Development Project and Control," WIP 88-5, CEC Contract EN3S-0109-D, July 1988.

[3] E.W. ter Horst et al, "The Terschelling PV/Wind System After Five Years of Operation," Proc. of the 8th European PV Solar Energy Conference, Florence, IT, May 1988.

[4] E.A. Alsema and W.C. Turkenburg, "PV Activities in the Netherlands," Proc. of the 7th European PV Solar Energy Conference, Sevilla, ES, October 1986.

[5] Terschelling Project Handout, Pilot Plant Optimization Meeting, Brussels, BE, 12-13 May 1986.

[6] P. Helm, "Manual of Photovoltaic Pilot Plants, Parts A and B," WIP report under CEC Contract EN 3 S-0007-D(B), Munich, FRG, February 1986.

[7] Terschelling Project Handout, Pilot Programme I Meeting, Tremiti Island, IT, 20-21 June 1984.

[8] R.P.M. Sonneville, "Solar-Wind Project Terschelling," Proc. of the EC Contractors' Meeting held in Hamburg and Pellworm, FRG, 12-13 July 1983.

[9] Terschelling Project Handout, Pilot Programme I Meeting, Mont Bouquet, FR, 12-14 April 1983.

[10] Terschelling Project Handout, Pilot Programme I Meeting, Crete, GR, 14-15 September 1982.

[11] Terschelling Project Handout, Pilot Programme I Meeting, Brussels, BE, 20-21 April 1982.

[12] R.P.M. Sonneville, "Solar-Wind Project Terschelling," Proc. of the Final Design Review Meeting on EC Photovoltaic Pilot Projects held in Brussels, BE, 30 November - 2 December 1981.

ANNEX A.   PERFORMANCE DATA FROM THE TERSCHELLING PILOT PLANT AS OF MID-1989 [1]

1. INTRODUCTION

The Terschelling PV/wind plant (see Fig. A-1) is used to supply electricity to a Marine Training School and for research at the system level. Ecofys has carried out the repair and optimization of this plant from 1987 to 1989 under DG XII sponsorship. All repair work has been done, and the system performance has improved considerably.

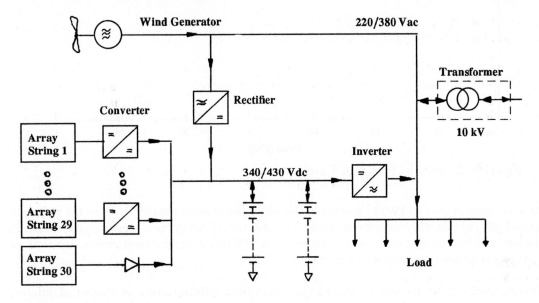

Figure A-1. Terschelling PV/Wind System Diagram

In July 1988, Ecofys started the experimental phase of the project. In this phase the control of the system was optimized for both grid-connected and autonomous operation. Special attention was given to the operation and performance of the converters and the battery. This annex presents a summary of the performance of the PV array, inverter, and converters available during the initial experimental work.

2. PV ARRAY

In June 1988, the Joint Research Centre (JRC) in Ispra, Italy, inspected the PV generator and conducted peak power measurements. These measurements were compared with those taken at the time of delivery of the system in 1983. The following value was found for the power under Standard Test Conditions (STC):

$$P_{STC} = 43.4 \pm 1.3 \text{ kW} \tag{1}$$

In the acceptance test carried out in 1983, a value of 44.2 kW under STC was measured. This implies that within the measuring error, no ageing effects on the peak power of the PV generator can be observed. The cell efficiency was estimated to be 7.9%.

3. INVERTER

For a number of weeks, input and output power of the inverter were measured and recorded every minute. Figure A-2 represents the efficiency of the inverter in operation based on these data. We see that the full-load efficiency is 93.6%. It can be deduced from the data that:

$$P_{out} = 0.95 \ x \ (P_{in} - 0.75) \tag{2}$$

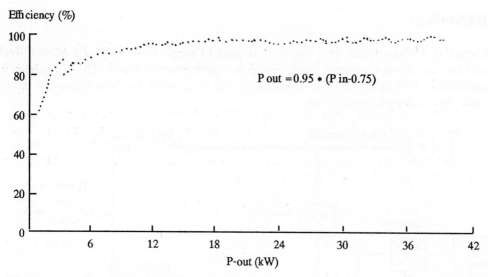

Figure A-2. Inverter Efficiency Curve

In the large range of over 15 kW the inverter has an efficiency of more than 90%. Even for an output power of 5 kW the efficiency still exceeds 80%. The inverter efficiency curve can be considered to be satisfactory and, according to the calculations made by SOMES computer programme, there will be an average annual inverter efficiency of 88.5%.

Ecofys examined whether the inverter size chosen would yield optimal results. A number of different situations were computed with the assistance of the computer programme SOMES, such as the use of a lower-capacity inverter. Figure A-3 depicts the results of these computations for smaller inverters.

Figure A-3 shows that the Terschelling inverter with a nominal power of 20 kW yields the highest output ___ 1,044 kWh/per year (3.7%) more than the inverter currently in use. Another, perhaps even more significant advantage is, of course, that a 20-kW inverter is much cheaper than a 50-kW inverter.

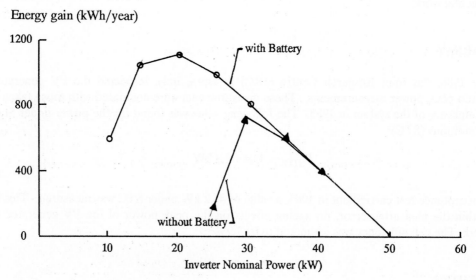

Figure A-3. Energy Gain of Smaller Inverters

One of the reasons why smaller inverters yield a higher annual output is that the inverter operates at a higher average efficiency. In the case without batteries, we observed (see Fig. A-3) that an inverter smaller than the current 50 kW inverter has a higher output. In this case, a 30-kW inverter produces 691.2 kWh more per annum.

4. CONVERTER

The converter with MPPT is designed to control the input voltage such that the PV array operates at its maximum power point. This MPPT operation is maintained as long as the battery is below its upper charge voltage limit.

For a number of weeks the converter input voltage and current of all subarrays were recorded every minute. In addition to this, the in-plane solar radiation was also monitored every minute. Figure A-4 represents the MPPT efficiency as a function of the input power. We see that the efficiency for 1250 kW at the outlet amounts to 89.3%, which is much worse than was reported in 1983. From the data we derived the following power output relationship:

$$P_{out} = 0.947 \; x \; (P_{in} - 78) \tag{3}$$

where P_{in} is the array input power.

The no-load dissipation of each converter was estimated to be 78 W.

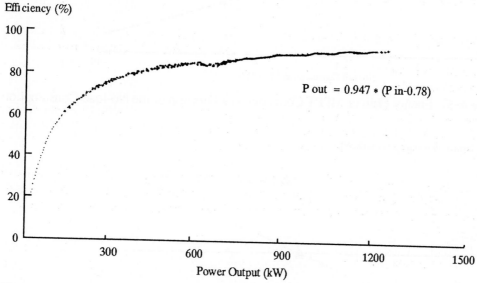

Figure A-4. Efficiency Curve of the MPPT Converter

Calculations for the Terschelling PV system (PV power 43.4 kW; battery 180 kWh) produced the results shown in Table A-1. It can be seen that the use of MPPT converters results in even lower energy production than is the case when the modules are linked directly to the batteries. Subsequently, a number of calculations were performed for the production of a 50-kWp PV generator for different converters. The coefficients α and β in the following equation were varied:

$$P_{out} = \alpha (P_{in} - \beta) \tag{4}$$

Table A-1. Net Production of the 43.4 kWp PV Generator in Terschelling System and the Energy Gain with Respect to an Identical System without MPPTs

Converter	PV Production (kWh/yr)	Energy Gain Relative to (a) (kWh/yr)
(a) without MPPTs	32,648	0
(b) MPPTs reported	35,050	2,402
(c) MPPTs measured	32.295	-352

Figure A-5 gives the energy gain from the application of MPPTs as a function of the MPPT consumption under no-load β (for a slope of 0.95) with respect to a system without MPPT. Figure A-6 represents the energy gain as a function of the marginal efficiency (the slope in equation 3) for a no-load consumption of 5% of the full-load.

From Figures A-5 and A-6, the conclusion can be drawn that the application of MPPTs does not produce any energy gain unless their efficiency is improved. A decrease in no-load β is more likely to produce more energy gain than an improvement of the marginal efficiency.

Energy Gain to Reference System (MWh/yr)

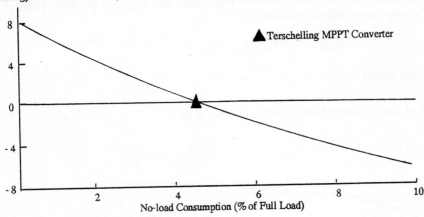

Figure A-5. Energy Gain of MPPT Converter as a Function of the No-load Consumption ß

Energy Gain to Reference System (MWh/yr)

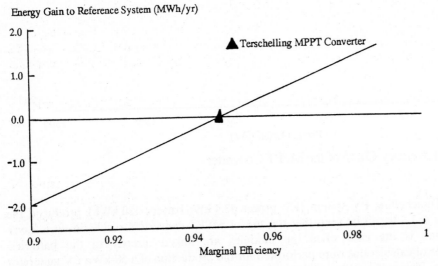

Figure A-6. Energy Gain of MPPT Converter as a Function of Marginal Efficiency

5. CONCLUSION

The inverter has operated well with a high yearly average efficiency during grid-connected operation. It appears that the inverter was not properly chosen for grid-connected operation. Although it is not necessary for improving the system performance, it would be interesting to install a 20-kW inverter in the Terschelling system to show the real energy gain compared with the currently installed 50-kW inverter.

The MPPT converter efficiency was found to be much lower than was expected. The high no-load losses appear to be too high. In addition, the initial data suggests that the MPPT efficiency itself is significantly lower than the other MPPT designs in use (normally about 95 to 98%).

Appendix 14

TREMITI PV PLANT

1. INTRODUCTION

This document describes the 65-kW TREMITI PV pilot plant and its initial operation which began in March 1984. Its purpose is desalination of sea water in order to supply drinking water to the island of St. Nicola in the Adriatic Sea. Figure 1 shows the array field and the control building.

The Tremiti Islands have no natural water sources. The water needed for the few dozen inhabitants and the nearly two thousand tourists during the summer months is transported by ship from the continent. The desalination process chosen, the reverse osmosis (RO), is capable of high efficiency and simple automatic operation.

The information contained herein was obtained from ref. [1 to 8] and individual contacts with the plant owner and designers.

2. PLANT DESIGNER, OWNER, AND OPERATOR

The plant in Tremiti was set up by a consortium headed by Italenergie. Tecmar was responsible for the water treatment plant. Rossetti in Milan, a subcontractor of Tecmar, built the reverse osmosis plant. The owner of the plant is Cassa per il Mezzogiorno. Plant improvement and maintenance activities are being done by Italenergie. Their address is as follows:

> Italenergie, SpA
> Via della Repubblica, 39
> I-67039 Sulmona
> Italy

3. SITE INFORMATION

3.1 Location

The plant is located on the southern coast of the island of St. Nicola which belongs to the Tremiti Island group in the Adriatic Sea. This island is 40 km from the east coast of Italy (see Fig. 2). The plant's coordinates are:

- Latitude: 42.1°N
- Longitude: 16.5°E
- Altitude: 58 m

View from the South. The 65-kW plant is installed over the rocky bluff,
about 58 metres above the water line

Dr. W. Palz (CEC-DG XII) on the left and Dr. K. Krebs (JRC)
inspecting the solar panels

View from the East

Figure 1. The TREMITI PV Pilot Plant

View of Isola di St. Nicola from the West (TREMITI Pilot PV Plant on the right side)

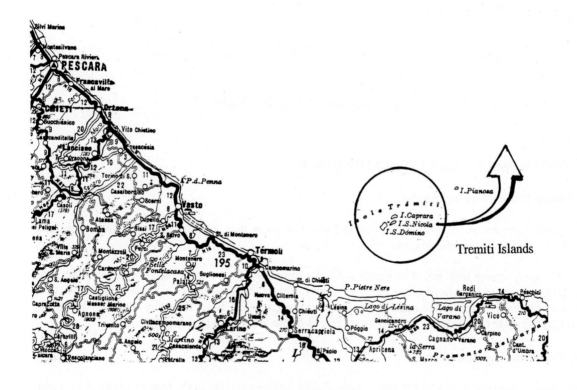

Figure 2. Location of the TREMITI PV Pilot Plant in the Adriatic Sea

415

3.2 Site Constraints

There were no major site constraints.

3.3 Solar Radiation and Climatic Conditions

The following data were obtained at Pescara (Aeronautica Militare), Amendola (Istituto Sperimentale Agrario) and Manfredonia (Aeronautica Militare) during the period 1913 to 1979:

- Insolation, global horizontal: 1,500 kWh/m^2 annual
- Ambient air temperature: 40°C maximum, -5°C minimum
- Wind speed: 130 km/h maximum

3.4 Load Demand

The total monthly energy demand based on an average water production of 30 m^3 per day is shown in Figure 3.

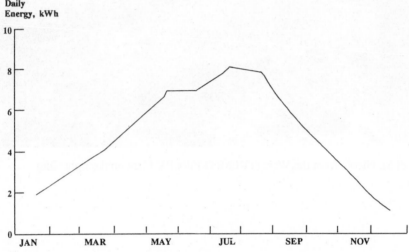

Figure 3. Daily Energy Demand During One Year

4. SYSTEM DESCRIPTION AND PERFORMANCE

4.1 Plant Architecture and Layout

The plant occupies an area of 4,000 m^2 on a cliff about 44 m high and the crest of a hill. The land surface is covered by fractured rocks and low vegetation.

The PV array covers 1800 m^2. The building for the remaining PV components, the power plant, and the RO equipment is situated along the northern border of the area and is less than 3 m higher than the surrounding ground. The natural slope of the terrain which is south facing and the cliff at the southern extremity permitted a very compact layout of the array field (see Fig. 1).

4.2 PV System Configuration

The plant consists of a 65-kW PV field, three 25-kW converters, one main battery (2,000 Ah), eight inverters (four 7.5 kVA and four 3 kVA), two UPSs (3.5 kVA and 15 kVA), an integrated control/monitoring system, and a reverse osmosis desalination system. Figure 5 shows a simplified system block diagram. Its major components and their key features are listed in Table 1.

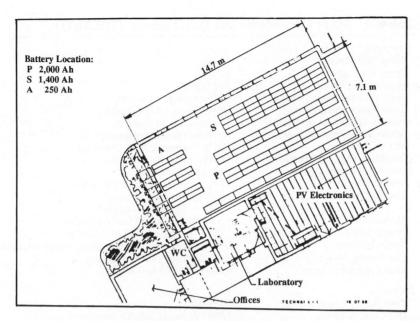

Figure 4. Layout of the Batteries and Electronics at TREMITI PV Plant

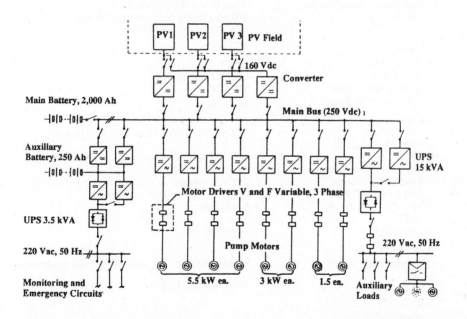

Figure 5. Simplified System Block Diagram of the TREMITI PV Pilot Plant

The RO system is designed to produce about 5000 m³ of water with an average PV energy requirement of 13 kWh per m³. A power flow supervisory system provides an optimum utilization of available energy and a display of system status.

4.3 System Operation and Management

The basic power management criteria were based on the following considerations:

- Minimize interruptions of the RO plant due to lack of array energy or pretreated water. These interruptions are limited to exceptionally long periods of bad weather;

Table 1. Major Components and Their Features (TREMITI)

Solar Array

Rated Peak Power:	65 kW
Array Tilt Angle:	22°
Nominal Bus Voltage:	160 Vdc
Number of Subarrays (Strings):	19 (Siemens and Ansaldo)
Total Number of Modules:	399 (Siemens), 627 (Ansaldo)
Number of Modules per String:	21 (Siemens), 33 (Ansaldo)
Module Manufacturer/Type:	Siemens/144-09; Ansaldo/AP 35
Module Rating:	120 W (Siemens), 35 W (Ansaldo)
Number of Cells per Module:	144 (Siemens), 36 (Ansaldo)
Number of Bypass Diodes:	1 per 18 cells in series
Module Weight:	27 kg (Siemens), 10.2 kg (Ansaldo)
Module Dimension:	1,470 x 1.020 mm (Siemens)
	1,298 x 343 mm (Ansaldo)
Module Area:	1.5 m^2 (Siemens), 0.45 m^2 (Ansaldo)
Active Solar Cell Area/Module:	1.13 m^2 (Siemens), 0.36 m^2 (Ansaldo)
Solar Cell Type:	Monocrystalline (Siemens), polycrystalline (Ansaldo)
Cell Packing Factor:	75 % (Siemens), 80 % (Ansaldo)
Total Land Area Used:	4,000 m^2 (1,800 for PV field)

Battery

Total Battery Energy Rating:	500 kWh
Cell Capacity Rating:	2,000 Ah (main), 250 Ah (auxiliary)
Voltage Range:	250 Vdc (main), 220 Vdc (auxiliary)
Allowable Depth of Discharge:	80 %
Number of Batteries:	2 (2,000 Ah and 250 Ah)
Number of Cells per Battery:	125 (main battery)
Cell Manufacturer/Part No.:	Varta Bloc VB 2420/VB 2305
Cell Weight:	199.4 kg (2,000 Ah), 29.1 kg (250 Ah)
	(with electrolyte)

Converter

Rated Power:	25 kW x 3
Manufacturer:	CO.EL (Pescara, IT)
Input Voltage Range:	160 Vdc
Output Voltage Range:	250 Vdc

Inverter

Rated Power:	30 kVA (7.5 kVA x 4), 12 kVA (3 kVA x 4)
Number of Inverters:	8
Type:	Self-commutated (all)
Manufacturer/Part No.:	CO.EL (Pescara, IT)
Input Voltage Range:	250 Vdc nominal
Output Voltage/Phase:	380 (220 Vac/3-phase (all)
Frequency:	50 Hz (all)
THD:	5 %
Efficiency:	94 % (full load)

Controller

Manufacturer/Model:	16-bit CPU LS 11/02, comprising 2 microprocessors

Data Acquisition: Siemens with ECMA

UPS

Number of UPS's:	2
Rated Power:	3.5 and 15 kVA
Input Voltage Range:	250 Vdc nominal (for 15 kVA),
	220 Vdc nominal (for 3.5 kVA)
Output Voltage Range:	220 Vac

- Maximize the operation time of the RO plant under direct supply from the PV array, thus avoiding the energy loss in the storage system;
- Optimize the ratio of water output to the input array energy on an annual basis.

The main battery was intended to assure the uninterrupted operation of the desalination system during periods of bad weather with no PV output.

Power for the auxiliary loads, chemical injectors, lighting, and fans are supplied by the single-phase inverter. Furthermore, the same UPS energizes all the starting switches of the PV plant. The central

418

control unit performs the automatic operation of the plant, the acquisition and processing of data, as well as on/off control of the three subarray groups and all power conditioning equipment.

4.4 Plant Control

The control is based on a 16-bit CPU LS 11/02. This system with two microprocessors has the following features:

- 64K RAM
- Fast parallel link for data acquisition
- 2 serial interfaces
- 20 MB Winchester
- 500-KB floppy disc, RX02 emulator
- RT11 V5 compatible operating system
- High level Basic language

The data collection and processing is controlled by a microprocessor and includes, besides the CPU, all analog and digital I/O, RAM, EPROMS, peripherals for data display, alarms, synoptic display, keyboard, and mass storage.

The second microprocessor system was intended for future on-line operation, with I/O peripherals, mass storage, and a firmware allowing operations through the use of a high level language.

Raw and processed data are displayed on a CRT colour monitor. Data for performance analysis are stored on cassette for further processing. The central computer issues diagnostic messages for possible operator corrective actions.

4.5 Plant Performance

The PV array output at 1000 W/m^2 and 25°C average cell temperature, based on the plant acceptance tests performed by JRC-Ispra in May 1984 is as follows:

- Array A (Ansaldo):   21.0 kW
- Array B (Siemens):   46.9 kW

The plant design was originally based on a water consumption rate of 4,100 m^3/year. During the first year of operation (June 1984 to June 1985), it produced 3,600 m^3 of potable water, which is 82% of the theoretical value. The difference is attributed mainly to the manual operating mode and to a lesser degree, the annual insolation level which turned out to be lower than the value used in the theoretical prediction.

In 1985, the water needs of this island had increased to 12,000 m^3 per year, due to a sharp rise in the number of local inhabitants and tourists. The plant improvement effort in 1988-1989 thus included an expansion in the desalination system capacity which required a corresponding adjustment of the PV size.

5. COMPONENT DESCRIPTION

5.1 Solar Array

5.1.1 Array and Cabling

The 65-kW PV array comprises three subarray groups, two with Siemens modules and the other with Ansaldo modules. These groups have the following characteristics:

	Siemens	Ansaldo
- Rated Power Output (kW):	45	20
- No. Modules in Series:	21	33
- No. of Strings:	19	19

The array strings are connected in parallel at the power distribution board inside the building. Each string is made up of 2-wire underground armoured cable which is connected to the junction box in the field. The module mismatch and the wiring losses were estimated to be 3 and 2%, respectively.

5.1.2 Modules

Two types of modules are employed:

- Siemens SM 144-09 module, with 144 monocrystalline silicon cells and two bypass diodes (one per 18 cells in series).

- Ansaldo AP 35 module with 36 polycrystalline silicon cells and two bypass diodes.

The main typical features of the modules at 1,000 W/m^2 and 25°C cell temperature are listed below:

	Siemens	Ansaldo
- Voc:	10.6 V	6.66 V
- Isc:	18 A	7.5 A
- Wp:	130 W	35 W
- Insulation Voltage:	2,000 Vdc	2,000 Vdc
- Weight:	27 kg	10.2 kg
- Dimensions:	1,470 x 1,020 mm	1,298 x 343 mm

5.1.3 Array Support Structures and Foundations

The basic structures for the modules are precast concrete blocks which support the horizontal galvanized steel beams to which the PV panels are attached (see Fig. 6). Furthermore, these concrete blocks rest on other pre-cast concrete blocks which serve as part of the foundation.

5.2 Batteries and Charge Control Method

Two batteries are provided, a main battery and an auxiliary battery. Their key features are as follows:

	Main Battery	Auxiliary Battery
- Battery Voltage (Vdc)	250 nominal	220 nominal
- Rated Capacity (Ah)	2000	250
- No. of Cells in Series:	125	110

Charge control is done by a dc-dc converter which essentially provides a constant voltage charging whenever the battery reaches a pre-set upper voltage limit.

5.3 Power Conditioning

5.3.1 Converter

The maximum power point tracking of the array and battery charge control is performed by three 25-kW dc-dc converters which operate in parallel. These converters boost the input voltage of the array from 160 Vdc to 250 Vdc, and have efficiencies of at least 90%.

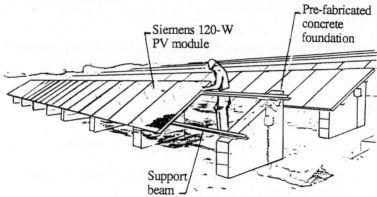

Figure 6. Details of Array Support Structures and Foundations at TREMITI PV Plant

5.3.2 Inverter

The inverters are of the variable frequency self-commutated type which use transistors with single-pulse PWM to give a square wave-shaped output. Step-up transformers provide a 380-Vac, 3-phase, 50-Hz output. The efficiency of the inverters including the transformers is about 80% at 10% full load, rising to 94% at full load.

The inverters are connected to eight 3-phase induction motors with rated power ranging from 1.5 to 5.5 kW. The inverter controls the output voltage and frequency during the turn-on phase of the motor in such a way that starting currents of the induction motors do not exceed the nominal rating of the inverters, while steady state voltage and frequency are maintained constant.

The reason for the use of eight inverters instead of a large one for supplying all pumps was to achieve better overall efficiency. The smaller pump motors are supplied with 380-V 3-phase ac through a single 1.5-kVA inverter, while the lighting load is supplied directly from the dc-busbar.

An additional single phase ac inverter supplies auxiliary loads of small motors, lighting and fans.

5.4 Control System

The plant control is based on a 16-bit CPU LS 11/02, a system comprising two microprocessors.

5.5 Data Collection

The monitoring system basically consists of analog and digital sensors, A/D multiplexer, cassette recorder, 8-colour CRT video console, printer, and a keyboard, all under one microprocessor control.

421

Measured and recorded parameters include solar and meteorological data and performances of PV components and RO system.

5.6 Lightning Protection

The probability of direct lightning strikes on the PV array was estimated to be very low (in the range of 0.5%/yr). Therefore, no protection against direct strikes was necessary. For protection against induced overvoltages, all array structures were strapped to the earth ground. In addition, metal-oxide varistors were installed on each power wiring.

5.7 Auxiliary Generators

The system has no auxiliary generators directly connected to the PV plant.

5.8 Load Equipment

The principal loads are the induction motors for pumping (1.5 to 5.5 kW), auxiliary loads (chemical injectors, lighting, and fans), as well as plant control and monitoring equipment.

6. MAINTENANCE

The maintenance of the PV plant consists primarily of the following activities:

- Checking the specific gravity and level of electrolyte in the battery cells periodically
- Cleaning of electronic cards and racks by blowing compressed air

7. SUMMARY OF PROBLEMS AND SOLUTION APPROACHES

The main problems encountered were as follows:

- The automatic control and data acquisition functions have not operated well and were shut down in early 1985. Since then, these systems have been under manual control. One of the reasons for this action was that the design of the automatic control system was too complex with too many requirements. For example, the data acquisition portion should be independent of the controller.

- To install the PV arrays, vegetation below the array panels was cleared. Since then, the plants have not regrown as expected, and consequently, soil erosion from rainwater has occurred. In the long term, the array structure foundations could be endangered.

- Frequent inverter malfunctions have occurred due to high starting currents of the induction motors for the pumps. The use of components with higher ratings apparently solved this problem.

- Corrosion in the electronic circuits has caused the control electronics and DAS to be unreliable. In addition, some of the electronic components and printed circuits were apparently sensitive to dust and humidity. Blowing these components with compressed air and replacing some parts allowed the manual operation of the power electronics and the DAS.

- Even with attempted modifications, the power management system has not been made operational. The problems were attributed to its complexity and unreliable control sensors and sensor wiring and calibration.

- The state of charge function on the Logistronic unit did not operate, and the unit has since been used only for visual monitoring of battery voltage and current.

- The membranes used in the RO system were found to be damaged several times. They were replaced with a new design and improved materials which have yielded satisfactory results.

- The piston pump valves may fail before their rated lifetime because of corrosion from frequent non-operating conditions.

- The data acquisition system and the power management system were switched off in early 1985. As a result, the key operation of this desalination plant could not be monitored properly.

- There is a lack of maintenance documentation on the PV system components for on-site trouble shooting and repairs.

- Since early 1985, the main battery has been used only as a bus voltage stabilizer, not for night-time operation. This is due to the manual mode of operation.

8. KEY LESSONS LEARNED

The key lessons learned are as follows:

- The available RO systems are still research items, requiring reliability improvements before they can be considered for large-scale commercial application. Particular design and selection of the membrane material appear to be the critical development needs.

- Protection from high humidity, temperature, and salty environments is a key requirement for all electronic components.

- Whenever computer-based control is used, appropriate software personnel must be available not only for the design but also for the long-term operational phase of the plant. Otherwise, a good mix of reliable hardware and computer approaches is a better design option than a computer-only approach.

- For expensive, high-capacity batteries, reliable back-up methods of battery charge/discharge protection should be provided.

9. CONCLUSIONS

The basic operation of the RO-type desalination system powered by the PV plant was demonstrated. The lessons learned from this plant are being put to good use in the plant improvement efforts scheduled to be completed in 1990.

10. REFERENCES

[1] Tremiti Project Handout, Pilot Plant Optimization Meeting, Brussels, BE, 12-13 May 1986.
[2] P. Helm, "Manual of Photovoltaic Pilot Plants, Parts A and B," WIP report under CEC Contract EN 3 S-0007-D(B), Munich, FRG, February 1986.
[3] Tremiti Project Handout, Pilot Programme I Meeting, Tremiti Island, IT, 20-21 June 1984.
[4] F. Fonzi, "Tremiti Desalination Plant," Proc. of the EC Contractors' Meeting held in Hamburg and Pellworm, FRG, 12-13 July 1983.
[5] Tremiti Project Handout, Pilot Programme I Meeting, Mont Bouquet, FR, 12-14 April 1983.
[6] Tremiti Project Handout, Pilot Programme I Meeting, Crete, GR, 14-15 September 1982.
[7] Tremiti Project Handout, Pilot Programme I Meeting, Brussels, BE, 20-21 April 1982.
[8] F. Fonzi, "Tremiti Desalination Plant," Proc. of the Final Design Review Meeting on EC Photovoltaic Pilot Projects held in Brussels, BE, 30 November - 2 December 1981.

Appendix 15

VULCANO PV PLANT

1. INTRODUCTION

This appendix describes the 80-kW VULCANO PV pilot plant and its initial performance. The plant has been operating since August 1984. This plant is intended to evaluate the technical and financial feasibility of supplying power from a PV solar energy system to small isolated villages not served by the utility grid. Figure 1 gives views of the plant.

The secondary aim was to investigate the economics of supplying power for use as a "fuel saver" for the existing Diesel power station.

The information contained herein was acquired from the plant operator, Ente Nazionale per l'Energia Elettrica (ENEL), direct contacts with Dr. Previ in 1988-1990, and ref. [1 to 17].

2. PLANT DESIGNER, OWNER, AND OPERATOR

The owner and operator of the PV plant in Vulcano is ENEL. Their address is as follows:

> ENEL - Centro di Ricerca Elettrica
> Attn: PV Dept.
> Via Volta, 1
> I - 20093 Cologno Monzese
> Italy

3. SITE INFORMATION

3.1 Location

Vulcano Island is part of the Aeolian archipelago north of Sicily (Italy). The island is mountainous, and typically volcanic. The vegetation (broom, cactus, wild olive, etc.) is typical of many

Figure 1. Views of the Array Field and Control Building at VULCANO PV Pilot Plant

426

Mediterranean coastal areas with little rainfall. The PV plant is located on a plateau (see Fig. 2). Its coordinates are:

- Latitude: 38.5°N
- Longitude: 15°E
- Altitude: 326 m

PV site is indicated by the circle

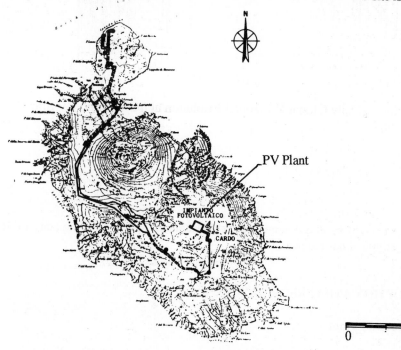

PV Plant

Figure 2. Location of PV Plant on Vulcano Island

3.2 Site Constraints

There were no major site constraints.

3.3 Solar Radiation and Climatic Conditions

- Insolation, global horizontal:
  - o Annual: 1,700 kWh/m^2
  - o Daily Average in June: 7.4 kWh/m^2
  - o Daily Average in December: 2.0 kWh/m^2
- Ambient air temperature: 40°C maximum, 0°C minimum
- Wind speed: 120 km/h maximum

Figure 3 shows a typical profile of daily global solar insolation on a horizontal surface for one year.

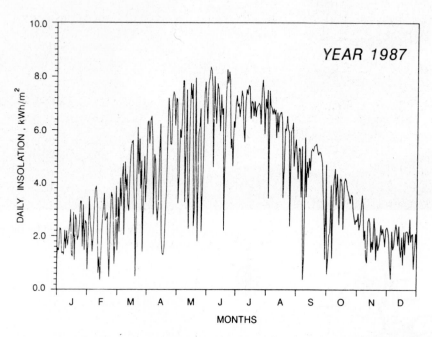

Figure 3. Profile of Daily Global Horizontal Insolation at the Vulcano PV Site in 1987

3.4 Load Demand

The plant supplies 55 residences and commercial establishments with a total capacity of 162 kVA. The maximum power supplied has reached 45 kW.

4. SYSTEM DESCRIPTION AND PERFORMANCE

4.1 Plant Architecture and Layout

The layout of the plant is shown in Figure 4. The storage battery and electrical equipment are housed in two separate prefabricated buildings with a total floor area of about 90 m^2. This relatively large surface is required to house the battery cells on seismic-resistant supporting structures.

428

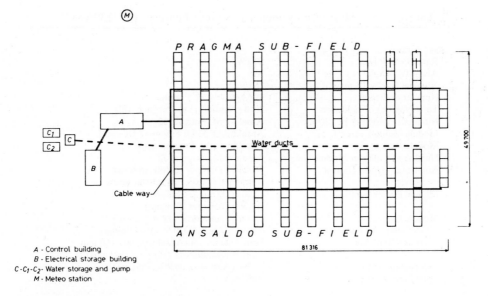

A - Control building
B - Electrical storage building
C-C_1-C_2- Water storage and pump
M - Meteo station

Figure 4. Layout of the VULCANO PV Plant

4.2 System Configuration

Figure 5 is a simplified block diagram of the PV system. The main components are the PV field, one battery, one battery charger, two inverters, system controller, and the data acquisition system. Its major components and their key features are listed in Table 1.

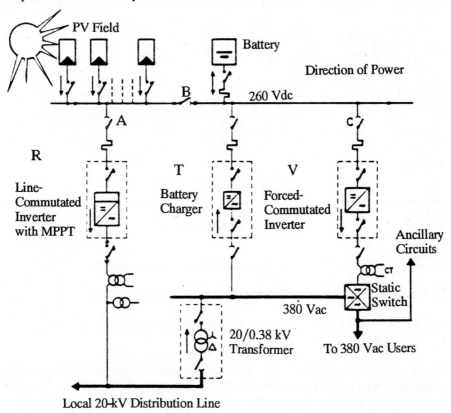

Figure 5. Simplified System Block Diagram of the VULCANO PV Pilot Plant

429

Table 1.   Major Components and Their Features (VULCANO)

Solar Array

Rated Peak Power:	80 kW
Array Tilt Angle:	35°
Nominal Bus Voltage:	260 Vdc
Number of Strings:	42 (Ansaldo: 21, Pragma: 21)
Total Number of Modules:	2016 (Ansaldo 1344, Pragma: 672)
Number of Modules per String:	Ansaldo: 64, Pragma: 32
Module Manufacturer/Type:	Ansaldo/AP 33 D; Pragma/2 DG/SOL
Module Rating:	Ansaldo: 33 W, Pragma: 55 W
Number of Cells per Module:	Ansaldo: 36, Pragma: 72
Number of Bypass Diodes:	Ansaldo: One per six cells in series
	Pragma: One per 24 cells in series
Module Dimension:	1,300 x 300 mm (Ansaldo)
	1,300 x 650 mm (Pragma)
Module Area:	0.44 m^2 (Ansaldo), 0.86 m^2 (Pragma)
Active Solar Cell Area:	0.39 m^2 (Ansaldo), 0.56 m^2 (Pragma)
Solar Cell Type/Size:	Polycrystalline Si (Ansaldo)/100 x 100 mm
	Monocrystalline Si (Pragma)/100 mm dia.
Cell Packing Factor:	92.3% (Ansaldo), 66.9% (Pragma)
Total Land Area Used:	4,350 m^2

Battery

Total Battery Energy Rating:	390 kWh
Cell Capacity Rating:	1,500 Ah
Voltage Range	260 Vdc (nominal)
Number of Batteries:	1
Number of Cells per Battery	130
Cell Manufacturer/Part No.:	Varta Vb 2415
Cell Weight:	152 kg (with electrolyte)
Cell Dimension:	550 x 307 x 383 mm

Battery Charger (Rectifier)

Rating:	10 kW
Manufacturer:	Borri
Input Voltage Range:	380 Vac
Ouput Voltage Range:	260 Vdc nominal

Inverter

Number of Inverters:	2
Rated Power:	40 kVA (s-c), 160 kVA (l-c)
Type:	Self-commutated (s-c), Line-commutated (l-c)
Manufacturer:	Borri S.p.A. (s-c)
Input Voltage Range:	Borri S.p.A. (s-c)
Manufacturer:	Marelli (l-c)
Output Voltage/Phase:	100-420 Vdc (l-c)
s-c:	380 Vac/3-phase + neutral
l-c:	480 Vac/3-phase
Frequency:	50 Hz
THD:	s-c: 3%, l-c: 5%
Efficiency:	90% minimum

Controller

Manufacturer/Model:	Hartmann & Braun, PLC (Contronic 3)

Data Acquisition:	ENEL-Marconi, Mini SAD/3 and cassette storage

Auxiliary Power Source:	Grid via 100-kVA back-up transformer

The PV field has 44- and 36-kW subfields. The first uses Ansaldo PV modules, the second Pragma modules. The storage battery has 130 Varta 1,500-Ah lead-acid cells in series. Two different inverters are provided: a 160-kVA line-commutated type and a 40-kVA self-commutated unit.

The plant is located in a region with low precipitation, so the supply of water is low. In order to provide water to clean the modules, rain gullets were placed at the lower edge of the panels and these channel the water to a storage tank. The sand settles in this tank and permits use of the water. In conjunction with this water collection system, sprinkler tubes were installed at the upper edge of the PV panel for module washing purposes.

4.3 System Operation and Management

Appropriate operation of power switches "A", "B", and "C" in Figure 5 allows two different modes of operation, stand-alone and grid-connected.

With "A" open, "B" and "C" closed: the PV array powers the segregated loads through the self-commutated inverter. This is the stand-alone mode with battery.

With "B" and "C" open and "A" closed: the PV array is tied to the grid network through the line-commutated inverter without the battery. In this grid-connected mode, the inverter is synchronized with the grid frequency. The inverter is designed with MPPT capability to continuously extract the maximum power available from the array field.

When power from the array field is insufficient or not available, the static switch automatically transfers the load to the grid via the back-up transformer.

4.4 Plant Control

The programmable logic controller (PLC) controls the plant in both stand-alone and grid-connected modes. The main automatic control functions are:

- Monitoring of the state of charge (SOC) of the battery
- Start-up of the line-commutated inverter
- Connection and disconnection of the stand-alone inverter according to the battery SOC
- Disconnection of the PV array at night in stand-alone operation
- Battery charge voltage limit control via switching of 10 array string groups
- State monitoring and control of switches

The data acquisition system collects solar, meteorological, and PV system performance data. A total of 48 measurement parameters are scanned and recorded, mostly with 15-min average values and some with peak and minimum values of scanned data.

4.5 Plant Performance

The plant has operated from October 1984 to February 1985 with some interruptions for small modifications and system adjustments. Since March 1985 it has worked with no interruptions in stand-alone mode and without any basic difficulties.

The line-commutated inverter was added in June 1986. Since that date the plant has been capable of either stand-alone or grid-connected mode of operation. The yearly performance of the plant is listed in Table 2. The ac energy output of the plant in stand-alone operation has ranged from about 4,000 kWh per month in winter to about 12,000 kWh per month in summer. Figure 6 shows the PV array output power vs. solar irradiance characteristics and daily energy profiles for the months of August and December 1985.

Tests performed on the PV field in September 1984 (acceptance test) and in May 1988 by JRC-Ispra [1] showed the following results, extrapolated to the standard reporting conditions:

	1984	1988	Difference
- Ansaldo Subfield (kW):	44.2	43.9	-1.0%
- Pragma Subfield (kW):	35.9	32.9	-6.0%

The uncertainty of the measurement is about $\pm 5\%$.

Table 2. Monthly Performance Summary, July to December 1988

PV Pilot Plant:	Vulcano	Date of Report:	January 1989	Array tilt (deg):	35
Site Manager:	V. Messina (ENEL)	Prepared by:	A. Buonarota, V. Piazza	Latitude (°N):	38.5
Project Manager:	G. Belli (ENEL)	Organization:	ENEL	Longitude (°E):	15
Organization:	ENEL (Milan, IT)	Tel.:	39-2-8847-5462	Altitude (m):	326 m

Parameter	Units	1988					
		Jul +	Aug	Sep	Oct	Nov	Dec
1. Insolation, plane of array, daily avg	kWh/m^2	6.91	6.06	4.88	4.68	3.32	2.5
2. Irradiance, Plane of Array, Sunlight Avg	W/m^2	550	500	470	480	420	360
3. Ambient Air Temperature							
a. Daily Average	°C	25.7	25.4	19.7	17.2	10.4	7.5
b. Maximum	°C	32.3	40.4	27.9	28.9	20	15.9
c. Minimum	°C	18.6	18.7	11.5	7.4	1.4	-2.9
4. Plant Energy Output							
a. Array Field	kWh	3998	2524	5598	6179	6073	4169
5. Plant Power Output							
a. Array Field, Daily Avg	kW	33.3	20.2	19.1	23.6	27.3	20.1
b. Array Field, Peak	kW	68.8	55.3	67.1	71.7	73.8	73
6. Batteries							
a. DOD*, % rated Ah, daily avg	%	67	35	24	37	76	86
b. Average daily SOC at sunrise	%	33	65	76	63	24	14
c. Average daily discharge current	A	45.1	23.9	20.7	24.1	25.4	16.9
d. Average daily charge current	A	34	43.1	39.6	50.5	67.4	55.8
e. Avg daily Ahi	Ah	770	453	379	449	494	401
f. Avg daily Aho	Ah	658	318	296	346	396	315
g. Avg recharge fraction (Ahi/Aho)	---	1.17	1.42	1.28	1.30	1.25	1.27

(∗) DOD - Depth of Discharge
+   Partial failure of the DAS occurred in July and August

5. COMPONENT DESCRIPTION

5.1 Solar Array

5.1.1 Array and Cabling

The photovoltaic array consists of 42 strings (21 with Ansaldo modules and 21 Pragma). Each Ansaldo string has 64 modules in series, and each Pragma string has 32 modules in series. Two strings are connected in parallel at the termination box which is mounted on the array structures. One termination box is provided per two strings. The modules are interconnected by cables via their single pole AMP connector plug and receptacles. Figure 7 shows the typical wiring of Ansaldo and Pragma array strings.

5.1.2 Modules

The 80-kW array comprises 2,016 modules (1,344 Ansaldo AP33D modules and 672 Pragma 2DG/SOL modules).

Typical characteristics of the two module types at 1,000 W/m^2 and 25°C cell temp rature are as follows:

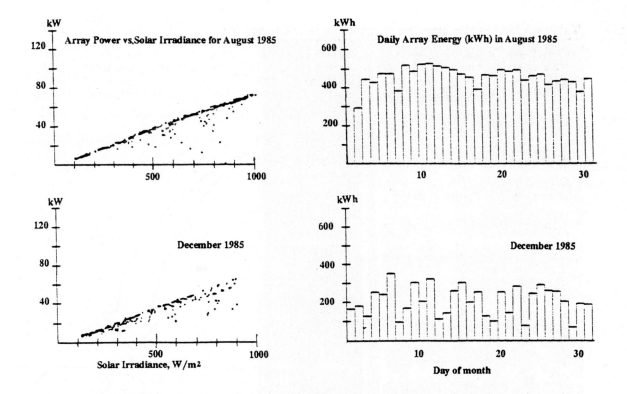

Figure 6. Array Power vs. Solar Irradiance and Daily Array Energy Output
in August and December 1985

	Ansaldo	Pragma
- Open-circuit Voltage (Vdc):	6.6	13.1
- Short-circuit Current (A):	6.9	6.4
- Maximum power (W):	33	55

5.1.3 Array Support Structure and Foundations

Figure 8 shows the modular type of supporting structure used. Mechanical differences between the Ansaldo and Pragma modules required a slightly different mounting configuration. Each structural segment, weighing 250 kg, can support either 16 Ansaldo or 8 Pragma modules. The segments are made of galvanized steel in conventional profiles. One hundred and sixty-eight of these module support segments are used in the PV array field. Each structure has its own prefabricated foundation, a steel-reinforced concrete slab weighing about two tons (see Fig. 9).

5.2 Batteries and Charge Control Method

The 1,500 Ah lead acid battery consists of 130 Varta Vb 2415 cells in series. The cells are installed in a ventilated building separated from the control building.

Battery charge control is provided by the system controller. For charge control the central controller provides a constant voltage charging whenever the battery reaches the pre-set voltage limit. It also considers the battery SOC which is calculated by the system controller.

5.3 Power Conditioning

Two inverters are provided, a self-commutated type for stand-alone operation and a line-commutated type for the grid-connected mode. The self-commutated inverter has the following characteristics:

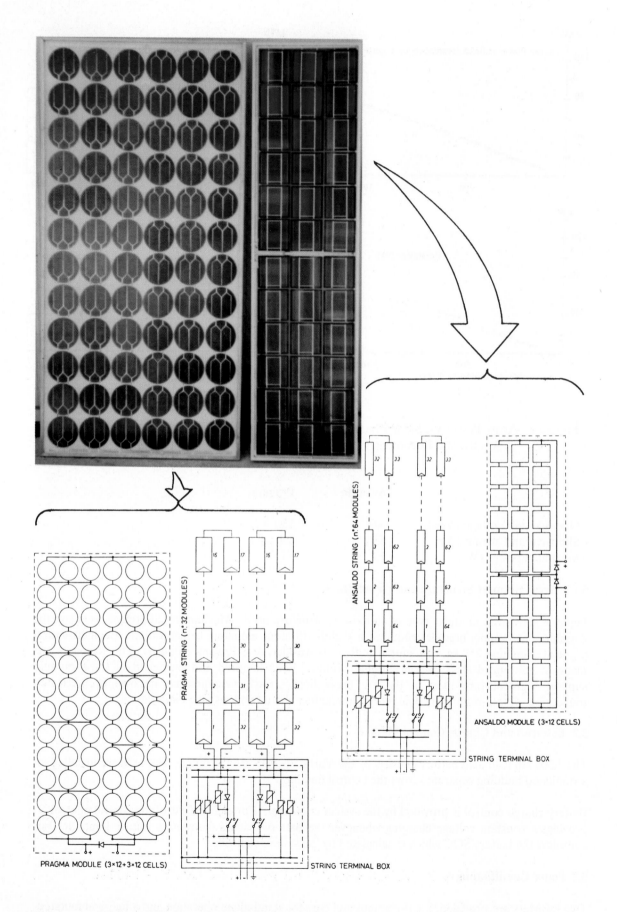

Figure 7. Wiring of Pragma and Ansaldo Modules

434

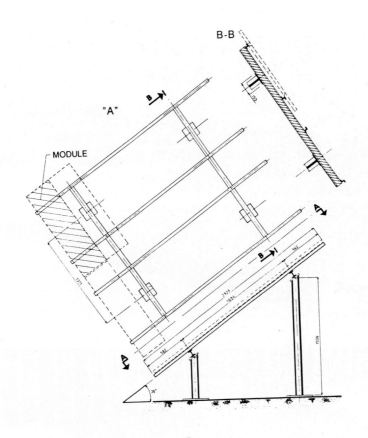

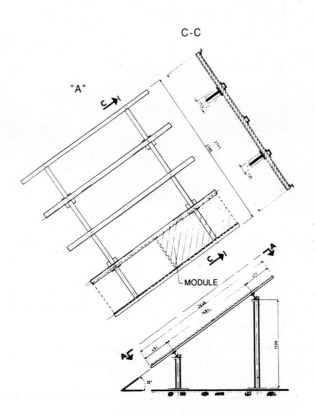

Figure 8. PV Module Support Structures at VULCANO PV Pilot Plant

Figure 9. Pre-fabricated Concrete Foundation for the Array Structures

- Rated power: 40 kVA
- dc input voltage: 230 to 310 Vdc
- ac output voltage: 380 V ± 1% 3-phase and neutral
- THD: Less than 3%
- Surge capability (10 s): 4 x nominal current (Inom) at voltage not less than
                          0.7 times nominal voltage
- Current limitation (30 mins): 1.5 times Inom

The inverter uses power transistors with pulse-width modulation. Its efficiency curve recorded at the VULCANO site is given in Figure 10.

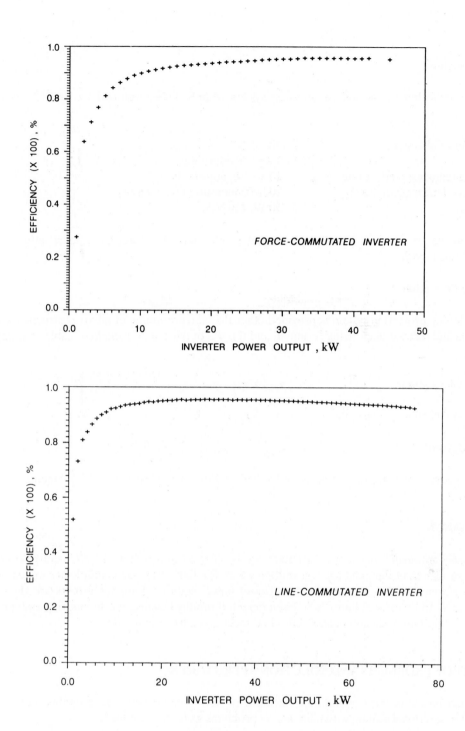

Figure 10. Efficiency vs. Output Power for Self-commutated and Line-commutated Inverters

The line-commutated inverter is designed to extract the maximum power from the PV array with a built-in MPPT. The MPPT operates two 3-phase bridge rectifiers connected in series. Its firing angle is controlled such that the THD of the output current is no more than 5%. A suitable filter is also provided to reduce the ripple on the array output voltage. The conversion efficiency of the inverter is at least 90% from 30 to 100% of rated power output (see Fig. 10).

5.4 Control System

The Contronic-3 H&B programmed controller is used in both stand-alone and grid-connected operation.

5.5 Data Collection

Plant data are collected by the data acquisition system Mini SAD/3 developed by ENEL. Its key features are:

- Total number of channels: 50
- Scan rate: 1 s - 1 h, adjustable
- Averaging, integrating, and storing: 10 s - 1 h, adjustable
- Cassette recorder/magnetic tape: 100,000 measurement capacity
- Power source: 50 W, 110 Vdc

If 25 measurements are stored at 15-minute averaging or integration intervals, the tape cassette would be filled up in one month.

5.6 Lightning Protection

No special lightning protection system is provided other than the grounding of all array structures, the use of varistors and fuses at each string terminal, and the use of twisted and shielded cables for signal transmission.

5.7 Auxiliary Generators

The plant has no auxiliary generators.

5.8 Load Equipment

The loads are conventional appliances in the individual residences and business establishments.

6. MAINTENANCE

The plant is fully automatic, but its operational state is checked in a remote ENEL control station, where the major alarms of the plant are transmitted via an RT link. Regular maintenance and simple repairs are performed by the personnel at the Diesel power station. Difficult repairs are done by ENEL specialists. In the past, failures have been detected rapidly because the monitoring system has some diagnostic and fault indication capabilities and local specialists are nearby.

7. SUMMARY OF PROBLEMS AND SOLUTION APPROACHES

The termination boxes on the Pragma modules have proven to be sensitive to humidity and water seepages. After additional sealing with silicone, no problems were encountered.

During the summer of 1985, some component breakdowns occurred due to overheating. As a simple fix, a temperature-controlled fan and a heat shield were installed in the inverter cabinet.

There is some indication that in stand-alone operation the battery may not be sufficiently charged during the winter months. This possible small problem can be remedied by giving priority to battery charging, by reducing the power fed to the grid or by using the rectifier.

8. KEY LESSONS LEARNED

The key lessons learned are as follows:

- Although the basic characteristics of the module were not affected, the terminal boxes attached to the modules were not designed properly to prevent water seepage.

438

- The natural washing of the PV modules by rain is sufficient to keep the modules clean. Thus, the water collection system installed in the array structures was considered unnecessary.

9. CONCLUSIONS

Since October 1984, the plant has operated very well with only a few shut-downs for the adjustment of power electronics and the installation of new hardware. With the installation of the line-commutated inverter in June 1986, the plant is now fully capable of both stand-alone and grid-connected modes of operation. The annual production of energy by the PV plant (June 1985 to May 1986) was about 90.5 kWh.

A reliable method of on-line evaluation of the battery SOC is useful for certain applications. The SOC calculation method installed in the H & B controller has shown that the math model is not accurate. A new math model will be installed, based on the test results of the ad hoc experimental facility (see Annex A).

10. REFERENCES

[1] A. Buonarota, P. Menga, P. Ostano, and V. Scarioni, "An Application of On-line Battery Monitoring to the Vulcano PV Plant," Proc. of the 8th European PV Solar Energy Conference, Florence, IT, May 1988.

[2] G. Blaesser, K. Krebs, and W.J. Zahiman, "Power Measurements at the 80-kW PV Plant Vulcano," Technical Note No I.88.54, JRC-Ispra, IT, May 1988.

[3] Vulcano Project Handout, Pilot Plant Optimization Meeting, Brussels, BE, 12-13 May 1986.

[4] P. Helm, "Manual of Photovoltaic Pilot Plants, Parts A and B," WIP report under CEC Contract EN 3 S-0007-D(B), Munich, FRG, February 1986.

[5] A. Iliceto, A. Invernizzi and A. Previ, "Design, Construction and Operation Experience of a Photovoltaic Plant for Village Power Supply," Proc. of CIGRE-UPDEA Symposium, Dakar, Senegal, Report No. 330-10, 25-27 November 1985.

[6] G. Belli, A. Iliceto, and A. Previ, "Design, Installation and Preliminary Operational Results of the Vulcano and Adrano Photovoltaic Projects," Proc. of the 6th European PV Solar Energy Conference, London, GB, 15-19 April 1985.

[7] G. Blaesser, K. Krebs, H. Ossenbrink and E. Rossi, "Acceptance Testing and Monitoring the CEC Photovoltaic Plants," Proc. of the 6th European PV Solar Energy Conference, London, GB, April 1985.

[8] A. Previ, "The Vulcano Project," Int. J. Solar Energy, Vol. 3, pp. 124-141, 1985.

[9] Vulcano Project Handout, Pilot Programme I Meeting, Tremiti, Island, IT, 20-21 June 1984.

[10] A. Iliceto, A. Previ, R. Buccianti and B. Morgana, "Metodi di caratterizzazione dei moduli fotovoltaici per la conversione diretta da energia solare in energia elettrica," Memoria No. 145, Riunione annuale AEI, Riva del Garda, IT, 1984.

[11] G. Ghiringhelli, F. Russo, "Acquisizione dati nell' ENEL," Automazione Oggi, No. 10, p. 34, IT, November 1983.

[12] Vulcano Project Handout, Pilot Programme I Meeting, Mont Bouquet, FR, 2-14 April 1983.

[13] Vulcano Project Handout, Pilot Programme I Meeting, Crete, GR, 14-15 September 1982.

[14] Vulcano Project Handout, Pilot Programme I Meeting, Brussels, BE, 20-21 April 1982.

[15] V. Arcidiacono, S. Corsi and L. Lambri, "Maximum Power Point Tracker for Photovoltaic Power Plants," Proc. of the 15th IEEE Photovoltaic Specialists Conference," San Diego, CA, USA, 1982.

[16] A. Brizzi, G.L. Fracassi, A. Invernizzi, G. Manzoni, "Alimentazione di un carico isolato con fonti di energia rinnovabili," Rassegna Tecnica ENEL, Vol. 2, IT, 1982.

[17] A. Taschini, "Alicudi Project," Proc. of the Final Design Review Meeting on EC Photovoltaic Pilot Projects held in Brussels, BE, 30 November - 2 December 1981.

ANNEX A. AN APPLICATION OF ON-LINE BATTERY MONITORING [1]

Reliable knowledge of the state of charge (SOC) of the battery of a PV plant can contribute to improvement of the system management.   Unfortunately, the technologies currently adopted to determine the battery SOC are not fully satisfactory. The experience obtained by ENEL (the Italian Electricity Agency) on traction lead-acid batteries, operating under cyclic conditions, led to the formulation of a simple model capable of describing the relationships between the operating parameters and the internal SOC of the battery.   This model was extended to the stationary accumulators at the VULCANO PV plant and checked by means of laboratory tests at CESI. On this basis, an automatic system for the on-line evaluation of the SOC of the battery has been recently set up and installed at Vulcano. This paper presents the basis of the methodology, the layout of the system, and the preliminary experience obtained so far. This work was co-financed by the CEC-DG XII (Contract EN3S-0135-I (A)).

1. INTRODUCTION

It is well recognized that knowledge of the battery SOC is necessary both for optimizing the management of the PV plant and for avoiding stresses on the battery which would result in reduction of its lifetime. Various methods are used for the determination of SOC in conventional applications. They include electrolyte density, open-circuit voltage, and ampere-hour balancing [A1, A2].

The electrolyte density of the lead-acid battery is in principle related directly to SOC. However, operation under cyclic conditions (typical of PV or traction) prevents the attainment of equilibrium in the mass of the electrolyte, thus reducing somewhat the reliability of density measurements for SOC predictions. Similarly, since accurate determination of the open-circuit voltage (Voc) at equilibrium would require a long rest period, use of the Voc for SOC estimation is not reliable.

Another method often employed is computing the ampere-hours balance of the battery. In this case limitations arise from the fact that the charge acceptance efficiency is lower than unity, especially when approaching the gassing threshold. This error integrates over time and leads to important inaccuracies in the SOC calculation. The measurement of battery voltage is a very simple and widespread practice. Unfortunately, the SOC is not a function of the terminal voltage alone. It is affected by temperature, DOD, discharge rate, charge rate, cycling durations, and ageing.

The difficulty of describing and forecasting the behaviour and the internal state of the lead-acid battery has stimulated in recent years the development of mathematical models as suitable tools for a better prediction of the battery response.

Many of the formulations proposed [A3, A5] are in the form of equivalent electrical circuits, which are particularly suitable for application in the electrical field. This tool can be easiliy oriented to the evaluation of the battery SOC.

2. BRIEF DESCRIPTION OF THE BATTERY MODEL

The discharge model considered here, which is described in detail in ref. [A3], is based on the equivalent circuit shown in Figure A-1. It consists simply of a generator with an emf E and two resistances R and R_d.

On the basis of experimental results it was found that:

- The quantity E is a function of extracted charge Q, temperature T and average discharge current I_m;

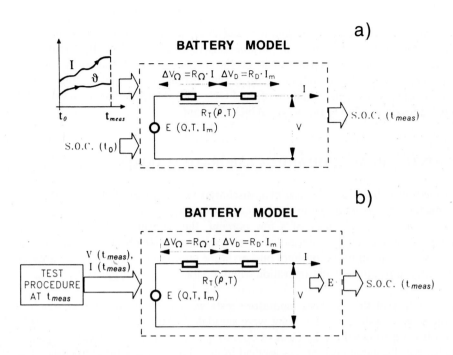

Figure A-1. Battery Models: a) Equivalent circuit with predetermined constants,
b) Equivalent circuit parameters adjusted wih actual V, I data

- The total resistance R is a function of the relative depth of discharge of the battery, taken as the ratio between the extracted charge Q and the battery capacity C at average regime I_m ;

- The component R takes into account the purely resistive voltage drop due to the instantaneous discharge current I;

- The component R_d represents electrolyte diffusion phenomena, whose corresponding voltage drop is evaluated on the basis of the current I_m.

To determine the values of these parameters, it is sufficient to perform simple tests, examining the response of the battery to a current step at various depths of discharge.

This procedure has been carried out on accumulators of the same type as those installed in the VULCANO PV plant; a series of laboratory tests simulating the operation of the battery in photovoltaic plants confirmed that the model gives a satisfactory representation of the behaviour of the storage system.

3. APPLICATION OF THE BATTERY MODEL FOR THE EVALUATION OF SOC

Different approaches can be adopted for evaluating the SOC by the use of the model. A first approach, shown in Figure A1a, consists of using the model in open loop, providing it with two basic inputs: the initial charge stored in the battery at time t_0, and the profile versus time of the operating quantities, namely current and temperature. The model processes these parameters in real time and evaluates the SOC. However, the open loop operation implies the integration of the unavoidable errors of the model, thus gradually degrading the accuracy of the SOC estimate.

A possible solution being investigated is to close the loop through a feedback, comparing the calculated and the actual value of an appropriate output quantity used as a calibrating parameter (generally the voltage at the terminals), and consequently modifying one or more internal elements of the model. The method proposed by the authors (Fig. A1b) uses the model in open loop, in such a way as to calculate the value of its internal parameters (resistances R and R_d , emf E, which are functions of SOC) through the measurement of the actual external quantities of the battery. Specifically, the knowledge of the output voltage, the instantaneous and average currents, and temperature allows the precise determination of the SOC. This principle has been validated through laboratory tests, and then applied to the VULCANO PV plant.

4. DESCRIPTION OF THE METHOD AND OF LABORATORY TRIALS

Our proposed method is based on the use of a discharge model of the battery. That implies that certain measurements are required while the battery is discharging. Moreover, in order to obtain a good degree of accuracy in the order of a few percent, the measurement of battery current and voltage has to be performed during a period of several minutes in steady-state conditions. For these reasons, it is convenient to impress, on-line, one or more controlled current steps large enough to mask the load fluctuations during night operation when the load current is generally small.

To check the reliability of this method, laboratory tests were carried out, during which operating conditions of the storage system of a PV plant were simulated. Figure A-2 gives an example of the daily current profile, reproduced on the basis of operating data from the VULCANO plant. For the sake of simplicity, the discharge current was applied in two steps of constant amplitude. The SOC is determined every night at a pre-established instant of the cycle (t_{meas} in Fig. A-2), by applying to the battery two current steps and evaluating by the model the emf E.

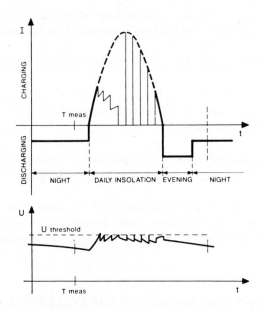

Figure A-2. Current Profile

Figure A-3 reports the results of the tests performed, showing the emf E versus the charge ΔQ extracted from the battery at determined values of I and T, starting from full charge. It shows that in the course of a succession (1,2,...,n) of operating cycles there is a progressive difference ΔQ between the charge value calculated by means of the emf (on the continuous curve) and that calculated on the basis of the Ah balance (abscissa of each dot). As expected, the latter shows an error that increases cycle by cycle.

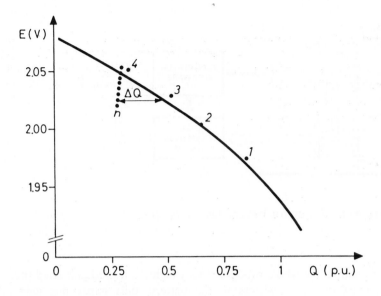

Figure A-3. Voltage vs. SOC

Various experimental checks, repeating the simulation several times for different test conditions (number of cycles, initial SOC) confirmed the reliability of the information supplied by the method, whose error remains within 3% of battery capacity.

These tests were performed on new and aged batteries. Figure A-4 shows that to a certain extent, and when the battery is sufficiently discharged, the prediction of SOC given by the method can take intrinsically into account the effect of battery ageing. In this condition, in fact, it can be seen that, although the evaluated value of the charge ΔQ extracted from the battery is erroneous, the estimate of the residual capacity ΔQ_N is sufficiently close to the actual value ΔQ_A.

5. IMPLEMENTATION OF AN EXPERIMENTAL TEST SYSTEM IN THE PV PLANT

In order to evaluate the reliability of the methodology previously described, an experimental test set-up has been temporarily installed in the PV plant of Vulcano. To avoid interference with the plant and its management, a separate test hardware has been developed and installed in a small prefabricated building. The basic layout of the system is given in Figure A-5.

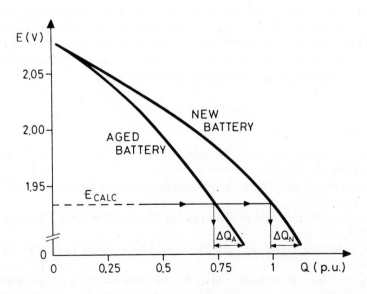

Figure A-4. Effects of Ageing of Battery on emf as a Function of SOC

443

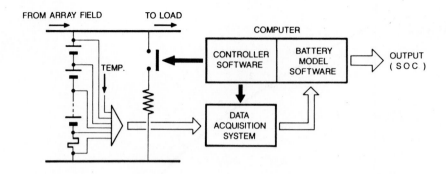

Figure A-5. Set-up for Battery Discharge Test

A personal computer controls the whole process. Every night, at a predetermined time, it closes the contactors which connect the test resistors to the battery, thus impressing current steps with appropriate intensity and duration, and records battery voltage and temperature by means of a data acquisition system. The same computer processes these data into the software that represents the battery model, and returns the SOC of the battery at that moment, as well as additional output information.

In the period between two successive SOC determinations, the evaluation of SOC is based on the use of the battery model in open loop, as no significant error occurs over a short time.

Actually, the individual voltages of several elements of the 130-cell battery are recorded, in order to study the scattering among them and provide a diagnosis of the battery condition. A comparison of cells with and without electrolyte agitation is also made.

The preliminary results after about two months of operation confirm that the system works correctly and returns reliable information. The trial will continue for 1 or 2 years, and will include periodic in-field measurements of the actual charge stored in the battery, in order to verify the accuracy of the SOC estimate given by the system.

6. CONCLUSIONS

A method for accurately evaluating the SOC, based on the use of a simple discharge model, has been developed. It has been implemented in the VULCANO PV plant, in the form of an experimental set-up which will be used for checking the long-term reliability of the approach. We expect to further simplify the method, tailored for industrial applications. We also plan to evaluate a better strategy for power system management on the basis of accurate knowledge of battery SOC.

7. REFERENCES

[A1] R. Giglioli, P. Pelacchi, M. Raugi, G. Zini: "A State of Charge Observer for Lead-acid Batteries," L'Energia Elettrica No. 9, 1987.
[A2] H.P. Schoner: "Battery State Monitoring Using the Voltage Response on Drive Currents," Drive Electric, Sorrento, IT, October 1985.
[A3] N. Polese, G. Betta: "An Equivalent Model of Batteries for Electric Vehicles," Drive Electric, Sorrento, IT, October 1985.
[A4] R. Buccianti, R. Giglioli, P. Menga, L. Thione: "An Electrical Model of the Lead-acid Battery," EVS-7, Versailles, FR, June 1984.
[A5] J. Meiwes, W. Schleuter: "Ein elektrisches Modell zur Beschreibung des Klemmen-spannungsverhaltens von Bleiakkumulatoren," Drive Electric, Amsterdam, NL, October 1982.

Appendix 16

ZAMBELLI PV PLANT

1. INTRODUCTION

This appendix describes the 70-kW ZAMBELLI PV pilot plant and its initial performance. The plant became operational in June 1984. The purpose of this plant is to supply power to the Zambelli drinking water pumping station near Verona and to demonstrate the viability of photovoltaic technology for industrial water pumping uses. Figure 1 shows the array field and the control building.

The information contained herein was obtained from ref. [1 to 12] and individual contacts with the plant owner and the subcontractor.

2. PLANT DESIGNER, OWNER, AND OPERATOR

Pragma supervised the design and installation of the plant and handed it over to AGSM, the municipal utility firm in Verona, in July 1984. Since then, AGSM has been the owner and operator of the plant. The following organization is responsible for its maintenance and operation:

> Azienda Generale Servizi Municipalizzati (AGSM)
> Attn: Renewable Energies Dept.
> Lungadige Galtarossa, 8
> I-37133 Verona
> Italy

3. SITE INFORMATION

3.1 Location

Zambelli is located near Cerro Veronese which is a small town approximately 35 km north of Verona, Italy (see Fig. 2). The PV plant is located nearby on a hilltop. Its coordinates are:

- Latitude: 45°N
- Longitude: 11°E
- Altitude: 830 m

The 70-kW Array field

Dr. Adami and Mr. Spagnolo (from left) of AGSM
inspecting the Pragma 72-cell PV modules

Mr Corradi (left), the local maintenance technician,
and Dr. Zamboni, the project leader from AGSM

Figure 1. Views of the ZAMBELLI Array Field and Control Building

446

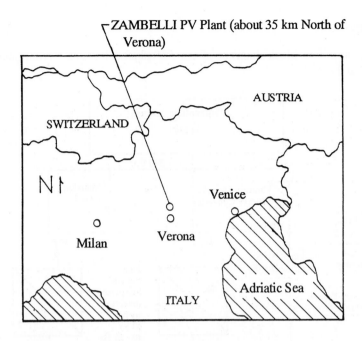

Figure 2. Location of the ZAMBELLI PV Pilot Plant

3.2 Site Constraints

There were no major site constraints.

3.3 Solar Radiation and Climatic Conditions

Typical solar energy and meteorological data at the PV site are as follows:

- Insolation, global horizontal: 1,300 kWh/m^2 annual
- Ambient air temperature: 35°C maximum
- Wind speed: 119 km/h maximum

3.4 Load Demand

The primary loads are two reciprocating pumps of 30 kW each.

4. SYSTEM DESCRIPTION AND PERFORMANCE

4.1 Plant Architecture and Layout

The PV-powered fresh water pumping station plays a key role in the water supply system for the Cessinia mountain area near Verona. The PV field is built near the pumping station building on top of a hill.

Electronic and electrical equipment are located at the station. The water tank has a capacity of 1,030 m^3. Originating from this station, a water pipe about 5 km long, 200 mm in diameter and with a capacity of 480 m^3 leads to another water basin at "Villa Ponti" (1,153 m above sea level). Therefore, a geodetic height of 323 m had to be considered by the pump designer in addition to the water flow resistance of the pipeline.

4.2 System Configuration

Figure 3 shows a simplified block diagram of the PV plant. Its major components and their key features are listed in Table 1.

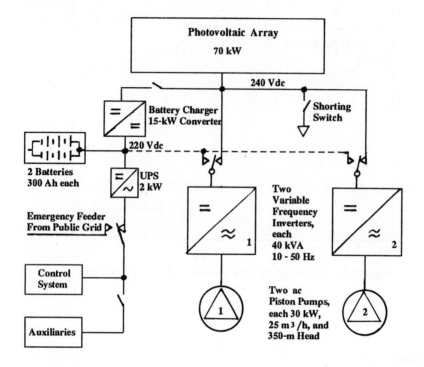

Figure 3. Simplified System Block Diagram of the ZAMBELLI PV Pilot Plant

4.3 System Operation and Management

The system can be operated in manual or automatic modes, entirely independently from the utility grid. The power management is designed to utilize all available array output. At low solar irradiance levels, only the battery is charged, but the pumps start operating one after the other when the solar irradiance increases above certain thresholds.

The plant has the capability of operating in several modes and of controlling the regulation devices for normal operation and special testing. The key ones are as follows:

Main Operating Modes- Battery charging with the UPS

- Pump operation with the UPS battery
- Pump operation with the PV array
- Single-pump and 2-pump operation
- Parallel operation of pumps and UPS battery charger

PV Array Regulation

- Maximum power point tracking
- Constant PV bus voltage operation
- I-V curve generation (entire array) for test purposes

Pump Inverter Regulation

- Inverter input voltage control
- Inverter output frequency control

Table 1. Major Components and Their Features (ZAMBELLI)

Solar Array
Rated Peak Power:	70 kW
Array Tilt Angle:	45°
Tilt Angle Adjustment:	None
Nominal Bus Voltage:	240 Vdc
Number of Subarrays (Strings):	27
Total Number of Modules:	1,296
Number of Modules per String:	48
Module Manufacturer/Type:	Pragma 72 DG/SOL
Module Rating:	56.5 W
Number of Cells per Module:	72 (12 in series, 6 in parallel)
Number of Bypass Diodes:	1 per 3 modules
Module Weight:	15.5 kg
Module Dimension:	1,308 x 658 mm
Module Area:	0.86 m^2
Active Solar Cell Area/Module:	0.565 m^2
Solar Cell Type/Size:	Monocrystalline silicon/10 cm dia.
Cell Packing Factor:	65.7 %
Total Land Area Used:	10,085 m^2

Battery
Total Battery Energy Rating:	120 kWh
Cell Capacity Rating:	600 Ah
Voltage Range:	216 Vdc (nominal)
Allowable Depth of Discharge:	25 %
Number of Batteries:	2
Number of Cells per Battery:	108
Cell Manufacturer/Part No.:	Tudor/4TF230/HC
Cell Weight:	15.4 kg
Cell Dimension:	134 x 222 x 410 mm

Inverter
Rated Power:	80 kVA
Number of Inverters:	2 (40 kVA each)
Type:	Self-commutated (PWM)
Manufacturer:	Silectron
Input Voltage Range:	100-300 Vdc
Output Voltage/Phase:	220 Vac/3-phase (both)
Frequency:	5-50 Hz (variable frequency)
THD:	15 %
Efficiency, % Full Load	
100:	97 %
10 :	90 %

Converter
Nominal Power:	15 kW
Number of Converters:	1
Type:	buck-boost
Manufacturer:	Silectron
Input Voltage:	210-273 Vdc
Output Voltage:	200-270 Vdc
Efficiency, % Full Load	
100:	90 %
10 :	97 %

Controller	Original System	New System
Model:	Z-80-based micro-processor	Since Feb. 1989 ORSI PMC with IBM PC Supervisor
Manufacturer:	Sincon-ROMA	
Features:	Provides MPT of PV array via voltage set point adjustment, manual and automatic	(Same features, with substantial improvements)
Data Acquisition:	Data logging soft-ware (Sincon-ROMA)	TEAM (on-site DAS) SFLTA (Remote monitoring)

Battery Charge Regulation

- Maximum power point tracking
- Maximum voltage limit, adjustable
- Battery state of charge indication

Battery Discharge Protection

- Minimum voltage limit, adjustable

This plant uses two variable frequency inverters. Frequency control allows the matching of power demand to the instantaneously available PV power. Moreover, it has the advantage of increasing the motor efficiency at low speeds. This principle is commonly used in some industrial applications where large motors are required to operate at variable speeds.

4.4 Plant Performance

Based on actual measurements of array string I-V curves in July 1987, JRC verified the PV array field output as 71.3 kW under standard test conditions (1,000 W/m^2, 25°C average cell temperature). No apparent power degradation in the array has been observed. The variable frequency inverter has demonstrated a peak efficiency of 96%.

The controller regulates the power consumption of the plant in such a way that the input voltage to the power conditioning system is held constant at a set point. This set point can be controlled automatically or manually. This is the approach taken to provide the MPPT feature; under this regulation, any sudden drop in the performance of the PV array, such as from a decrease in solar irradiance, causes an immediate deceleration in the speed of one or both pumping units, resulting in the recovery of the input voltage to the pump inverter. This allows a stable operation of the pumping units according to the available PV power.

Table 2 shows monthly performance data for January to June 1990. The information and format of this table are in accordance with the CEC/DG XII criteria established in 1988.

Figure 4 is an efficiency plot of the major elements in the pumping system vs. the inverter output frequency. The largest losses are due to the pump itself while the inverter losses are generally less than 4%.

Figure 5 shows 1-day profiles of solar irradiance on the plane of array, actual PV output power, inverter output power, equivalent power used in pumping water, and battery charge/discharge power for 20 October 1989 and 20 February 1990. The data points are 1-h averages. As indicated by the positive inverter output power (see Fig. 5) at least one pump starts operating when the solar irradiance reaches 100 W/m^2. Also note that inverter output is roughly proportional to the solar irradiance. The plant start-up and shut-down thresholds are actually 50 W/m^2.

Figure 6 shows plots of battery parameters for several days in December 1989 and February 1990. An interesting parameter is the battery SOC. A variety of SOC calculation equipment exist in many other PV plants in Europe, but not much real data have been published. ZAMBELLI's SOC calculator is apparently working as intended, and it can be effective in providing an additional signal for battery protection (i.e., for deep-discharge cutoff).

5. COMPONENT DESCRIPTION

5.1 Solar Array

5.1.1 Array and Cabling

The array has a total of 1,296 modules connected into 27 strings. Each string is composed of 48 modules in series. Modules are mounted in nine rows at a tilt angle of 45°. Array performance specifications are:

Table 2. Monthly Performance Summary, January 1990 to June 1990

PV Pilot Plant:	**ZAMBELLI Pumping Station**	Date of Report:	**07-05-90**	Latitude (°N):	**45**
Site Manager:	**G. Corradi**	Prepared by:	**A. Sorokin**	Longitude °E):	**11**
Project Manager:	**G. Zamboni**	Organization:	**TEAM (Rome, IT)**	Altitude (m):	**830**
Organization:	**AGSM, Verona, IT**				

		1990					
PARAMETER	Units	Jan	Feb	Mar	Apr	May	Jun
1. Solar/Temperature/Wind Information:							
a. Insolation, plane of array	kWh/m^2	102.2	119.1	139.1	104.9	154.6	139.7
b. Insolation, horizontal surface	kWh/m^2	49.8	70.1	110.6	111.7	180.8	164.4
c. Irradiance, plane of array, sunlight avg	W/m^2	.372	.418	.386	.262	.343	.302
d. Irradiance, plane of array, daily peak	W/m^2	1.003	1.118	1.096	1.157	1.181	1.174
e. Irradiance, horizontal, sunligh avg.	W/m^2	.181	.245	.305	.279	.402	.355
f. Temperature, PV module	°C	15	16.6	16.1	10.7	20.8	24.1
g. Temperature, ambient air, peak	°C	11.2	16.3	18.2	17.2	20.8	25.4
h. Temperature, ambient air, daily avg	°C	4.3	7.0	9.3	7.5	14.6	16.3
i. Temperature, ambient air, lowest	°C	-4.7	-2.1	- 0.3	.4	5.4	7.6
j. Wind speed, daily avg.	m/s	2.5	3.9	5.1	6.8	7.1	5.3
k. Wind speed, daily peak	m/s	23.9	25.9	32.0	30.6	31.8	25.6
2. Plant Output Information:							
a. Energy array field	kWh	2183	3738	4694	2196	4458	6567
b. Energy, inverters	kWh	1920	3193	3571	1913	4354	6303
c. Power, array field, sunlight avg	kW	7.9	14	15	5.6	10	14.3
d. Power, array field, daily peak	kW	41.8	65	62.2	63.3	64.7	65.2
e. Julian day of peak occurrence	1-365	21	51	63	118	130	162
3. Plant Performance Indices							
a. PV array energy capability	kWh	7194	8323	9555	7317	10664	9557
b. Array utilization factor	%	30.3	44.9	49.1	30	41.8	68.7
c. Array energy efficiency	%	1.91	2.8	3.01	1.87	2.57	4.2
d. Inverter energy efficiency	%	88	85.4	76.1	87.1	97.7	96
e. System energy efficiency	%	1.68	2.39	2.29	1.63	2.51	4.03
f. Plant availability	%	100	96.2	76.7	60.5	84.9	91.7
4. Battery Information							
a. DOD, % rated Ah, daily avg	%	14.3	13.7	12.8	17.6	12.8	14
b. Nol. cycles*	--	29	23	23	23	27	32
c. Cell temperature, avg.	°C	6.3	7.5	10.1	10.1	16.8	18.5
d. Daily avg. recharge factor	--	.84	.87	.94	1.13	1.04	1.05
e. Lowest state of charge reached	%	51.7	63.5	50	40.3	66.5	75.4
5. Operation & Maintenance Information:							
a. % total time DAS operational	%	99.3	94.5	89.7	96.8	95.6	98.9
b. % total time inverter oper.: auxiliary	%	99.3	94.5	89.7	96.8	95.6	98.9
pumping	%	8.1	20.9	25.5	10.7	24.2	42
c. Number of key failures	--	--	--	--	--	1	--
d. MTTR	h	--	--	--	--	240	--
e. Sunlight hours:							
1) Daily avg, actual	h	7.8	8.5	10.4	10.8	12.3	12.6
2) Daily avg., theoretical	h	8.9	10.1	11.7	13.3	14.6	15.4
3) Total actual	h	242	238	322	324	381	378
4) Total, theoretical	h	276	283	363	399	453	462
6. Appliction Information							
a. Consumer load energy	kWh	1920	3193	3571	1913	4354	6303
b. User energy supplied by PV plant	kWh	1920	3193	3571	1913	4354	6303
c. Total energy from grid to PV plant	kWh	--	--	--	--	--	--
d. Total water volume pumped (pumping head 37 bar)	m^3	1099	2308	2998	1055	2712	4290

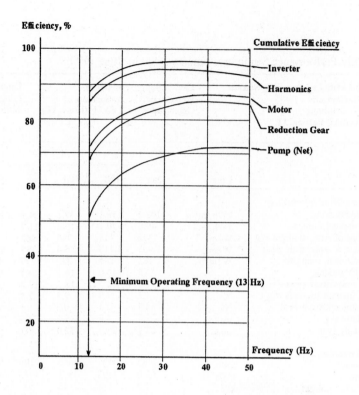

Figure 4. Efficiency of Pumping System Elements as a
function of Inverter Output Frequency

- Rated power: 70 kW
- Nominal voltage: 240 Vdc
- Nominal current: 290 Vdc

Total power losses due to module mismatch, diodes, and cables were estimated at 4.5%. Each array string is connected to the array power collection panel in the control building.

5.1.2 Modules

The modules manufactured by Pragma contain 72 circular monocrystalline Si cells of 10 cm diameter. The specifications for each module at 1,000 W/m^2 and 25°C cell temperature are:

- Rated peak power: 56.5 Wp
- Voltage at peak power: 5 Vdc
- Current at peak voltage: 12 Adc

5.1.3 Array Support Structures and Foundations

The material used in the panel frame is hot galvanized steel (GS) according to the Italian standard CNR UNI. The array structures and foundations are shown in Figure 7. Mounting costs were minimized as far as possible by using standard steel components which can be easily assembled.

5.2 Batteries and Charge Control Method

Two batteries are connected in parallel. Each battery consists of 108 Tudor 600-Ah cells in series. The batteries supply power to the auxiliary loads and the peak inverter demand during the pump start-up.

452

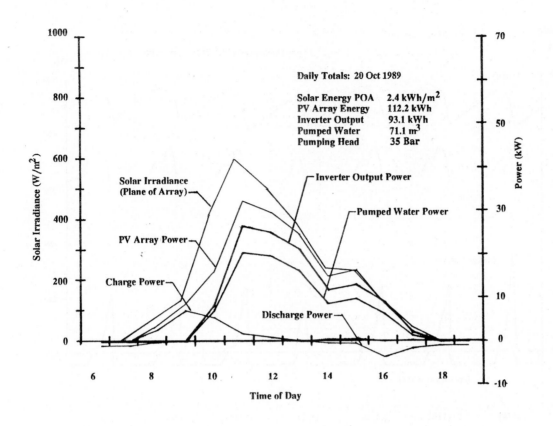

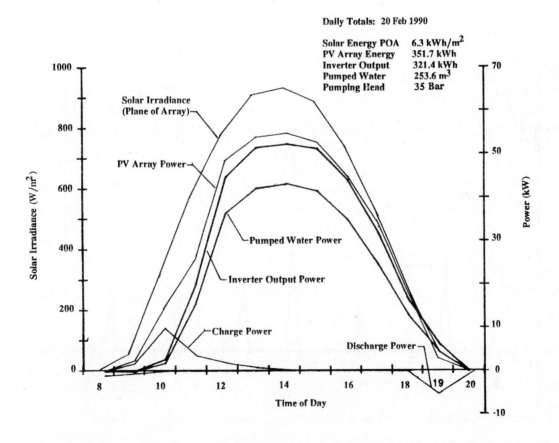

Figure 5. Plant Performance Profiles for 20 October 1989 and 20 February 1990.
Data points are 1-h averages.

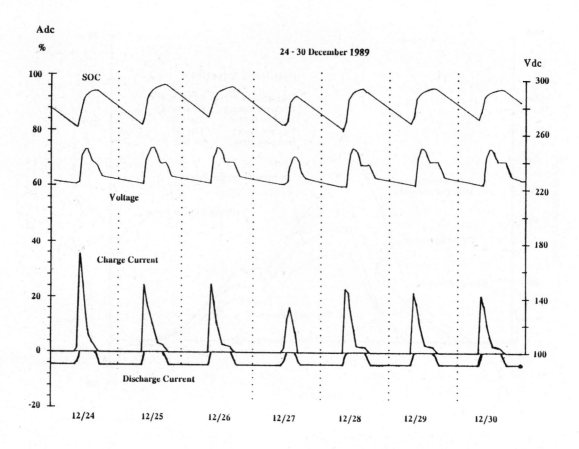

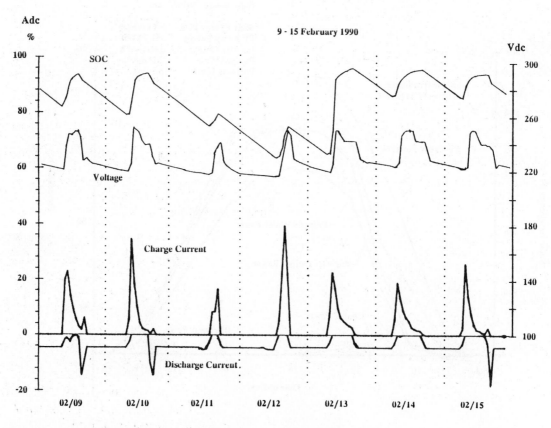

Figure 6. Profiles of Battery SOC, Voltage, and Current for Several Consecutive Days in December 1989 and February 1990.

Mr. Sorokin (TEAM) pointing at the
temperature sensor (PT 100)

Electrical connection box on top of
the concrete foundation

Figure 7. Array Structures and Foundations at the ZAMBELLI PV Pilot Plant

Battery charge voltage regulation is performed by the dc-dc converter. The automatic plant supervisor controls the converter which charges the two batteries in either MPPT or constant voltage mode (the charge current can vary until batteries reach their upper voltage limit). It also provides deep discharge protection by cutting off the load at a predetermined low-voltage limit. Both the charge voltage and discharge voltage limits are manually adjustable at the control panel.

5.3 Power Conditioning

The 15-kW converter used for battery charging has both constant voltage and current limits which are both adjustable.

Two 40-kVA self-commutated inverters use high power transistors and operate with pulse-width modulation. They have variable frequency output which can be adjusted manually or automatically via the central controller.

The motor speed, and consequently the power demand, can be adjusted by changing the inverter output frequency. The power consumption of the pumps is matched to the PV array output via this control. An internal control circuit is also provided to operate the solar generator at the maximum power point. If a defect occurs in the pump motor, the inverter is automatically shut down.

5.4 Control System

The original automatic control system was replaced in early 1989 (see Fig. 8) by the following industrial process controller and peripheral equipment:

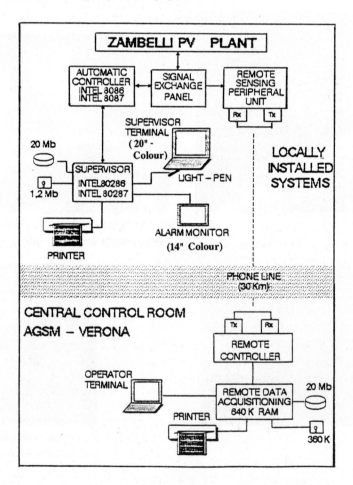

Figure 8. New Controller and Remote Monitoring Systems

456

- TEAM-ORSI PMC Controller
- IBM-PC supervisor
- Video Display Terminal
- Winchester
- Floppy Disk
- Printer

5.5 Data Collection

Since the beginning of 1989, data acquisition is performed by two separate monitoring systems: the local automatic control and supervisor system and a remote sensing and data monitoring system. The latter is connected to the AGSM central control room in Verona via a modem (see Fig. 8).

5.6 Lightning Protection

Since the area is not subject to lightning according to the available information, the plant is not protected against direct lightning strikes. Varistors provide protection against induced overvoltage.

5.7 Auxiliary Generators

The system has no auxiliary generators. Grid power is available for the critical control and monitoring equipment in the event of loss of PV power.

5.8 Load Equipment

The principal loads on the PV plant are two piston pumps driven by standard ac motors. Pump features are:

- Delivery head   37 bars
- Delivery flow rate (at 50 Hz)   27.5 m^3/h each
- Pump speed (at 50 Hz)   400 rpm
- Motor type   asynchronous (squirrel cage)
- Power consumption (at 50 Hz)   30 kW
- Pump efficiency   83%
- Motor efficiency   94%
- Gear efficiency   97%

To reduce its peak current demand during start-up, each pump is provided with a recycling bypass, which is kept open until the pump has reached minimum operating speed. Furthermore, the pumping system is equipped with a fault detection and interlock system, which turns off the inverters and signals the process controller in case of a malfunction.

6. MAINTENANCE

The technical personnel of AGSM, who is responsible for the operation and maintenance of the plant, checks the system on a weekly basis. The maintenance of the components is performed by AGSM, TEAM, and equipment suppliers, as needed.

7. SUMMARY OF PROBLEMS AND SOLUTION APPROACHES

During the first period of operation, the plant suffered frequent failures of the power transistors in the variable frequency inverters. These failures were due to uneven current distribution among paralleled inverter stages and back-emf from motors during deceleration. However, Silectron (power conditioning manufacturer) managed to solve the problem in the spring of 1985.

457

The original automatic plant control and data monitoring system proved to be unreliable with frequent hardware failures. As a result, most of the plant performance data were lost. Furthermore, during unattended automatic operation, failures tended to trigger off alarms resulting in automatic plant shut-downs, requiring reset/restart by the maintenance staff. However, since no data were transmitted from the plant to the AGSM headquarters (30 km away), the operational status of the system was not immediately available to the AGSM staff. As a result, it became difficult to run the plant on a continuous basis.

To increase the reliability of the system the entire automatic plant control and data monitoring system was replaced in early 1989 by a new INTEL 8086/8087 μp-based industrial local controller and a separate remote sensing and data acquisitioning system. This allowed the monitoring of the plant directly from the AGSM headquarters in Verona via modem.

An indirect lightning strike in the spring of 1989 caused severe damages to the electronics (inverter, converter, and data acquisition). Complete mechanical overhaul of the piston pumps was done in January 1990. In the past only small preventive maintenance had been performed on these pumps.

Four battery cells all indicated partial internal shorts and/or low capacity. The manufacturer rejuvenated these cells by replacing the electrolyte and subjecting the cells to slow charge/discharge cycles. They were then reinstalled into the battery banks.

The performance results of the improved system configuration in 1989 and 1990 indicated that most of the system operational problems and monitoring difficulties have been resolved. However, some instability in the voltage regulation and maximum array power tracking of the variable frequency inverters still occurs. The investigation of this problem will be done in the CEC JOULE programme.

8. KEY LESSONS LEARNED

Following initial enthusiasm relating to the new photovoltaic technology, the manufacturer of the power conditioning system (Silectron) lost interest in continuing developments in these components for PV application. Accordingly, it became very difficult to obtain the necessary technical assistance for maintenance and improvements in the power conditioning system.

To avoid similar problems in the future, it is recommended that standard equipment be used wherever possible. In the case of components designed specifically for PV application, only manufacturers showing a strong interest in maintaining their equipment during the long operational periods should be selected. Also, detailed documentation, especially for maintenance use, should be requested at the time of contract award.

For unmanned automatic installations, it is advisable to provide remote monitoring of alarms in case of malfunction. Otherwise, plant stoppages can go undetected and operation inevitably becomes discontinuous.

Tests carried out by JRC [3] in June 1987 (three years after installation) indicated that the only measureable degradation in the PV array was due to contact corrosion of electrical parts in the terminal boxes located outdoors.

9. CONCLUSIONS

Since the main hardware for the ZAMBELLI PV pumping station was designed and selected in 1983, several PV equipment manufacturers have developed standard commercial PV pumping systems of similar conceptual design (directly connected variable frequency inverter, such as Grundfos, AEG Solarverter etc.). Accordingly, there is little doubt that this basic concept deserves attention and industrial exploitation.

The system was designed as a pilot plant for experimental and demonstration purposes, and for that reason, adequate scientific evaluation of its performance is still needed. A significant improvement in the data monitoring system was made in 1989. This will permit a detailed assessment of plant performance in the future. Some of the unique plant design and performance features are:

- Battery state of charge (SOC) which is calculated in real time by the computer. It is used by the controller to avoid deep discharging of the battery. The SOC software has proven to be reliable so far.

- Acquisition of the I-V curve of the entire PV array is automatical done by operator command.

- The plant has a low start-up threshold (50 W/m^2).

- System and component efficiencies are very high (inverter efficiency > 96%, pump efficiency > 75%, overall pumping efficiency > 70%).

- The system design allows the operation of inverters tied directly to the PV array, the battery, or the converter/battery, so a comparison of these operating modes is possible.

- MPPT (Maximum Power Point Tracking) may be performed by either the converter (UPS battery charger) or one of the two variable frequency inverters.

- The following operating modes are possible both automatically and manually:

  o Inverter frequency regulation to control pump motor speed
  o PV bus and battery voltage regulation
  o Constant current discharge of battery for test purposes (the discharge rate can be adjusted manually)

10. REFERENCES

[1] A. Sorokin and G. Adami, "Zambelli PV Plant Optimization," Proc. of the 9th European PV Solar Energy Conference, Freiburg, FRG, September 1989.
[2] D. Braggion and A. Sorokin, "Zambelli PV Plant Optimization," Contractors' Meeting, Brussels, BE, January 1988.
[3] G. Blaesser, W. Zaaiman, "Power Measurements at the 70-kW PV Plant Zambelli," Technical Note I 87.95, CEC-JRC, Ispra, IT, July 1987.
[4] D. Braggion and A. Fonzi, "Photovoltaic Pumping Station for Fresh Water Supply," Project Paper, Pilot Programme Meeting, Brussels, BE, 12-13 May 1986.
[5] Zambelli Project Handout, Pilot Programme I Meeting, Brussels, BE, 12-13 May 1986.
[6] P. Helm, "Manual of Photovoltaic Pilot Plants, Parts A and B," WIP report under CEC Contract EN 3 S-0007-D(B), Munich, FRG, February 1986.
[7] S. Merlina, G. Peluso, C. Rossati, and A. Sorokin, "Pumping Station for Fresh Water Supply," Int. J. Solar Energy 1985, Vol. 3 pp. 164-172.
[8] Zambelli Project Handout, Pilot Programme I Meeting, Tremiti Island, IT, 20-21 June 1984.
[9] Zambelli Project Handout, Pilot Programme I Meeting, Mont Bouquet, FR, 12-14 April 1983.
[10] Zambelli Project Handout, Pilot Programme I Meeting, Crete, GR, 14-15 September 1982.
[11] Zambelli Project Handout, Pilot Programme I Meeting, Brussels, BE, 20-21 April 1982.
[12] G. Germano, "Pump Station for Fresh Water Supply," Proc. of the Final Design Review Meeting on EC Photovoltaic Pilot Projects held in Brussels, BE, 30 November - 2 December 1981.

Appendix 17

ADRANO PROJECT

A. ILICETO and A. TASCHINI
ENEL
Via A. Volta 1
I - 20093 Cologno Monzese (Milan)
Italy

Abstract

The purpose of the ADRANO Photovoltaic Project was to permit direct on-site testing and comparison of different PV generators. The project includes the design, fabrication, installation, and testing of four 2.5 - to 3-kW blocks, representing two different types of fixed flat-plate array and one type of sun-tracking array. Each block consists of PV modules, sun-tracking devices, support structures, wiring, protection devices, the power conditioning system (MPPT-inverter), and the related instrumentation. Therefore each block is a fully independent PV power source tied to a 380-Vac, 50-Hz three-phase line. Furthermore, two experimental systems, one a 200-Wp module with GaAs cells and the other a 200-Wp panel with amorphous silicon cells are installed in the same area for the testing of these particular technologies.

1. INTRODUCTION

The photovoltaic plant at Adrano is a test installation for photovoltaic generators. Constructed on an area adjacent to the Eurelios heliothermal power station, which is equipped with a tower and an array of mirrors, the plant was intended to compare six different photovoltaic array configurations, cell technologies (monocrystalline, polycrystalline, amorphous silicon, and gallium arsenide), as well as their support structures (fixed, with seasonal adjustment, or with single- and double-axis tracking).

Despite the difference in power between the Eurelios station and the photovoltaic test installation, it will also be possible to compare the conversion efficiency, the annual utilization hours of the installed generating capacity, and the reliability of both systems.

Furthermore, the new plant includes an experiment in the "residential" application of photovoltaic technology, both in the size of the PV array and the method of direct hook-up to the utility grid via an MPPT-inverter.

461

The plant is designed for fully automatic operation. The data acquisition system monitors the performance of the various systems and collect the operating data on the maintenance and reliability of the different systems.

2. GENERAL PLANT DESCRIPTION

The plant consists of six completely different systems, installed over an area of about 4,000 m^2 (see Fig. 1). Four of these systems, which have a 2.5 to 3-kW peak power, comprise different flat-plate modules with silicon cells. The modules are electrically interconnected in series and parallel, and each string is connected to a terminal box in the field, containing power distribution devices and protection equipment. Bipolar cables connect each terminal box to a general electrical control and measurement panel. Connected to this panel are four sets of power-conversion and conditioning equipment that feed power into the local 20-kV grid via a transformer. The cross-sections of the power conductors were chosen to handle a current density up to 0.8 A/mm^2.

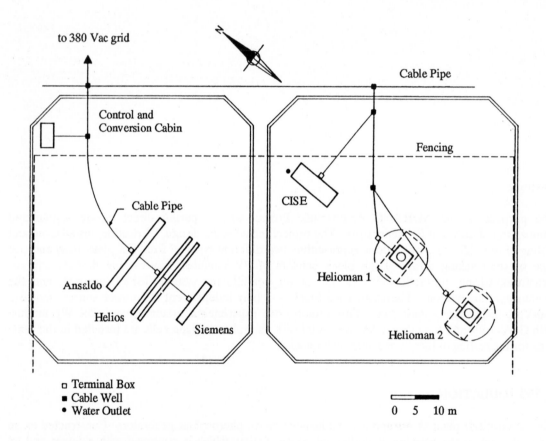

Figure 1. General Layout of the Plant

The electrical control panel, the conversion equipment, and the cassette data acquisition system are installed in a cabin behind the array area.

The plant's overall wiring system for these four systems is given in Figure 2. There are also two other experimental systems in the plant, one consisting of a parabolic mirror concentrator, with a 200-Wp gallium arsenide photovoltaic receiver, and the other, an array of modules with amorphous silicon cells, totalling 200 Wp. The six systems have 220/380 Vac operating outlets, and a water faucet for washing the panels. The plant is equipped with meteorological and other sensors, to record electrical and thermal parameters.

462

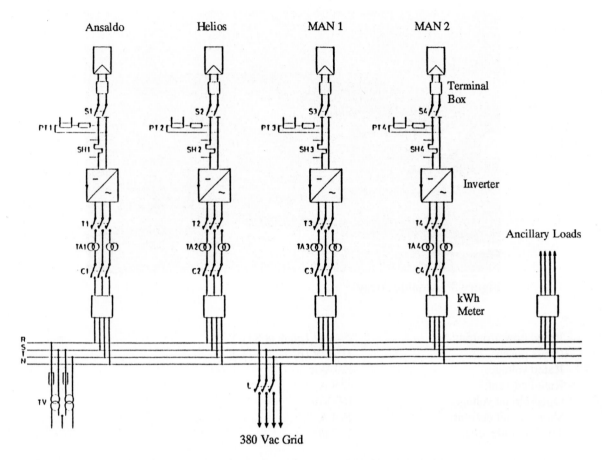

Figure 2. Interconnection of the Four PV Systems to the Utility Grid

3. TECHNICAL PLANT DATA

A block diagram of the four power systems, referred to as the Ansaldo, Helios, MAN 1, and MAN 2, is shown in Figure 2.

3.1 Ansaldo Photovoltaic Generator

The Ansaldo array comprises 88 modules with polycrystalline silicon cells (see Fig. 3). Each module has the following characteristics:

- Dimensions:	1298 x 343 x 41 mm
- Weight:	10 kg
- No. of 100 x 100-mm cells, (connected in series):	36
- Encapsulating material:	front glass/silicon/rear glass with A1 frame
- Rated power:	36 W (peak at STC)
- Rated voltage:	16.3 Vdc
- Rated current:	2.15 A
- Open-circuit voltage:	20 Vdc
- Short-circuit current:	2.5 A

Eight modules are connected in series per string. Each string leads to a block diode and then to a pair of busbars, which connects the 11 strings in parallel. Surge-arrestors (varistors) and a bipolar output switch are also included. These devices are housed in a waterproof box behind the array. The photovoltaic array has the following electrical characteristics:

Figure 3. Ansaldo Array

- Rated power output: 2941 W (peak at STC)
- Gross power: 3168 W (peak at STC)
- Rated voltage: 129 Vdc
- Rated current: 22.8 A
- Open-circuit voltage: 160 Vdc
- Short-circuit current: 26.4 A
- Total module area: 39.5 m^2

The modules are assembled on 11 flat-plate panels on a single structure with tilt angle adjustment capability (see Fig. 3). The total weight of each supporting panel (for 8 modules) is 180 kg. Both the modules and structures were supplied by Ansaldo.

3.2 Helios Photovoltaic Generator

This consists of a fixed flat array of 88 modules, with monocrystalline silicon cells (see Fig. 4). Each module has the following characteristics:

Figure 4. Helios Array

- Dimensions: 1041 x 406 x 47 mm
- Weight: 5.6 kg
- No. of 4" (diameter) cells,
  (connected in series):   36
- Encapsulating material: front glass/EVA tedlar with A1 frame

- Rated power:	32 W (peak at STC)
- Rated voltage:	15.5 Vdc
- Rated current:	2.13 A
- Open-circuit voltage:	19.5 Vdc
- Short-circuit current:	2.5 A

The modules are connected in series and in parallel in accordance with the same wiring arrangement as the Ansaldo generator, with terminals in a similar box containing disconnector and protection devices. The photovoltaic array has the following electrical characteristics:

- Rated power output:	2612 W (peak at STC)
- Gross power:	2816 W (peak at STC)
- Rated voltage:	130 Vdc
- Rated current:	20 A
- Open-circuit voltage:	164 Vdc
- Short-circuit current:	26.4 A
- Total module area:	37.2 m^2

The mechanical support structure allows for periodic adjustment of the angle of inclination. The total weight of 8 modules is 60 kg. The modules were subdivided into strings in the same way as the Ansaldo array. Both the modules and supporting structures were supplied by Helios Technology.

3.3 Helioman Photovoltaic Generators

The two identical arrays are made up of 144 flat modules each, mounted on a 2-axis sun-tracking structure (see Fig. 5). The modules, supplied by AEG, use polycrystalline silicon cells. Each module has the following characteristics:

Figure 5. MAN Heliostat

- Dimensions:	563 x 459 x 11 mm
- Weight:	3.85 kg
- No. of 100 x 100-mm cells, (connected in series):	20
- Encapsulating material:	glass/PVC/glass, with stainless-steel frame
- Rated power:	19.2 W (peak at STC)
- Rated voltage:	8.73 Vdc
- Rated current:	2.2 A
- Open-circuit voltage: ·	11.2 Vdc
- Short-circuit current:	2.41 A

Sixteen modules are connected in series per string. Each string with a blocking diode is paralleled with all others at the dc bus and protected against overvoltages. The resulting photovoltaic array has the following electrical characteristics:

- Rated power output: 2466 W (peak at STC)
- Gross power: 2765 W (peak at STC)
- Rated voltage: 132.2 Vdc
- Rated current: 18.6 A
- Open-circuit voltage: 169 Vdc
- Short-circuit current: 21.7 A
- Total module area: 37.2 m^2

The mechanical support and handling structure consists, for each of the two heliostats, of a column mounted on foundations, a rotating head, a leverage mechanism for raising the panels, and a platform for fixing the modules. The platform can rotate on the vertical and the horizontal axes (azimuth and elevation). The automatic tracking system consists of two solar sensors, an electronic control switchboard, and two 24-Vdc motors powered by an external battery.

In the evening the structure automatically moves into the rest position (horizontal), with the front of the modules pointing downwards. The same procedure happens whenever the wind speed exceeds the safety level. Furthermore, high diffused irradiance levels often cause the platform to move automatically into the horizontal position with the modules pointing to the sky. The structure was supplied by MAN (F.R. Germany) and the modules by AEG (F.R. Germany).

4. POWER CONDITIONING AND CONVERSION EQUIPMENT

Each of the four generators is connected independently to a general electrical switchboard, where control measurements and disconnectors are installed.

Electrical power is routed from the general control board to a line-commutated inverter. This inverter extracts the maximum power available from each photovoltaic array by an appropriate firing of the thyristors forming the three-phase inversion bridge.

The equipment inside the cabin is shown in Figure 6. Figure 7 shows the schematic diagram of the inverter, while its working principle is illustrated in Figure 8.

The rating of each of the four inverters is as follows:

dc input

- Rated power: 3 kW
- Rated voltage: 120 Vdc
- Rated current: 25 A
- Input voltage range: 60 - 180 Vdc

ac output

- Rated power: 6 kVA
- Rated voltage: 380 Vac $\pm$ 15%, 3-phase
- Frequency: 50 Hz $\pm$ 2%

The conversion efficiency of the equipment exceeds 95% between 20 and 100% of its rated power. The maximum power point tracking operates in the range of 2 - 100% of the rated irradiance.

Recording of the entire current-voltage characteristics of the PV generator, with a local and remote control for automatic sweeping has also been provided.

Figure 6. Control/Display Panel and Power Conditioning Equipment

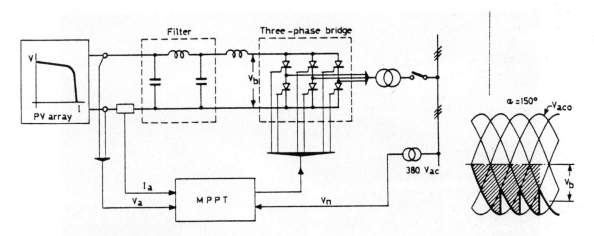

Figure 7. Inverter Circuit Diagram

5. EXPERIMENTAL SYSTEMS

5.1 The CISE System

For experimental purposes a 200-Wp concentrator system has been installed. The system was designed and manufactured by CISE (Italy), with financial contributions from both the CEC and ENEL. It consists mainly of a parabolic mirror concentrator and a receiver in focus with it (see Fig. 9).

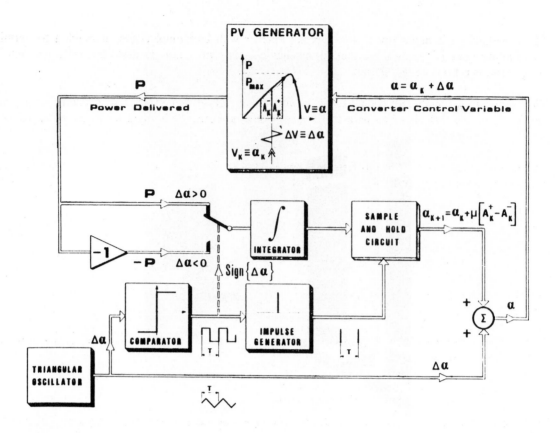

Figure 8. Operating Principle of the Inverter

Figure 9. CISE GaAS Photovoltaic System

The concentrator is made up of 46 spherical mirrors with hexagonal edges, assembled to permit micro-adjustment of a parabola-shaped mechanical support 1.5 m in diameter. Its geometric concentration ratio is about 600 suns.

This concentrator assembly tracks the sun with an accuracy of 5 mrad, and is controlled by an electronic device that also returns it to the resting position during the night, or in the event of bad weather or equipment malfunction.

The receiver shown in Figure 10 has 30 GaAs cells with a total area of 19.5 cm^2. The electrical characteristics of the array at STC are as follows:

- Open-circuit voltage: 9.4 Vdc
- Short-circuit current: 50.4 mA

The following performance data have been recorded at about 600 suns:

- Open-circuit voltage: 11.2 Vdc
- Short-circuit current: 22.9 A

In addition, a water circulation system from the receiver to a heat exchanger enables the thermal energy to be recovered.

5.2 The Siemens System

A fixed flat-plate array of modules with amorphous silicon cells of about 200 Wp was installed to evaluate the performance of this cell technology which has a lower-cost potential in comparison with mono- or polycrystalline silicon cells.

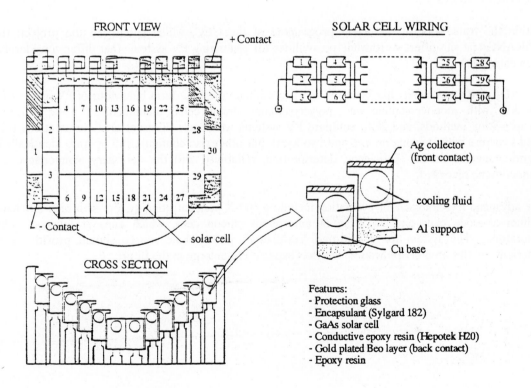

Figure 10. Receiver for the CISE GaAs Array

6. DATA ACQUISITION SYSTEM (DAS)

All electrical, thermal, mechanical, and meteorological data affecting the operation of the PV systems are recorded by special-purpose instrumentation, (miniSAD/SOLE) designed by ENEL as integrated equipment. This integrated system which is simple to use includes measurement transducers and a digital magnetic tape cassette recorder (capacity: 100,000 measurements).

All the electronic transducers give continuous measurements. The DAS calculates and records the averages and integrals of selected data. The integration period may be preset within a range of 10 s-1 h (the tape cassette lasts one month for 25 parameters at a standard integration period of 15 min).

Data collected on ECMA 34 cassette tape are transferred to a computer-compatible magnetic tape by a centralized reader-translator to enable off-line data processing by a digital computer.

7. FIRST RESULTS AND OPERATING EXPERIENCE

During the initial test period, the plant was tested to check the performance of the various systems, correct operation of the tracking mechanisms and the inverters, the effectiveness of the protection devices installed, and the accuracy of the monitored measurements. Moreover, the following corrective actions were taken: installation of a voltage relay for each phase in the 400-Vac grid to protect the conversion equipment, replacement of a module with a shattered glass back substrate (presumably due to thermal stress), installation of bypass diodes on the gallium arsenide cell panel, search for a system to provide a more uniform illumination of the receiver of the CISE system, and replacement of some components in the electrical protection devices. Generally, after the initial corrective measures had been taken, the systems performed reasonably well, as expected.

8. CONCLUSIONS

Within the framework of the R&D programme of the CEC, which sponsored the project, the ADRANO test site offers worthwhile possibilities for evaluating PV systems that differ considerably from each other.

The ADRANO test facility has been set up 1) to evaluate the long-term behaviour of PV arrays of a few kW with monocrystalline and polycrystalline silicon cells, manufactured using different encapsulation methods, and 2) to compare PV systems which use different array orientations (i.e., fixed structure with tracking on one and two axes), but otherwise configured to operate identically in a grid-connected mode. Efficiency, deterioration, reliability, and the necessary maintenance are planned to be recorded.

The advanced technologies for evaluation include a 600-X concentration system with high-efficiency gallium arsenide cells and an array of amorphous silicon cells, which aims at low-cost solar generators. The system for a grid-tied "residential" photovoltaic source should provide some experience in this area of application, which is expected to enlarge in the future.

Appendix 18

FLEXIBLE CABLE-MOUNT
ARRAY STRUCTURES

The cost of conventional fixed tilted PV array structures is now about 100 ECU/m². The PV module price continues to decline, hence the cost of array structures including the foundations will play a major role in the economics of large PV plants. The new idea of a tensile structure supported by poles was conceived by the Commission on the basis of the wire structures customary for vineyards. Cost reductions were anticipated by using low-cost steel cables and wooden poles locally available and plantable with unskilled labour.

A prototype array consisting of 4.3 kW of PV modules was designed and installed in 1983 at a vineyard in Lamole, south of Florence (Italy) (see Figs. 1 and 2). This project was completed in an earlier programme (1981-1985) sponsored by the Commission. In 1989, as part of the array structures concerted actions, Solarconvert of Florence (Italy) was subcontracted by WIP to evaluate the previous configuration designed by CRITI with a view to reducing the hardware cost and on-site assembly work. The key results of this study are as follows:

- Considerable improvements in design and cost were achieved. The number of parts has been reduced, resulting in about 50% reduction in the number of parts from the first generation structure.

- Assembly procedures have been simplified, and assembly time reduced to less than one hour per square metre of PV active area. Simple tools have been identified for fast assembly, for example, for drilling holes and for cable tensile adjustment and maintenance.

- Improved design of the cable-mounted structural parts has been defined and a second generation array prototype is ready for manufacture and installation.

- A PC-based analytical tool suitable for use in design process and economic estimates has been developed and utilized. After inputting module type, soil and climatic features, and mechanical configurations, the programme readily calculates site dimensions, geometrical configuration of the spans, mechanical operating conditions of each component under stress, and the list of materials. It also performs cost analyses.

- The cost analysis carried out for the sample improved configuration with a redesigned tensile support structure resulted in a total installed cost of about 25-35 ECU/m² of PV module area.

Two upper rows contain live modules

Mr. A. Hänel (WIP) and Dr. M. Maltagliati (consultant)

PV modules are attched to three cables

Array output monitoring and on-off relays

Figure 1. Lamole 4.3-kW PV Plant with Flexible Array Structures

Immediately after installation

1 year later

View from the west

Cable stabilizers

PV modules mounting brackets and cables

Module fasteners

Figure 2. Construction Details of Cable-mounted PV Modules

473

Appendix 19

PV-POWERED HOUSE
IN BRAMMING

B. MORTENSEN
Jydsk Telephone
Sletvej 30
DK - 8310 Aarhus-Tranbjerg J.
Denmark

1. INTRODUCTION

The project entitled "Photovoltaic Solar Energy Supply for a Danish Standard House," under CEC-contract No. ESC-R-097-DK was initiated in February 1984 and completed in June 1984. It was officially put into operation in August 1984.

A new contract, No. ENS-0211-DK(SP) with the title "PV Energy Supply of a Danish Standard House - Experiments on Battery Storage" was carried out between February 1989 and late 1989. Because of the unique monitoring and battery experiments set up earlier, a special importance was given to the analyses of battery performance.

The project was intended to stimulate the interest of Danish industries in the application of PV systems as an energy source, primarily for export markets. The PV house has operated with a load profile controlled by a computer. This profile simulates the load of a family of 4 persons.

The following four Danish companies implemented this project:

Prime Contractor:

Jutland Telephone
T-Power Supply Section
Sletvej 30
DK-8310 Aarhus-Tranbjerg J.
(Denmark)

Subcontractors:

Accumulatorfabrikken Lyac
Lyacvej 16
DK-2800 Lyngby (Denmark)

Alex Grosman
Transformervej 13
DK-2730 Herlev (Denmark)

Kalmargarden (House Owner)
Hellesvej
DK-6740 Bramming (Denmark)

2. DESIGN CONSIDERATIONS

The design criteria for this PV residence were established as follows:

- The PV source shall be integrated in the house both technically and architecturally.

- The PV array shall have a peak power output of 5 kW at a solar irradiance of 1,000 W/m^2 and an average cell temperature of 25°C.

- The system shall be designed to operate in a normal grid-connected mode.

- The PV system shall be sized for "stand-alone" operation under climatic conditions corresponding to those in the Mediterranean area.

- The utility grid shall provide the back-up power needed, especially during the winter months.

- The regulation of the PV array voltage shall be done by switching on or off the necessary number of array strings.

- The PV system shall be modular and flexible to facilitate easy adjustments to various load profiles, climatic conditions, layout of the house, etc.

- The know-how and/or products of the participating companies shall be developed, put into operation, and evaluated.

3. DESCRIPTION OF THE PV HOUSE

3.1 The Photovoltaic Residence

In the construction of the photovoltaic house, great attention has been paid to the creation of a compact but well-arranged house. The solar cell modules are integrated into the south slope of the roof. Figure 1 shows the layout and view of the house from the south. For functional reasons and to achieve a presentable internal layout, the technical equipment has been placed in the utility room which is referred to as the "technical room".

Most construction and mounting materials used in the house are based on wood. The wooden frame construction also provides a better heat insulation for a given thickness than an ordinary brick construction. The house is built up of modular units which facilitate shipment and assembly. The units do not involve cold bridges in joints, etc. The roof uses standard trusses, and its tilt angle was selected to be the same as the site latitude.

In connection with the design and arrangement of the house, it was very important for it to function as an ordinary single-family standard house. Thus, apart from the size and construction of the roof, no special measures have been taken which differentiate it from a standard house. For the internal arrangement, the room for the PV electronic components was intentionally kept small. The house has been equipped with all the conventional single-phase appliances of a one-family house. The capability to turn these appliances on or off remotely was included in the plant controller.

3.2 PV Power Plant Description

The simplified system block diagram and power/signal distribution diagram are shown in Figures 2 and 3, respectively. A brief description of the key components follows.

View from the south

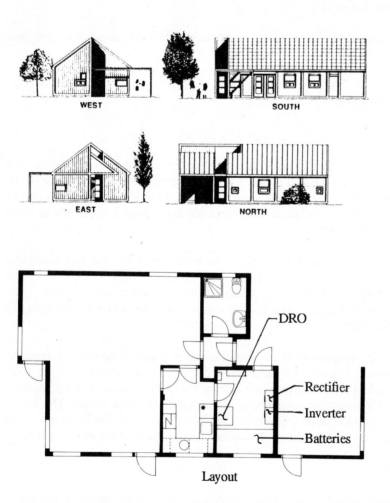

Layout

Figure 1. PV House at BRAMMING

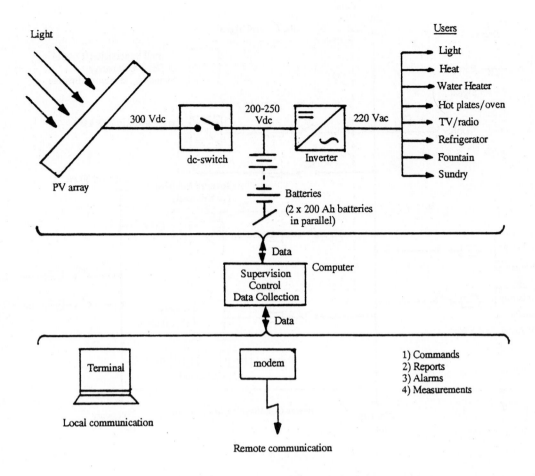

Figure 2. Simplified Block Diagram of the BRAMMING PV Powered House

3.2.1 PV Array

The array consists of 9 parallel strings, each made up of 29 AEG 19-W modules. The number of modules in series has been chosen such that the nominal array voltage at the maximum power point is roughly equal to the upper charge voltage limit.

3.2.2 Rectifier

A 5-kW thyristor type rectifier was installed in order to maintain the battery above the minimum capacity level. This rectifier is considered as an auxiliary·equipment which is not required for normal operation of the PV system. Directly connected to the utility grid, it can supply 20 A at 240 Vdc. The energy drawn from the grid when charging the battery is monitored and stored, along with the other plant data.

3.2.3 Inverter

One key objective of the project was to develop and produce an inverter suitable for household application. The basic requirements and features were established as follows:

- Output power: 10 kVA
- Input voltage: 200-260 Vdc
- Output voltage: 220 Vac, single-phase, 50 Hz
- Modular design
- High efficiency
- Switch-mode

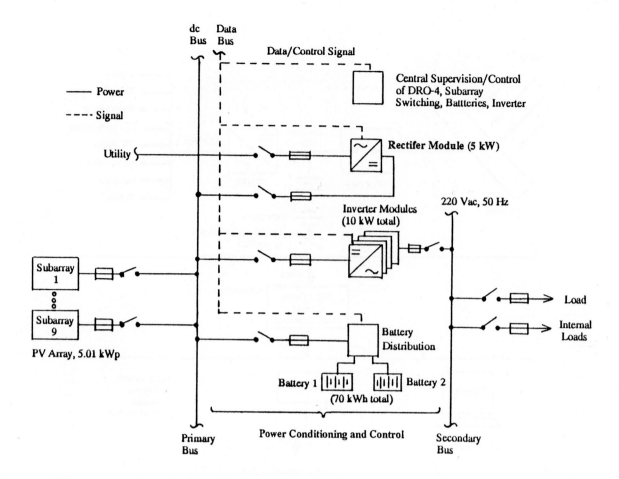

Figure 3. Power and Signal Distribution Diagram

The supervision and control of the operation of the inverter have a higher software priority than the array string switching because of the speed consideration. The inverter makes use of a crystal-controlled frequency generator. The supervisory control function has been divided so that the four subsystems can be supervised individually. When the line frequency and voltage are out of tolerance, the inverter disconnects the load automatically.

The inverter is capable of handling 10 kW continuously. In the case of an overload of up to 160%, the inverter maintains frequency and voltage for approximately 2 s, after which the individual inverter modules disconnect. Any faulty module can be detected by means of light-emitting diodes. Restart of the inverter can be effected manually or automatically via remote control. With an overload capability of 160% for 2 s it is possible to restart the refrigerator and other loads even though the inverter may be already at full load.

The inverter, developed by Alex Grosman, uses PWM technique with a switching frequency of 20 to 40 kHz. In order to supply the required 10 kW, the inverter was built up of 32 modules each containing a half-bridge connection (see Fig. 4). The 32 modules are divided into 4 groups of 8 modules of 625 W. Each of the four groups is connected to its own transformer. The secondary sides of the transformers have been connected in parallel to give a total power of 10 kW. The PWM system uses MOSFET power transistors which resulted in a low power consumption for the control modules. The inverter emits no audible noise because of the high switching frequency.

3.2.4 Battery

The energy storage consists of two 200-Ah batteries connected in parallel. Each battery has 108 cells in series made up of 36-cell monoblocks, each with three cells in one case. Table 1 lists the basic specification from Lyac Power. Figure 5 shows the battery installation.

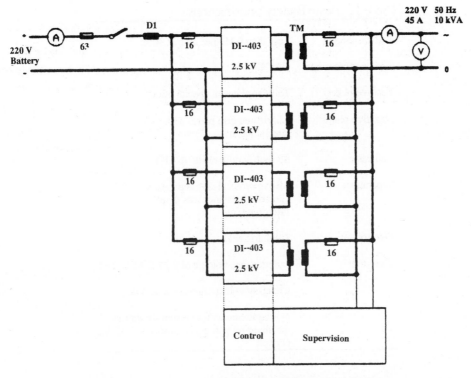

Electrical configuration of the inverter

UPS and inverter consoles (on the left), and battery racks (on the right)

Figure 4. Configuration of the inverter (top), and interior of the technical room (bottom)

The battery size was chosen for an expected load of a daily average of about 12 kWh. This would permit operation for approximately six days without recharging at an allowable DOD of 80%.

Table 1. Basic Battery Specifactions

Self-discharge:	Less than 0.1% per 24 h at 30°C
Temperature range:	5-35°C
Postiive plates:	Tubular plates of special alloy
Negative plates:	Lubricated grid plates with Lyac's special paste
Cell case:	Styrene acrylic nitrite (SAN)
Separators:	Microporous plastic separators. Electric resestance less than 0.2 microohm/cm
Electrolyte:	Sulphuric acid
Charging:	Maintenance voltage of 2.22-2.24 V/cell
	Charging voltage limit of 2.35 Vdc
	No requirements for current limitation if the charging voltage is limited to 2.35 Vdc per cell

3.2.5 Microprocessor-based Supervisory System, DRO-4

The supervision and control equipment for this project was initially based on DRO-1, developed by Jutland Telephone. This system was intended for supervision of power supplies for the telephone exchanges.

Based on the initial operating experience of the DRO-1, further development resulted in the new design, DRO-4 (see Fig. 5). This new unit has the following functions which are implemented by a 16-bit 12 MHz computer:

- Supervision and control of the PV installation
- Supervision of battery storage
- dc and ac voltage supervision
- Supervision and control of the load
- Power-flow control
- Alarm supervision and communication
- Data acquisition and collection

A key function of the DRO-4 is the scanning and collection of data. It has the following features:

- 128 analog inputs
- 32 digital inputs
- 32 digital outputs
- 40 kB RAM
- 24 kB PROM
- 2 floppy disks of 800 kB each
- 16-bit/12 MHz processor

One of the drives contains the software for the DRO-4. If a total reset of the software is needed, the DRO-4 re-inputs the programme into the RAM. This makes it possible to make changes, reset, and test the changes implemented, without being present at the site. The DRO-4 also has the capability to initiate a telephone call automatically, via modem, to the central office. Thus, it is possible to determine when the alarm occurred. Furthermore, via the modem connection, the analyst can acquire tabular reports, change parameters, reset alarms, etc. remotely.

The second drive is used for storage of collected data from the plant. Every five minutes all data are registered on the disk. Once every 24 hours calculations are made, e.g., energies and mean values, which are stored on the disk.

Air pumping motor

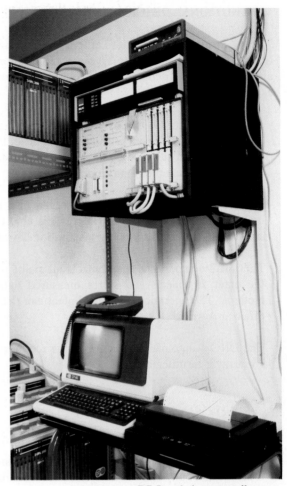

DRO and plant controller on top

Ducting for pumping air into the cells

Electrolyte level sensor

Figure 5. The Battery Installation and the DRO Supervisory Controller. A forced-air Bubbling System was Installed in the Cells of One Battery

481

4. FUNCTIONAL DESCRIPTION

The PV arrays are divided into nine parallel strings each supplying a voltage of 0-300 Vdc, depending on load and solar intensity. To maintain an appropriate battery voltage level, the individual strings are connected or disconnected by remotely controlled dc switches. The main functional features are as follows:

- If the battery storage is fully charged, strings will be connected or disconnected depending on the load requirement. By measuring the output of the strings that are still connected, the power which might have been supplied by the disconnected strings is calculated. This figure is stored in the disk together with the other data from the plant.

- The single-phase ac power is distributed in a group panel to the low voltage installation of the house. The supervision unit (DRO-4) is able to connect or disconnect the individual groups for simulation of a normal family consumption day and night.

- The DRO-4 supervises and controls all major plant components. Besides real-time supervision of the plant, it collects and stores measured data. Communication with the plant is possible via modem. The operation of the PV plant can thus be observed at the video terminal from the home office at Aarhus.

- The DRO-4 automatically connects the rectifier to the battery whenever the power supplied by the PV array is insufficient.

5. PERFORMANCE EVALUATION

The PV plant has operated very well after the initial period of adjustments and problem solving. Table 2 lists the monthly performance data for January - June 1989. It is interesting to note that the PV plant availability and the array utilization factors remained very high throughout this period.

The following paragraphs summarize the performance of the PV system and key experiences during the operational period (June 1984 to December 1989).

5.1 House Load and Application

During the operational period the house has been used two to three times a week for meetings. Thus, as the house has not been permanently occupied during the operational period it has been attempted to simulate a normal family's consumption pattern by means of controlled group outlets in the low voltage installation. For various reasons this simulation pattern has been changed during the initial period.

Originally the intention was to supply the water heater, fountain pump, and electric heating as well as electric lighting and kitchen installations in a real-time consumption pattern day and night. However, this consumption would be too heavy for a pilot plant located in Denmark. Furthermore, hot-plates as well as the cooker cannot be switched on without supervision.

Over the past few years the load consisted of internal lighting as well as a radiant-heating unit placed in the carport. By connecting these comsumers at appropriate intervals the load has been varied day and night. The nominal load was about 12 kWh daily.

Technically, the house has functioned extremely well. Apart from the cracked cover glasses on PV modules, there have been no technical problems. Moreover, the architecture of the house and its usability make it quite suitable for possible export to the southern regions of Europe and to developing countries.

Table 2. Monthly Performance Summary, January 1989 to June 1989

PV Pilot Plant: PV-house Bramming	Date of report: 20-09-89	Latitude (N): 56
Site Manager: M. Jorgensen	Prepared by: M. Jorgensen	Longitude (E): 9
Project Manager: B. Mortensen	Telephone: 45-86-293366	Altitude (m): 30
Organization: JYDSK Telephone, Aarhus, DK		

Parameter	Units	1989					
		Jan	Feb	Mar	Apr	May	Jun
1. Solar/Temperature							
a. Insolation, plane of array	kWh/m^2	19	41.9	62.7	102.2	152	133.5
b. Irradiance, plane of array, day avg	W/m^2	25.5	62.4	84.3	141.9	211.1	185.4
c. Temperature, back of module, day avg	°C	10.4	10.9	12.9	14.9	23.5	27.8
at 6 pm	°C	9.2	8.8	8.5	7.6	11.1	14
at 12 am	°C	13.7	17.4	20.6	26.1	40.9	47
at 6 am	°C	9.5	11.8	12.8	18.8	28.7	33.6
d. Temperature, ambient, day avg	°C	6	6	7.4	7.3	13.9	18.1
e. Temperature, air ambient, day low	°C	-7.7	1.3	1.2	-2	5.7	6.3
2. Plant Output Information:							
a. Energy, array	kWh	98	197	287	463	635	641
b. Energy, inverter	kWh	463.9	469.9	509.7	492	453.7	390.4
c. Power, array day avg	kW	0.1	0.3	0.4	0.6	0.9	0.9
d. Power, Inverter day avg	kW	0.6	0.7	0.7	0.7	0.6	0.5
e. Energy, array field (unused)	kWh	--	1	1	2	28	37
3. Plant Performance Indices:							
a. PV array energy capability	kWh	98	198	288	465	663	678
b. Array utilization factor	%	100	99.5	99.7	99.6	95.8	94.5
c. Array energy efficiency (Irrad. >50)	%	5.5	7	8.4	8.8	9.4	13.1+
d. System energy efficiency day avg	%	58	58	59.9	61.7	57.2	54.1
e. Plant availability	%	100	100	100	100	96.8	100
4. Battery Information:							
a. DOD, % rated Ah, daily avg	%	79.7	78	79.2	84.2	85.4	86
b. No. cycles *	-		23	21	28	22	18	18
c. Cell temperature, daily avg.	°C	23.4	23.6	23.8	23.9	28	30.3
d. Level, electrolyte, daily avg	%	63.6	57.7	52.5	46.3	71.3	68.3
5. Operation & Maintenance Information:							
a. Total time DRO operational	%	100	100	100	100	96.8	100
b. Total time inverter operational	%	100	100	100	100	96.8	100
c. Number of failures, major components							
1) MTIR	h	--	--	--	--	--	--
6. Application Information:							
a. Consumer load energy, total	kWh	463	469.9	509.7	492	453.7	390.4
b. System consumption, total	kWh	96.7	87.4	96.7	93.6	93.6	93.6
c. Energy from grid to PV plant, total	kWh	701.2	612.9	563.8	333.9	158	80.7

* DOD >5% counts as one charge/discharge cycle
+ Erroneous data, caused by shadows on solar sensor

5.2 PV Array

The PV modules were installed on the roof of the house according to the instructions of the supplier. The mounting was easy, and their replacement, if required, is possible without too much trouble.

The mounting approach based on two clamping points, the non-planarity of the mounting surface, and mounting on wood structures probably contributed most to the cracking of the cover glass.

The first cover glass damages on PV modules appeared immediately after the plant was put into operation, and they were replaced with new ones. At the end of the project a total of 15 modules were found with cracked glasses. Even though the cracks apparently did not cause a lower energy production from the individual strings, energy output can decrease in later years.

Based on measured data, the array power efficiency has been calculated to be between 7 and 12% on an average, dependent on the period of the year and other external factors. An abnormally high efficiency (18% in spring 1988) was indicated at times. This was due to periodic shadowing of the solar irradiance sensor from a tree branch that eventually grew taller and obstructed the sun. The PV array has not shown any apparent deterioration over the 5-year period.

5.3 Rectifier and Inverter

The rectifier charges the battery when the start signal is issued by DRO. This occurs when the battery state of charge (SOC) reaches a preset limit (20% initially, increased to 50% later).

The inverter, developed specially for this project, has functioned satisfactorily during the entire operational period. The inverter has had no major faults and has fulfilled our anticipations during the operational phase. However, the inverter has had a tendency to disconnect the load when sudden load changes occur. After a minor modification, the inverter has been able to withstand certain high load changes without disconnecting, and has, on the whole, functioned satisfactorily.

5.4 Batteries

The key results of the battery performance are as follows:

- The two batteries charge at 21 A maximum and discharge at approximately 3 A per battery on an average.

- In May 1987, a discharge test of the batteries was made. This test showed that it is possible to discharge approximately 80% of the rated capacity before the battery reached its low voltage limit.

- During the deep discharge tests some of the battery cells apparently suffered damage in the form of sulphation and loss of active material. It was not possible to recharge these cells to the same level as the other cells in the battery.

- At the subsequent laboratory tests on the replaced cells, it was found that these cells contained only 50-60% of their nominal capacity after several cycles of charge and discharge. Furthermore, it was found that all the positive tubular plates were almost empty of active material. This may be due to the deep discharges encountered in the initial periods.

- In May 1989, another deep discharge test was made, corresponding to the one made in 1987. This test revealed 6 cells in the batteries with very low voltages, and it was possible to obtain only 60% of the nominal capacity.

- Because of the acid stratification in the batteries and the resulting problems in the specific gravity readings, a system which pumps air into the cells of one of the batteries was installed (see Fig. 5). After each charge and at least every 100 hours, this device bubbles air into the bottom part of the cells in order to agitate the electrolyte; this procedure could also minimize sulphation. This experiment was intended to provide some answers to questions such as:

o Does the forced-air bubbling improve the life of the cell and consequently of the battery?

o After five years of operation, are the batteries in this application in a poorer state than batteries of the same age applied in a conventional stand-by application?

The key conclusions of battery post-mortem analyses are:

- It was fairly clear from the initial examinations that the amount of plate deterioration was greater for the cells which had been installed during the entire operational period than in the recently installed blocks. Furthermore, the amount of material emptied as sediment in the cells mounted with the air bubbler was lower than in the cells without the bubbler.

- No indications of serious corrosion were found on any of the cells. The "new" cells showed pronounced indications of the effects of heavy stratification because only the lower part of the plates had participated in the chemical reaction in the battery. For this reason, a periodic "exercise" of the battery (i.e., controlled cycling) is recommended if an air bubbling system is not installed.

5.5 Supervision Equipment DRO-4

The first DRO-4 model which was put into operation in 1984 had a relatively poor electrical noise immunity. Furthermore, the data collection units, especially the disk drive, were of poor quality. These characteristics have caused a few problems relating to the accuracy of battery capacity calculations and the collection of performance data. As a result, a new version of the DRO-4 was developed and installed at the beginning of 1986, which essentially corrected the hardware problems. The present version of the DRO-4 is now a reliable unit which is able to supervise and control the PV pilot plant continuously and collect data over the modem virtually with no losses.

6. CONCLUSIONS

The significant results and conclusions of this project are as follows:

- Great importance has been attached to integrating the PV modules architecturally in the roof construction, and the result was quite satisfactory. The cover glasses of 15 modules cracked during the operational period. The tensions in the wooden roof structure, combined with the poor module mounting method, contributed to this problem.

- On the whole, the PV modules have functioned well during the operational period and module power efficiencies of 7% to 12% have been measured, depending on the time of the year.

- The 10-kVA switch mode inverter has operated without problems during the entire monitoring period. Because it operates at a 20-kHz switching frequency, it emits no audible noise. Consequently, it is quite acceptable for household use.

- The microprocessor-controlled unit, DRO-4, developed for the project, has operated with only minor technical problems. It has unique capabilities for the monitoring and control of stand-alone systems. It was also adapted later in PV hybrid plants.

- The DRO has been designed to calculate the battery capacity and display it in real time. This SOC parameter serves an important function in maintaining a reliable battery and for periodic assessment of battery performance.

- The operational period has shown that the battery is the most complicated component to monitor properly in practice, and it is the most important component in a PV plant which is expected to function satisfactorily in a stand-alone system.

- Forced-air bubbling clearly demonstrated that it improves the battery performance and can extend the battery life.

- The PV energy production was approximately 25% below the originally calculated value.

- Our investigation of the battery performance has helped answer some of the general questions about how to extend battery life. However, there are many other unknowns which should be studied further.

- The capacity test showed very clearly that the remaining Ah capacity of the cells, without the bubbler, was considerably less than those with the bubbler.

Appendix 20

PV-POWERED HOUSE
IN MUNICH

G. BOPP, R. KAISER, R. SCHÄTZLE and J. SCHMID
Fraunhofer Institute for Solar Energy Systems
Oltmannsstrasse 22
D - 7800 Freiburg

Abstract

The first German utility interactive photovoltaic residential system was installed in a private residence in Munich by the Fraunhofer Institute for Solar Energy Systems (ISE), Freiburg (F.R. Germany) in 1982. This project was sponsored by the Commission of the European Communities, and several German firms. The integration of the solar panels into the roof by replacing existing glass panes proved to be successful. Since the installation in November 1982 no problems concerning water-tightness or electrical performance have been encountered. The inverter, specifically developed for photovoltaic applications, confirmed the high efficiency (> 90% in the range 5-100% of output power) and low harmonic distortion (less than 5%).

1. INTRODUCTION

The photovoltaic plant was installed in a residential house in Munich in 1982 [1 to 3]. The project was sponsored by the Commission of the European Communities and the German photovoltaic industry represented by Siemens, AEG, and Varta. The goals of the experiment were to gain experience with small-scale grid-connected photovoltaic systems, the integration of solar panels into the roof, safety aspects of the electrical system, and acceptance of such systems by the inhabitants of the house. The system was put into operation in September 1983. The private residential building shown in Figure 1 was chosen for the experiment due to its unique design. Its cross-section is a right triangle resting on one of the short sides, while the hypotenuse carrying the glass-covered roof faces the south. The glass roof is inclined at an angle of 45°. This constitutes nearly an optimum angle, taking into account Munich's latitude of 48°. In general, the building has been optimized for passive solar heat gain.

Figure 1. Photovoltaic Residence with Roof-integrated PV Modules

2. PV ARRAY AND SYSTEM CONFIGURATION

The basic idea for integrating the photovoltaic system in the house was to replace the existing glass panes with frameless PV modules. Thus, no additional costs for mounting structures would be incurred and costs for photovoltaic components could be limited to PV panels and wiring. Over half of the solar cell modules (30 m²) are integrated directly into the roof, replacing the glass panes (see Fig. 2), and the others (24 m²) are mounted beneath the glass roof because of their larger dimensions (see Fig. 3).

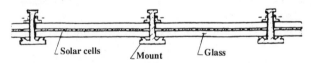

Figure 2. Direct Integration of AEG PV Modules

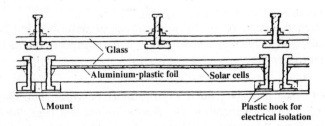

Figure 3. Indirect Integration of Siemens PV Modules

The manufacturer AEG was able to meet the requirements for a suitable frameless module of 8 mm thickness to fit the outer dimensions of the panes to be replaced (736 x 1263 mm) as shown in Figure 2.

The AEG frameless PV modules are of double-glass construction (i.e., glass-solar cell-glass). Each module contains thirty-six 100 x 100 mm polycrystalline Si cells. Thirty-two modules were integrated in this way without difficulties.

The dimensions of the Siemens modules (1020 x 1470 mm) were such that direct integration was not possible without altering the building's roof structure. Therefore, they were installed in a different manner, i.e., below the glass roof (see Fig. 3). The advantage of this method is the access to the entire panel, allowing simple installation and easy replacement. The inconvenience is radiation losses due to reflection and absorption on the outer glass and shading from the support structure.

The PV modules are interconnected to form five groups with the nominal voltages, 11, 22, 44, 88 and 176 Vdc, respectively. The dc power they supply is converted by a newly developed inverter to 220 Vac and fed into the house grid. Excess power is fed into the public grid, and if the household consumption exceeds the photovoltaic system capability, the difference is drawn from the grid. Figure 4 shows the system power flow diagram and the key characteristics of the PV array.

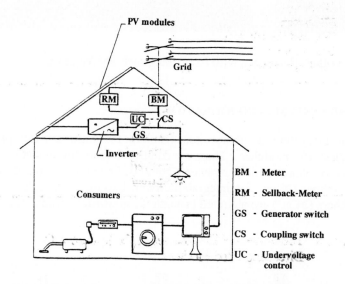

Figure 4. Utility-connected Residential PV System

3. INVERTER

The inverter was designed especially for PV applications as described in [2]. Some of its unique features are: extremely high efficiency, low harmonic content, and small volume resulting from its transformerless working principle.

Due to the special working principle, the PV array must be wired into five independent groups according to the inverter's input requirements. For the initial test runs, 24 AEG and 12 Siemens modules were used to form the subfields. The surplus modules were used to vary the subfield's voltage and power in order to achieve optimal configurations. Table 1 shows the subfield characteristics.

In 1985, detailed performance measurements were made, and the efficiency values of the individual system components were determined. In order to analyse the total system efficiency, the system was modelled with individual component efficiencies as illustrated in Figure 5. The analysis of the system efficiency, according to the component efficiencies, was restricted to those PV modules directly integrated into the glass roof, thus eliminating shadowing effects.

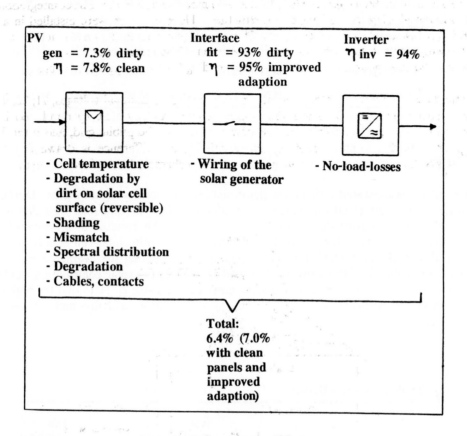

Figure 5. Breakdown of Efficiencies of the System with Direct-mounted Modules

The average efficiency of the PV array at the maximum power point was 7.3%. This value, in comparison with the manufacturer's specification (10%), resulted from the higher operating temperature (40°C to 60°C) compared with the standard temperature (25°C), gradual soiling of the solar cells, and increased reflection of light falling at non-normal incidence angles. After cleaning the PV modules, two years following their installation, the array efficiency increased from 7.3% to 7.8%.

The average energy efficiency for the combined array-inverter was around 93%. This value can be increased to 95% by optimizing the wiring of the PV modules.

The average inverter energy efficiency was 94.0%. Even at only 1/15 of the nominal load (3 kW), the efficiency of the inverter exceeded 90% (see Fig. 6). This is attributed to very low no-load losses. Thus, the total system energy efficiency of the PV array directly integrated into the glass roof along with the transformerless inverter is 7%.

Table 1. Modules Used to Form the Subfields

| | Subarray | | | | | |
	1	2	3	4	5	Total
No. of Modules in Series	0.50	1.00	1.50	3.00	6.00	12.00
Module Area (m^2)						
- Siemens	0.75	1.50	2.25	4.50	9.00	18.00
- AEG	0.95	1.90	2.85	5.70	11.40	22.80
Cell Area (m^2)						
- Siemens	0.57	1.13	1.70	3.40	6.80	13.60
- AEG	0.72	1.44	2.16	4.32	8.64	17.28
Vmp (60°C) (Vdc)	14.50	29.00	43.50	87.00	174.00	348.00
dc Input to Inverter (Vdc)	11.00	22.00	44.00	88.00	176.00	341.00

4. RESULTS

The entire plant was operated without any significant interruption for eight months. During the first eight months, 1,014 kWh electrical energy was produced, while the total solar radiation recieved on the plane of the array during this period was 19,425 kWh. This resulted in a system efficiency of 5.22%. If this is extrapolated for the whole year 1984, and the entire solar cell area is assumed to be connected, this results in 2,700 kWh electrical energy. Considering the average annual household electricity consumption in F.R. Germany of around 3,000 kWh, the solar contribution would be 90% based on the above estimates. The results are summarized in Table 2. Note that the high electricity consumption in this particular household is due to a film-cutting machine and several additional electric heaters.

Table 2. Energy Balance

	1984 (direct and under-glass-mounted panels in parallel	1986 (only direct mounted panels)
PV module area	30 m^2	23 m^2
Energy produced by PV modules in 8 months (kWh)	1,014	1,190
Solar irradiation in 8 months (kWh)	19,425	18,594
System efficiency (%)	5.22%	6.4
Energy delivered to the grid (kWh)	170	160
Energy consumption of the household (kWh)	16,000*	16,000*
Average energy consumption of households in Germany (kWh)	3,000	
Solar energy produced at 100% availability and with the whole module area (kWh)	2,700	
Contribution from solar System (%)	90	

* The very high energy consumption in this household is caused by a film-cutting-machine and several electricadditional-heaters.

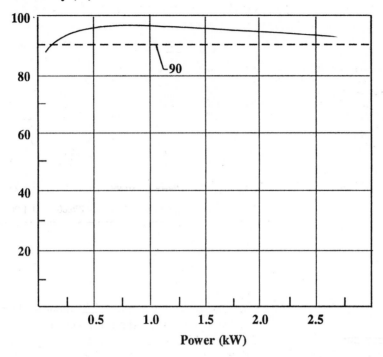

Figure 6. Power Efficiency Characteristics of ISE's Transformerless Inverter

5. CONCLUSION

With the system efficiency reaching 7%, the Munich PV plant compares favourably with other grid-connected systems in many European countries. This is largely due to the very high power conversion efficiency of the transformerless inverter inthe partial load range. The average energy efficiency of the inverter is 94.0%. This high efficiency is due to the low no-load losses of the particular inverter used. The roof section with the directly integrated solar cell modules remained waterproof for over four years.

6. REFERENCES

[1] J. Schmid, Design Considerations and Economic Aspects of Photovoltaic Modules for Architecture, Proc. of the 1st EC Conf. on Solar Collectors in Architecture, Venice, IT, March 1983.

[2] J. Schmid and R. Schätzle, "Simple Transformerless Inverter with Automatic Grid Tracking and Negligible Harmonic Content for Utility Interactive Photovoltaic Systems," Proc. of the 4th European PV Solar Energy Conference, Stresa, IT, May 1982.

[3] J. Schmid, G. Bopp, R. Kaiser, and R. Schätzle, "Experiences with a Grid-connected Residential Photovoltaic System," Fraunhofer Institute for Solar Energy Systems, Freiburg, FRG, April 1985.

Appendix 21

PV-POWERED HOUSE
VILLA GUIDINI

M.S. IMAMURA
WIP - Munich
Sylvensteinstr. 2
D - 8000 München 70
F.R. Germany

The PV plant at Villa Guidini was intended to provide power to an 18th century residential villa which has been converted to a public garden and museum. Figure 1 shows the layout of the villa and the PV array. The secondary aim was to demonstrate the use of lightweight PV module design and to conduct experiments to evaluate the effects of various array tilt angles.

The plant was designed and installed by Helios Technology of Galliera Veneto (Padova, Italy). It has a 2.5-kW PV array, one 24-Vdc battery rated at 1,600 Ah, a 1-kVA inverter, and a monitoring system consisting of a data logger, digital cassette recorder, video terminal, computer, and modem. Figure 3 is a block diagram of the plant. Figure 2 shows the balance-of-system hardware.

The PV array contains 80 PV modules mounted on two rows of array structures. It consists of two module types (Helios H40 and AL40), each rated at about 34 Wp. Twenty modules of each type are mounted on one array plane tilted at 27 degrees simulating the mean inclination of roofs in this region. The other 40 PV modules (20 of each module type) can be inclined manually at different angles in a single group of two subarrays. The differences in output at various angles will permit a comparison of the two autonomous groups. The module supports are mobile and are arranged so as to permit the replacement of PV modules by types from other potential suppliers. A double-array configuration was chosen because the AL40 module, which weighs about 3.9 kg is considered particularly suitable for this application. The lightweight modules can be mounted on a lighter support structure which is an advantage for roof-mounting on older buildings that cannot support heavy structures.

The battery cells, supplied by Fiamm of Montecchio Maggiore (Vicenza, Italy), are of the tubular positive plate type. The data acquisition system was designed to collect the plant's performance data as well as the solar meteorological data.

View of the principal antique building from the east

PV Arrays

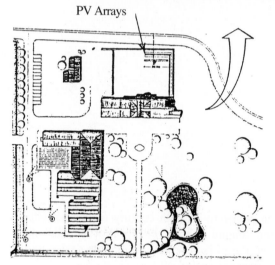

Two rows of PV panels

Tilt angle is adjustable

Figure 1. Layout of the Villa Guidini House and the 2.5-kW PV Array

494

Cc Control / display Panel

Solar / Meteorological Sensors

Monitoring System

1,6 1,600-Ah 24-Vdc Battery
(Supplier: Fiamm)

Figure 2. Balance of System Hardware

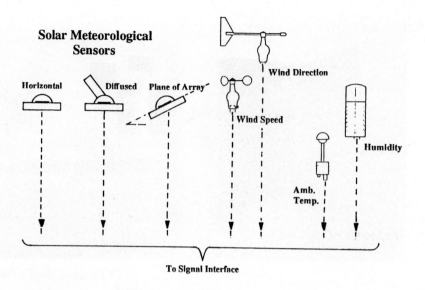

Solar Meteorological
Sensors

Horizontal Diffused Plane of Array

Wind Direction

Wind Speed

Humidity

Amb.
Temp.

To Signal Interface

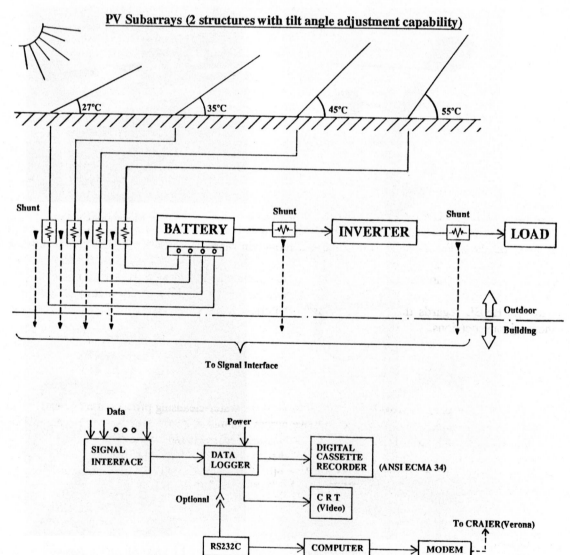

PV Subarrays (2 structures with tilt angle adjustment capability)

27°C 35°C 45°C 55°C

Shunt

Shunt

Shunt

BATTERY INVERTER LOAD

Outdoor

Building

To Signal Interface

Data

Power

SIGNAL
INTERFACE

DATA
LOGGER

DIGITAL
CASSETTE
RECORDER (ANSI ECMA 34)

Optional

C R T
(Video)

RS232C COMPUTER MODEM

To CRAIER(Verona)

Figure 3. Functional Block Diagram of PV System at Villa Guidini

496

Appendix 22

ANCIPA PV PLANT

A. PREVI
ENEL-CREL
Via A. Volta
I - 20093 Cologno Monzese (Mi)
Italy

B. MORGANA
Conphoebus Scrl
Passo Martino-Zona Industriale
Casella Postale
I - 95030 Piano d'Arci (Catania)
Italy

D. LICATA
ENEL
Via Marchese Villabianca, 121
I - Palermo
Italy

Abstract

The primary aim of the ANCIPA Project is the establishment of a PV plant to supply the electric power necessary to drive one trash-rack rake and four sluice gates. The plant is located at the water spillway on the River Bracallà in the Nebrody mountain range in northeast Sicily. The trash-rack rake removes large stones and debris from the metal screen, while two gates are used for safety reasons to prevent water from flowing to the Ancipa dam. The total electric power required daily by the hydraulic system driving the rake and the gates is handled by the 1.8-kWp PV array with a 540-Ah battery. ENEL regards this pilot plant as a prototype for similar future applications in remote mountainous locations.

1. INTRODUCTION

The ANCIPA PV project provides for automation of the water-cleansing process at a spillway. The mechanical trash-rack rake is designed to move across a stainless-steel screen over the water intake opening to free the water intake from large stones and other coarse debris brought down by the torrent. In addition, the raising and lowering of the water control gates have been automated by the addition of a hydraulic actuator driven by dc motors. This plant was installed and started up in November 1989.

2. WATER CONTROL AND CLEANING SYSTEM

The Bracallá River spillway is situated in the Nebrodi range, in northeast Sicily, nearly 1000 m above sea level at 37.86°N, 14.65°E. It is one of the five spillways which channel and control the water flow into the Ancipa reservoir (see Fig. 1). This reservoir and the dam serve one of Sicily's biggest hydroelectric plants.

ANCIPA Spillway

Artist's sketch of the surrounding region. ANCIPA is located to the West of the Etna Volcanic Mountain.

View of Etna from the South

Figure 1. Location of the ANCIPA Site

View from the road (Dr. Previ of ENEL)

PV Array

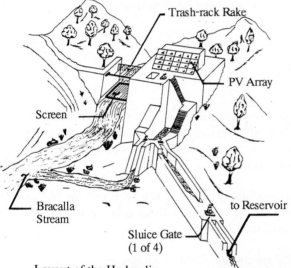

Layout of the Hydraulic
System and PV Array

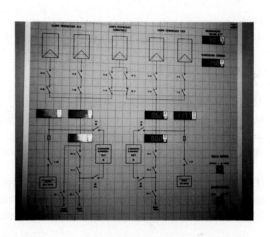

Mosaic design display panel

View southward from the Control Building

Figure 2. Views of the ANCIPA Project Site

Figure 2 shows the general layout, the PV array, and the hydraulic elements involved. Figure 3 presents a schematic diagram of the spillway and photos of the major components. The water in the river flows from north to south down the mountainside. At point "A" in Figure 3, the water flows over a stainless steel screen which is one of the five intakes of the underground canal leading into the Ancipa reservoir. The screen prevents large stones from being carried into the reservoir.

Sluice gate "B" (see Fig. 3), located at the input of the gravel cleansing bed, is normally open, but can stop the water flow when it is so violent that the hydraulic systems are considered inadequate to contain all the material carried by the torrent.

The bed has two output gates: "C" and "D". When gate "D" opens, the in-rushing water cleans the bed by diverting the gravel into the river. Gate "C" provides for a second cleaning of the sand-trapping bed. This, in turn, has an output gate "E" that, when open, enables the bed to be cleaned in the same manner as above. However, gate "E" is generally closed, and clean water overflows into the canal and finally, into the Ancipa reservoir, about 10 km away.

Prior to this project, the hydraulic system was hand-operated; the screen was cleaned periodically with a manual rake, and the gates were moved up and down by coaxial Archimedean screws.

3. PV PLANT DESIGN AND OPERATION

3.1 PV Array and Batteries

Figure 4 shows the block diagram of the plant. The PV array is made up of state-of-the-art flat-plate modules. Two different power sources at 110 and 24 Vdc are provided to supply power for the motor and valves and the control devices, respectively. Each circuit has its own PV array and battery. Back-up dc power is available from a Diesel generator via two rectifiers.

The 110 Vdc PV array has 24 Ansaldo monocrystalline modules of about 50 Wp each, and the 24 Vdc array has 8 modules of the same type. The batteries are made up of 540-Ah cells. The 110 Vdc battery has 55 cells in series and the 24 Vdc system has 12 cells in series.

The design was based on the following data, using a PV system simulation programme:

- Meteorological data acquired at Messina (nearest meteorological station):
  - Daily insolation, global horizontal: 1.5 to 1.9 kWh/m^2 in winter
         4.6 to 6.5 kWh/m^2 in summer
  - Precipitation: 750 mm avg
  - Snowfall: up to 1.5 m
  - Temperature (ambient):   -10° to 30° C
- Tilt: 60°/S-facing, optimized for winter output performance and to permit snow slippage
- Expected load (power and ancillary circuits):   2.5 kWh/day
- Required autonomy: 10 days (for sizing the batteries)

3.2 Hydraulic Converter

The pump in the hydraulic converter is powered by a dc compound motor. There are differing points of view regarding the choice of a dc motor to power an electro-hydraulic pump. A squirrel-cage asynchronous ac motor is favoured for reliability, maintenance, and economy. Hence, we gave due consideration to the possibility of supplying the motor through a static variable frequency inverter fed by the 110-Vdc battery. After a careful survey, we found that commercially available inverters required ac power input, and specially designed converters fed directly by a battery are not easily available. They are very costly, and would not have the easy maintenance and high reliability characteristics which are among the main reasons for choosing an ac motor.

The rake above the screen

Raking the screen (A)

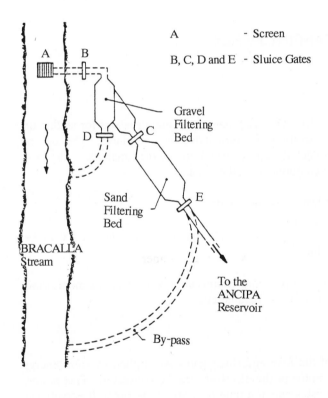

A   - Screen

B, C, D and E - Sluice Gates

Gravel
Filtering
Bed

Sand
Filtering
Bed

BRACALLA
Stream

To the
ANCIPA
Reservoir

By-pass

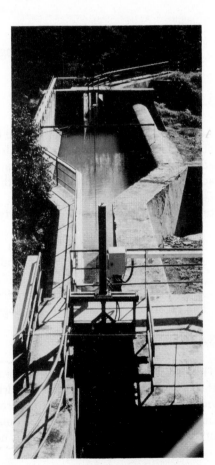
Two Sluice Gates (C and E)

Figure 3. Details of the Hydraulic System

501

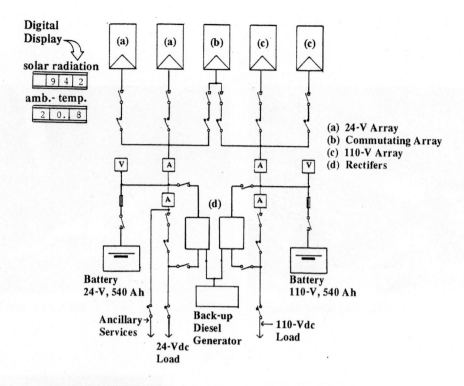

Figure 4. Schematic Diagram - ANCIPA PV Plant

3.3 Data Acquisition System (DAS)

As in previous R&D work co-financed by the CEC (DG XII), one of the main reasons for setting up pilot plants is to acquire experience for the operation of future industrialized plants. This can be achieved only if sufficient plant data are collected and recorded for future analysis. The data acquisition system is therefore one of the main components of the plant.

In the ANCIPA project, two different types of DAS were installed:

- Energy meters (ENEL/Advel-Milan design) to supply complete data on solar energy, PV array power output, and the power supplied to the load;

- A DAS (MiniSAD, designed by ENEL/Marconi) for more comprehensive continuous monitoring and recording of about 15 meteorological and electrical parameters.

3.4 Plant Operation

The power needed for automatic operation of the rake and sluice gates is supplied by solar energy through photovoltaic conversion. It drives a motor in the electro-hydraulic converter. This motor, rated at 3.4-kW power, is used to operate the gates one at a time or to drive the rake. It is switched on for no more than 15 minutes at a time. The load profile depends on the meteorological conditions at the site. The total estimated load is no greater than 2.5 kWh/day.

4. CONCLUSIONS

The chief motivation of the project was to set up a PV plant which supplies electric power for the automatic operation of a water spillway serving a large reservoir. This pilot plant is of interest to ENEL and other utility companies as well.

Appendix 23

BERLIN PV PLANT

T. MIERKE and B. VOIGT
EAB Energie-Anlagen Berlin GmbH
Flottwellstrasse 4-5
D - 1000 Berlin 30
F.R. Germany

R. HANITSCH
Technical Univ. Berlin
Einsteinufer 11-15
D - 1000 Berlin 10
F.R. Germany

K. BÜRGEL
Berliner Kraft- und Licht (BEWAG)-AG
Stauffenbergstrase 26
D - 1000 Berlin 30
F.R. Germany

Abstract

In early 1989 BEWAG/EAB, in collaboration with the Technical University of Berlin, installed a PV pilot plant in a residential district of West Berlin. The plant has a total PV power of about 10 kWp and a battery with a capacity of about 100 kWh. It is used to supply the electric demand of a heat pump, household appliances, and some lighting for the building, and to serve as a charging station for electric car batteries. Additionally, the system is capable of feeding solar energy into the local grid. Both the microprocessor-based controller and data acquisition system utilize state-of-the-art technologies (1988). The testing phase began in late 1989. The project was sponsored by the Commission of the European Communities (CEC) and the Federal Ministry for Research and Technology (BMFT), F.R. Germany.

1. INTRODUCTION

The photovoltaic technology may help solve some of our future energy problems. In order to investigate and demonstrate the operation of a PV plant in an urban area, BEWAG/EAB installed a 10-kWp plant in West Berlin. The primary concern was to optimize the interaction of all components.

2. SYSTEM DESCRIPTION

The installed PV array of 9.4 kWp is made up of two subarrays mounted on fixed tilted support racks (see Fig. 1). These are referred to as the Siemens and AEG subarrays. Figure 2 shows the block diagram of the entire plant. Each subarray is connected to a separate dc-dc converter. The converters boost the array voltage from a nominal 220-Vdc to 350-Vdc level. Moreover, the converters have a common MPPT control, and their outputs are connected in parallel.

Until the battery voltage reaches the upper charge limit, the MPPT controller automatically searches for the maximum array power point, which is temperature dependent. It adjusts the array output

Siemens Array (left) and AEG Array (right)

Solar irradiance sensor, Matrix 1G (Si) on top and K&Z CM 5/6 (thermopile)

Back view of Siemens array

Back view of AEG array

Figure 1. The BERLIN Array is Made up of 5.6-kW Siemens and 3.8-kW AEG Modules

504

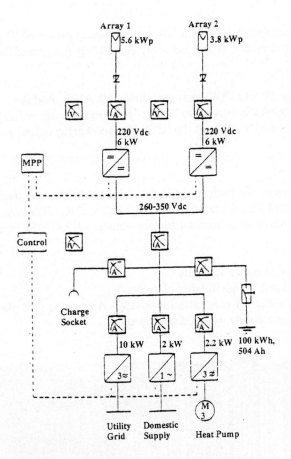

Figure 2. Block Diagram of the BERLIN PV Plant

voltage to the battery voltage which is determined by the battery state of charge. The function of the battery is to allow for fluctuations in the daily solar irradiance and the consumer power demand.

The consumers include a heat pump (2.2 kW, 3-phase inverter, frequency controlled), house appliances (1-phase inverter), lighting appliances, and charging station for an electric car, all connected to the 280-V bus.

If the needs of the consumers are below the power output of the solar generator and the battery is fully charged, energy is fed into the public grid. The battery charge voltage of 320 Vdc or over is used as the indication of surplus power condition, and the switch-over to the grid-connected mode occurs automatically. If the battery voltage falls below the cut-off voltage (245 V) due to insufficent solar power, the battery is charged by the public grid. A three-phase inverter provides for energy feed-back and battery charging. The reactive power is compensated for a load of 50%. A transformer provides a galvanic separation and an operation at different voltage levels. The transformer enables power transfer in either direction.

A standardized BMFT/KFA-supplied measurement system, designed and installed by WIP-Munich has operated successfully since mid-1989.

3. COMPONENTS

3.1. PV Array

The PV array consists of two subarrays (see Fig. 1). One of these subarrays contains 112 Siemens SM-50 modules. Each string consists of 14 modules in series and 8 strings are connected in parallel.

505

The module which uses monocrystalline Si cells has a rated output power of 50 Wp each at 25°C and 1000 W/m². The gross output power of this subarray is 5.6 kWp (i.e., rated module power x total number of modules).

The other subarray contains 98 PQ 10/40 polycrystalline-cell AEG modules. It is divided into 7 strings in parallel and 14 modules in series per string. The modules use polycrystalline Si cells, and each module has a rated power of 38.4 Wp at the STC. The total array output power is 3.8 kWp.

3.2 Battery

The rated capacity and energy of the battery is 504 Ah and 100 kWh, respectively, at the C/120 rate. The battery cell, Type 4 OCSM 320, is manufactured by Hagen, F.R. Germany. A total of 140 cells are connected in series, resulting in a nominal battery voltage of 280 Vdc (see Fig. 3). The cell's construction features are:

- High efficiency and long maintenance intervals
- Positive armour plate to achieve a high number of cycles
- Use of a lead alloy with an antimony concentration < 3% - A negative grid plate with an expanded-copper metal grid for voltage stability during discharging

Battery Rack (Cell Manufacturer: Hagen)

Figure 3. Battery Consisting of 140 Cells in Series (280-V Nominal Voltage, 100-kWh Capacity)

3.3 Converter

The dc-dc-converter is designed to operate as a voltage "boost" regulator. The outputs of the two 6-kW units are connected in parallel. The two converters are controlled by a 16-bit processor which generates staggered clock pulses, resulting in low output-voltage ripple. Through use of the SIPMOS field-effect power transistors, a power conversion efficiency of over 95% has been achieved. The converter and other electronics are mounted in the cabinets shown in Figure 4.

506

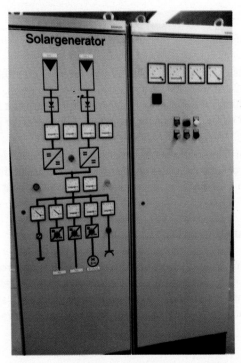

Figure 4. Cabinets Containing the Converters,
Inverters, and Control Elements

View of the PC and a box containing
power and current integrators

COMPAQ PC

Current and Power Integrators
with mechanical counter

Behind the Integrator is an IMP which
interfaces with the sensor outputs

Figure 5. Monitoring Equipment Designed and Installed by WIP-Munich

3.4 Inverter

Two different inverter designs are used. The 10-kW inverter manufactured by Siemens and referred to as "Simoreg" contains 2 three-phase bridges which use thyristors. Its functions are to feed power into the utility grid and charge the battery. Synchronization with the 50-Hz grid occurs automatically. The inverter is controlled by the supervising processor. The inverter power efficiency exceeds 90%.

The second inverter, also from Siemens and known as "Simovert", is a three-phase PWM-transistor unit. It powers the control unit (stepless speed) of standard induction motors. Its nominal power is 2.2 kW at frequency from 1 to 100 Hz. It supplies power to the heat pump, controlled by an internal computer.

3.5 On-line Monitoring and Control

The Siemens "Simatic" unit provides on-line monitoring and control. A portable IBM-compatible 16-bit computer controls the interaction of all components. The serial data transfer (current-loop, TTY) ensures trouble-free operation over long data lines. This systems displays all critical data concerning the plant conditions. Key control operating parameters can be changed on line via the portable PC.

3.6 Data Collection

In order to implement a standardized monitoring approach, the BMFT provided a unique system for data acquisition and storage. This system shown in Figure 5 consists of a Compaq 386 computer (IBM PC-compatible), Schlumberger IMP, Sun Power current integrator (solar irradiance integrator and mechanical energy counter), Si-based solar irradiance sensors, and WIP's data processing and storage software.

4. REFERENCES

[1] K. Bürgel, R. Hauk, R. Hanitsch, T. Mierke, B. Voigt, "10-kWp PV Plant in Berlin," Proc. of the 8th European PV Solar Energy Conference, Florence, IT, 9-13 May 1988, pp. 229-231.
[2] K. Bürgel, R. Hanitsch, T. Mierke, B. Voigt, "10-kWp Solaranlage in Berlin," Proc. of the 6th International Solar Forum, Berlin, FRG, 1988, pp. 507-511.

Appendix 24

POZOBLANCO PV PLANT

J.R. SANTOS, A. GONZALEZ MENENDEZ and
M. SIDRACH DE CARDONA
CIEMAT - IER
Avenida Computense, 22
E - 28040 Madrid
Spain

D. MAYER, G. ELALAOUI
Centre d'Enegétique ARMINES
Rue Claude Daunesse,
F - 06560 Valbonne
France

Abstract

The purpose of the POZOBLANCO PV plant is to evaluate the effectiveness of a stand-alone PV power source for an isolated dairy farm in Pozoblanco, Southern Spain. The consumer loads include the milking and cooling machines, small ac motors for milk shaking and an electric fan, domestic lighting, and water pumping. The plant was installed in late 1988 and put into operation by mid-1989. A simulation programme for system sizing has been developed in collaboration with the Centre d'Energétique of Armines (France). The data acquisition system was installed at the beginning of 1989 and the first results were available by March 1989. Data evaluation continued into 1990. The main technical work included computer simulation and analysis of plant performance, design and analysis of power management, failure detection and prevention, and economic analyses. A 17-kVA Diesel generator was added in late 1989 to meet the total demand of the farm, which had increased more than originally anticipated.

1. INTRODUCTION

The POZOBLANCO project is an application of PV systems in Southern Spain, where a considerable number of isolated farms with no utility grid are located. The PV plant supplies a 48-cattle dairy farm with an average milk production of 750 l/day. Figure 1 shows the farmhouse and the PV array.

The total energy demand is about 27 kWh/day, which is practically constant throughout the year. The installed PV peak power is 12.6 kW and the total rated battery storage capacity is 2,800 Ah. The project carried out the development of a new type of transformerless inverter for the main load (milking and cooling machines) using an electronic dc voltage converter working at high frequency. All system components, including the PV modules, were tested prior to installation at the laboratories of the Renewable Energies Institute (CIEMAT) in Madrid. Figure 2 shows the layout of the farmhouse, consumers, and the PV system components.

Figure 1. The PV Array and the Farmhouse (POZOBLANCO Project)

510

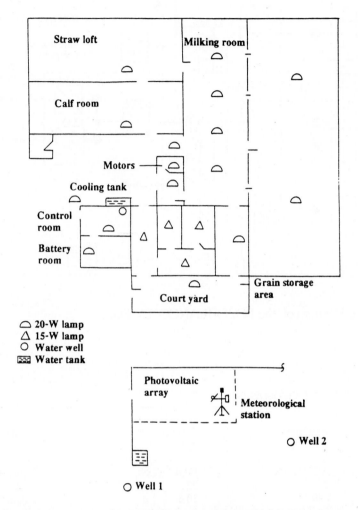

Figure 2. Layout of POZOBLANCO Farmhouse and PV Plant

2. SYSTEM CONFIGURATION

The PV system is divided into three independent subsystems as illustrated in Figure 3. Subsystem A with a PV array and a battery supplies power to the domestic lighting, milk shaker, and the monitoring equipment. Because of the load types used, this subsystem provides 24-Vdc and 220-Vac outputs.

Subsystem B, also with a battery, satisfies the highest power demands at the site, which are the milking and cooling machines, and the electric fan.

Subsystem C, consisting of a PV inverter set, supplies only the water pumping loads.

2.1 PV System

Figure 3 shows the three independent PV sources. The main components in these subsystems and their features are listed in Table 1. The basic ratings of the subsystems are as follows:

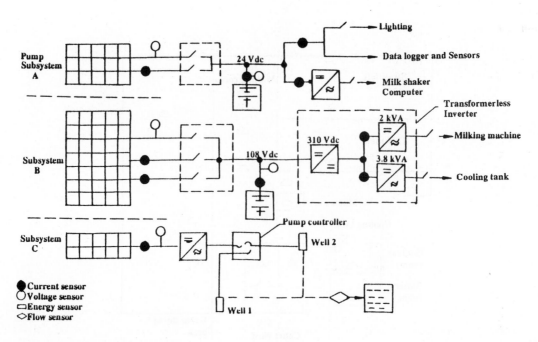

Figure 3. Block Diagram of the POZOBLANCO PV Plant

Table 1. List of PV System Components

Component	Subsystem A	Subsystem B	Subsystem C
PV Array			
- Module Manufacturer:	ISOFOTON	ISOFOTON	ISOFOTON
- No. of modules	30	225	14
- Peak power (kW)	1.40	10.6	0.66
- Module Wiring:	2 x 7 (series x parallel	2 9 x 8 (series x parallel)	7 x 2 (series x parallel)
	2 x 9 (series x parallel)	9 x 9 (series x parallel)	
Battery			
- Manufacturer:	TUDOR	TUDOR	
- Cell Type	7 EAN 100	10 EAN 100	
- No. of cells in series:	12	54	
- Capacity:	1,120 Ah	1,600 Ah	
Regulator			
- Manufacturer:	ATERSA	ATERSA	
- Model:	RS-130T	CRMCM-03	
- No. of Steps:	2	3	
Inverter (Auxiliary)			
- Manufacturer:	AMBAR	JEMA, S.A.	GRUNDFOS
- Rated Power:	300 W	5.8 kVA total	1,000 W
- Type:		Transformerless	Variable Frequency
- Input:			
o Min. Voltage:		97.5 Vdc	
o Max. Voltage:		130 Vdc	
o Rated Voltage:		110 Vdc	100 Vdc
- Output:		Three-phase, 220 Vac, 50 Hz	Variable voltage, three-phase 60 Vac, 10-A, 6-60 Hz
- Technology:		dc-dc converter (chopper) Intermediate voltages: 310 Vdc FUJI dc/ac converter PWM	

Subsystem A

- PV array: 1.4 kWp (30-Vdc)
- Battery: 24-Vdc, 1,120 Ah
- Inverter: 300 W, 220-Vac

Subsystem B

- PV array: 10.6 kWp (230-Vdc)
- Battery: 108-Vdc, 1,600 Ah
- Inverter: 220-Vac (both inverters)

Subsystem C

- PV array: 0.66 kWp (120-Vdc)
- Inverter: 1.0 kW, 60-Vac, 10 A, 6-60 Hz

2.2 Load Description

The loads in Subsystem A consist of 24-Vdc and 220-Vac consumers. Table 2 is a list of these loads along with their power requirement and estimated "on" times.

Table 2. List of Loads, Subsystem A

Loads	Power (W)	Daily On-Time (h)
24-Vdc		
Lighting:		
- 4 lamps (15 W each)	60	1
- 12 lamps (20 W each)	240	4
- 4 lamps (20 W each)	80	2
Sensors:	73	24
220-Vac		
Agitator (Milk Shaker):	250	2 min/0.5 h
Computer:	200	5 min

The daily predicted load profile of Subsystem A is shown in Figure 4. The loads in Subsystem B, listed in Table 3, all use motors with a 0.8-factor. The daily predicted load profile for this system is shown in Figure 5.

The principal load in Subsystem C is the 10 m^3/day water pumping system at the two wells located at 63 m and 88 m from the storage tank. The depths are 21 and 22 m, respectively. Under worst-case conditions, the peak power needed is 448 W. To satisfy this demand, two 658-W nominal power SP4-8 Grundfos pumps were selected.

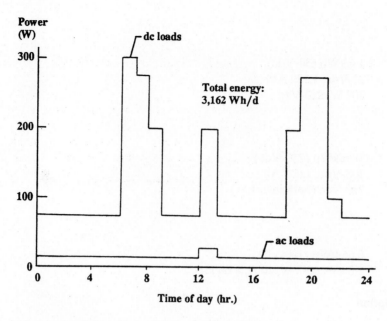

Figure 4. Daily Load Profile of Subsystem A

Table 3. List of Loads, Subsystem B

Consumer	Power (W)	Daily On-Time (h)
Milking machine:	1472	3
Cooling machine:	2944	6
FAn:	240	6
Milk Pumping:	2220	5 min every 2 days

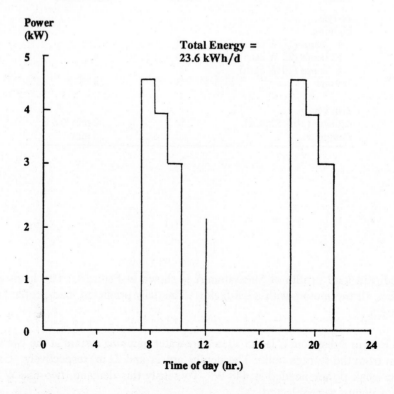

Figure 5. Daily Load Profile of Subsystem B, ac Power

3. SYSTEM SIZING

A computer analysis for the system sizing was carried out by the Centre d'Energétique of Armines (France), who collaborated in the project.

The global horizontal irradiance was available for the Cordoba site from the National Meteorological Institute in Madrid between the years 1976-1982. Different methods were used to estimate the values of the global irradiance at the tilt angle of 60°. Values of monthly average ambient air temperature for the same location were also used. This information is presented in Table 4. The irradiance at the

Table 4. Global Irradiance, Measured and Calculated, and Ambient Temperature

| Month | Measured Irradiance Cordoba Airport (W/m²) | Calculated Irradiance at 60° Tilt Angle (W/m²) | | Average Ambient Air Temperature (°C) |
		Method of Frutos	Method of Liu and Jordan	
January	2500	4700	4720	9.1
February	3000	4000	4390	10.7
March	4800	5400	5680	13.5
April	4900	4300	4280	16.3
May	6400	4700	4630	19.4
June	6400	4700	4250	24.4
July	7200	5200	4870	27.9
August	6200	5600	4980	27.6
September	5300	6000	5550	24.3
October	3880	5400	5260	18.6
November	2500	4600	4310	13.6
December	2000	4000	3830	9.6

* Horizontal surface

60° tilted surface was calculated by the Frutos correlation method which is the most appropriate for the Iberian Peninsula. In order to assure the maximum autonomy and reliability of the system throughout the year, the solar radiation received on the plane of the array in December, the most unfavourable month, was taken as the basis for array sizing. This analysis resulted in the following sizes of the PV array and batteries:

Subsystem	Peak Power Required (W)	Storage Capacity Required (Ah)
A	1122	1120
B	10152	1600
C	658	

4. DATA MONITORING SYSTEM

The monitoring has been carried out, taking into account the requirements given in the "Plan for Monitoring of R&D Photovoltaic Power Plants," for PV-powered applications sponsored by CEC/DG XII. The main characteristics of the DAS are as follows:

- Data logger (Datataker from Australia): very low consumption (0.48-W in active mode) with 11-K RAM battery back-up storage measured data. It is programmable by a PC compatible computer with an RS-232C interface.

- Number of sensors: 23

- Computer: PC with 20-MB hard disk and one 1.2-MB floppy disk. Daily measured data are stored in both disks.

- To reduce the energy consumption, the PC is turned on a total of 20 minutes maximum per day.

- Some of the sensors include galvanic isolation and 0-20 mA output to facilitate connection to the data logger.

- Two software packages have been developed. One is used for the data logger management, including channel programming, measuring time, alarms and control signals. The other programme is for the storage of data-logger information for the real-time display of system performance data.

5. PLANT PERFORMANCE

Table 5 lists the actual performance of the 270 PV modules delivered initially and tested by CIEMAT (44.3-W average). Upon replacement of the modules that did not meet the minimum power specification, the average power increased to 46.1 W as shown in Table 6. Figure 6 is a histogram of the initial and final distributions of module output based on 270 modules.

Table 5. Performance Spread of Initial 270 PV Modules (M75L) Delivered by Isofoton

	Isc (A)	Voc (Vdc)	FF	Vm (Vdc)	Im (A)	Pm (W)
Maximum	3.32	20.00	0.79	15.79	3.09	46.88
Minimum	2.93	18.90	0.69	14.10	2.76	42.22
Average	3.14	19.45	0.73	15.04	2.95	44.30
σ	0.08	0.22	0.02	0.30	0.06	0.82

No. Modules: 270
No. Cells in Module: 33
Cell Type: Monocrystalline Si, 100-mm square
Light Source: Simulator

Table 6. Performance Spread of PV Modules (M75L) form Isofoton After Out-of-specification Modules were Replaced

	Isc (A)	Voc (Vdc)	FF	Vm (Vdc)	Im (A)	Pm (W)
Maximum	3.32	19.94	0.79	16.16	3.10	47.78
Minimum	3.01	18.82	0.69	14.43	2.84	44.45
Average	3.18	19.55	0.75	15.40	2.99	46.11
σ	0.05	0.20	0.02	0.33	0.04	0.94

The installation of the PV system and connection of the loads were completed in early 1989. Various subsystems were started up for checkout and verification in March 1989. The monitoring system was turned on in June 1989.

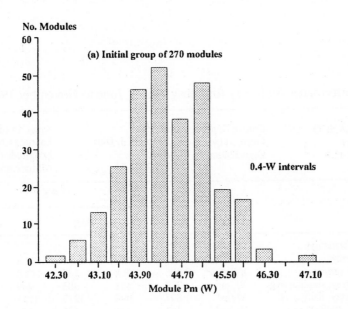

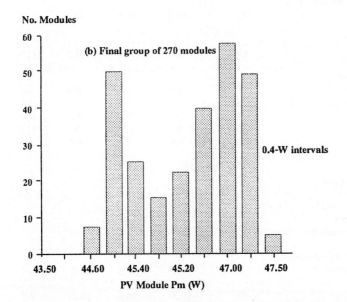

Figure 6. The Spread in Maximum Power on (a) Initial and (b) Final Groups of PV Modules for POZOBLANCO

Table 7 lists the key performance parameters from June to November 1989. The significant results and problems experienced during this period are as follows:

- Two failures in the output fuses of the 300-W inverter (Subsystem A)

- One failure of the overvoltage protection device from an indirect lightning strike

- One failure of the milk agitator machine (Subsystem A)

Table 7. Monthly Performance Summary for Subsystem A, June to November 1989

PV Pilot Plant:	**POZOBLANCO**	Date of Report:	**20-12-1989**	Array tilt (deg):	**60**
Project Leader:	**M. Sidrach**	Prepared by:	**M. Sidrach, J. Diaz**	Latitude (°N)	**38.42**
Organization:	**CIEMAT-IER**	Organization:	**TEAM**	Longitude (°W)	**4.87**
		Tel.	**34-1-346 62 54**	Altitude (m):	**651**

Parameter	Units	1989 Jun	Jul	Aug	Sep	Oct	Nov
1. Solar/Temperature/ Information:							
a. Insolation, plane of array	kWh/m²	114.7	170.9	159.9	153.4	153.4	83.7
b. Insolation, horizontal surface	kWh/m²	208.6	238.8	187	141.8	106.4	52.3
c. Irradiance, plane of array, sunlight avg	W/m²	329	351	383	449	460	382
d. Irradiance, plane of array, daily peak	W/m²	897	988	1213	1211	1149	1435
e. Irradiance, horizontal, sunlight avg.	W/m²	464	501	424	396	319	249
f. Temperature, ambient air, daily avg.	°C	28.2	29.7	28.1	24	21.3	16.7
g. Temperature, ambient air, lowest in daytime	°C	19.7	17.8	19.1	16.9	14.2	9.8
2. Plant Output Information:							
a. Energy array field	kWh	101.6	116.5	108.5	111.2	114.3	76.2
3. Plant Performance Indices:							
a. Array utilization factor	%	64	64	60	65	69	77
b. Array energy efficiency	%	5.8	5.7	5.7	5.8	6.2	7.6
c. Plant availability	%	100	100	100	100	100	80
d. Tilt factor	%	69	113	86	113	114	160
e. Specific energy output	%	2.5	5.6	4.6	5.6	7.8	13.9
4. Battery Information:	(Not reported)						
a. DOD, % rated Ah, daily avg							
b. No. of cycles							
c. Cell temperature, avg.							
d. Daily avg. recharge factor							
e. Lowest state of charge reached							
5. Operation & Maintenance Information:							
a. Total time DAS operational	%	40	90.3	64.5	66.7	74.2	76.7
b. Sunlight hours:							
1) Daily avg, actual	%	10.3	10.5	10.2	9.7	8.7	5.8
2) Daily avg., theoretical	%	14.7	14.5	13.6	12..4	11.1	10
3) Total actual	%	310	325.5	315.2	290	268.7	175
4) Total, theoretical	%	441	449.5	421.6	372	344.1	300
6. Appliction Information: (as applicable)							
a. Consumer (load) energy	kWh	62.7	98.9	102.9	95.8	99.9	87.1
1) Consumer (Light) energy	kWh	5.9	6.0	7.7	14.0	18.8	17.9
2) Consumer (D.A.S.) energy	kWh	30.3	37.7	44.2	41.1	42.5	39.5
3) Consumer (inverter) energy	kWh	26.5	55.3	51.1	39.9	38.7	29.7

- The failure of the Subsystem B inverter (transformerless) caused damage in the dc side because of high voltage surges from ac to dc side; this was attributed to inverter design problems. As a result, Subsystem B has been inoperative for a few months. To meet the increased load demand, a 17-kW Diesel generator was added. The inverter was redesigned using inductive filters to protect the dc side.

- Data collection stopped once with little loss of data; this was caused by the Datataker computer's failure to command the data scanning cycle.

- The respective energy consumptions of the cooling and milking systems from March to November were higher than expected. This is because the cooling system efficiency varies exponentially with the ambient air temperature.

- The number of cows has increased from 48 to 70, which substantially increased the milk production.

6. CONCLUSIONS

The initial 6-month results obtained on the POZOBLANCO installation are insufficient to draw general conclusions on the system performance. Optimization steps involving the milk cooling system and the PV power source were found to be necessary in view of the increase in the cattle population from the time of system installation.

Appendix 25

PV APPLICATION IN
PASSENGER CARS

A. KÖNIG and E. GRUNDMANN
Volkswagen AG
Postfach
D - 3180 Wolfsburg 1
F.R. Germany

Abstract

PV applications in passenger cars are defined and evaluated, and improvements of the cost/benefit ratio of such applications are discussed. The project was co-financed by the Commission of the European Communities, DG XII. In cooperation with the car accessories industry, design solutions for the integration of PV panels in a car body are investigated. A particular consideration stems from the fact that all areas of the car body are curved, mostly in two axes, and that vibrations and dynamic deformations have to be taken into account. One application of solar energy, which is already feasible nowadays is solar ventilation. This is a very effective means to decrease the overheating of a car parked in the sun. In winter time, the PV panel can be used for trickle charging of the battery. Special design of the system is necessary, because these two applications require different voltages. Vehicle propulsion by solar energy is not a realistic prospect for a general purpose passenger car. However, it is technically feasible to supply the electrical energy required for a car temporarily from a solar-charged battery instead of the alternator. The fuel economy improvement thus achieved ranges between 2 and 6%.

1. GENERAL ASPECTS

Although the direct drive of vehicles by means of solar power is technically possible, as demonstrated by the large number of automobiles built for solar rallies, it will remain restricted to special vehicles and will not be introduced into large-scale car production. The reason for this is simply the large disparity between the power density required by present-day cars and the power which can be generated by the solar panel mounted on such cars.

Even the possibility of storing solar energy in an on-board battery when the car is not being driven fails to produce a concept fulfilling the utility criteria for present-day cars in general use. This fact, however, does not mean that solar cells will play no role in future vehicles. A practical approach is to define auxiliary functions which could be powered by solar cells and to optimize appropriate equipment on the basis of cost-utility factors.

2. POSSIBLE APPLICATIONS FOR SOLAR CELLS IN PASSENGER CARS

The most advanced application for solar cells in cars is the ventilation of stationary vehicles to avoid overheating of the interior which occurs rapidly in summer. This is one of the few applications where the energy supply and demand are synchronous. Indeed, the efficiency of this application is very high since heat flows in the order of several kW can be conveyed with very few watts of electrical power. Furthermore, whenever the car is stationary and air-conditioning is not switched on, the solar ventilation system is a useful supplement. Table 1 lists some of the applications for solar cells in cars and the size of the momentary electric power requirements.

Table 1. PV Applications in Cars

Application	Power (W)
Battery trickle charge	< 1
Solar ventilation	20
Solar-powered coolbox	40
Minimum car electrical power requirements	
- Diesel engine	60
- Petrol engine (carburettor)	90
- Petrol engine (fuel-injection)	140

A good application of solar cells is for trickle charging of the battery. Internal battery self-discharge can have noticeable consequences, especially with older batteries. When the vehicle is left idle for long periods of time, battery self-discharge can be compensated for by charging from a solar panel. Also, during cold winter months, the trickle charging current helps keep the battery warm when the car is parked and its engine is not running.

Large solar panels combined with a suitably designed battery are capable of supplying the electrical power requirement of a car independently of the on-board generator. In this case, the generator can be disengaged from the engine, thus reducing its mechanical power output requirements and resulting in a corresponding saving in fuel.

3. DESIGN CONSIDERATIONS

Passenger cars usually have curved external contours, most of which map into 3-dimensional surfaces. Solar panel design must take this fact into consideration in order to follow the contours of the car body. A relatively simple technical method is to attach crystalline solar cells to sunroofs or sliding roofs. Roofs measuring between 0.2 and 0.3 m^2 are already available on the automotive parts market. The attachment and connection of each of the cells to a panel and the process to fabricate a solar roof are not simple, and only limited production and cost reduction possibilities exist. A major breakthrough will be possible when stable amorphous silicon cells become available. Because of their flexibility, these cells can be attached directly to the curved contour of a sliding roof or sunroof.

The methods described above are less suitable for attaching solar cells to large areas, such as the whole of the car roof. The best method for large areas is to use flexible, preassembled solar panels which can be bonded to the roof. The flexible panels themselves have a very thin stainless steel foil substrate on which multi-layer amorphous silicon is applied in a continuous process. The top- and under-sides of these panels are encapsulated in a plastic foil.

Figure 1 shows some of the test cars on which solar cells have been integrated on the roof. Polycrystalline solar cells are attached to the standard optional extra steel sliding roof on the vehicle in the foreground and the Multivan. A flexible matrix with Si amorphous (multi-layer) cells was attached to the roof of the Passat Variant in the middle of the picture.

Figure 1. Test Vehicles

4. TEST RESULTS

4.1 Solar Ventilation

When a solar panel (roof-mounted) is mainly used to operate the car fan, it is not practical to design the panel's operating voltage to be identical with the car's electrical system voltage. It is better to use the power-voltage characteristic of the fan motor. In practice, this figure is between 4 and 5 V. If the solar panel is also used from time to time to charge the battery, a voltage regulation is required, either in the form of a dc-dc converter or by switching sections of the roof (e.g., 3 x 4.5 V) in series.

Figure 2 shows the efficiency of the ventilation. The tests were performed in a climatic wind tunnel with solar simulation to ensure the reproducibility of the tests from one test point to another. The test car was a Passat Saloon with a sliding roof fitted with polycrystalline solar cells. The global test conditions led to a slight underassessment of the effect of solar ventilation since there is a greater rise in the temperature of the solar cells in the climatic chamber than on a vehicle in the open air with a corresponding efficient air flow around the car. The result is that solar ventilation causes a temperature drop in the car interior of approximately 18 K. This effect lies within the same order of magnitude as when the car fan runs continuously at stage 1.

The effect of solar ventilation on the time period after the start of a vehicle until the air conditioning has lowered the interior temperature to an acceptable value is shown in Figure 3. The basic tests were conducted on consecutive days during a very stable fair-weather period at the end of May 1989. The test car was a Jetta equipped with a sunroof covered with monocrystalline cells. The car was parked in the sun for several hours until the interior temperature had reached an approximately steady condition. The car was then started with the air-conditioning switched on and operated for some time at a raised idling speed of 2000 rpm. At the same time as the engine was started, the line-recorder paper for plotting the temperature was speeded up. It again shows that the solar stationary ventilation reduces the overheated condition of the car (difference between headroom temperature and ambient temperature) by over 50%. The time period until the temperature reached an acceptable level, i.e., 30°C, was shortened by several minutes.

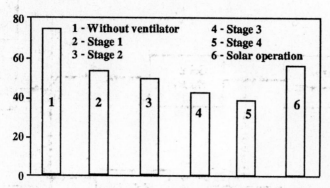

Figure 2. Headroom temperatures at various ventilator operating conditions. Wind speed 6 km/h; outside temperature 25°C; and solar irradiance (simulated) 800 W/m²

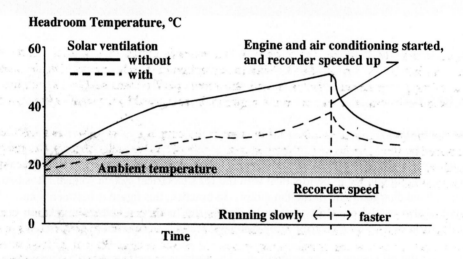

Figure 3. Influence of Solar Ventilation on Temperature in Headroom with and without Air Conditioning

4.2 Solar Power to Cover the On-board Electrical Power Requirement

Figure 4 shows in a very simplified form what the electrical system of a car would look like if its electrical power requirements were covered either by a solar panel or by the conventional car generator. In the practical implementation, some special requirements must be considered to ensure the reliability of the car. Standard car starter batteries and solar batteries are optimized in different ways. Starter batteries are designed to be capable of supplying high current flows for short periods of time and are not designed for low discharges since they are normally recharged immediately by the car generator. Batteries for solar systems are capable of deep discharging and a very large number of charge/discharge cycles. On the other hand, they are not capable of delivering a high short-term current like starter batteries.

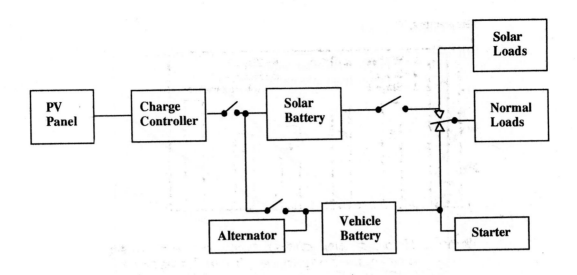

Figure 4. Block Diagram for the Test Multivan

In this specific case, the test car was fitted with two batteries of different designs. The solar panel charges a solar battery via a control unit. With the exception of the starting process, the solar battery is capable of supplying normal car loads, and also some specific loads such as a coolbox. The car battery is used for starting the car and can, if required, be recharged by the standard generator.

However, the car generator is switched off by a magnetic coupling if the power requirements of the car are covered by the solar battery. When operating with dipped headlight beams, the car generator is switched on continuously for safety reasons since its lights are designed for the higher generator voltage and not battery voltage.

When the generator is switched off, the load on the car engine decreases, resulting in a corresponding reduction in fuel consumption. The fuel consumption in practice is dependent on a number of parameters, such as car power consumption, generator power output, the partial load consumption characteristic of the engine, generator efficiency, and car weight. The data shown in Figure 5 were obtained in stationary operation on the fuel-injected Multivan used in the test stand. Taking the average of all measured points, there was a reduction in consumption of about 3% when the car generator was switched off.

During a practical test, the vehicle was driven for several consecutive days on a circular course around Wolfsburg, mainly on country roads. The total distance around the course was 77 km and the average travel time was 78 min. For the rest of the test day, the vehicle was parked in the open air to recharge the solar battery.

Figure 6 shows a typical curve of solar radiation power, solar cell power and the power extracted from the solar battery during the test time. The average electrical power required for the travel distance was calculated as being 247 Wh. During one day an aggregate of 346 Wh was charged into the solar battery. After four days, the solar battery which was originally largely discharged was fully charged up to gassing voltage. Further charging was then interrupted by the regulator. Compared with the test run with the car generator switched on, an average fuel saving of 4% was determined for operation using the solar panel.

Table 2 shows the approximate charging level of the solar battery and the battery open-circuit voltage at the beginning and at the end of each driving period.

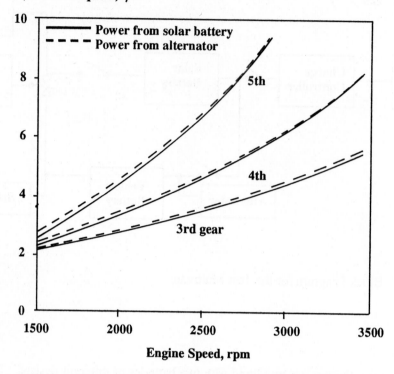

Figure 5. Fuel Consumption Improvement by Solar Power.
Multivan, 82 kW with Electronic Injection Engine

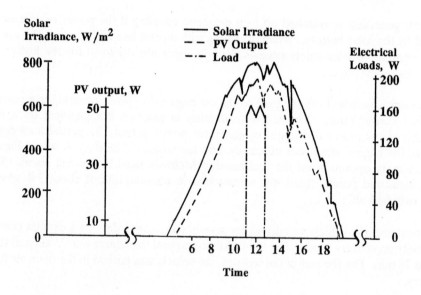

Figure 6. Vehicle Test Multivan with Solar Roof

Greater saving potentials can be obtained for car/engine combinations with extremely low specific fuel consumption for which the generator drive requires a relatively higher power component. Thus, fuel savings of 4-8% were obtained on stand tests using a Diesel engine at a load corresponding to the partial load of a Volkswagen Jetta.

Table 2. Charging Level and Battery Open Circuit Voltage at Beginning and End of Each Driving Period

Day	Charge Level %		Battery Voltage, Vdc	
	Start	End	Start	End
1	50	30	12.7	12.2
2	48	27	12.7	12.1*
3	60	40	13.0	12.2
4	75	55	13.1	12.2
5	(100)		14.0**	

* Low voltage warning
** Charging interrupted

5. CONCLUSIONS

Solar cells for cars can offer valuable services for auxiliary functions. They can also be used for purposes for which they have a quantifiable utility, e.g. fuel saving and improvement of passenger comfort and car reliability. The main factors governing their success in the car market will be:

- A further cost reduction of solar cells and panels,

- Attractive visual appearance of solar panels, and

- The availability of suitable battery charge control electronics which are reliable in typical car conditions, featuring extremely low internal consumption.

Appendix 26

RECENT ADVANCES IN
SOLAR IRRADIANCE MONITORING
DEVICES AND CALIBRATION METHODS

H. OSSENBRINK
CEC - JRC
Ispra Establishment
I - 21020 Ispra VA
Italy

G. BEER and S. GUASTELLA
Conphoebus Scrl
Passo Martino - Zona Industriale
Casella Postale
I - 95030 Piano d'Arci (Catania)
Italy

M. IMAMURA
WIP - Munich
Sylvensteinstrasse 2
D - 8000 Munich 70
F.R. Germany

Abstract

Within the concerted actions of the Commission's PV Pilot Plant R&D Programme a study was undertaken to determine and validate a cost-effective on-site calibration method for silicon solar sensors used in PV plants and to identify a reliable sensor design with a stable lifetime comparable to that of the modules.

This paper summarizes 1) the present state-of-the-art devices used for monitoring solar irradiance mostly by the PV community, 2) practices employed by device suppliers and users in calibrating such devices, and 3) preliminary results of concerted action studies being carried out by several test centres in Europe regarding the use of Si-based devices and their calibration methods.

Our preliminary observations are that 1) the simplified calibration method is effective and suitable for system designers, installers, and operators, and 2) the Si-based devices evaluated (mainly monocrystalline) are cost-effective substitutes for the thermopile pyranometers.

1. INTRODUCTION

Since the early 1970s there have been many discussions and publications on the calibration of silicon (Si) sensors in natural sunlight [1 to 4]. The methods described and standardized are relatively complex and expensive to implement. What is needed is a simple but reliable method of calibrating solar irradiance sensors, which the solar energy monitoring communities and systems throughout the world can agree to and use at the installation site without sending every sensor to a calibration agency.

The method proposed in this paper is a slightly modified version of new approaches proposed [5, 6] but not yet widely used. The calibration constant is the slope of the linear regression line over a range of intensity, including the origin. This method does not require special skills or expensive equipment and can be applied easily by field workers either when they receive the sensors, or at the time of sensor installation and during periodic site inspections. In the strict sense, this method is not intended as a primary reference cell calibration method as it relies on a relative reference instrument, i.e., a thermopile pyranometer.

Another point of concern in solar irradiance monitoring is the suitability of the sensor itself. It should keep its calibration precision as long as possible without any environmentally induced degradation. From the point of view of PV users, a spectral and optical match with the photovoltaic array has been generally favoured.

The paper first describes the solar irradiance sensors and calibration methods in use and presents a simplified calibration approach and a review of state-of-the-art sensors. The second part focuses on the experimental approach taken in collaboration with a number of experienced laboratories, to assess both calibration methods and sensor design and concludes with preliminary results. This investigation is being continued in the JOULE programme, and final results should be available by late 1991.

2. CALIBRATION METHODS

Standards for the calibration of a solar sensor differ according to the level of the instrument, i.e., primary or secondary sensors. A primary reference sensor is one which has been calibrated outdoor against a cavity radiometer, which in turn is traceable to the World Radiometric Reference (WRR). A secondary reference sensor is one which is calibrated against a primary reference cell or an ISO secondary standard [7] reference pyranometer.

2.1 Standards for Outdoor Calibration Methods

The standardization of reference Si-cell calibration methods for terrestrial applications is under discussion at the international level. In the USA, two outdoor calibration standards, one for direct beam and the other for global irradiance, have been issued by the American Society for Testing and Materials (ASTM) [2]. The International Electrotechnical Commission is considering their adoption in a slightly modified form within the activities of its Technical Committee 82 [7]. A description of similar techniques can be found already in Specification 101 of the Joint Research Centre [8]. The criticism of these outdoor methods concerns their reproducibility under different atmospheric conditions [9] and the reliability of a reference pyranometer [10]. We also found that the calibration precision obtainable with outdoor methods depends on the construction of the reference sensors.

To minimize problems with thermopile pyranometers, the existing standards recommend that each user select and maintain in a relatively unused state two or more thermopile working standards for the purpose of calibrating other thermopile instruments and Si sensors used for continuous outdoor measurements. These working standards are usually periodically checked against an absolute cavity radiometer traceable to the WRR.

2.2 Proposed Method

A basic issue in the use of radiation sensors is to what extent the calibration factor is valid in terms of initial and temporal precision. If installed in a field, a simple but reliable method should be applied in order to verify sensor sensivity at regular intervals. The method described here was intended to be executed by plant installers and operators, by simply correcting the manufacturer's or previously established calibration factor to that of a precision pyranometer (ISO secondary standard) used as the reference.

Basically, the calibration run consists of simultaneous readings of the reference pyranometer and the sensor to be calibrated at irradiance levels between 200 and 1,000 W/m^2. The measurement is done only on one clear day in contrast to three days specified in available procedures [2, 5, 7]. The calibration factor is calculated from a straight-line fit through the data representation of sensor voltage vs. irradiance as measured with the reference instrument. The fit procedure should be applied in a manner that the resulting straight line is made to pass through the origin. This is done to provide repeatability in the calibration factor and to minimize errors due to the slow-response characteristics of the reference thermopile sensor, combined with the rapidly changing solar flux and cosine effects of the sensors, especially at low irradiances (below 500 W/m^2).

The simplified calibration approach requires a sun-tracking platform (manual or automatic), an ISO-secondary standard thermopile pyranometer selected as the calibration reference, and the sensors to be calibrated. We have found that a Kipp & Zonen CM11 or Eppley PSP pyranometer is perhaps the best choice for the reference sensor. Another variation of this approach is to calibrate the sensors in the tilted or horizontal position, for cases where the sensors should not be disturbed from their mounted position. For this purpose, we recommend use of solar irradiance data between 11:00 and 13:00 h to minimize their cosine errors.

The method: 1) assumes a linear relationship between the Si cell's short-circuit current and the thermopile pyranometer output, especially at high irradiance levels (e.g., greater than 800 W/m^2); 2) it neglects Si cell temperature effects; and 3) it disregards atmospheric and air mass constraints. Moreover, it does not require cell spectral response or sun spectral irradiance measurements. It is, therefore, more simplified than other outdoor methods in use or in discussion. However, partially sunny days with clouds obscuring the sun periodically should not be considered as optimum for calibration purposes. Data obtained at several locations in Europe (Southern France, Sicily, Spain, and Germany) indicate that under different site conditions one can achieve acceptable results (within 2% of the thermopile working reference) with properly constructed Si sensors. This remains to be proven for other locations in the world, so we encourage others to investigate it at their geographical locations.

3. SOLAR SENSORS IN USE

3.1 Thermopile Sensors

This type of sensor comprises pyranometers and pyrheliometers based on a thermal absorption principle. The main instrument manufacturers are Eppley Laboratories, Kipp & Zonen, and Schenk. The temperature difference between the hot and cold junctions of the thermocouple produces a voltage which is proportional to the solar flux. Various approaches have been implemented to provide for linearity and stability over the whole range of solar irradiance and ambient air temperature by optimizing the coating of the receiver surface, balancing thermal mass, and designing optimum thermocouple layout. The pyranometer covered by a glass dome has a full hemispherical field-of-view, whereas pyrheliometer has a 5.6° field-of-view in order to measure the direct sunlight component only. Figure 1 shows examples of these instruments.

For a majority of PV projects in Europe, only pyranometers are used; depending on the information needed they are mounted horizontally or tilted to the plane of PV array. The principle of measurement has two major drawbacks for monitoring purposes, one being the rather low signal output (about 10 mV for full irradiance), which imposes high requirements on the data acquisition system, the other being the temporal stability of the back paint on the receiver surface. Noticeable degradation can occur even after a half-year exposure [10].

3.2 Silicon Sensors

Silicon solar sensors consist basically of a monocrystalline or polycrystalline PV cell with a sufficiently small resistor to provide a reading of its short-circuit current (Isc). The linearity of the device Isc

Kipp & Zonen CM11

Eppley PSP on a sun tracker

Kipp & Zonen CM11 (without cover)

Eppley NIP on a sun tracker

Figure 1. Commonly Available Thermopile Sensors for Monitoring Global
Irradiance (CM11 and PSP) and Direct Normal Irradiance (NIP)

with the solar irradiance is very good. This feature, as well as the fact that they are usually cheaper and easier to maintain than the thermopile sensors, provide good argument for choosing Si-based sensors for the continuous monitoring of solar radiation.

An important feature of these sensors is their spectral sensitivity which is limited to that of silicon (0.4 to 1.1 micrometre). The thermopile sensor's spectral window is between 0.4 and 2.8 micrometre, which is the characteristic of the cover glass. Depending on the application and the needs of the user this can be an advantage or disadvantage. Usually, in PV monitoring applications one prefers to have a sensor which indicates the irradiance level only of that part of the solar spectrum which is converted into electricity. Unlike the thermopile sensors, silicon sensors have the same fast reaction time to irradiance changes as the photovoltaic array. This is of importance not only for the accuracy of solar flux and energy data, but also for system control purposes.

The silicon sensor cell has to be protected from the environment. Silicon sensors in use differ mainly in their encapsulation design, the cover glass, and shunt resistance type and value.

3.2.1 Commercial Silicon Sensors

Only a few silicon sensors are commercially available for monitoring purposes. The cost of these sensors ranges from DM 600 to 800 in Germany. Figure 2 shows commonly used devices with significant field data. They are briefly described below.

Matrix

Haenni 130

Dodge Product SS100

Licor 200SZ

Figure 2. Commercially Available Si-based Sensors Used for Global Irradiance Monitoring

Haenni Solar 130

This sensor consists of a small (less than 0.5 cm^2 area) monocrystalline solar cell (Siemens BPX91B) incorporated in an aluminium housing, a curved white plastic diffusor (Perspex Opal 040) as a light window, a shunt resistance between 250 and 400 ohms, and a 2-wire cable. It is calibrated by Haenni to yield an output voltage of 100 mV at 1000 W/m^2.

As an option, model 118 comes with the sensor (Solar 130) and a battery-powered portable digital unit which integrates the solar irradiance data from the Solar 130 sensor and displays both energy and flux values on its LCD. Spectral response measurements show that the milky white plastic window absorbs certain components of the spectral irradiance, effectively resulting in a flat response curve. Obviously, the manufacturer's intention is to compromise between the flat but broader response of a pyranometer and that of silicon.

Li-Cor 200 SZ

The construction of this sensor is very similar to that of the Haenni, except that it uses a smaller monocrystalline solar cell made by EG&G and a white plastic (acrylic) cover with a flat surface. The cell's spectral response is similar to the conventional Si cell.

The manufacturer calibrates the sensors indoors under a preadjusted light source and provides a calibration coefficient for each cell individually (usually in units of microamps per 1000 W/m^2).

Matrix 1G

The Matrix 1G sensor has a 4-cm^2 monocrystalline solar cell on an aluminium base, covered by a pyrex glass dome to optimize its cosine response. The design has also the advantage of being used in a horizontal position while preventing dust accumulation (self-cleaning effect). The output voltage can be optionally 50 mV or 100 mV at 1200 W/m^2. The Matrix uses manganin material for its shunt resistance with a $\pm$ 0.1% tolerance and provides 2-wire shielded electrical output leads. The spectral response of the silicon cell is not affected by the glass dome.

The manufacturer calibrates the sensors outdoor against an Eppley PSP and provides a calibration coefficient for each cell individually (in microvolts per W/m^2).

Dodge SS100

This unit is completely encapsulated in white acrylic material, including the cover for the Si sensor. The sensor is a 5.7-cm diameter PV cell. Each sensor is individually calibrated in sunlight against a thermopile sensor before shipment. The manufacturer adjusts the electrical output signal to 100 mV for a 1000 W/m^2 solar irradiance via a potentiometer accessible from its base.

3.2.2 Custom-made Silicon sensors

Custom-made devices are encapsulated solar cells made on special request by the system users, module manufacturers, or calibration agencies. Thus they are normally not widely distributed. They can be purchased in Germany for about DM 1,200. Figure 3 shows some of the devices in use and a brief description of the main sensors is given below.

DSET Reference Cell

The layout of the cell holder follows the ASTM standard [11], incorporating a 4-cm^2 solar cell, which is covered by a quartz or pyrex glass window. The device is not particularly designed for long outdoor exposure. In contrast with the commercial sensors, it is usually provided with a thermocouple for cell temperature monitoring, and no shunt resistor is attached to the output of the cell.

Reference Cell in Module Package (RCMP)

This reference cell is usually a crystalline solar cell encapsulated with the same materials used for the PV module production of the respective manufacturer (glass front, EVA, Tedlar or glass back cover, and aluminium frame). Some manufacturers incorporate a shunt resistance within the laminate or a terminal box. To achieve the best thermal match with modules [12, 13, 14], dummy cells or parts of them are included in the laminate around the active and instrumented cell. This design is usually constructed to meet the proposed international standard [15]. A typical example fabricated by Siemens is shown in Figure 3. As in the Matrix unit, the Siemens RCMP uses a 0.1% precision resistor made of manganin material connected across the cell terminals. The main advantage of the RCPM design is that its encapsulation technology is identical to the PV module which has already been proven in various qualification tests, and therefore it is environmentally qualified for long-term outdoor use.

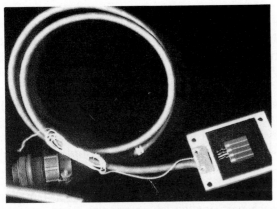

Reference cell in ASTM package (Available from Univ. Politecnica, Madrid, ES)[4]

Mesa (Konstanz, FRG)

AEG (Wedel, FRG)

GBI (Berlin, FRG)

Siemens (Munich, FRG)

Italsolar (Rome, IT)

Figure 3. Custom-designed Reference Si Sensors for Global Irradiance Monitoring

4. DESCRIPTION OF THE SENSOR CALIBRATION STUDY TASK

To investigate the calibration and design issues of silicon sensors, a special analysis task was initiated in 1988 in the frame of the Concerted Action on Modules and Arrays in the Pilot Plant R&D Programme. This task will continue into 1992, but key results are expected by 1991.

The structure of the study carefully considered the problems involved in sensor calibration, i.e., sites of different geographical location have to be involved and the problem of sensor stability has to be assessed by an initial and final calibration. In order to refer to a common radiation standard, all reference pyranometers were initially calibrated by a single agency (JRC).

Five European laboratories are involved in the calibration study: IER-CIEMAT (Madrid, Spain), Fraunhofer Institute ISE (Freiburg, F.R. Germany), Conphoebus (Catania, Italy), CEA (Cadarache, France), and Joint Research Centre ESTI (Commission of the European Communities, Ispra, Italy). The first four laboratories are following the outdoor calibration, whereas ESTI performs the initial and final laboratory checks.

The main purpose of this study is to develop a simplified but acceptable method of performing recalibration of sensors in the field. Therefore, each participating laboratory was asked to make calibration runs following the simplified procedure in December, March, June, and September (i.e., at approximately yearly solstices and equinoxes) after the baseline calibration. In the course of the programme, stability and repeatability of the calibration factor, cosine and temperature effects, and endurance of the sensors were to be investigated. Basically, all sensors described earlier are included in the study.

4.1 Baseline Calibration

The baseline calibration of the silicon sensors was done by means of an indoor calibration using two methods: the solar simulator method with a primary reference cell as described in [16], and the absolute spectral response method which is in principle a calculation of the calibration factor based on absolute spectral response and standard spectral data [17]. The baseline calibration value was calculated as the mean of the results. The reference pyranometers were calibrated before the tests against an absolute cavity radiometer according to ISO/DIS 9060 [18].

4.2 Sensors

The question of a suitable Si-sensor construction was assessed by including various kinds of sensors in the calibration task presented. The sensors brought into the campaign represent the variety of silicon sensors in use: Haenni, Li-Cor, Matrix, DSET, and Reference Cells in Module Pack (RCMP). RCMPs from a single manufacturer (Siemens, Munich, FRG) have been given to the participants in order to assess the site-dependence of the calibration method and to evaluate sensor construction differences.

At all calibration sites, the reference pyranometer used was either a Kipp & Zonen CM11 or Eppley PSP which was initially calibrated by JRC.

4.3 Data Analysis and Presentation

In accordance with the proposed method, readings of the test sensor(s) and the reference pyranometer have been taken throughout the day. A typical plot of these data is shown in Figure 4, where the pyranometer reading is plotted versus time of day. Analysis is performed on the plot sensor reading versus irradiance, as demonstrated in Figure 5. The straight line fit through the data points and origin yields the calibration value for this day. Some laboratories repeated the measurements on one or two other days. In the final results, the mean of these values has been taken. Table 1 summarizes these calibration data for the participating calibration sites with their respective sensors measured at different periods of the year; the values have been normalized to the baseline calibration and are presented as percentage deviations from this initial value. The last column averages the data over the periods and shows the value for twice the standard deviation for each sensor.

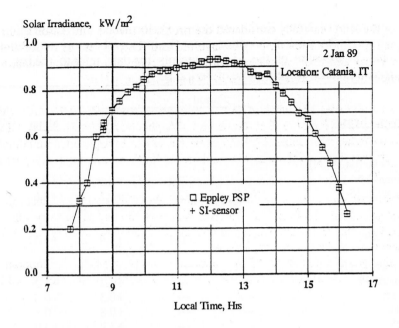

Figure 4. Typical Calibration Measurements: Solar
Irradiance vs. Time of Day

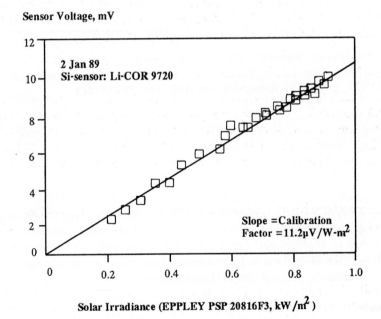

Figure 5. Test Sensor Output Voltage vs. Reference
Pyranometer Solar Irradiance

5. RESULTS

As the calibration exercise was under way at the time this paper was completed (late 1989), only preliminary results are described here.

Table 1. Summary of Initial Results on Sensor Performance and Calibration Method. Shown is the percent deviation from the indoor laboratory calibration carried out. In the case of the Li-Cor sensors the data refer to the first available measurements, namely those of Period II.

	Period				
Laboratory and Sensor	II	III	IV	Mean	2 Sigma
CEA Cadarache (lat. 44 N)					
Siemens 19	---	- 2.3	+2.6	+0.2	4.9
Photowatt 15	+5.7	+2.8	+5.2	+4.6	2.5
ISE (lat. 48 N)					
Siemens 9	+2.8	+0.4	+1.4	+1.5	2.0
CIEMAT (Lat. 41 N)					
Siemens 17	---	---	+1.2	+1.2	---
Solarex	---	---	+0.3	+0.3	---
PT04	---	---	+0.8	+0.8	---
PT05	---	---	+3.9	+3.9	---
Conphoebus (Lat. 37 N)					
Siemens 12	---	+0.3	- 0.7	- 0.2	1.0
Siemens 13	---	+0.1	- 0.9	- 0.4	1.0
LiCor 9720	+0.0	- 4.5	- 6.3	- 3.6	5.3
LiCor 9721	+0.0	- 3.7	- 4.6	- 2.8	4.0
Haenni 10185	+7.6	+2.7	- 0.5	+3.3	6.7

Note:
1) Period II: Dec/Jan 89, Period III: Mar/Apr 89, Period IV: Jun/Jul 89.
2) Commercial Sensors are Solarex, Haenni, and LiCor; the others are custom-made sensors. PT04 and PT05 are ASTM designs fabricated by Univ. Politecnica.

5.1 Site Effects

All laboratories reproduced the indoor calibration values of the Siemens RCMPs to a high degree of accuracy (within $\pm 2.5\%$ of the initial calibration). This indicates first, that the calibration method is quite independent of the site and geographical location, and second, as standard instrumentation is used, that errors due to sophisticated equipment and procedures are minimized. However, one should bear in mind that all reference pyranometers have been previously calibrated by a single agency, effectively eliminating pyranometer calibration errors.

5.2 Seasonal Effects

The Conphoebus data seem to indicate a degradation effect over the three periods. A confirmation can only be given after the final calibration of the sensors. The Period III (March) data of CEA and ISE are lower by 1% to 5% than at other (winter amd summer) periods. However, these data are closer to the laboratory values, a tendency which is also supported by the Conphoebus data. Presumably low and high elevations of the sun influence the result, as can be expected because the indoor laboratory calibration is valid for AM 1.5 (sun elevation 48.2°) [19].

5.3 Sensor Differences

As can be seen from the column listing the standard deviation of the data, some of the sensors have yielded more consistent results throughout the year than the others. The worst case was the Haenni sensor, where the spread of calibration results from period to period is the largest. Reasons could be the shunt resistance accuracy and/or the quality of its connection, or spectral mismatch between reference pyranometer and sensor. The Li-Cor sensor calibration values scatter with the calibration period similar to those of the Haenni.

On the other hand, the Siemens RCMPs show little spread in data and also small deviations from the laboratory indoor calibration. The largest absolute deviation from the baseline calibration is found with the Photowatt RCMP. The reason for this is unclear, but we suspect that the precision and type of shunt resistance used is one of the primary contributors.

6. CONCLUSIONS

Our preliminary observations are that 1) the simplified calibration method is effective and suitable to system designers, installers, and operators, and 2) the Si-based devices evaluated (mainly monocrystalline) are cost-effective substitutes for thermopile pyranometers.

The calibration results using the simplified method proposed seem to depend on the construction of reference sensors: there are sensors which read the same calibration factor as in the laboratory, independent of location and season. The large deviations of some of the sensors were already known and can be interpreted as effects due to the plastic diffusor and/or the accuracy of the shunt resistor used. More data will be necessary to establish the proper design criteria for stable and reliable reference sensors. Taking into account the simplicity of the calibration method applied, the sensor precision is quite high. In nearly all instances the Siemens RCMP reproduces its indoor calibration factor within $\pm 2\%$. The repeatability of sensor sensitivity for Matrix 1G calibrated against CM11 has also been found to be excellent (within $\pm 2\%$ based on comparison with outdoor measurements).

The reliability of the sensors involved should be checked after the calibration exercise by a final calibration and an accelerated ageing test similar to that of modules.

7. ACKNOWLEDGEMENTS

The authors wish to thank other participants of this concerted action campaign for their support, which include Mr. Chenlo of IER-CIEMAT, Mr. Heidler of ISE, Mr. Ragot of CEA Cadarache, and Mr. van Steenwinkel and Mr. Rau of JRC Ispra. The Siemens RCMP was designed by Mr. Bednorz of Siemens Munich, and we would like to thank him for his special efforts. This work has been supported by DG XII (Science, Research and Development) of the Commission of the European Communities.

8. REFERENCES

[1] R.D. Whitaker, A.W. Purnell, G.A. Zerlaut, "Progress in the Development of Standard Procedures for the Global Method of Calibration of Photovoltaic Reference Cells," Solar Cells, 7 (1982-1983) 135.

[2] "Standard Method for Calibration and Characterization of Non-concentrator Terrestrial Photovoltaic Reference Cells Under Global Irradiation," ASTM Standard E 1039-85.

[3] K.A. Emery and C.R. Osterwald, "Measurement of Photovoltaic Device Current as a Function of Voltage, Temperature, Intensity and Spectrum," Solar Cells, 21 (1987) p. 313.

[4] K.A. Emery and C.R. Osterwald, "Solar Cell Measurements," Solar Cells, 17 (1986) p. 253.

[5] F.C. Treble, "The Calibration of Primary Terrestrial Flat-plate Reference Devices - A Proposed International Standard," Proc. of the 7th European PV Solar Energy Conference, Seville, ES (1986) p. 89.

[6] M.S. Imamura, M.B. Mahfood, and M. Hussain, "Simplified Calibration Method for Silicon Solar Irradiance Sensors," Proc. of the 7th European PV Solar Energy Conference, Seville, ES (1986) p. 54.

[7] "Primary Reference Device Calibration Methods," International Electrotechnical Commission, Technical Committee 82, Document 82/ WG2 (Secretary) 101, (unpublished).

[8] K. Krebs, "Standard Procedures for Terrestrial Photovoltaic Performance Measurements" - Specification No. 101 - Issue 2, EUR 7078 EN, CEC/DG XII Joint Research Centre Ispra (1980, Reprinted 1989).

[9] H.B. Curtis, "Global Calibration of Terrestrial Reference Cells and Errors Involved in Using Different Irradiance Monitoring Techniques", Proc. 14th IEEE PV Specialists Conference, San Diego, CA, 1980, p. 500.

[10] M.B. Mahfood, M.S. Imamura, and M. Al-Khaldi, "Long-term Outdoor Performance Characteristics of Silicon and Thermopile Solar Irradiance Sensors," Proc. of the 7th European PV Solar Energy Conference, Seville, ES (1986), p. 349.

[11] "Standard Specification for Physical Characteristics of Non-concentrator Terrestrial Photovoltaic Reference Cells," ASTM Standard E 1040-84.

[12] F.C. Treble, "The Selection of Reference Devices for Photovoltaic Performance Measurements," Proc. of the 8th European PV Solar Energy Conference, Florence, IT (1988), p. 53.

[13] R. Shimokawa, F. Nagamine, Y. Hayashi, "Photon Collection Enhancement by White Rear Cover Reflection and the Design of Reference Cells for Module Performance Measurement," Jap. Jour. Appl. Phys., Vol. 253, 1986, L165.

[14] R. Shimokawa, Y. Miyake, Y. Nakanishi, Y. Kuwano, Y. Hamakawa, "Possible Errors Due to Deviation from the Cosine Response in the Reference Cell Calibration under Global Irradiance," Jap. Jour. Appl. Phys., Vol. 252, (1986), L105.

[15] "Requirements for Reference Solar Cells," International Electro-technical Commission, Technical Committee 82, Central Office 11, Geneva, Switzerland (in print).

[16] "Standard Practice for Determination of the Spectral Mismatch Parameter Between a Photovoltaic Device and a Photovoltaic Reference Cell," ASTM Standard E 973-83.

[17] "Standard Test Method for Calibration of Primary Non-concentrator Terrestrial Photovoltaic Reference Cells Using a Tabular Spectrum," ASTM Standard E 1125 - 86.

[18] Specification and Classification of Instruments for Measurements of Hemispherical Solar and Direct Solar Radiation, ISO/DIS 9060 Standard, 1988.

[19] R. Shimokawa, Y. Miyake, Y. Nakanishi, Y. Kuwano, Y. Hamakawa, "Effect of Atmospheric Parameters on Solar Cell Performance under Global Irradiance," Sol. Cells, 19, 1986-1987, p. 59.

Appendix 27

A SIMPLIFIED METHOD FOR DETERMINING THE AVAILABLE POWER AND ENERGY OF A PHOTOVOLTAIC ARRAY

M. S. IMAMURA
WIP- Munich
Sylvensteinstrasse 2
D - 8000 Munich 70
F.R. Germany

Abstract

A practical method of calculating the power and energy capabilities of a PV array source is presented in this paper. The method is based on power flow relationships with average values for constants and actual measurements of key variables, solar irradiance and module temperature on an hourly basis or other discrete time intervals. Included in this paper is an example for a PV plant and a simple procedure for constructing and validating the math model.

The method is applicable to any PV array configuration which is operating normally, i.e., when no PV modules are bypassed by the shunt diodes. The array can comprise one or more PV modules. The accuracy of this method is dependent upon proper selection of the values of various constants in the math model as well as how well the model has been validated. The calculation and display of maximum power capability in real-time have been implemented in several PV plants. Some of them have found it to be very useful for operation, maintenance, performance analysis, and energy utilization improvement purposes.

1. INTRODUCTION

To optimize the PV array design or to determine how well the capability of the installed PV array is being used (i.e., the array utilization factor), a knowledge of the power and energy capability of the array is essential. However, existing math models for such parameters, depending on the approach taken, are usually very complicated because they resort to the current-voltage model of the PV array rather than the power relationship and/or involve a heat-transfer model [1 - 5]. Most of the computer programmes [3 - 6] were developed for system sizing or design analysis purposes. Hence, PV designers and operators have not exploited the use of these specific parameters for operational optimization or system performance analysis purposes.

541

If a simplified calculation procedure is available to calculate these parameters using a certain set of measurements, more plant designers and operators will make greater use of these parameters for real-time display and/or subsequent performance analysis. This calculation capability is especially important for stand-alone systems because they do not operate at the array maximum power point most of the time. Even for the grid-tied PV plants, these parameters are very useful for detection of major faults in the inverter and the array subfields.

2. DERIVATION

For the purpose of this paper the power capability of a PV array is defined as the maximum power available from the array at the array dc bus at a given time.

There are basically two approaches in formulating a steady-state math model of a flat-plate PV array operating in a terrestrial environment. One is to use an equivalent circuit containing all current sources and discrete devices. The current sources can be at the individual solar cell or PV module level. Discrete devices in each array string are usually the line resistances and diodes. The solution of this PV array equivalent circuit for output voltage from zero to the maximum value (or array current equal to zero) yields the complete I-V curve of the total array. For each given computational time interval, the maximum available power can then be calculated from the array I-V curve. Compared to the power flow representation, the formulation of the math model for such a V-I equivalent circuit is more complex and not warranted for the purpose intended here.

Another approach which is easier and faster to calculate makes use of a power relationship as a function of PV module temperature and solar irradiance. This approach is presented in this paper as the "simplified method."

The power capability of any PV module type at a given time or time interval is basically a function of module temperature and solar irradiance as follows:

$$P_{m1} = P_{mo} [1 - \beta (T_1 - T_o)] H_1/H_o \tag{1}$$

These terms are defined in Table 1. The power capability for an array string in the array configuration shown in Figure 1 with N number of modules in series is:

$$P_b = N K_b P_{m1} \tag{2}$$

where $K_b = n_d n_{mm} n_{fw} n_{bw} K_d$

Other terms are defined in Table 1 and illustrated in Figure 1.

The resulting array field power capability at a given time for M number of strings in parallel is:

$$P_{sa} = M n_{bw} P_b \tag{3}$$

When eq. 1 to 3 are combined, P_{sa} can be expressed in the following form with only three variables, P_{mo}, T_1, and H_1 (the other terms are constants):

$$P_{sa} = \rho K_{sa} P_{mo} [1 - \beta (T_1 - 25)] H_1 \; 10^{-3} \tag{4}$$

where ρ = Correction factor to be established when validating the above equation from actual measurements, and

$K_{sa} = M N n_{bm} n_d n_{mm} n_{fw} n_{bw} K_d$

Note that the product M x N is the total number of modules in the array.

Table 1. PV Array Power Capability Determination Equation and Definition of Its Terms

Terms used in the basic PV module power and array relationships (eq. 1 and 4) are:

P_{mo} = Average module maximum power at a known or reference solar irradiance, H_o and reference cell temperature T_o (usually 1000 W/m^2 and 25°C, respectively. P_{mo} should be the mean of the modules in the field if such a value is available; if not, the rated value is acceptable)

P_{sa} = Array maximum power at measured intensity H_1 and cell temperature T_1, respectively

ß = Temperature coefficient of module maximum power expressed in % per °C (a negative constant, usually available from the module supplier for the cell used)

N = Total number of modules connected in series in a string

M = Total number of strings paralleled at the array bus

n_d = Diode efficiency

n_{mm} = Mismatch factor for series-connected modules

n_{bm} = Mismatch factor for parallel-connected strings

n_{bw} = Intra-module wiring loss factor

n_{fw} = Field wiring loss factor

K_d = Power loss factor for dust and dirt on module cover glass and/or power degradation factor

The array energy capability, E_{sa}, is the integral of P_{sa} profiles during a sunlight period. When discrete measurements of H_1 and T_1 are available, E_{sa} can be expressed as:

$$E_{sa} = \rho \, K_{sa} \, 10^{-3} \, P_{mo} \sum_{i=1}^{n} \{[1 - \beta(T_1 - 25)]H_i \, \Delta t_i\} \tag{5}$$

3. CALCULATION EXAMPLE

The method may be illustrated by calculating the maximum power and energy capability for one discrete time interval for which appropriate data are available. For this example, we will use the VULCANO PV pilot plant located on Vulcano Island north of Sicily. It has the following characteristics:

- PV array rating: 80 kW
- Array bus voltage: 300 Vdc nominal
- Number of modules: 1,344 Ansaldo and 672 Pragma

Further, let us assume the following measured values (H_1, T_1), predefined constants and values of parameters as listed in Table 2:

H_1 = 820 W/m^2 average for a 1-hour Δt as measured by the plane-of-array irradiance sensor

T_1 = 45°C average for a 1-hour Δt for Ansaldo modules (as measured by the back-of-module sensor)

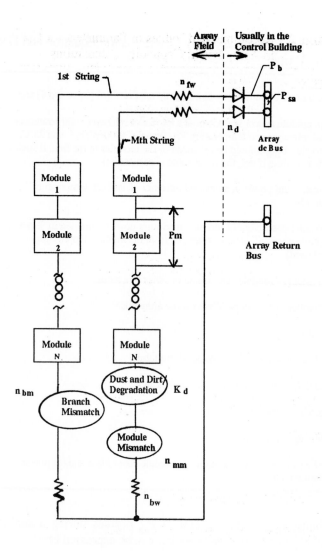

Figure 1. Simplified Power Flow Diagram of the PV
Array Field, Showing Typical String Wiring

T_1 = 42°C average for Pragma modules

P_{mo} = 33 W for Ansaldo modules, and 55 W for Pragma modules (for other modules, derate 5 to 15% of rated value, depending on the manufacturer)

ρ = 0.996 for the Ansaldo subfield, and 0.971 for the Pragma subfield determined from validation data

(Note: ρ should be calculated from eq. 4 when the inverter MPPT is operating such that maximum P_{sa} can be actually measured along with H_1 and T_1)

n_d = 0.995

β = 0.0050 for Ansaldo modules, and 0.0045 for Pragma modules

K_d = 0.95 for either module

Then the total array field power capability is the sum of P_{sa} for the Ansaldo and Pragma subfields using equation (4):

Table 2. Recommended Values of Parameters for Use
in Array Power Capability Calculations

Parameter	Value
n_d	0.98 for 28 Vdc Array Bus 0.99 for 120 Vdc Array Bus 0.995 for 200 - 300 Vdc Array Bus
n_{mm}	0.98
n_{bm}	0.98
n_{bw}	0.99
n_{fw}	0.99
ß	0.0045 for monocrystalline 0.0050 for polycrystalline Si module

For the Ansaldo array,

$$P_{sa} = (0.996)(1,344)(0.98)(0.995)(0.98)(0.99)^2(0.95)(33)\ [1 - 0.005\ (45 - 25)]\ 820\ \ 10^{-3}$$

$$= 29,007\ W$$

For the Pragma array,

$$P_{sa} = (0.971)(672)(0.98)(0.995)(0.98)(0.99)$$

$$x\ \ (0.99)(0.95)(55)\ [1 - 0.0045(42 - 25)]\ 820\ \ 10^{-3}$$

$$= 24,181\ W$$

Hence, the total average maximum power available for the 1-hour period is:

$$29,007\ W + 24,181\ W\ =\ 53,188\ W$$

and the array energy capability E_{sa} during this one hour is 53.2 kWh.

4. ACCURACY IMPROVEMENT METHOD

It is possible to further improve the accuracy of P_{sa} and E_{sa} by finding a range of values for ρ in equation (4) as a function of seasonal periods like summer and winter and adjustments of T_1 and β based on actual data. It is advisable to first correct the values of β and T_1 before making any changes to ρ. The following steps are suggested:

1) Correct β - Validate the value of the temperature coefficient of maximum power, β, at the module level via your own measurements. If not, use the manufacturer's β for the cells used in the module and calculate it for the module, considering the cell interconnection IR losses. The best method is to use actual I-V curves on the PV module obtained at different temperatures (e.g., 0°C and 60°C) because the I-V curves include the cell interconnection losses.

2) Correct T_1 - The value of T_1 should theoretically be the average temperature of all cells in the array. In reality, there is a temperature gradient across the entire array, so one cannot make precise measurements to arrive at an average value. Therefore, a single temperature transducer measurement on one or more modules may not necessarily be representative of the average module temperature. However, this difference can be minimized by measuring the open-circuit voltage (Voc) of the entire array or several individual array strings and the module temperature transducer (T_1). The new T_1 is the average array temperature determined from the Voc vs. module temperature relationship.

3) Correct ρ - With the system operating at the array peak power point, make measurements of the three parameters, P_{sa}, H_1, and T_1 at different times during one or more sunny days during the summer. From a few selected measurements of P_{sa}, H_1, and T_1, determine the value for ρ which gives the best fit using eq. (4). Repeat this procedure during the winter months to determine the value of ρ for the winter months.

The basic math model has been evaluated at several PV plants [7 - 10]. The preliminary observations are that the accuracy of the power and energy capability calculation using the simplified model is between 2-4% for solar irradiances above 300 W/m^2 [7]. Sorokin of TEAM reported that the power capability parameter has been extremely valuable for the real-time assessment of the ZAMBELLI PV pumping system performance. ZAMBELLI operates in the stand-alone mode and autonomously with only periodic visits by the operator.

5. CONCLUSIONS

A simple method of determining the power and energy capabilities of a PV array is described. It is based on power flow relationships and a minimum set of real-time measurements (plane-of-array irradiance and back-of-module temperature). This method is applicable to any size and configuration of PV array. Its accuracy is dependent on how the validation is accomplished. A real-time calculation of maximum array power can serve very effectively as a "gauge" for sensing the actual PV array capability (both instantaneous power and energy). These parameters are also very useful for system efficiency optimization and real-time operation improvement.

6. REFERENCES

[1] Evans, D.L., "Simplified Method for Predicting Photovoltaic Array Output," Solar Energy (27:6), pp. 555-560, 1981.
[2] Siegel, M.D., Klein, S.A., and Beckman, W.A., "A Simplified Method for Estimating the Monthly Average Performance of Photovoltaic Systems," Solar Energy, V. 26, pp. 413-418, 1981.
[3] Menicucci, D.F., and Fernandez, J.P., "User's Manual for PVFORM: A Photovoltaic System Simulation Program for Stand-alone and Grid-Interactive Applications," Sandia National Laboratories, SAND 85-0376, 1988.
[4] Chapman, R.N., and Fernandez, J.P., "A. User's Manual for SIZEPV: A Simulation Program for Stand-alone PV Systems," SAND 89-0616, March 1989.
[5] "Analysis Methods for PV Applications," Solar Energy Research Institute (Denver, CO.), SERI/SP-35-231, 1984
[6] Wrixon, G.T., McCarthy, S., and Keating, L., "Computer Modelling and Simulation," CEC Contract EN3S-0129-IRL, National Microelectronics Research Centre (Cork, IRL), 1989
[7] Private communication with Mr. A. Sorokin of TEAM (Rome, IT) on Zambelli 65-kW PV plant monitoring system performance, July - September 1990.
[8] Private communication with Mr. B. Mortensen of Jydsk Telephone (Aarhus, DK) on Bramming 5-kW PV House, comparison with that of the existing software in their monitoring system, May 1990.
[9] Private communication with Dr. A. Previ of ENEL (Milan, IT) on Vulcano 80-kW PV Plant performance, December 1989.
[10] Private communication with Mr. E. Ehlers of WIP (Munich, FRG), January - August 1990, on several PV projects in the German National PV programme.

Appendix 28

ELECTRICITY PRODUCTION COSTS FROM PHOTOVOLTAIC SYSTEMS AT SELECTED SITES WITHIN THE EUROPEAN COMMUNITY

W. PALZ
CEC, DG XII
Rue de la loi, 200
B - 1049 Brusseis
Belgium

J. SCHMID
Fraunhofer Inst. ISE
Oltmannstrasse 22
D - 7800 Freiburg
F.R. Germany

Abstract

The costs for photovoltaics, as calculated in this paper, are based essentially on experience gained in Europe with prototype systems. Social and other external costs in electricity production are not taken into account. In addition, an estimate for the year 1995 is projected. This is based on the extrapolation of existing and proven technology, and does not assume any technological breakthroughs. In the longer term, one can assume that further significant cost reductions will be achieved as a result of the continuing research activity.

1. SYSTEM PRICES PER INSTALLED POWER UNIT

1.1 Simplest Case

For the simplest case, we will consider PV modules integrated into the roof of a grid-connected building, without storage. The best possible system price in Europe in 1990 for large production volumes, in ECU per peak watt is:

- Module:	4.40
- Support structure:	1.50 (not applicable for roof integration)
- Inverter:	1
- Installation:	0.50
Total:	5.90 (excluding support structure)

The French company, Photowatt, was the reference source for the module price. Unlike most other producers of silicon PV modules, Photowatt benefits from the cost advantage gained by producing its own silicon wafers sliced by wire sawing. In addition, it has low overhead costs. Prices for the inverter and the support structure were obtained from Fraunhofer Institute ISE, Freiburg (F.R. Germany).

By 1995, Photowatt expects to achieve a module price of 1.40 ECU per peak watt. Corresponding R&D work is under way, supported financially by the Commission of the European Communities (DG XII) and others. In our opinion, this price estimate is credible, as it can be based solely on

547

further development of the processes used today. Furthermore, it is assumed that the costs for the inverter and the array support structure listed above will be reduced by half. Thus, the result is a system price of 2.15 ECU per peak watt for 1995 (based on 1990 monetary values).

1.2 PV Power Stations

No commercial PV power station yet exists in Europe. Thus, costs cannot be calculated on the basis of practical experience. Certainly, an attempt has been made in the past to deduce the costs from development projects such as the 300-kW plant in PELLWORM (F.R. Germany), which was also supported by the CEC/DG XII. However, this can lead to serious misunderstandings. It must be assumed that power stations with separate array fields are generally more expensive than PV systems installed on buildings, because significant additional costs for the land area, site preparation and support structure must be expected. Simply to give an idea of the order of magnitude, the cost of a 3-MW PV power station in Italy is given here as an example. This plant, scheduled for installation in 1992 as a grid-connected system, will be the largest in Europe. Planning is based on experience with the 80-kW plant, which ENEL installed on the Vulcano island in 1984 under contract to the CEC, and which has delivered excellent results since then.

The ENEL estimate, converted to ECU/Wp, is as follows:

- Modules:	4.5
- Land, support structure:	2.5
- Inverter, grid connection:	0.6
Total:	7.6

1.3 Special Applications

Developing countries offer important application areas for photovoltaics, and PV water pumps are of particular interest. The CEC/DG VIII initiated a programme in 1990 to introduce more than 1000 PV-powered water pumps in the Sahel region. The total costs for such systems, including the pumps and adaptation, are expected to be around 20 ECU/Wp. This is a comparatively high value. However, it should be noted that these PV systems are still competitive with Diesel generators.

2. CALCULATION OF THE ANNUAL ENERGY PRODUCTION OF A PV GENERATOR

To give an example of the calculation process, we have developed the following equation:

$$E \quad = \quad Iy \; x \; A \; x \; Q \; x \; U \tag{1}$$

where E   =   annual energy produced by 1 kWp of array

Iy   =   total annual solar radiation (here global radiation on a south-oriented surface of 1 m^2)

A   =   availability of the generator. As accidents and maintenance periods are negligible for stationary systems, A = 1 is valid for most plants.

Q   =   quality factor. For example, in systems without batteries it includes the inverter losses and losses in the modules. A good inverter today has an energy efficiency of about 93%; the energy yield from the modules will be reduced in operation by lower efficiency at lower radiation levels, losses related to non-perpendicular incidence of solar irradiance, temperature increase under strong irradiance, and variation in the solar spectrum. Experimental plants developed by the European Communities give values for Q in the range 0.6 - 0.7.

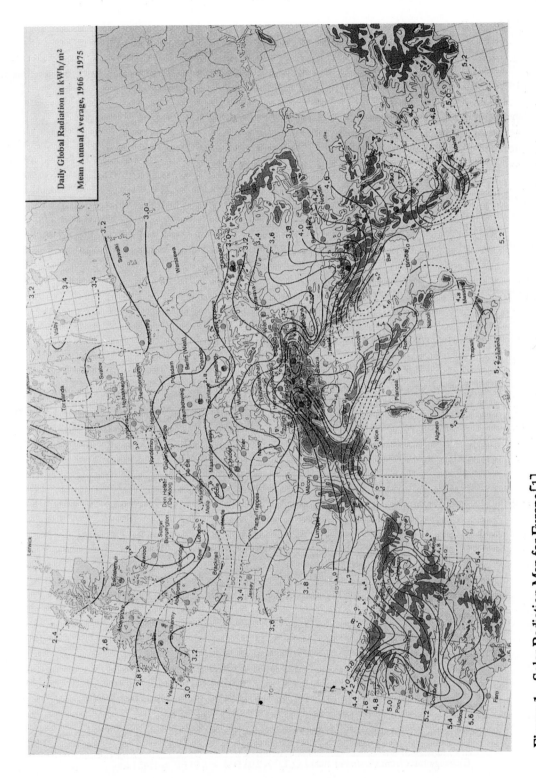

Figure 1. Solar Radiation Map for Europe[1]

U = utilization factor for the system. This term includes the storage efficiency and the mismatch between energy production and consumption. For grid-connected systems without storage, U = 1 can be assumed.

3. ENERGY COSTS FOR SELECTED EXAMPLES IN EUROPE

For the sake of simplicity, we have restricted the cost assessment to basic grid-connected systems with roof-integrated PV array. The following assumptions have been made:

- Operating costs: none
- Interest on capital: 10% or 4%
- Lifetime: 30 years
- Availability: 100%

Equation (1) has been used to calculate the energy production, assuming that 1, U = 1, Q = 0.65. The corresponding energy prices have been calculated for F.R. Germany, France, and Italy. The regional range of global irradiation, incident on a surface oriented to the South and tilted appropriately for the latitude of the site, as listed below, is taken from the European Solar Radiation Atlas (see Fig. 1).

Country	Annual Radiation on 1 m^2 of array
F.R. Germany	1100 - 1400 kWh
France	1100 - 1900 kWh
Italy	1300 - 1900 kWh

The annuity factor f was used to calculate the discounted kWh cost. It amounts to 9.43 for a 10% interest rate and 17.29 for a 4% interest rate. As explained earlier, a system price of 5.9 ECU/Wp is assumed for 1990 and 2.15 ECU/Wp for 1995. Using equation (1), together with

$$\text{kWh price} = \frac{\text{installation costs}}{f \cdot E}$$

results in the values which are presented in Table 1.

Table 1. PV System Energy Costs in Three European Countries

	1990		1995	
	with 10% interest	with 4% interest	with 10% interest	with 4% interest
F.R. Germany				
ECU/kWh	0.87-0.69	0.47-0.37	0.32-0.25	0.17-0.13
DM/kWh*	1.76-1.40	0.95-0.75	0.65-0.51	0.35-0.26
France				
ECU/kWh	0.87-0.51	0.47-0.28	0.32-0.19	0.17.0.10
FF/kWh*	6.04-3.54	3.26-1.94	2.22-1.32	1.18-0.69
Italy				
ECU/kWh	0.74-0.51	0.40-0.28	0.27-0.19	0.14-0.10
Lira/kWh*	1111-766	600-420	405-285	210-150

* Exchange rates from December 1989 1 ECU ≈ 2.03 DM ≈ 6.94 FF ≈ 1502 Lira

4. REFERENCES

[1] European Solar Radiation Atlas, Volume 2: Inclined Surfaces. Verlag TÜV Rheinland, Cologne, FRG, 1984

Appendix 29

RELIABILITY AND AVAILABILITY ASSESSMENT METHODS

M. S. IMAMURA
WIP - Munich
Sylvensteinstrasse 2
D - 8000 Munich 70
F.R. Germany

Abstract

This paper provides a definition of basic reliability terms most commonly used and describes general methods of determining the reliability and availability of components, subsystems, or systems. Simple procedures are outlined and examples given on PV components for use by system design engineers in making reliability estimates for the purpose of improving their systems by selecting better components and/or adding redundancies. High reliability cannot be achieved without a cost penalty. But it can decrease the overall life cycle cost through reduced maintenance effort and increased system availability.

1. INTRODUCTION

Fifty years ago, reliability was a term seldom applied to engineering products. Today, it is commonly used in all manufactured items. Reliability requirements and techniques are being specified regularly, and they are quite often the basis of how various items of equipment are to be designed and configured.

A majority of the equipment items used by PV plants installed in the EC are commercial types, i.e., off-the-shelf or catalogue items available from several sources. A small number of them are custom-designed or of a special design and not widely used. Failure rates of these custom-designed equipment items are largely unknown, but they are beginning to accumulate in the pilot plant and demonstration projects. To improve the reliability of individual pieces of equipment, and hence, the total plant reliability, it is important that field performance data be collected properly and timely feedbacks be given to the suppliers of such equipment.

Many project and system engineers are surprised to find their systems and subsystems failing immediately or shortly after the installation of hardware. In fact, many pilot plants along with their monitoring equipment had frequent failures or stoppages, necessitating repairs more often than anticipated. In such cases, a basic knowledge of the reliability estimation and design procedures would have helped prevent short equipment lifetimes. This paper is intended to give a basic understanding of the reliability terminologies and how to detect unreliabilities in the electronic hardware as well as how to design a system which can perform successfully for a desired duration of time.

The actual lifetimes of many PV components operating outdoor or indoor are basically not known at this time. Improvements in equipment reliability can lower the overall cost of the PV plant via a reduction in maintenance costs and can also avoid loss of revenue in the case of grid-connected plants. High reliability cannot be achieved without a cost penalty. Thus, the designer must carefully weigh cost constraints against reliability needs in the subsystems and systems.

1.1 Why Reliability?

It is important to understand the basic ideas and methods involved in reliability so one can quantitatively study why the product or a system does not last very long, and how to improve the designs. Equipment is required to work in a variety of environments and is becoming more complex. As the severity of the working environments increases, so does the complexity of the design.

The high cost of the equipment, its lifetime, and the labour and spare parts required to maintain it are other justifications for emphasis on high reliability. It would be extremely unfortunate if the failure of a 100-DM part in the inverter caused a complete stoppage of a multi-million DM PV plant operation, or the failure of a small 50-DM bias power supply to the data acquisition equipment wiped out data recording for several months.

1.2 How Do We Achieve Reliability?

The generally applicable methods for achieving high reliability of all systems include the following steps:

- Proper choice of configurations and parts during the design phase to achieve the expected level of reliability.
- Failure modes and effects analysis (FMEA); timely FMEA and redesign effort during the development phase can minimize the chances of subsystem and system level failures.
- Adequate testing and inspection.
- Good quality control throughout product manufacturing and installation.
- Monitoring and proper failure reporting, e.g., an adequate test programme and timely dissemination of information.
- Configuration or procedural changes based on performance and trend data.

The first four steps above dictate the inherent reliability designed into the equipment or an entire system. The last two determine the potential for achieving high reliability and availability of the PV power plant and subsystems. Based on the monitoring results, certain changes can be implemented in the system configuration or operational mode in an attempt to extend the life of components or to prevent further premature failures.

It must be emphasized that reliability cannot be tested into a system or added in any way if the design itself is inadequate to begin with.

2. DEFINITION

The reliability of a system is simply the probability that the system (or an item of equipment) will complete its intended mission successfully. More precisely, we can define it as the probability that the system will work as designed for a specified duration of time. The environments which the system will be exposed to throughout its lifetime include manufacturing, shipping, assembly, and operational conditions after installation.

3. RELIABILITY RELATIONSHIPS

To understand how to make use of reliability assessment techniques, we need to review the basic concepts and relationships, focusing on the basic reliability equation for non-redundant and redundant system elements and the failure rates. A "redundant system element" is one whose function can be taken over by another element in the system.

The probability that an item of equipment will work successfully is described by

$$P = e^{-\lambda t} \tag{1}$$

where
P = success probability in % or decimal fraction
λ = total failure rates of electrical parts in the equipment
t = total time of successful operation of the equipment

The reciprocal of the failure rate is the MTBF (mean time between failures), or

$$MTBF = \frac{1}{\lambda} \tag{2}$$

If a system consists of several subsystems or items of equipment in series, each having its own probability, the total system reliability is equal to the product of all individual probabilities. Thus, if a system is made up of two items of equipment having P_1 and P_2 probabilities (see Fig. 1), the system success probability is $P_1 \times P_2$. This extends to an item of equipment with n serial parts that are all used at 100% duty cycle:

$$P_n = P_1 \times P_2 \times P_3 \times \ldots = \prod_{i=1}^{n} P_i \tag{3}$$

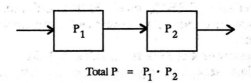

Total P $= P_1 \cdot P_2$

Figure 1. A System Made UP of Two Subsystems or Parts in Series with P_1 and P_2 Probabilities

It can be seen that as more and more items are added in series, the system reliability decreases since the individual probabilities are all less than unity. Substitution of Eq. (1) into Eq. (3) results in

$$P_s = e^{-(\lambda_1 + \lambda_2 + \ldots)t}$$

$$= e^{-\left[\sum_{i=1}^{n} \lambda_i\right]t} \tag{4}$$

The reliability of a system, therefore, can be found by summing up the failure rates of the component parts. This holds true for a simple non-redundant system as illustrated in Figure 1. For a system that has a redundancy as shown in Figure 2 in which block P1 or P2 can successfully transfer power or signal from point A to B, the probability of this system is

$$P_s = 1 - (1 - P_i)^2 \qquad (5)$$

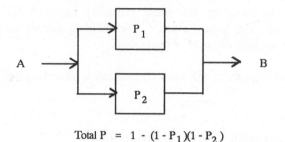

Total P $= 1 - (1 - P_1)(1 - P_2)$

Figure 2. Redundant Path in which Either Block Can Successfully Transfer Power or Signal from A to B, Independently of the Other

For n parts each with equal reliability:

$$P_s = 1 - [(1 - P_i)]^n \qquad (6)$$

where P_i is the reliability of each part, and P_s is the system reliability at time t.

Eq. (5) is no longer in the simple form of Eq. (4), so it does not permit the simple addition of failure rates for a system with redundancy. However, an equivalent failure rate for a redundant pair can be calculated by an approximation using the exponential expansion of the basic reliability equation.

$$P = e^{-\lambda t}$$

$$= 1 - \lambda t + \frac{\lambda^2 t^2}{2!} - \frac{\lambda^3 t^3}{3!} + ... \qquad (7)$$

By substituting Eq. (7) into P_i in Eq. (5), the above series can be approximated by

$$P \approx 1 - \lambda t \qquad (8)$$

Then using Eq. (8) for P_i in Eq. (5), the reliability of a system with redundancy becomes

$$P_s = 1 - \lambda^2 t^2$$

$$= 1 - (\lambda^2 t)t \qquad (9)$$

A comparison of Eq. (9) and (8) shows that a parallel redundant path can be replaced by an equivalent single path or block that has an effective failure rate

$$\lambda_{eff} = \lambda^2 \qquad (10)$$

and its effective MTBF is

$$MTBF_{eff} \approx \frac{1}{\lambda^2 t} \qquad (11)$$

To illustrate the result of using redundancy, consider a device with a failure rate of 500 failures per 10^6 hours required to operate for 1,000 hours. If a redundant device is used, the effective failure rate is 250 failures per 10^6 hours, and the system reliability (using Eq. 8) is 0.75 as compared to 0.5 for the non-redundant case.

The basic reliability equations cited above assume a constant failure rate, resulting in the exponential failure probability. This is a valid assumption in most mature equipment over a portion of its life. Figure 3 is a plot of failure rates as a function of time for most mature designs. The first portion of the curve, 0 to t_1, illustrates the high failure rates inherent in the infant mortality phase of the equipment. Included here are poor workmanship, defective or marginal components, and procedural failures. The portion of the curve beyond t_2 ($t_2 < t < \infty$) illustrates the increase of failure rates resulting from the onset of equipment wearout. This would include fatigue-type failure which is very difficult to predict or detect. The region between t_1 and t_2 illustrates the exponential failure rate zone of constant λ. In this area, we say that failures occur randomly and at a constant rate.

Note that Figure 3 is idealized but is commonly accepted for most electronic equipment. Quite often equipment and parts "burn-in" times are sufficient to discover all the infant mortality failures, and life tests are necessary to assure that normal equipment operation times are well within the t_2 - t_1 region.

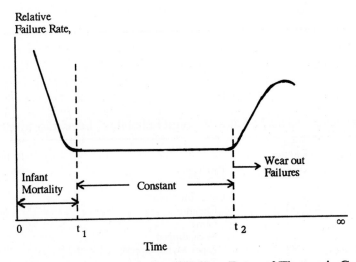

Figure 3. Characteristics of Failure Rates of Electronic Components

4. AVAILABILITY

System availability is defined by the following equation:

$$A = \frac{MTBF}{MTBF + MTTR} \qquad (12)$$

where A = availability

    $MTTR$ = mean time to repair

The values of MTTR and MTBF must be determined by the user from actual field experience on each major replaceable unit. However, MTBF can initially be estimated using standard tables available from the major manufacturers of electrical parts, national laboratories or ESA (European Space Agency). Examples of such data are given in Tables 1 and 2 for two reliability levels. These tables are usually available in documentation such as reference [1].

Table 1. Some Examples of Failure Rates of "Standard Reliability" Level Electronic Parts

Part Type	Failures per 10^9 h	Part Type	Failures per 10^9 h
Diode		Capacitor, fixed	
Silicon	84	Ceramic	214
Germanium	120	Glass	143
Transistor		Mica, dipped	143
Silicon	250	Paper-Mylar, dipped	429
Germanium	358	Tantalum	
Integrated Circuits	168	Foil or Wet Slug	286
Resistor		Solid	143
Carbon Composition		Transformer, Low voltage, Class H	21
General Application	58	Inductor, Low Voltage, Class H	21
Digital Application	16	Relay	
Metal Film	58	General Purpose	220
Wirewound, Precision	142	Magnetic Latching	400
Wirewound, Power	174	Connector, per Active Pin	40
Variable		Connection, Soldered or Welded	4
Composition	122		
Wirewound	206		

Source: Ref [1]

Table 2. Some Examples of Failure Rates of "High Reliability" Level Electronic Parts

Part Type	Failures per 10^9 h	Part Type	Failures per 10^9 h
Diode, Silicon		Capacitor, Fixed	
General Application	33	Ceramic	86
Digital Application	7	Glass	57
Transistor, Silicon		Mica, dipped	57
General Application	49	Paper-Mylar, dipped	143
Transistor, Germanium		Tantalum	
General Application	33	Foil or Wet Slug	72
Integrated Circuits	84	Solid (Series resistance 3 ohms /V)	43
Resistor		Connection	
Carbon Composition		Soldered	4
General Application	16	Welded	4
Digital Application	3		
Metal Film	16		

Source: Ref [1]

When evaluating the system design, one can determine an item's contribution to system availability via Eq. (12) or reliability via Eq. (1), (5), or (6). To arrive at these system-level values, the individual item's availability and reliability number must first be calculated. Examples are given in the next section.

5. EXAMPLES OF RELIABILITY CALCULATIONS

There are several ways to compute the reliability of an equipment or a "black box" and likewise a system consisting of a number of these black boxes. A first-order approximation is to sum up the failure rates of all component parts, assuming a 100% duty cycle on these parts. Table 3 shows an example of this approximation on a computer. Using Eq. (4), the estimated probability of this computer operating successfully for 1,000 hours is: The MTBF for the small high-quality computer example in Table 3 is:

$$P = e^{-(239,165)(1,000)/10^9}$$

$$= 0.787$$

Table 3. Reliability Assessment of A Small Computer from Parts Count

Description	Parts Quantity	Unit Rate (Failures per 10 9 h)	Subtotal, Failure Rates
Integrated Circuits	520	168	87,360
Diodes			
Silicon	500	84	42,000
Transistors			
Silicon	5	250	1,250
Germanium	5	358	1,790
Resistors			
Composition	700	16	11,200
Precision Wire	55	142	7,810
Wirewound (Power)	105	174	18,270
Capacitors			
Paper-Mylar, dipped	45	429	19,305
Tantalum (Solid)	130	286	37,180
Connections (Solder)	3,400	4	13,600
TOTALS	5,465		239,765

The MTBF for the small high-quality computer example in Table 3 is:

$$MTBF = \frac{1}{\lambda}$$

$$= \frac{1}{(239,165)10^{-9}}$$

$$= 4,181 \text{ hours}$$

Now consider a dc regulator with a simplified reliability model shown in Figure 4 with the failure rates of individual blocks already calculated. The problem is to determine the probability of the total dc regulator. Let us assume a total operational time of one year or 8,760 hours.

blocks 3 and 4:

$$P_{3-4} = e^{-(1,767 + 485)(8,760)10^{-9}}$$

$$= 0.9805$$

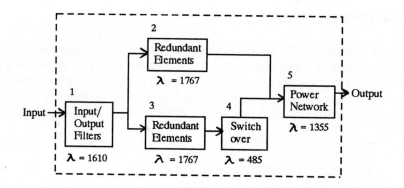

Figure 4. Simplified Reliability Model of a dc Regulator Showing
Calculated Failure Rates for Each Functional Block

block 2:

$$P_2 = e^{-(1,767)(8,760)10^{-9}}$$

$$= 0.9846$$

blocks 2, 3, and 4:

$$P_{2\text{-}3\text{-}4} = 1 - (1 - P_2)(1 - P_{3\text{-}4})$$

$$= 1 - (1 - 0.9846)(1 - 0.9805)$$

$$= 0.9996$$

The λt corresponding to 0.9996 probability is 0.0004. Thus, the equivalent λ for blocks 2, 3, and 4 is

$$\lambda_{2\text{-}3\text{-}4} = 45.7 \text{ failures}/10^9 \text{ h}$$

block 1:

$$P_1 = e^{-(1.61)(8.76)10^{-3}}$$

$$= 0.986$$

block 5:

$$P_5 = e^{-(1.355)(8.76)10^{-3}}$$

$$= 0.9882$$

The total failure rate for the dc regulator is then,

$$\text{total} \quad \lambda = \lambda_1 + \lambda_{2\text{-}3\text{-}4} + \lambda_5$$

$$= 1616 + 45.7 + 1355$$

$$= 3011 \text{ (per } 10^9 \text{ h)}$$

and its probability of successful operation for one year:

$$P = e^{-(3.011)(8.76)10^{-3}}$$

$$= 0.974$$

An example of availability calculation using Eq. (12) for a typical PV system component is given in Table 4.

Table 4. Availability of Typical PV Plant Components from Known MTBF and MTTR

Equipment	Failure Rate/ 10^9 h	MTBF (h)	MTTR (h)	Availability
PV Field	2660	376	1.5	0.996
dc Regulator	3	333333	20	0.994
Inverter	156	6410	20	0.997

Note that both MTBF and MTTR can be established using actual data. As far as practicable, they should be determined on a line-replaceable item such as the entire black box, e.g. a dc regulator or a printed circuit board in the inverter. The MTBF of an equipment is determined from the field data simply by dividing the total number of all failures encountered on this equipment by the total hours between these failures. Likewise, the MTTR is the total time required to repair or replace the unit and get the system operating again.

6. CONCLUSIONS

Estimates of reliability, MTBF, and availability of an item of equipment can be made using simple procedures. This assessment can be verified and updated if new values of MTBF's and MTTR's can be established from actual system operation. Using the procedure defined, a system engineer can easily estimate the total reliability of his system or any of its components from a knowledge of the type and quantity of parts used in the component. He can then make the necessary adjustments to the system configuration and layout during the design phase on the basis of both reliability and cost constraints.

7. REFERENCES

[1] MIL-HDBK-217B, "Reliability Prediction of Electronic Equipment," 1982 and later revisions.
[2] Drake, A.W., "Fundamentals of Applied Probability Theory." New York: McGraw-Hill, 1967.
[3] "Reliability Applications and Analysis Guide," The Martin Company, Denver, CO, Internal Report MI-60-54, July 1961.

Appendix 30

TRACKING VS. FIXED FLAT-PLATE ARRAYS: EXPERIMENTAL RESULTS OF ONE YEAR'S OPERATION AT THE ADRANO PILOT PLANT

S. GUASTELLA
Conphoebus Scrl
Passo Martino-Zona Industriale
Casella Postale
I - 95030 Piano d'Arci (Catania)
Italy

A. ILICETO and V. PIAZZA
ENEL-CREL
Via A. Volta, 1
I - 20093 Cologno Monzese (Mi)
Italy

Abstract

To investigate the cost/benefit analysis of sun-tracking flat-plate PV systems as compared with fixed arrays, an experiment was conducted at the ADRANO PV plant to look into the performance of currently operating systems. This paper reports both theoretical predictions of energy gain achievable with partially and fully sun-tracking PV arrays and actual data collected over a year of continuous operation. Such data permit an analysis of financial viability for a particular PV installation.

1. INTRODUCTION

As part of the ADRANO PV Project [1], which was co-financed by the Commission of the European Communities, ENEL and Conphoebus carried out a research programme to look into the benefits of partially and full *sun-tracking* as opposed to *fixed* flat-plate PV arrays, under real operating conditions. This research also made it possible to test three PV/inverter systems, each with about 2.5-kWp PV arrays connected to the inverter. It enabled a thorough comparative analysis and evaluations to be made on the effectiveness of such grid-connected systems.

This paper presents the analytical and experimental work done concerning the benefits of different types of tracking array surfaces. Experimental results were obtained using one fixed and two tracking arrays, and a comparison of five different array orientations was done using computer predictions. An analysis of their financial viability is also presented.

2. COMPUTER MODEL EVALUATIONS

Figure 1 shows the five array orientations considered. The calculations that follow refer to the solar energy data recorded in Sicily over a period of 20 years from which the "average" year was selected. The use of computer codes [2], run on an hourly calculation basis, made it possible to process the data available (daily global horizontal irradiation and number of daily hours of sunshine) and to compute the integrated values of the solar irradiance incident on the different fixed and tracked surfaces, corresponding to each of the five systems considered.

System	Azimuthal Orientation	Zenithal Orientation	Rotation Axis
a	South	30°	None
b	South	Tracking	E - W Horizontal
c	Tracking	30°	N - S 30° Tilted
d	Tracking	30°	Vertical
e	Tracking	Tracking	2 axes

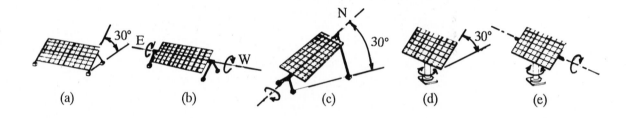

(a)   (b)   (c)   (d)   (e)

Figure 1. The Five PV Array Configurations Studied by Computer Analysis

The above calculations produced the annual values listed in Table 1, where Δ % represents the gain in annual energy collection of the tracking systems as compared with the fixed system "a". The same calculations based on different solar data led to quite different absolute figures, but Δ % values remained substantially unchanged. Note that these figures, while relating to a real meteorological year, represent "solar geometry" data, and therefore, do not take into account the real performance characteristics of the devices required for the tracking systems mentioned above (e.g., the operating PV module temperature).

Table 1. Annual Insolation Estimates for Various Tracking and Fixed Surfaces

System[1]	Insolation (kWh/m^2-y)	Δ%
a	2050	0.0
b	2269	+10.7
c	2955	+44.1
d	2665	+30.0
e	3079	+50.2

[1] See Fig. 1

3. ONE YEAR'S OPERATING RESULTS

As illustrated in Figure 2, the three PV systems compared have about the same rated power and are connected to the grid through three identical MPPT inverters (see characteristics in Table 2) [3]. This layout was intended to provide perfectly symmetrical operation for the various PV arrays. Thus, these systems are three separate small grid-connected systems.

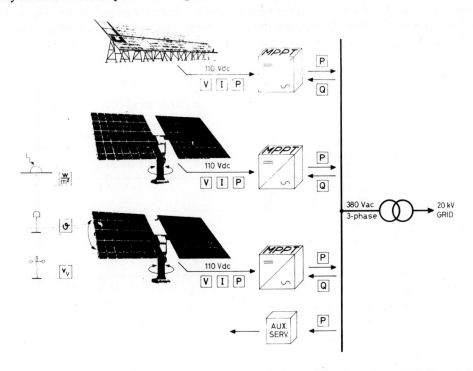

Figure 2. Functional Diagrams of the Three PV Systems at the Adrano PV Plant

Table 2. Main Characteristics of the Three PV Systems Installed at the ADRANO Plant

System*	Orientation	Module Type**	System Mftr.	Total Module Area [m²]	Rated Voltage [Vdc]	Rated Current [Adc]
a	az. South el. 30°	36 cells 2 glass	Ansaldo	39.5	129	22.8
d	az. track. el. 30°	20 cells 2 glass	MAN	37.2	132	18.6
e	2 axis track.	20 cells 2 glass	MAN	37.2	132	18.6

*   See Fig. 1
**   All contain 10 x 10 cm polycrystalline cells

The two automatic tracking systems [4] have the following main functions:

- Tracking start-up with irradiation threshold equal to 200 W/m²
- Tracking of sun with precision of ± 5°
- Horizontal positioning under cloudy conditions for maximum collection of diffuse irradiation
- Horizontal positioning for protection against strong winds

A high-precision data acquisition system [5] makes it possible to record all the electrical, thermal, and meteorological quantities that characterize operation of the plant (37 measurements recorded on cassette every 15 min. integration period). After one year of operation, the data effectively available cover about 90% of total time.

Figure 3 shows the weekly integrated energy values (52 data) for each of the three systems (expressed in peak units, i.e., related to 1 kWp power for each system, for comparison purposes). Figure 4 shows the percentage of energy gain in the tracking systems as compared with the fixed array, according to the data reported in Figure 3. The complete data for a whole year are given in Table 3.

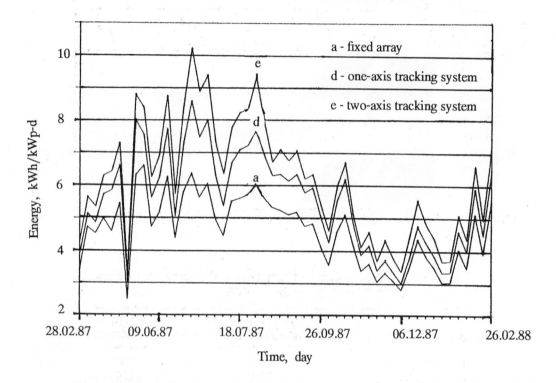

Figure 3. Daily Average Energy Production of the Three PV Systems Based on Weekly Measurements

The real operating results of the three systems (Table 3) are considerably lower than the theoretical energy gain values for moveable surfaces (Table 1), since they were determined by the operating characteristics of the systems, such as tracking limit thresholds, protections, variations in performance with module temperature, effects of dust and pollution, etc. However, the theoretical values remain valid for the purpose of comparing the different systems, and may be regarded as limits for the evaluation of array orientation differences.

4. COST EVALUATION

Because of the higher cost of installation and maintenance, it is not cost effective to set up tracking systems in photovoltaic plants for small isolated users or communities. One of the main advantages of photovoltaics over other possible energy sources is that the PV plant can be made completely static. In the case of large power plants, however, it is worth assessing the financial viability of tracking, as well as fixed flat-plate systems.

The cost comparison requires the calculation of the cost based on energy produced. This is difficult because it depends on many different factors, such as latitude, meteorological conditions (solar

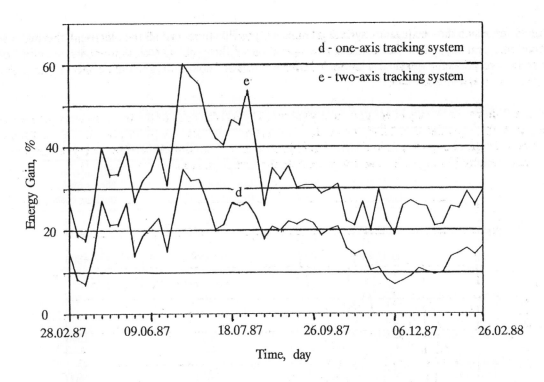

Figure 4. Weeky Energy Gain in Tracking PV Systems as Compared with Fixed Array

Table 3. One Year's Energy Production of the Three PV Systems (From 28/02/87 to 26/02/88)

System	Energy Production (kWh/kWp-y)	Difference (%)
Fixed (a)	1683	0.0
1 - axis tracking (d)	2010	+19.4
2 - axis tracking (e)	2257	+34.1

energy availability on different surfaces), cost and type of land, mass production of systems, cost of PV modules, cost of reflectors and/or concentrators, O&M costs, reliability, discount rate, lifetime, etc. This requires a large number of assumptions that limit the validity of the conclusions arrived at in each individual case.

Our cost assessment considered PV plants with the following features and factors:

- Number of PV generation units installed: 1000
- Peak power per generation unit: 5 kW
- Type of structures: one column flat-plate panels
- Conversion efficiency of modules: 10%
- Site: flat uncultivated land
- Location: Sicily
- Daily mean global irradiation on a horizontal surface: 4.5 kWh/m^2-d
- Fixed system: 30° tilted
- Tracking systems: two-axis tracking
- Cost of land and clearance: negligible
- Lifetime: 20 years
- Discount rate: 5%

Based on all of the above and given the same installed peak power, two realistic assumptions of additional tracking costs have been considered (see Table 4, a and b), combined with two assumptions of energy gain in tracking vs. fixed systems (see Table 4, a1, a2 and b1, b2).

Table 4. Tracking vs Fixed PV Systems: Results of Cost Analysis

			Tracking Systems	
Item	Units	Fixed Array	Assumption A: Low additional tracking costs	Assumption A: High additional tracking costs
Modules	ECU/kWp	4000	4000	4000
Structures & Mounting	ECU/kWp	1000	1000	1500
Tracking Electronics and Sensors	ECU/kWp	---	200	500
Mechanical Devices	ECU/kWp	---	650	1500
Total annual capital cost	ECU/kWp	400	470	600
Maintenance	ECU/kWp	---	50	100
Total annual cost*	ECU/kWp	400	520	700

Item	Units	Fixed Array	Assumpt.a1: high gain	Assumpt.a2: low gain	Assumpt.b1: high gain	Assumpt.b2: low gain
Tracking Energy Gain	%		+50	+30	+50	+30
Cost of Energy (without power conditioning)	ECU/kWh	0.22	0.19	0.22	0.26	0.29

* Maintenance plus capital cost

5. ACKNOWLEDGEMENTS

The authors wish to thank MAN of Munich for their co-operation in the experimental work done and for the data they provided.

6. REFERENCES

[1] A. Iliceto, "The Adrano Project," Int. J. Solar Energy, 1985 Vol. 3, pp 174-190.

[2] P.Bullo, M. Gasbarra, and G. Emanuele, "A Detailed Package of Digital Codes Specially Developed for the Array Design," Proc. the 4th European PV Solar Energy Conference, Stresa, IT, 10-14 May 1982.

[3] V. Arcidiacono, S. Corsi, L. Lambri, " Maximum Power Point Tracker for Photovoltaic Power Plants," Proc. of the 15th IEEE Photovoltaic Specialists Conference, S. Diego, USA, 1982.

[4] Pfeiffer, "Solar-Guided Photovoltaic Generators," Proc. of the 6th European PV Solar Energy Conference, London, GB, 15-19 April 1985.

[5] G. Ghirighelli, F. Russo, "Acquisizione Dati dell'ENEL," Automazione Oggi - Anno I - no 10, November 1983.